Non-Conventional Copyright

Non-Conventional Copyright

Do New and Atypical Works Deserve Protection?

Edited by

Enrico Bonadio

Senior Lecturer in Intellectual Property Law, City, University of London, UK

Nicola Lucchi

Associate Professor of Law, Jönköping International Business School, Sweden

Edward Elgar PUBLISHING

Cheltenham, UK • Northampton, MA, USA

Cover image: 'Holding Hands' by Stik and LA2.

Published by
Edward Elgar Publishing Limited
The Lypiatts
15 Lansdown Road
Cheltenham
Glos GL50 2JA
UK

Edward Elgar Publishing, Inc.
William Pratt House
9 Dewey Court
Northampton
Massachusetts 01060
USA

A catalogue record for this book
is available from the British Library

Library of Congress Control Number: 2018944733

This book is available electronically in the **Elgar**online
Law subject collection
DOI 10.4337/9781786434074

ISBN 978 1 78643 406 7 (cased)
ISBN 978 1 78643 407 4 (eBook)

Typeset by Servis Filmsetting Ltd, Stockport, Cheshire

Contents

Figures

Contributors

Enrico Bonadio, Senior Lecturer in Intellectual Property Law, The City Law School, City, University of London, UK.

Shane Burke, Lecturer in Law, Cardiff University, Law School, UK.

Charles Cronin, Lecturer, University of Southern California, Law School, Adjunct Professor, Keck Graduate Institute, Claremont Colleges, US.

Teshager Dagne, Associate Professor, Faculty of Law, Thompson Rivers University, Canada.

Tim W. Dornis, Professor of Law, Leuphana Law School, Germany.

F. Jay Dougherty, Professor of Law, Loyola Law School, Los Angeles, US.

Trevor M. Gates, Attorney, K&L Gates LLP, Seattle, Washington, US.

Mathews P. George, IP lawyer and blogger. He practises in the High Court of Kerala and other forums, India.

Eldar Haber, Senior Lecturer, Faculty of Law, University of Haifa; Haifa Center for Law and Technology, Faculty of Law, University of Haifa, Israel.

Tom Iverson, Partner, Media and Entertainment/Defamation and Privacy Disputes Lawyer at Simkins LLP, UK.

Stavroula Karapapa, Professor of Law, School of Law, University of Reading, UK.

Yolanda M. King, Interim Assistant Dean for Student Affairs and Associate Professor of Law, Northern Illinois University College of Law, US.

Nicola Lucchi, Associate Professor of Law, Jönköping International Business School, Sweden.

Massimo Maggiore, Founding Partner of Maschietto Maggiore Beseghini law-firm; Adjunct Professor of Marketing Law at Bocconi University of Milan, Italy.

Giuseppe Mazziotti, Assistant Professor in Intellectual Property Law, Trinity College Dublin, Ireland.

Jani McCutcheon, Associate Professor, University of Western Australia Law School, Australia.

Luke McDonagh, Senior Lecturer, The City Law School, City, University of London, UK.

Péter Mezei, Associate Professor, University of Szeged, Hungary.

Marc Mimler, Lecturer in Law, Bournemouth University, UK.

Arul George Scaria, Assistant Professor of Law and Co-Director of the Centre for Innovation, Intellectual Property and Competition (CIIPC), National Law University, Delhi, India.

Cathay Y. N. Smith, Assistant Professor, University of Montana Blewett School of Law, US.

Xiyin Tang, Lecturer, Yale University; Visiting Fellow, Yale Law School, Yale Information Society Project; Associate, Mayer Brown LLP, US.

Foreword

The past two decades have seen policy makers, academic commentators and industry experts advancing a plethora of law and policy proposals to recalibrate the copyright system. As they observe, the internet, social media and other new communication technologies have raised novel questions, posed unforeseen challenges and generated new stakeholders. In view of this transformation of the copyright landscape, reform is needed to ensure the adequate and appropriate protection of creativity. Such reform will not only enable creators to recoup the time, effort and resources invested in the creative process, but will also allow them to obtain much-deserved rewards and recognition.

One area that scholars have underexplored in intellectual property literature concerns the protection of non-conventional subject matters under copyright law. These subject matters often lie outside the copyright system. Even when they fall within, they tend to lurk at the fringes and receive no or very limited protection.

This book examines three distinct groups of non-conventional subject matters. The first group covers those subject matters that have received considerable or growing attention in the copyright arena. Examples are graffiti artworks, tattoo designs, comedic routines, typeface appearances and fragrance formulae. Thus far, copyright disputes involving these subject matters have remained infrequent, and commentators have treated these disputes as rarities, if not oddities. In intellectual property literature, these subject matters have also raised more questions than answers. The chapters in this book are no exception. They provide helpful assessments of whether the courts or the parties have drawn the right conclusions – and, if so, whether the arguments and rationales involved could be extended to cover other non-conventional subject matters.

The second group of non-conventional subject matters comprises those having longstanding existence yet receiving no or limited copyright protection. Examples are culinary presentations, jazz improvisations, magic performances and sport moves. The chapters discussing these subject matters are highly insightful because legal experts continue to question their fit within the copyright system. If they are asked about copyright protection for these subject matters, it will take them time and effort to formulate thoughtful answers. Such answers will require not only a deep understanding of copyright law and policy, but also creative legal analyses in areas that do not lend themselves to easy answers.

The last group of non-conventional subject matters includes novel issues that have slowly begun to garner attention from copyright experts from around the world. Providing cases of first impression, these subject matters are the fruits of new technological advances, such as 3D printing, synthetic biology and artificial intelligence. The chapters examining these subject matters are timely, important and of great practical value. They provide useful insights into the ongoing and future development of copyright law and policy.

Taken together, all of the chapters in this book have addressed important questions concerning the role, objectives and boundaries of the copyright system. They also bring to mind the distinction the US Supreme Court made between traditional and non-traditional contours of copyright protection in *Eldred v. Ashcroft*. Although that decision focused on US law, the Court's failure to provide additional guidance foreshadows a worldwide challenge in the copyright field. Since the establishment of the modern copyright system more than three centuries ago, protection has been greatly expanded from books and other printed matter to cover photographs, motion pictures, architectural plans, sound recordings, computer programs and now 3D printed artefacts. What constitute 'traditional contours' – or, for our purposes, conventional subject matters – remains highly debatable.

In the past few decades, copyright law and policy has evolved considerably, thanks to the active and continuous demand by creators and their investors for new or increased protection. As a result, the boundaries of the copyright system have been vastly expanded to cover non-conventional subject matters that ordinarily would not receive protection. As the system continues to expand, its coverage will begin to encompass other non-conventional subject matters.

There is an irony in this seemingly endless expansion, however. What is deemed non-conventional depends largely on what falls within the convention. The more protection the copyright system affords to non-conventional subject matters, the more of these subject matters will eventually become conventional. Thus, whether intentional or not, the present treatment of non-conventional subject matters will colour the future development of copyright law and policy. The current developments and reform in the copyright area will also directly influence the future protection of these subject matters.

In this volume, Enrico Bonadio and Nicola Lucchi have put together an excellent collection of chapters examining the treatment of non-conventional subject matters in copyright law. Written by authors from around the world and covering laws in multiple jurisdictions, the individual contributions are intellectually stimulating, legally creative and generally fun to read. They call on us to think more deeply about the nature and scope of copyright protection, the justifications behind such protection and the interplay between copyright law and other areas of intellectual property law. They also invite us to explore the core question in today's copyright debate: should protection be expanded – and if so, how and why?

This book has covered a wide variety of non-conventional subject matters. It has also explored many important questions concerning the protection or non-protection of these subject matters. To deepen our understanding of copyright law and policy, I invite you to join the volume's editors and contributors to embark on an exciting and rewarding journey traversing the landscape of non-conventional copyright.

Peter K. Yu
Professor of Law and Director, Center for Law and Intellectual Property;
Texas A&M University School of Law
April 2018

Introduction: setting the scene for non-conventional copyright

Enrico Bonadio and Nicola Lucchi*

Informal conversations and chats sometimes give rise to new ideas and trigger adventures. This is exactly what happened when we met for a 'fika' (Sweden's ritual coffee break) on a cold February afternoon in Nicola's university office in the Swedish town of Jönköping and started talking about bizarre and to some extent esoteric forms of creativity in the fields of art, music, popular culture, technology, science and beyond.

The chat took a playful turn from the very beginning. Enrico asked: 'Nicola, what if more and more graffiti writers start enforcing their copyright against corporations that commercially exploit their tags? . . . And if disc-jockeys decide to complain about party-goers that attend their set and use Shazam [the smartphone application that can identify music] to learn about their tunes and copy the entire compilation? . . . Also, last week I was in a restaurant in East London and had a wonderful and quite eye-catching radish salad. Could the chef claim copyright in such culinary presentation?'. Nicola seemed interested and enthusiastically entered the game: 'Yeah. That's intriguing. And what about Michael Jordan's classic spin move under the basket? Could he protect such an athletic movement? . . . Or could comedians and magicians hold and enforce a valid copyright over their jokes and tricks?'. The chat went on and on.

After mentioning other extravagant products of human ingenuity, including tattoos, bio, conceptual and land art, typefaces and the scents of perfumes, we both wondered if it was worth spurring a comprehensive academic debate about whether copyright laws should accommodate the interests of such an increasingly broad category of creators by guaranteeing them protection.[1] We agreed it was. That belief has not

---

* Both authors contributed equally to this introductory chapter and are listed alphabetically.

[1] Some scholars have already speculated about whether some non-traditional works could be protected by copyright. See for example R. Anthony Reese,

only been fuelled by our acknowledging that artists, musicians, writers and other creative people are continually exploring new ways of expression; we also started puzzling over whether copyright laws – in a world characterised by fast social and technological changes – are too obsessed with categorising works and a rigid application of the requirements for protection.

Our investigation aims therefore at offering a constructive view of the changing nature of contemporary creative efforts and verifying whether it is possible to imagine a more open and flexible copyright system which is capable of welcoming new forms of creativity.

It is not just art, music and culture which are evolving and challenging the traditional principles and comfortable contours of copyright laws. Recent technological developments are also playing a crucial role. Take synthetic biology, bioengineering, artificial intelligence and 3D printing techniques, which are increasingly helping to address some of today's most pressing industrial, technological and biomedical challenges, and are producing outputs that might soon be recognised as meriting copyright protection. It could be argued that this is just 'history repeating'. Indeed, copyright laws have always needed to adapt to and catch up with new technologies and their societal consequences; this occurred for example with the advent of photography, motion pictures, computer programs and several other 'new' works.[2]

We therefore believe it is time to offer a new set of perspectives on, and views of, copyright regimes, with a view to investigating whether the latter should be more flexible regarding whether new or non-conventional forms of expression in atypical fields deserve to be protected. The issues that will be addressed by the contributors to this volume, and are briefly summarised in this introduction, predominantly refer to copyrightable subject matter, the originality and fixation requirements, the functionality exception, the authorship requirement, the idea/expression dichotomy and to the debate whether illegal and immoral works should be denied protection.

Copyrightable Subject Matter in the 'Next Great Copyright Act', 29 Berkeley Tech. L.J. 1489 (2014); Pamela Samuelson, Evolving Conceptions of Copyright Subject Matter, 78 U. Pitt. L. Rev. 17 (2016); Christopher Buccafusco, A Theory of Copyright Authorship, 102 Va. L. Rev. 1229 (2016).

[2] See also Brad Sherman and Leanne Wiseman (eds), *Copyright and the Challenge of the New* (Wolters Kluwer, 2012) 1 (noting that 'one of the most challenging things about copyright law is that it is constantly subject to change').

1. COPYRIGHTABLE SUBJECT MATTER

As mentioned, a first hurdle for new and non-traditional works to be protected is the rigid categorisation embraced by several copyright laws. One of these is the UK copyright act, which contains a closed list granting protection to only eight categories of works, namely literary, dramatic, musical and artistic works, films, sound recordings, broadcasts, and typographical arrangements of published editions.[3] Only works which happen to fit within these available boxes are protected, thus creating what can be labelled as a 'pigeon-hole' system. While this regime may be praised for providing legal certainty, criticisms have also been voiced. First, closed list systems may leave several works unprotected, as confirmed by the British case *Creations Records Ltd v. News Group Newspapers Ltd.*[4] In that dispute it was held that no copyright subsists in a specially prepared scene (consisting of a white Rolls Royce in a swimming pool of a hotel and an arrangement of other objects and props), as this scene could not be categorised as a dramatic work or as an artistic work, or sculpture, collage or work of artistic craftsmanship. It has been suggested that such an outcome would create an unjustifiable discrimination, and in particular offend public policy, as the intellectual and manual efforts used to come up with the scene in *Creations Records* (the ensemble of object and props was photographed, and the official picture was used as a cover for an LP album of the band Oasis) reflected a significant amount of creativity which would deserve protection.[5]

Categorisation is not completely rigid, though. Take laws like the 1790 US copyright act which granted protection only to a very few works such as books, charts and maps; or the UK Statute of Anne of 1710 which covered just books. Both laws featured closed lists of copyrightable works, which have however expanded over time, through both the introduction of new subject matter into copyright laws[6] and a (broad) interpretation of existing

---

[3] Other common law jurisdictions such as Australia and New Zealand have also a closed list system.

[4] *Creation Records v. News Group Newspapers* [1997] EMLR 444.

[5] See Andrew Christie, A Proposal for Simplifying United Kingdom Copyright Law, 23 Eur. Intell. Prop. Rev. 26 (2001).

[6] In the US for example, after the 1790 Act, the copyrightable subject matter was expanded by Congress by adding engravings, etchings and prints in 1802, then musical compositions in 1831, dramatic compositions in 1856, photographs in 1865, paintings, drawings and statuary in 1870, motion pictures in 1912, and sound recordings in 1971; see Christopher J. Buccafusco, A Theory of Copyright Authorship, 102 Va. L. Rev. 1229, 1238, 1242 (2016).

categories by judges and copyright offices.[7] It has also been argued that current closed systems like the 1988 UK copyright act might be interpreted as encompassing new forms of creative expressions not explicitly mentioned in the eight categories.[8] Indeed, following the *Infopaq* decision by the Court of Justice of the European Union (CJEU)[9] (and even after the UK leaves the EU in March 2019), the British copyright act might be read to reflect the EU principle that anything that constitutes an 'intellectual creation' should be protected, and accordingly the fact that a work does not fall within those eight categories would not exclude protection.[10]

While the rigidity of closed list systems may still be tempered by lawmakers' interventions or judicial interpretation – because the existing categories can actually operate in an open-ended manner[11] – jurisdictions that have adopted open and illustrative lists of copyrightable works, such as France, Germany and the Netherlands, at least in theory should leave the 'copyright door' more open to new and non-traditional works.[12] The

---

[7] For example, in *Burrow-Giles Lithographic Co. v. Sarony*, 111 US 53 (1884), the US Supreme Court upheld the power of Congress to extend copyright protection to photography, by holding that photographs could be considered as 'writings' pursuant to Section 1(8) of the US Constitution (comparing them to engravings, etchings and prints), and noting that they embodied 'the intellectual conception of the author'. Later, *in Edison v. Lubin*, 122 F. 240 (3d Cir. 1903) the concept of photographs was interpreted extensively to encompass films as they were considered as moving pictures. Also, under the 1909 Act the US copyright office registered computer programs as 'books', considering them as 'how to do it' books; see R. Anthony Reese, Copyrightable Subject Matter in the 'Next Great Copyright Act', supra note 1, at 1519. And in the UK 'printed sheet music' was considered as included in the sub-category of books in *Bach v Longman* (1777) 98 Eng Re p. 274, 1275 (K.B.).

[8] See Lionel Bently and Brad Sherman, *Intellectual Property Law* (4th edn, 2014) 60–61.

[9] Case C-5/08, *Infopaq Int'l A/S v. Danske Dagblades Forening*, 2009 E.C.R I-6569 (setting out the EU originality standard for copyright protection).

[10] See the British case *SAS Institute Inc. v. World Programming Ltd* [2013] EWHC 69 (Ch), [27]. At the time of writing it cannot be ruled out that the UK, even after leaving the EU, will still need to observe several laws governing the EU single market, including copyright-related directives and CJEU decisions. And even in the case of the departure of the UK from EU single market rules, the 'intellectual creation' test for copyright subsistence might still be kept by British courts.

[11] Tanya Aplin, *Copyright in the Digital Society: The Challenges of Multimedia* (Hart Publishing, 2005) 249. See also the recent decision of the English High Court in *Banner Universal Motion Pictures Ltd v. Endemol Shine Group Ltd & Anor* [2017] EWHC 2600 (Ch), which suggests that the format of a TV game show or quiz show can be protected by copyright as a dramatic work.

[12] It is also because of the fact that the Dutch copyright act contains an open list of copyrightable works that the local Supreme Court found that the scent of a

open list approach would have the advantage of injecting greater flexibility into the system and aim at guaranteeing comprehensiveness of protection (for example, new works will not be denied protection merely because they do not fall within any category of the list).[13] What the US Congress noted when passing the 1976 Act and justifying the open list of protectable works it includes, is quite eloquent and relevant here: '[a]uthors are continually finding new ways of expressing themselves, but is impossible to foresee the forms that these new expressive methods will take'.[14]

Yet, the propensity for 'categorisation' has remained even in countries that have adopted open list systems. In the US, for instance, new works like computer programs and video games were considered copyrightable since they had been deemed as falling within already identified categories of subject matter (literary and/or audiovisual works),[15] and not because they had been treated as original works of authorship belonging to a new category.[16] That categories of copyrightable subject matter remain relevant in the US copyright system has led some commentators to argue that the list provided by the act is effectively a closed one.[17] Also, grey areas surrounding the scope and width of these categories remain. Take the lack of definition of choreographic works and pantomimes within the current

---

perfume can be eligible for protection: see *Kekofa BV Lancome Parfums et Beauté et Cie SNC* [2006] ECDR 26.

[13] See Tanya Aplin, Subject Matter. In *Research Handbook on the Future of EU Copyright* (Estelle Derclaye, ed., Edward Elgar Publishing, 2009) 49. An extensive but non-exhaustive list of works is also provided by Article 2(1) of the Berne Convention (Berne Convention for the Protection of Literary and Artistic Works, Sept. 9, 1886, S. Treaty Doc. 99-27, 1161 U.N.T.S. 3 (amended Sept. 28, 1979) (hereinafter also: Berne Convention), which clarifies that the expression 'literary and artistic works' includes 'every production in the literary, scientific and artistic domain'.

[14] 1976 House Report No. 94-1476, at 51 (also noting that '[t]he history of copyright law has been one of gradual expansion in the types of works accorded protection . . . The historic expansion of copyright has also applied to forms of expression which, although in existence for generations or centuries, have only gradually come to be recognized as creative and worthy of protection').

[15] The definition of 'computer program', for instance, was added in 1980 by the US Congress to 17 U.S.C. § 101. This legislative intervention, and a court decision in *Apple v. Franklin* in 1983, clarified that the Copyright Act treated computer programs as literary works.

[16] See Aplin, *Copyright in the Digital Society*, supra note 11.

[17] One of these is our contributor Xiyin Tang, who has authored Chapter 1 on 'Copyright in the Expanded Field: On Land Art and Other New Mediums'. See also (again) the 1976 House Report No. 94-1476, at 51, stressing that the 1976 US Copyright Act does not aim to 'allow unlimited expansion into areas completely outside the present congressional intent'.

US copyright Act; and the ambiguity such a lack has created in relation to, for example, the protectability of sport moves as choreographic works and of magicians' performances of tricks as pantomime.[18] The state of uncertainty created by such grey areas is epitomised by the 'rule of doubt' certificates occasionally issued by the US copyright office in relation to unconventional subject matter. These certificates show the office's scepticism about the copyrightability of a certain work, but give the applicant an opportunity to convince courts that the claimed subject matter is indeed protectable.

2. ORIGINALITY

Falling within the copyrightable subject matter is not enough. Whether or not forms of creativity can be protected also depends on whether the work in question meets a sufficient level of originality. Although the Berne Convention does not explicitly state that copyrightable works must be 'original', most countries do provide so. The originality test varies across jurisdictions, ranging from the US law 'minimal degree of creativity' test (as affirmed in *Feist v. Rural*)[19] to the EU 'intellectual creation' requirement.[20] Despite such differences, the originality threshold has traditionally been low in many countries.

Requiring that copyrightable works simply be authors' intellectual creations or the result of minimum level of creativity may make the categorisation of eligible subject matter less relevant. The CJEU decisions which affirmed the copyrightability of works such as 11-word extracts from protected newspapers,[21] a user interface in a computer program,[22] or simple portrait photographs,[23] seem to confirm this point. In light of such case law – as mentioned above – it could be argued that where there is an intellectual creation, it is unnecessary to further categorise it as a particular type of work in order to consider it copyrightable. Whereas these findings may not particularly affect those EU countries that have

[18] See R. Anthony Reese, Copyrightable Subject Matter in the 'Next Great Copyright Act', supra note 1, at 1526.

[19] *Feist Publications, Inc., v. Rural Telephone Service Co.*, 499 U.S. 340 (1991).

[20] *Infopaq v. Danske*, C-5/08, supra note 9. In this case the CJEU found that anything that constitutes an intellectual creation should be protected by copyright.

[21] See again *Infopaq v. Danske*, C-5/08, supra note 9.

[22] See Case C-393/09 *Bezpečnostní softwarová asociace – Svaz softwarové ochrany v. Ministerstvo kultury*, 2010 E.C.R. I-13971.

[23] See Case C-145/10, *Painer v. Standard Verlags*, 2011 E.C.R. I-12533.

an open-ended list approach, it may instead push jurisdictions like UK (even after Brexit)[24] to expressly drop their closed list system and link the concept of work to originality.[25] This would not really come as a big surprise as also the *travaux préparatoires* for the (Berne Convention) Brussels Revision Conference of 1948 indicated that the requirement of 'intellectual creation' is implicit in the concept of 'literary and artistic work'.[26]

Against this backdrop, this volume will investigate whether various forms of expression are capable of satisfying originality tests in several jurisdictions. Can for example culinary creations (which often borrow elements from other chefs' presentations) or certain esoteric graffiti and street artworks such as tags and short sentences written on walls be considered original? What about sport moves like those performed in figure skating? Can they be protected as choreographic works where substantial creativity exists on the athletes' and their trainers' side? And CAD (computer-aided designs) files which are used to print 3D products? Can we argue that a CAD file depicting an object that is simply scanned or whose picture is taken demonstrates creative and free choices? And what about DJs' sets? It does not seem heretical to maintain that disc-jockeys display genuine artistic creativity in compiling and executing their sets and that their compilations can be considered original. Contributors to this book will expand on these and similar issues.

3. FIXATION

As copyright protects just expressions of ideas, and not ideas themselves, in several jurisdictions copyrightable works need to be fixed in a tangible medium.[27] Indeed, despite the Berne Convention protecting 'literary and artistic works . . . whatever may be the mode or form of its expression',[28] it also allows countries to determine that works are ineligible for protection 'unless they have been fixed in some material form'.[29] The fixation require-

---

[24] See supra note 10.

[25] On the CJEU case law suggesting that the notion of 'originality' and 'work' conflate see Eleonora Rosati, *Originality in EU Copyright: Full Harmonization through Case Law* (Edward Elgar Publishing, 2013).

[26] See Sam Ricketson and Jane Ginsburg, *International Copyright and Neighbouring Rights: The Berne Convention and Beyond* § 8.03 (Oxford University Press, 2006).

[27] See also Megan M. Carpenter, If It's Broke, Fix It: Fixing Fixation, 39 Colum. J.L. & Arts 355 (2016) (noting that fixation 'is often the judicial threshold over which idea becomes expression').

[28] Berne Convention, Article 2, Sec. 1.

[29] Berne Convention, Article 2 Sec. 2.

ment is thus allowed by Berne – and while copyright laws such as the US and UK statutes have incorporated it, others have not (this is the case with several civil law jurisdictions).

That said, some authors in this volume wonder whether certain copyright regimes are too obsessed with the tangible and physical embodiment requirement. Do contemporary copyright laws – in the words of a contributor[30] – remain very much focused and dependent on a specific concept of medium purity? Are these laws still focused, for example, on a kind of traditional art which is just displayed in museums and dependent on judges' subjective aesthetic judgments about what can be considered 'artistic',[31] with the result that many pieces of contemporary art may be excluded from protection?

Old and recent cases have highlighted how a rigid application of the fixation requirement may leave new forms of art without protection. Take the garden 'Windflower Works' devised by painter and gardener Chapman Kelley, an important example of landscape art. In the seminal US case *Kelley v. Chicago Park District*, [32] it was held that such a living garden lacks the kind of stable fixation normally required to support copyright: 'a garden is simply too changeable to satisfy the primary purpose of fixation; its appearance is too inherently variable to supply a baseline for determining questions of copyright protection'.[33] A not very different outcome occurred in the British case *Merchandising Corporation of America v. Harpbond*,[34] where it was found that the facial make-up of the pop artist Adam Ant did not constitute a painting for the purpose of copyright ('if the marks are taken off the face there cannot be a painting. . . . a painting without a surface is not a painting'). This finding attracted criticism.[35] It has been noted that it is difficult to see why a pop star's face is less of a surface than a piece of canvas and could not be protected by

---

[30] See Xiyin Tang's Chapter 1 'Copyright in the Expanded Field: On Land Art and Other New Mediums'.

[31] See Anne Barron, Copyright Law and the Claims of Art, 4 Intell. Prop. Q. 368, 371 (2002).

[32] *Kelley v. Chicago Park District*, 635 F.3d 290, 303 (7th Cir. 2011).

[33] Ibid.

[34] *Merchandising Corp. of America Inc. & ors v. Harpbond Ltd & ors* [1983] FSR 32.

[35] See also Anne Barron, Copyright Art and Objecthood. In *Dear Images: Art, Copyright and Culture* (Daniel McLean and Karsten Schubert, eds, Ridinghouse, 2002) 277, 293, 304 (noting that art practices nowadays are often based on 'the dematerialization of the art object: the production of art that yielded no object; or in which process, context-dependence, chance or randomness [are] prioritised over form').

copyright.[36] After all, in the US make-up designs have been considered as satisfying the fixation requirement.[37]

Issues of fixation have also arisen in relation to more esoteric, and non-strictly artistic, matter, such as the taste of a cheese. In a case which at the time of writing still needs to be decided by the CJEU (*Levola Hengelo*), Advocate General Melchior Whatelet noted that the fact that tastes are ephemeral, volatile and instable prevents them from being precisely and objectively identified and thus considered copyrightable works.[38]

Not only certain forms of artistic works may be left unprotected by a rigid application of the fixation requirement (the book will expand on subject matter as diverse as graffiti and street art, tattoos and bio art). Music can also be affected. This would be the case with jazz performers, especially when they come up with arrangements directly on stage and do not record their improvised performances. The same is true of comedians' performances. Comedians' routine may indeed depend in large part on audience interaction, such as responding to heckles from an audience member, reading subtle cues, and adapting a pre-planned routine as the performance goes on: with the result that, if comedians fail to fix their joke, they cannot rely on copyright protection.[39] Could we then argue that these outcomes openly challenge the paradigm in copyright laws that are predominantly based on the primacy of notated or fixed musical compositions or scripts?[40] This volume will also shed light on these points.

4. FUNCTIONALITY

Works that have prominent functional features may be considered uncopyrightable in several jurisdictions. That works must not conform to some

---

[36] See Bently and Sherman, *Intellectual Property Law* supra note 8, at 73–74.

[37] Make-up designs for the characters in the Broadway show 'Cats' were found to be copyrightable in *Carell v. Shubert Org., Inc.*, 104 F.Supp.2d 236 (S.D.N.Y. 2000).

[38] *Levola Hengelo* (Case C-310/17), Opinion of 25 July 2018, para 60. From the above comment it seems clear that a fixation-related issue has been raised, although the Advocate General took the pain to stress that EU copyright law does not provide a fixation requirement (para 59). The Advocate General also held that the only categories of works included in the definition given by Article 2(1) of the Berne Convention ("every production in the literary, scientific and artistic domain") are those that can be perceived through sight or hearing, but not other senses such as taste, smell or tactile sense (para 51).

[39] See Chapter 10, 'Copyright Protection for Modern Comedic Material', by Trevor Gates.

[40] See Chapter 8, 'Music Improvisation and Copyright', by Giuseppe Mazziotti.

technical necessities is a requirement which is somehow related to originality. Indeed, any work whose relevant aspects are dictated by functionalities may at the same time lack originality. Where the work also meets the requirements for patent protection, this exception serves the important policy of keeping patent and copyright regimes separate by preventing creators and innovators from relying on copyright to obtain a 'backdoor patent' on a functional article that cannot be patented (or whose patent protection has expired).[41] Yet, where artistic features are incorporated into a functional product (for example, a garment), these features are still eligible for protection in many countries subject to various conditions: in several jurisdictions including the US such artistic features are protectable provided that they are separable from the underlying product.[42]

A wide variety of works can be affected by the functionality exception. It is because of this exception that, for instance, the selection, ordering and arrangement of a Bikram yoga sequence was considered by a US court as a 'process' for achieving good health, which precludes copyright protection.[43] But what about other sequences, such as movements in sports like ice dancing, figure skating, synchronised swimming, skateboarding, or even wrestling? Can these moves be considered as being developed for a purely aesthetic purpose, and not as attempts to directly increase the chances of prevailing in an athletic contest (in which case they would not be functional for copyright purposes)? And tricks performed by magicians? Are they functional methods and systems for creating the illusions, and therefore not protectable by copyright? What about a device that, for example, merely appears to cut a person in half? Is this device really 'useful' when its only use is to entertain and create an illusion as well as evoke emotions?[44] Analogous issues are raised by food recipes (while one may consider them as functional directions for achieving a result,[45] recipes for many dishes

---

41 See the US case *Smith & Hawken, Ltd. v. Gardendance, Inc.*, 75 U.S.P.Q.2d 1853 (N.D. Cal. 2005).

42 In other countries requirements for protection of useful products seem less strict. In France, for example, it is accepted that an artistic expression should not be disqualified and deprived of copyright protection simply because it is fixed or embodied in a utilitarian article (this is the so-called theory of 'unity of art').

43 *Bikram's Yoga Coll. of India, L.P. v. Evolation Yoga*, LLC, 803 F.3d 1032 (9th Cir. 2015).

44 See Chapter 11, 'Now You Own It, No You Don't – Or Do You? Copyright and Related Rights in Magic Productions and Performances', by F. Jay Dougherty.

45 See the US case *Publ'ns Int'l, Ltd. v. Meredith Corp.*, 88 F.3d 473, 481 (7th Cir. 1996) (with the court defining a recipe as 'a set of instructions for making something . . . a formula for cooking or preparing something to be eaten or drunk: a list of ingredients and a statement of the procedure to be followed in making

may also be determined by creative expressions independent of most functional influences);[46] scents and tastes (do feelings equal function?);[47] typefaces and graffiti lettering (are they objects whose utility outweighs their artistic elements or rather stylish renditions of letters that deserve copyright protection?). This volume will expand on most of these issues.

5. AUTHORSHIP

The concept of authorship is relevant in copyright law, as without an author there cannot evidently be a copyrightable work. While the Berne Convention determines the minimum protections to be granted by countries to the works of their authors, it leaves states free to interpret the concept of authorship – and many copyright acts still do not provide a definition of authors. Indeed, few judicial decisions address what authorship means, and who an author is.[48] In the US for example the Supreme Court has defined an author as 'he to whom anything owes its origin; originator; maker; one who completes a work of science or literature'.[49] Lower courts have also noted that an author must be more than one who contributes creativity or originality to a work;[50] basically it must be one 'who superintended the whole work, the "master mind"'.[51] In other words, authorship could be

---

an item of food or drink . . . a method of procedure for doing or attaining something'). The US Copyright Office seems to confirm the exclusion by stating that 'mere listing[s] of ingredients or contents' are amongst the materials not subject to copyright protection: see 37 C.F.R. § 202.1.

[46] See Christopher Buccafusco, Making Sense of Intellectual Property Law, 97 Cornell L. Rev. 501, 508 (2012). The same could be said of the creative way food is presented on plates. In Chapter 6 of this volume Cathay Smith notes that after the US Supreme Court ruling in *Star Athletica, L. L. C. v. Varsity Brands, Inc.* (580 U.S. 2017), which held that the pictorial and graphic features of cheerleaders' uniforms could be applied to other medium and therefore could be protected by copyright, culinary presentations that present artistic qualities might also get the same protection.

[47] A case about copyright protection of tastes is the already mentioned *Levola Hengelo* (Case C-310/17), where the Advocate General Melchior Whatelet considered a taste of a cheese as not copyrightable (although he did not expand on functionality related grounds).

[48] See Jane C. Ginsburg, The Concept of Authorship in Comparative Copyright Law, 52 DePaul L. Rev. 1063, 1066 (2002).

[49] See again *Burrow-Giles Lithographic Co. v. Sarony*, 111 U.S. 53, 58 (1884).

[50] See *Aalmuhammed v. Lee*, 202 F. 3d 1227, 1233 (9th Cir. 2000).

[51] Ibid. (citing *Burrow-Giles Lithographic Co.*, 111 U.S. at 53, 61).

defined as a requirement that no creation that does not entail at least some degree of human intervention is eligible for copyright protection.

One may therefore legitimately doubt whether works produced without a predominant human 'touch' can be considered protectable. Let us think about works created by animals (for instance, the sound of toads or frogs) or growing plants.[52] A recent US case which has attracted considerable attention has dealt with whether an animal can be considered a copyright holder. A macaque monkey had used a camera belonging to British photographer David Slater in Indonesia to take a selfie, which then became famous worldwide. The judge dismissed the argument brought by the plaintiff, an animal rights organisation, which had filed a lawsuit requesting that the monkey be assigned the copyright. The argument was that US copyright law does not prohibit an animal from owning a copyright. Yet, such argument was rejected by the court which held that while the animal had constitutional standing it 'lacked statutory standing to claim copyright infringement of photographs'.[53]

The issue of authorship and ownership of copyright in works created by non-humans is not limited to artworks created by animals; and the debate around the required degree of human intervention in a work is not novel in copyright law. As far as computer-produced works are concerned, for instance, Australian courts have often been reluctant to affirm copyright due to a focus on requiring a human author (in other countries such as UK, Ireland and New Zealand, however, these works do attract copyright instead with the author being considered the person that has made the arrangement necessary to produce the work).[54] The debate has recently been revamped in light of the rise of creations made artificial intelligence and robots, without substantial human involvement: here the focus is not on whether 'computers are enabling people' to create works, but on whether 'people are enabling computers' to do that.[55]

Authorship- (and ownership-) related issues arise in several chapters of the book. What about (again) a living garden designed by an environmentalist artist but which grows by itself, or biological resources manipulated as a consequence of genetic engineering? And what about traditional

---

[52] These examples are given by Jani McCutcheon in Chapter 3, 'Copyright in Bio Art'.

[53] See *Naruto v. Slater*, 2018 U.S. App. LEXIS 10129 (9th Cir. Cal., April 23, 2018).

[54] See for example Section 9(3) of the UK Copyright, Designs and Patents Act 1988.

[55] See Annemarie Bridy, Coding Creativity: Copyright and the Artificially Intelligent Author, 2012 Stan. Tech. L. Rev. 5 (2012).

folk music where a 'chain' of musicians have added their own creativity and variations to older tunes and songs, many of which may be orphan works or in the public domain? Also, what happens when tattoo artists receive specific instructions from their client about the style and details of the tattoo? Is there an, or who is the, author (and copyright owner) in these circumstances? Our contributors will come back on these and similar points.

6. PUBLIC INTEREST CONCERNS

Copyright laws do not only protect the interests of creators, but are also concerned with public policy. They aim in particular at guaranteeing that certain subject matter and elements are not protected (and thus cannot be taken away from the public domain) because they need to remain available for, and usable by, anybody. Judges adjudicating copyright disputes might also be inclined to exclude from protection works that display an unlawful or immoral content, or that have been illegally produced.

A key copyright principle which aims at boosting the public domain is the idea–expression dichotomy. It ensures that the manifestation of an idea is protected rather than the idea itself and therefore reconciles the interests of copyright owners with those of down-stream creators, users and the public at large, who can use and exploit non-copyrightable elements for their own purposes.[56] While this principle is codified in US copyright law ('in no case does copyright protection for an original work of authorship extend to any idea . . . [and] concept . . .')[57] and in the TRIPS Agreement[58] (the first international treaty to incorporate such a principle), in other jurisdictions it has been developed through case law. In theory, the

---

[56] Users' interests in copyright law have often been closely looked at by scholars. For recent perspectives see Christophe Geiger, Copyright as an Access Right, Securing Cultural Participation through the Protection of Creators' Interests. In *What if We Could Reimagine Copyright?* 73, 97 (R. Giblin and K. Weatherall, eds, ANU Press, 2017) (stressing the need to introduce a sort of three-step test in order to obtain access to copyright protection and arguing for the centrality of copyright's role in permitting access to culture through the recognition of both specific obligations for copyright holders and rights for users).

[57] 17 U.S. Code § 102.

[58] See Article 9(2) of the Agreement on Trade-Related Aspects of Intellectual Property Rights, April 15, 1994, Marrakesh Agreement Establishing the World Trade Organization, Annex 1C, 1869 U.N.T.S. 299 (hereinafter: TRIPS).

principle looks relatively easy to grasp.[59] Yet, verifying in practice whether a specific element falls within either of these categories is definitely more difficult. The crucial question is: where does the idea stop and the expression start? This dilemma is particularly evident when it comes to certain forms of art and creativity where the line distinguishing ideas and expressions is rather blurred. For example, what about conceptual art, for which the concept behind the artwork is more important than the finished art object? And bio art, which is often (again) fundamentally conceptual and process-focused? Or jokes, which seem to be based on merely funny ideas, but could also be personalised by comedians that change the arrangement of the words and perform them in a different style? These and similar issues will be dealt with by several contributors to the volume.

The often blurred line between ideas and expressions is confirmed by the doctrines of 'merger' and *scènes à faire*, which are sometimes referred to by US courts. The merger doctrine argues that where the idea and expression are intrinsically connected, and the expression is indistinguishable from the idea because they have merged, such an element cannot be protected by copyright. The *scènes à faire* doctrine (literally meaning 'scenes that must be done') refers to elements that are standard, or indispensable in the type of work at issue or sequences of events that necessarily result from the choice of a setting or situation (in an old movie focusing on cops in the Bronx, for example, it is inevitable that the scenery would include drunk people, stripped cars, prostitutes, and rats;[60] and in a spy movie we should expect to see secret agents hiding self-defence gadgets in their clothes). The rationale of both doctrines is clear. Granting the first comer monopolistic rights over these elements would significantly hinder the creation of follow-on works, and thus restrict the public's enjoyment of further creative expressions. The application of these doctrines to specific works, such as jokes, magic tricks, sport moves and pornographic works, will be discussed in this book.

Not only ideas. Facts cannot be copyrighted either. In the (already mentioned) case *Feist v. Rural*[61] the US Supreme Court confirmed that copyright can only apply to the creative aspects of a collection of data

---

[59] A very recent EU case where the idea-expression dichotomy has been dealt with is the above mentioned *Levola Hengelo* (Case C-310/17) on the protectability of a taste of a cheese. In his Opinion of 25 July 2018, Advocate General Whatelet was quite clear and straightforward, as he considered tastes of food as non protectable ideas (para 55).

[60] This was actually noted by a US court in *Walker v. Time Life Films, Inc.*, 784 F.2d 44 (2d Cir. 1986).

[61] *Feist Publications, Inc. v. Rural Telephone Service Co.*, 499 U.S. 340 (1991).

(namely, the creative choice of what data to include or exclude and the order and style in which the information is presented), but not to the information itself. Many other jurisdictions have followed this approach, thus respecting the spirit of the Berne Convention, which does not extend protection to news of the day or to miscellaneous facts having the character of mere items of press information.[62] It is therefore understandable that several commentators have criticised recent legislative moves, for example by Germany and Spain, introducing a *sui generis* right to press publications that cover short snippets of news made available by news aggregators and web search engines.[63] A chapter of this book will expand on what is considered by many as too broad a right which may end up covering facts such as news items and lead to an unwelcome monopolisation of information and a consequential restriction of free speech.[64]

What about works that display immoral or illegal contents (including adult entertainment), or that are illegally created (for example, graffiti)? Could they be considered uncopyrightable on the assumption that they do not advance knowledge or positively contribute to society? A positive answer to this question could be justified on the ground that denying these works protection would serve overriding public interests, such as the defence of common morality and of public order in general. On the other hand, however, it could also be argued that refusing to protect these controversial forms of expression may amount to governmental censorship, which in turn would create a chilling effect on creativity as well as put pressure on people (mainly judges) who are required to decide which works deserve to be protected by copyright. Contributors to this volume will expand on these issues and in particular on the benefits (and drawbacks) of a 'neutral' approach to the illegality and morality aspects surrounding a work.

---

[62] Article 2(8), Berne Convention.

[63] See Lionel Bently et al., Response to Article 11 of the Proposal for a Directive on Copyright in the Digital Single Market, entitled 'Protection of press publications concerning digital uses' on behalf of 37 professors and leading scholars of Intellectual Property, Information Law and Digital Economy (5 December 2016), available at https://www.cipil.law.cam.ac.uk/sites/www.law.cam.ac.uk/files/images/www.cipil.law.cam.ac.uk/documents/ipomodernisingipprofresponsepresspublishers.pdf (last accessed on 5 February 2018); European Copyright Society, Opinion on the Proposed Press Publishers Right (April 25, 2018) available at https://europeancopyrightsocietydotorg.files.wordpress.com/2018/06/2018_european-copyrightsocietyopiniononpresspublishersright.pdf.

[64] See Chapter 14, 'The Press Publisher's Right in the European Union', by Stavroula Karapapa.

7. THE STRUCTURE OF THE BOOK

The volume is structured in four parts, each of which highlights a broad category of works. Contributors will look at specific jurisdictions, including – depending on the chapter – the US, the UK, the EU and some of its Member States as well as Australia and India.

The investigation focuses on forms of expression which have recently attracted attention amongst courts and scholars. The subjects have been selected with the purpose of identifying trends and highlighting possible commonalities and differences in the progressive expansion of the range of protectable works. We have focused on topics that have had or may have a lasting impact on case law or policy. In providing a direct and concise insight in a new area of copyright-related research, the selection of topics for comment and analysis may appear uneven or occasionally idiosyncratic. In order to overcome this limit, such a selection follows a deliberate pattern: we have tried to identify topics that are particularly capable of stimulating academic discussion and the contamination of legal approaches amongst different legal systems, with the aim of promoting exchanges and dialogues between legislators, judges and legal scholars.

Part I analyses whether new or non-conventional forms of art can, and should be, protected by copyright. The art which is looked at in this section is not the kind of art that is usually found in traditional galleries or museums.

Chapter 1, authored by Xiyin Tang, focuses on land art. This form of art uses materials found in nature, such as the soil and rocks, vegetation and water found on-site. The author casts doubt about whether land art can satisfy the current requirements for copyright protection in the US, especially fixation, and explores other possible forms of protection when traditional copyright claims may not be feasible or successful.

In Chapter 2, Shane Burke focuses on language as a signature style of conceptual art, arguing that copyright law struggles to protect the artistic authorship inherent in text instructions used to communicate the work to those who execute its final visual form. He makes the point that, irrespective of whether more traditional or progressive approaches to the legal definition of art are employed, more systemic forms of artistic production, particularly those that involve a literary to visual 'translation' of the work as part of that process of production, currently struggle to be fully protected by copyright law.

Jani McCutcheon then describes, in Chapter 3, the broad spectrum of creations that might arguably fall within the definition of bio art, including art usually generated in laboratories using biotechnical media and tools, such as tissue, blood, bacteria, plants or other organisms. While explaining the potential misalignment between bio art and copyright, the chapter explores the practical and normative ramifications of this dissonance.

Chapter 4, authored by Enrico Bonadio, addresses issues related to copyright protection of street art and graffiti. After focusing on whether these forms of urban and public art can satisfy the requirements for protection under US and UK laws, he expands on the possible clash between the artist's moral right of integrity and the right of the owner of the property upon which the artwork is placed, as well as on whether illegally created street artworks can and should be protected by copyright.

In Chapter 5 Yolanda King deals with tattoo art. She notes that tattoo artists increasingly claim copyright protection for subject matter that – the author argues – both fits within traditional conceptions of creativity and satisfies the fixation requirement. This form of art – she stresses – pushes the boundaries of copyrightable subject matter due to, among other factors, the unique nature of the physical support upon which it is placed, i.e. the human body.

With Chapter 6 Cathay Smith concludes this section by focusing on culinary presentations. She explores each element of copyrightability – originality, work of authorship, fixation, and useful article separability – to detail how copyright law treats such presentations. She argues that as a consequence of the recent developments in US case law, one of the biggest hurdles to copyright protection of culinary presentations – the useful article exclusion – may actually not be as big a hurdle as it has been in the past.

Part II of the volume addresses a wide category of music-related creations and other cultural endeavours.

Luke McDonagh, in Chapter 7, highlights what traditional music is and why it poses challenges to copyright. The author analyses how musicians who add originality to the arrangement of a recorded song may be entitled to copyright over that arrangement as a musical work in its own right; and makes the point that it is possible to protect traditional music under copyright – but that this might not be in the best interests of the process of music creation.

Chapter 8, authored by Giuseppe Mazziotti, focuses on improvised jazz music. It explains why copyright systems have ended up disfavouring extemporaneous authorship, in spite of an international legal framework that is strongly protective of authors' rights. The author argues that a rigid application of the fixation requirement may leave jazz performers unprotected, especially when they come up with arrangements directly on stage and often do not record their improvised performances.

In Chapter 9, Tom Iverson deals with compilations of musical works made by disc-jockeys. He wonders whether play lists created by DJs, either at parties or for radio stations, may be protected by copyright as original selections and arrangements of songs. He does so by also analysing a recent case where a music streaming service company was sued in the UK, for alleged copyright infringement, by a company that manages a well-known London music club.

Chapter 10 is authored by Trevor Gates. It examines the requirements for and difficulties in protecting jokes under copyright law, how a court would handle a copyright infringement action between comedians, and potential solutions aimed at finding a balance between protection and enforcement. Gates concludes there are ways copyright law could be more flexible in its application to modern comedic material and proposes some solutions to assist comedians enforce their rights in their works, including one that may not require any formal legal changes.

Jay Dougherty, in Chapter 11, delves into copyright and related rights in magic productions and performances, primarily under US law. He argues that while copyright law might in theory provide magicians with a property right, its limiting doctrines preclude protection in many instances. After exploring both limits on copyright and potentially protectable subject matter, the chapter discusses two judicial decisions that confronted claims for protection of magic productions – one where the magician failed to receive protection and a more recent case where the copied magician was successful.

With Chapter 12 Péter Mezei concludes this section by focusing on the copyrightability of sport moves. Examples of famous moves are given, including Bob Cousy's behind-the-back pass in basketball, Antonín Panenka's penalty kick in football, Werner Rittberger's loop jump in ice skating and Dick Fosbury's flop in high jump. The author wonders at what juncture sporting moves may become creative pieces in their own right and thus protectable as unique artistic expressions.

Part III of the volume shifts the analyses towards works created on an industrial scale, in particular to address market demands, and in the context of scientific research.

In Chapter 13 Arul George Scaria and Mathews George expand on whether typefaces – the designs for fonts – are and should be copyrightable. By looking at US, UK and Indian scenarios, the authors look into what may be protectable (the aesthetic and ornamental features not dictated by functionality) and non-protectable (utilitarian aspects) of typefaces. They stress that legibility is clearly a utilitarian function associated with the design choices for typefaces, and this poses challenges for their acceptance as a protectable subject matter.

Stavroula Karapapa then explores in Chapter 14 the right to press publications, which has been introduced in countries like Germany and Spain, but which has also provoked harsh criticism. Devised to provide news publishers with an additional layer of protection, this sui generis right aspires to offer a solution to the so-called 'newspaper crisis'. The author argues however that such a new right does not constitute an appropriate measure for addressing declining revenues in the press publishing sector.

Chapter 15, authored by Charles Cronin, focuses on whether fragrances can be protected by copyright or other forms of intellectual property. The author reminds us that perfumes' manufacturers have recently attempted to protect against the copying of their fragrances by claiming they are copyrightable works of expression – and looks at some interesting and to some extent surprising decisions by Dutch and French courts on this issue.

In Chapter 16 Teshager Dagne looks into CAD (computer-assisted) files for 3D printing and related copyright issues. CAD files are necessary components of the 3D printing process, as the file holds the instructions on which the printer relies. The author then delves into thorny questions, including whether copyright subsists in such files under Canadian, US and EU laws.

Nicola Lucchi then expands in Chapter 17 on copyright as a possible form of intellectual property protection for engineered DNA sequences. He argues that copyright might turn out to be not only flexible enough to handle contemporary technologies producing living organisms, but also socially preferable to patent protection, especially in terms of enhanced chances for users to access and use essential public knowledge assets in the life sciences sector.

Chapter 18 is authored by Massimo Maggiore, who speculates on whether computer generated works such as *The Next Rembrandt* (an artwork produced by a computer that had analysed many paintings by the Dutch maestro) can and should be protected. While pushing for a refocusing of the rationale for copyright protection from the 'author' to the 'work' as such, the author calls for a reform of copyright laws in jurisdictions which, unlike the UK and a few others, do not expressly protect this category of works.

The last part of the volume sheds light on another peculiar category of creative outputs – works that display an illegal or immoral content.

This section starts by hosting a contribution by Eldar Haber on illegal works. In Chapter 19 he wonders whether copyright law should be neutral as to content and other forms of illegality surrounding the work, on the assumption that it is beyond this branch of law to act as a censor and create a chilling effect on creators. This chapter offers a taxonomy for illegal works and lays foundations for the normative discussion on whether the law should grant rights and remedies for the makers of works which – many stress – do not advance knowledge or contribute to society.

In Chapter 20 Enrico Bonadio and Nicola Lucchi analyse the intersection between pornography and copyright. After commenting on US, UK and French cases on the copyrightability of obscene works (especially, porn movies), the authors highlight the arguments both in support and against copyright protection of this controversial subject matter. They conclude that the overriding need to protect free speech makes the arguments supporting the copyrightability of pornography stronger.

Chapter 21, authored by Marc Mimler, then addresses the question of how to deal with works that promote hate by focusing on books authored by German Nazi leaders. While noting that in some circumstances copyright in these works has been relied on to generate income, the author notes that copyright has also been applied as a tool to restrict access to and dissemination of controversial works such as the book *Mein Kampf* by Adolf Hitler.

This book is brought to an end by a concluding chapter, written by Tim Dornis, which provides an economic analysis of non-conventional works. Many chapters in this volume – Dornis stresses – address the demarcation line between non-protectable and protectable subject matter and it is in this regard that a careful handling of non-conventional copyright issues proves to be a question of fundamental economic policymaking. The author concludes that the current copyright system offers numerous ways to 'thin out' – and thereby economically fine-tune – rights protection.

As mentioned, the analysis offered by the volume aims at spurring an academic debate about whether certain new and non-conventional forms of creativity can, and should, attract copyright protection. While many chapters conclude that the works in question are, or should be, protectable, a few of them make the point that the opposite is the case (for example, because the requirements for protection cannot be met) and highlight instead that other legal or self-regulatory solutions may be followed to protect and reward these forms of creativity: such alternative solutions include – again, depending on the type of works analysed in each chapter – patent protection, the law of confidence, rules against unfair competition, the right of publicity as well as trespass and nuisance laws and even free and open source style licences.[65] After all, that more appropriate legal tools could

[65] Finally, a few contributors, while dealing with their own category of non-traditional works, not only look at copyrightable subject matter, requirements for protection and the other issues briefly discussed in this Introduction; they also expand on other legal specificities typical of the range of works analysed in their chapters, for example the impact of the exclusive rights of authors and publishers on derivative works (see the chapters authored by Luke McDonagh (Chapter 7 on traditional music) and Giuseppe Mazziotti (Chapter 8 on improvised jazz perform-

be available to protect creations that cannot be identified with sufficient precisions under copyright law (especially the taste of food products such as cheeses) was recently noted by a group of copyright law scholars.[66]

The chapters of this book therefore not only make the point that it may be time to challenge some basic tenets of copyright laws by embracing more flexible and objective ways to identify protectable works and interpret the requirements for protection. Some contributors also cast doubts about whether copyright is the right instrument to address and regulate new and non-traditional forms of expression. This is not an entirely new issue. Several scholars have already identified a negative intellectual property (IP) space where various creative works, including some explored in this volume, are often produced without the lure of monopolistic rights.[67] In other words, making copyright available in these cases is not the main motivational trigger that pushes people to generate works. But is copyright *always* unattractive to these categories of creators? If so, will this lack of interest in copyright last *for ever*? Or could such creators instead acquire a strong interest in copyright, especially if and after their work has been imitated and exploited economically by someone without their authorisation? After all, as will be examined by several contributors in this volume, some of these creators have recently started taking copyright-related actions against imitators. Thus, the application of the IP negative space theory to several of the works highlighted in this book may need to be further tested in the not so distant future as more and more creators in the areas in question may soon nurture expectations of economic return and look for (copyright-focused) ways to protect their works and turn them into an even more profitable source of revenues.

---

ers)) or the need to find a balance between the moral rights of artists (especially the integrity right) and the property rights of others (see Chapter 4 'Street Art, Graffiti and Copyright', by Enrico Bonadio).

66 See European Copyright Society, Opinion on Reference to CJEU in Case 310/17, Levola Hengolo (February 19, 2018) available at https://europeancopyrightsocietydotorg.files.wordpress.com/2018/03/ecs-opinion-on-protection-for-tastes-final1.pdf.

67 See for example Karl Raustiala and Christopher Springman, *The Knockoff Economy* (Oxford University Press, 2012); Elizabeth Rosenblatt, 'A Theory of IP's Negative Space', 34 Colum. J.L. & Arts 317 (2011); Rochelle Cooper Dreyfuss, 'Does IP Need IP? Accommodating Intellectual Production Outside the Intellectual Property Paradigm', 31 Cardozo Law Review 1437 (2010); Kate Darling and Aaron Perzanowski, *Creativity Without Law: Challenging the Assumptions of Intellectual Property* (NYU Press, 2017). All these studies highlight scenarios where creativity and innovation flourish and prosper without legal incentives.

PART I

Art

1. Copyright in the expanded field: on land art and other new mediums

Xiyin Tang*

1. INTRODUCTION

It is about fifteen hundred feet in length and fifteen feet wide, built from 6,650 tons of basalt rocks and earth.[1] It sits just south of Rozel Point, in Utah's Salt Lake, next to swirls of black oil from abandoned rigs. Thousands of people a year make the pilgrimage to see it as they are photographed from above, standing like tiny ants atop the salt as if minute stars dotting a constellation. For the rest of us, we see it only through photographs, and they are almost always aerial views, in which the swirl cuts across its site like a 'strange footprint on the shores of the unknown'.[2] As it turns out, 'even the artist is startled to discover that this footprint is our own'.[3]

Is it architecture? Is it a larger-than-life sculpture? Is it nature? Is it a man-made imitation of the former? Is it site? Is it life itself?

It is Robert Smithson's Spiral Jetty, an iconic earthwork or piece of 'land art' that has, since its creation, been the stuff of art historical myth and fable – not least of which was its submersion under water at the same time that Smithson died in a fatal plane crash, creating a sort of double

* This chapter takes its name from the 1978 essay by Rosalind Krauss titled 'Sculpture in the Expanded Field', which perfectly documents the phenomenon of post-1960s artists working in an expanded medium that encompasses elements of architecture, landscape, nature, and traditional sculpture. See, Rosalind Krauss, Sculpture in the Expanded Field, 8 October, 30–44 (1979). An earlier version of this essay was originally printed in the Hofstra Law Review. Reprinted with permission.

[1] Lynne Cooke, A Position of Elsewhere. In *Robert Smithson: Spiral Jetty* (Lynne Cooke et al. eds., University of California Press, 2005) 53, 53.

[2] Ann Reynolds, At the Jetty. In *Robert Smithson: Spiral Jetty*, supra note 1, at 73, 73.

[3] Ibid.

negative or disappearance.[4] While the jetty is perhaps the most famous earthwork, it is not alone. Walter de Maria's Lightning Field in New Mexico, Michael Heizer's Effigy Tumuli in Illinois, and James Turrell's Roden Crater in Arizona are just a few of the many examples of artists working in expanded mediums that implicate site, architecture, sculpture, and nature-as-material.

US copyright law, on the other hand, only extends protection to eight categories of works: literary works, musical works, dramatic works, choreographic works, pictorial/graphic/sculptural works, motion pictures, sound recordings, and architectural works.[5] Numerous legal scholars have already noted that evolving forms of art – not just appropriation art (for which fair use remains a field ripe with possibilities and problems)[6] but conceptual art, performance art, and temporal 'happenings', which are notable for their ephemerality[7] – seem incompatible and fundamentally unprotectable under the very laws intended for their protection. Land art, specifically, does not neatly fit into any of the above-named categories. However, few solutions outside of legislative reform have been posed.[8]

Yet the fast-changing nature of art likely renders legislative reform impractical, and would leave new legislation largely obsolete in its wake. What is needed, then, are creative solutions within already-existing law – solutions that may lie outside of the copyright statute itself. Artists asserting copyright claims already frequently assert trademark and right

---

4 Ibid.

5 17 U.S.C. § 102 (2006).

6 See, e.g. Alex Kozinski and Christopher Newman, What's So Fair About Fair Use?, 46 J. Copyright Soc'y U.S. 513 (1999); Gideon Parchomovsky and Phillip J. Weiser, Beyond Fair Use, 96 Cornell L. Rev. 91, 100–1 (2010); Xiyin Tang, That Old Thing, Copyright . . .: Reconciling the Postmodern Paradox in the New Digital Age, 39 AIPLA Q.J. 71, 86–94 (2011).

7 See, e.g., Randall Bezanson and Andrew Finkelman, Trespassory Art, 43 U. Mich. J. L. Reform 245 (2010); Charles Cronin, Dead on the Vine: Living and Conceptual Art and VARA, 12 Vand. J. Ent. & Tech. L. 209 (2010).

8 See, Virginia M. Cascio, Hardly a Walk in the Park: Courts' Hostile Treatment of Site-Specific Works Under VARA, 20 DePaul J. Art, Tech. & Intell. Prop. L. 167 (2009); Bryan J. Hopping, A Proposal for the WTO to Enhance Moral Rights Protection, 11 Gonz. J. Int'l L. 2 (2007–2008); Lauren Ruth Spotts, Phillips Has Left VARA Little Protection for Site-Specific Artists, 16 J. Intell. Prop. L. 297 (2009). However, Randall Bezanson and Andrew Finkelman have applied real property principles to 'trespassory' art (i.e. art that involves trespassing in spaces) – including endowing artists with a limited 'easement' to use private property for artistic purposes. See Bezanson and Finkelman, supra note 7 at 277–9.

of personality claims as well, suggesting an awareness that personhood, branding, and artistic creation are all interlinked.[9] But as evolving forms of art begin to look less comfortably like traditional artworks, perhaps it would serve us well to also expand our thinking beyond intellectual property law. This is where focusing on the specific problem of land art or earthworks can prove instructive in demonstrating that copyright need not be the only answer to the exasperating problem of a copyright doctrine hostile to contemporary art.

Art historian Rosalind Krauss has traced art history's suspicions of the new hybridity of sculpture to 'the modernist demand for the purity and separateness of the various mediums'.[10] While land art may seem 'rigorously logical' from 'within the practice of postmodernism', where 'practice is not defined in relation to a given medium – sculpture – but rather in relation to the logical operations on a set of cultural terms, for which any medium – photography, books, lines on walls, mirrors, or sculpture itself – might be used',[11] contemporary copyright law still remains very much focused and, might I add, dependent on, very specific notions of modernist medium purity, autonomy, and specificity. Indeed, throughout the history of copyright law, judges have engaged in exactly the kind of modernist definitional line-drawing that attempts to reduce, via a single set of finite, discrete criteria, a specific medium to its 'inherent' end logic.[12] Thus if legal scholarship from the past decade has insisted on copyright's outdated reliance on the notion of the 19th-century romantic author,[13] I argue instead that it is precisely the 20th-century avant-garde that copyright law owes its debt to in its insistence on the purity of the medium itself.

Further, land art proves a timely and especially interesting case study for our purposes, as two circuit court decisions have essentially deemed this hybrid art form unprotectable – either under copyright law or its

---

[9] See, e.g., Stacey L. Dogan and Mark A. Lemley, What the Right of Publicity Can Learn from Trademark Law, 58 Stan. L. Rev. 1161 (2006); Xiyin Tang, The Artist as Brand: Toward a Trademark Conception of Moral Rights, 122 Yale L.J. 218 (2012).

[10] Rosalind Krauss, Sculpture in the Expanded Field. In *The Originality of the Avant-Garde and Other Modernist Myths* (MIT Press, 1985) 277, 288.

[11] Ibid.

[12] See infra, notes 21–23 and accompanying text.

[13] See, e.g., Peter Jaszi, Toward a Theory of Copyright: The Metamorphoses of 'Authorship', Duke L.J. 455 (1991); Martha Woodmansee, The Genius and the Copyright: Economic and Legal Conditions of the Emergence of the 'Author', 17 Eighteenth-Century Stud. 425, 428–30 (1983–1984) (both arguing that authorship is a socially-constructed category reliant on the Romantic notion of an inspired genius).

newer moral rights component, the Visual Artists Rights Act of 1990 (VARA).[14] I focus the first part of this chapter on these two decisions and their implications for earthworks. The second part then links this legal reasoning to a surprising predecessor: the 20th-century avant-garde insistence on medium purity, not the 19th-century Romantic author construct that most legal scholars have focused on in their analysis of copyright doctrine's limitations. Applying this reasoning, I come, in my last part, to the seemingly disappointing conclusion that this new form of hybrid art is, logically, uncopyrightable. Yet, I propose two new forms of protection under real property and tort law – specifically, the claims of trespass and nuisance – which seem to most aptly reflect land art's site-specific aspirations. Above all, I suggest that we may think of legal protection for artworks not as an either-or (either it is uncopyrightable and hence unprotectable or vice versa), but as existing in an expanded field that encompasses other viable options for redress like real property or tort law.

2. THE LEGAL DILEMMA OF NON-TRADITIONAL ART FORMS

This section begins by reviewing a few of the fundamental principles of copyright law that this chapter is concerned with, before examining how recent court decisions have applied these requirements to works of art that move beyond the walled-in ethos of the museum.

2.1 Copyright Basics

There are a few strange requirements of copyright law that will form the chief concerns of this chapter and its case law analysis. I will review them each briefly in this section, but they can be summed up into three major themes: fixation, idea/expression dichotomy, and medium categorization.

The first two are relatively straightforward. One of the most basic requirements for copyright protection is that an 'original work . . . of authorship [be] fixed in a tangible medium'.[15] The fixation-in-a-tangible-medium requirement interacts with and informs the other major requirement of copyright law: that only expression, not ideas, can be

---

[14] The two cases are *Kelley v. Chicago Park Dist.*, 635 F.3d 290 (7th Cir. 2011) and *Phillips v. Pembroke Real Estate, Inc.*, 459 F.3d 128, 131 (1st Cir. 2006).

[15] 17 U.S.C. § 102 (2006).

copyrighted.[16] After all, if an idea is abstract, expression is concrete, manifesting itself in something physical and tangible. Indeed, the two requirements are frequently discussed together by courts in examining whether a work satisfies this most basic of requirements.[17] But, of course, the two present their own, unique problems. Even presuming that an 'idea' were 'expressed' in a concrete way (in a performance, for example), if there is no tangible record of it (in a recording, for example), that expression would not be copyrightable, either.[18]

The third major concern of this chapter is a surprising one, for the very word 'medium' seems to be more of an art historical preoccupation than a legal one. For modernism, which reached its apex in the 1960s with the dominance of such art critics as Clement Greenberg, medium specificity was an imperative that drove critics to attempt to distill just what made a painting a painting, a sculpture a sculpture, a photograph a photograph, and art art. Greenberg, the king of modernist art critics, penned his most famous work on medium specificity – 'Modernist Painting', in which he defines what painting is – a purely visual experience that is first and foremost flat, rendered so by the flatness of its pictorial surface.[19]

Yet copyright, too, relies on assumptions about the inherent characteristics of a specific medium, even if such assumptions are implicit rather than made explicit. For example, in deciding that an object which has utilitarian concerns simply cannot qualify for copyright protection,[20] courts are doing more than adhering to the letter of the law, which states that useful articles belong to the realm of patents[21] – they are saying, rather, that

---

[16] Ibid. ('In no case does copyright protection for an original work of authorship extend to any idea, procedure, process, system . . . concept, principle, or discovery, regardless of the form in which it is described, explained, illustrated, or embodied in such work').

[17] See, *Dollcraft Industries, Ltd. v. Well-Made Toy Mfg. Co.*, 479 F.Supp. 1105, 1113 (E.D.N.Y. 1978) ('[C]opyright law protects an individual's concrete expression of his own idea').

[18] See, e.g., *Hoopla Sports and Entertainment, Inc. v. Nike, Inc.*, 947 F.Supp. 347, 354 (1996) (noting that even if a basketball game is copyrightable, the fact that it was not recorded renders it ineligible for copyright).

[19] Clement Greenberg, Modernist Painting. In *The Collected Essays and Criticism: Modernism with a Vengeance, 1957–1969*, Vol 4 (1960) (University of Chicago Press, 1995) at 85.

[20] Consider the famous case of *Brandir Int'l, Inc. v. Cascade Pac. Lumber Co.*, 834 F.2d 1142 (2d Cir. 1987), in which the court had to decide whether a bike rack was sufficiently artistic to be considered sculpture or else influenced by utilitarian concerns and thus a product of industrial design.

[21] 17 U.S.C. § 101 (2006).

artworks, like Greenberg's flat picture plane, should be inherently useless, and the more useless it is, the more likely it is to be eligible for copyright. Then again, some might object that this leaves us merely with a distinction between copyrightable artworks and non-copyrightable non-art, without necessarily defining mediums within the copyright statute.

Yet land art, in traversing the boundaries of sculpture, architecture, land, or site, brashly confronts judges with the problem of medium definition within the copyright statute itself. Even more importantly, consider the Visual Artists Rights Act of 1990 (VARA), which gives the author of a painting, drawing, print, sculpture, or photograph, existing in a copy of 200 or less, the right to prevent the destruction, distortion, mutilation, or other modification of a work that would be prejudicial to the author's reputation.[22] A work labelled 'architecture' instead of 'sculpture', then, would not be eligible for VARA protection. Judges could either decide, for example, that land art is more like 'architecture' and hence not subject to VARA, or else that it is not art at all.

As we will see below, the few courts that have confronted the issue of land art have also implicitly confronted the issue of medium specificity – what a medium is, what it should be, what it contains, what it does. But their failure to acknowledge the very assumptions they are making about medium has led to much confusion and, as others have pointed out, poor case law.[23] For as it turns out, the two courts that have most notoriously opined on land art have done so from a highly modernist-inflected view of sculpture: that it should be man-made, static and unchanging, and, above all, that it is a self-contained monument that should sit on a pedestal within the white walls of the museum.

2.2 *Phillips v. Pembroke*: Stripping Land Art of VARA Protection

Yet earthworks do not fit comfortably into the self-contained, walled-in ethos of the museum and its pedestals. Minimalism, with its emphasis on atmosphere and environment, was the natural predecessor of these large-scale site-specific works that cannot be easily divorced from their surroundings. In a problematic case for this new hybrid sculpture, the court created a judicially-imposed VARA carve-out for site-specific art, holding that VARA's 'integrity' right (the right to prevent modifications

---

[22] 17 U.S.C. § 106(A) (2006).

[23] See, e.g., Rachel E. Nordby, Off of the Pedestal and Into the Fire: How Phillips Chips Away at the Rights of Site-Specific Artists, 35 Fla. St. U. L. Rev. 167, 186 (2007); Cascio, supra note 8, at 184–5.

and mutilations of the art work) did not apply to removals of site-specific art from its, well, site. In *Phillips v. Pembroke Real Estate*, the work in question was one by an artist named David Phillips, who designed a series of stoneworks for Boston's Eastport Park. These sculptures, which depicted Zodiac signs and marine animals, contained granite stones that mirrored those stones located along the Boston Harbor.[24] In 2003, the defendant, Pembroke Real Estate, moved to relocate some of Phillips' sculptures off of the park grounds. Phillips sued under VARA, claiming the integrity right and arguing that the park itself, including the granite paths and the sea walls of Boston Harbor near which it was located, was integral to his work. The First Circuit disagreed – to say the least. Rather than simply holding that the sculpture could be dissembled and removed without violating VARA rights so long as the integrity of the sculptures themselves were kept intact, the court went further in holding that no work of site-specific art is protected under VARA.

Their reasoning relied on the 'public presentation' exception of VARA, which states that 'the modification of a work of visual art which is the result of conservation, or of the public presentation, including lighting and placement, of the work is not a destruction, distortion, mutilation, or other modification'.[25] The court below had interpreted this to mean that VARA allows for the removal of the work from its site but requires that the work itself be kept intact.[26] There is evidence that this was in line with Congress's wishes when drafting VARA, as they noted that 'the removal of a work from a specific location comes within the [presentation] exclusion because the location is a matter of presentation'.[27]

But the First Circuit went farther. It first acknowledged Phillips' argument that the site itself is integral to site-specific art, helped along by a few testimonies from those within the art historical community. One testimony, for example, opined to the court that '[m]uch of modern sculpture does not exist separate from its context, but rather integrates its context with the work to form, ideally, a seamless whole'.[28] The First Circuit, applying this logic, deemed it an illogical reading of VARA that the act would both protect site-specific art against modification or mutilation and yet allow for precisely its modification or mutilation by divorcing it from its site, which operates as part of the work.[29]

---

24 *Phillips v. Pembroke Real Estate, Inc.*, 459 F.3d 128, 131 (1st Cir. 2006).
25 17 U.S.C. § 106(A)(c)(2) (2006).
26 See, *Phillips*, 459 F.3d at 140.
27 H.R. Rep. No. 101-514, at 12 (1990).
28 Phillips, 459 F.3d at 134.
29 Ibid. at 140.

Needless to say, the reasoning is slightly attenuated, though best elucidated in this one either-or: '[e]ither VARA recognizes site-specific art and protects it, or it does not recognize site-specific art at all'.[30] The court's ultimate decision on the latter reflects a deep antagonism to land art's uncomfortable position between that of traditional art objects, governed by the law of chattels (i.e. movable property), and land, governed by the law of real property.

American law is already deeply protective of the rights that attach with any piece of property, including chattel – VARA has been roundly criticized for its limitations on what an owner may do with his own painting, including destroy it, alter it, and paint over it.[31] The First Circuit's refusal to grant VARA protection to land art, then, shows judges' reluctance to extend VARA any farther than the traditional categories of 'painting, prints, sculpture, and drawing' that it protects. (Further, both the *Phillips* and *Kelley* court, discussed below, have interpreted these categories as autonomous, separate mediums, rather than loose categories within which hybrids may be constructed.) Real property – land – on the other hand, presents graver concerns, as the court suggests that 'owners of nearby property who had nothing to do with the purchase or installation of Phillips' works would be subject to claims that what they do with their property has somehow affected the site and has, as a result, altered or destroyed Phillips' works'.[32]

What *Phillips* leaves open, then, is whether site-specific art or land art is copyrightable at all. One possible reading of *Phillips* suggests that the court's concern solely with modification or destruction – and ownership, in general – could nonetheless qualify land art as an architectural work, for which 'the owners of a building embodying an architectural work may, without the consent of the author or copyright owner of the architectural work, make or authorize the making of alterations to such building, and destroy or authorize the destruction of such building'.[33] This exception nicely satisfies the First Circuit's concern with property ownership, as the statute's divorce of building from copyrightable work (in which a building owned by another may contain a copyrightable architectural work owned by the artist) would transfer over nicely to the land art context. Thus we could analogize the copyrighted architectural work to the artwork itself, and the building to the site that surrounds it.

Phillips, then, was at least hinting towards one viable option for placing

---

30 Ibid.

31 See, Amy Adler, Against Moral Rights, 97 Cal. L. Rev. 263 (2009).

32 Phillips, 459 F.3d at 142.

33 17 U.S.C. § 120 (2006).

land art more firmly within the architectural medium – for legal purposes, at least. But in a more recent case involving land art, the court firmly moved in an even more radical direction – of declaring the uncopyrightability of land art.

2.3 *Kelley v. Chicago Park District*: Stripping Land Art of all Copyright Protection

If *Phillips* presented a question of medium categorization, in which defining land art as architecture (or at least as not-sculpture) may entail losing certain rights that sculptures are subject to, *Kelley* eschews the issue for the very basics of copyrightability instead – the idea/expression dichotomy and the fixation-in-a-tangible-medium requirement. The case centred around a wildflower garden created by the artist Chapman Kelley. 'Wildflower Works' consisted of 'two enormous elliptical flower beds, each nearly as big as a football field, featuring a variety of native wildflowers and edged with borders of gravel and steel'.[34] 'Wildflower Works' immediately calls to mind other 'expanded field' artworks like Maya Lin's Wave Field at Storm King, a wavy expanse of lush green grass framed by the forest beyond.[35]

Chapman had planted his work in 1984 in the Chicago Park District, but the condition of the flower beds began deteriorating by the early 2000s. By then, the city decided to modify the garden, 'substantially reducing its size, reconfiguring the oval flower beds into rectangles, and changing some of the planting material'.[36] Chapman sued under VARA, claiming that the city had engaged in an unauthorized modification and mutilation of his work.

The Seventh Circuit began by immediately calling into question whether 'Wildflower Works' can be thought of as a painting or a sculpture at all – a question of medium. 'To qualify for moral-rights protection under VARA, Wildflower Works cannot just be "pictorial" or "sculptural" in some aspect or effect, it must actually be a "painting" or "sculpture". Not metaphorically or by analogy, but really', the court writes.[37] Unfortunately, the court never articulates just what about a work makes it 'really' a painting or sculpture, rather than having mere 'sculptural' or

---

34 635 F.3d at 291.

35 See, Holland Cotter, Where the Ocean Meets the Mountains, N.Y. Times (7 May 2009), http://www.nytimes.com/2009/05/08/arts/design/08lin.html?pagewanted=all.

36 635 F.3d at 291.

37 Ibid. at 300.

'pictorial' qualities. Perhaps in order to avoid the more vexing issue of determining just what each medium is, the court moves on to a more basic (and seemingly simpler) question: whether the work can be copyrighted at all, as 'a living garden lacks the kind of authorship and stable fixation normally required to support copyright'.[38] The first, the court maintains, is true because much of 'Wildflower Works' owes its form to nature – 'the colors, shapes, textures, and scents of the plants originates in nature, not in the mind of the gardener [the derivative term they have now bestowed upon Chapman]'.[39] The second, that the work is not fixed, is also true, the court opines, because '[a] garden's constituent elements are alive and inherently changeable, not fixed'.[40] Because the work is unable to cross even the most basic of all copyright hurdles – originality and fixation – Chapman does not own a copyright, let alone moral rights, in the work.

Yet this reasoning is somewhat attenuated, suggesting the deep discomfort with medium that led the court to address a more straightforward question about fixation instead. However, the idea that a living garden lacks fixation would also call into question much of sculpture made from materials meant to degrade or wear over time – even man-made ones like steel. The sculptures of Richard Serra, for example, are not constant, stable, unchanging –but are intended to register the marks and wear of life as the steel ages, changing in color and texture throughout the years.[41] In this sense, even more 'traditional' sculpture is not 'fixed' under the court's definition. And ruling out flowers as merely an 'idea' of nature seems silly to say the least – plenty of sculptures depend on natural materials (sand, for example, in the sculptures of Joseph Beuys) for their existence.[42]

All of which suggests that what we are really getting at in both these opinions are basic assumptions about what sculpture – and really, art – is. As I argue in the following part, judges' natural distrust of land art owes much to modernism's insistence on a purity of art form – meant to be

---

[38] Ibid. at 304.

[39] Ibid.

[40] Ibid.

[41] Deborah Solomon, Our Most Notorious Sculptor, N.Y. Times (8 October 1989), http://www.nytimes.com/1989/10/08/magazine/our-most-notorious-sculptor.html?pagewanted=10&src=pm ('Cor-Ten steel [the material Serra makes his sculptures from] was designed to acquire a dark, even patina of rust over time, and this natural process endows Serra's sculptures with a handsome brown hue. The rust can function metaphorically, too, reminding the viewer of his own inevitable decay').

[42] Mary and Leigh Block Museum of Art, Joseph Beuys: Sand Drawings, Block Museum of Art, http://www.blockmuseum.northwestern.edu/view/exhibitions/past-exhibits/2012/joseph-beuys-sand-drawings.html.

placed in the museum, free from the every day, and transcendent in its aspirations.

3. COPYRIGHT IN THE EXPANDED FIELD: WHAT COPYRIGHT LAW OWES TO MODERNIST MEDIUMS' SPECIFICITY

I have already suggested that copyright law, like modernism, focuses on distilling a medium down to its inherent end logic.[43] Yet both *Phillips* and *Kelley* seem less guided by what sculpture or painting is than by an almost intuitive, implicit idea of what it isn't. And yet what are both judges so convinced that art shouldn't be? For this, we turn to another great modernist critic, Michael Fried, who famously set out to define what sets 'art' apart from mere 'objects' in his 1967 essay 'Art and Objecthood'. For if this chapter had begun by positing that perhaps it is not just the 19th-century notion of the romantic author-genius that copyright law owes its debt to, but its 20th-century avant-garde successors, a re-reading of Fried's article garners unexpected parallels between the rhetoric of the Kelley court and the high modernist mantra.

Fried, in his polemic against minimalist art, termed the works of well-known artists like Donald Judd, Carl Andre, and Robert Morris as objects, not artworks. Inspired by Greenberg's earlier rant against minimalism, in which he argued that '[m]inimal works are readable as art, as almost anything is today—including a door, a table, or a blank sheet of paper',[44] Fried takes this new 'condition of non-art' and calls it 'objecthood'. It is as though objecthood alone can, in the present circumstances, secure something's identity, if not as non-art, at least as neither painting nor sculpture; or as though a work of art – more accurately, a work of modernist painting or sculpture – were in some essential respect not an object'.[45] Here, we see Fried and Greenberg visibly working to categorize the new sculpture as something other than – other than painting, other than sculpture, maybe non-art, at best an object – just as the *Kelley* court expressed suspicion at granting VARA protection to something that did not fit neatly into the traditional category of painting qua painting or

---

43 Supra, notes 20–23 and accompanying text.

44 Clement Greenberg, Recentness of Sculpture. In *The Collected Essays and Criticism: Modernism with a Vengeance, 1957–1969* (University of Chicago Press, 1995) at 250, 253–4.

45 Michael Fried, Art and Objecthood. In *Art and Objecthood: Essays and Reviews* (University of Chicago Press, 1998) 148, 152.

sculpture qua sculpture. The court's categorization of 'Wildflower Works' as merely having 'sculptural' or 'painterly' qualities turns what Rosalind Krauss had termed the expanded sculpture's negation-as-positive into a negation-as-negative (surely how Greenberg and Fried would see it): 'not-sculpture' or 'not-painting' does not merely expand into a quaternary field of 'not-sculpture as painting or not-painting as sculpture'.[46] Rather, 'not-sculpture' or 'not-painting' becomes, simply, not-art.

Yet why is this so? Why is Fried so eager to categorize minimalism as not-art, as condemned to 'objecthood' status? Similarly, how does the *Kelley* court come to the conclusion that 'Wildflower Works', too, is not art, but at best a garden, an ordinary work of landscaping in the middle of a public park? Surprisingly, we can view the justifications both provide as one and the same. That is, we can read the Seventh Circuit's anxiety over the 'not fixed', intangible nature of 'Wildflower Works' as parallel to Fried's distrust of the 'endlessness' of minimalist sculpture.

'Endlessness', Fried writes, or the condition of 'being able to go on and on, even having to go on and on, is central both to the concept of interest and to that of objecthood'.[47] And so in many ways, experience blurs art and life, just as minimalist objects are not self-contained, but rely on the viewer, on the room, on atmosphere and the sounds and air of light itself for its effect. Certainly, Smithson's 'Spiral Jetty' takes the idea of art as experience or art-blurred-into-life to its next logical progression, as viewers travel along the length of the jetty one step at a time in the long journey towards its centre. In person, the jetty is not viewable in an instant, not graspable as a whole. And no two individuals' experience of the work will be the same – the jetty, whether submerged, re-emerged, laced with rain or dry from drought, is inextricably linked with atmosphere, with the time of day or the workings of the weather. As such, it, like the works of its minimalist predecessors that Fried railed against, lacking the 'presentness . . . that one experiences as a kind of instantaneousness, as though if only one were infinitely more acute, a single infinitely brief instant would be long enough to see everything, to experience the work in all its depth and fullness, to be forever convinced by it'.[48]

It is this instantaneousness that modernist art aspires towards (as in the works of Kenneth Noland or Jules Olitski or David Smith or Anthony Caro, in which 'at every moment the work itself is wholly manifest',

---

[46] Rosalind Krauss, Sculpture in the Expanded Field. In *The Originality of the Avant-Garde and Other Modernist Myths* 277, 283 (MIT Press, 1985).

[47] Ibid. at 166.

[48] Ibid. at 167.

contends Fried),[49] and this instantaneousness that minimalist objects lack. For the latter, 'the experience in question persists in time, and the presentment of endlessness . . . is essentially a presentment of endless or indefinite duration'.[50] So we return to the *Kelley* court's emphasis on 'fixation in a tangible medium', which now manifests itself as an understanding of art as precisely Fried's modernist notion of instantaneousness. A work of art, in other words, should not persist in time, but should be fixed in one moment, a brief instant that captures the work in all its depth and fullness, in its oneness and wholeness, forever. A work of art cannot go on and on, ever-changing, inherently alive as the flowers in Chapman's work. It is this suspicion of the intangibility of experience itself that the *Kelley* court seems to abhor, the idea that art may expand beyond the object itself to encompass the conditions of life and time. Somewhere we may read the mystic's hope for immortality in here, for transcendence. Art should endure even as we must die. Hence why Fried disdains minimalist art, calling it 'literalist' instead, in what could be a direct evocation of the drabness of life that art should aim to transcend.[51] 'We are all literalists most or all of our lives', Fried writes. But 'presentness is grace'.[52]

4. A CREATIVE SOLUTION FOR SITE-SPECIFIC ART: LOOKING OUTSIDE INTELLECTUAL PROPERTY LAW

The phrase 'beyond IP' has, in recent years, come into fashion. The premise that other forms of law, regulation, or social norms could govern, stimulate, encourage, or incentivize knowledge and information production is nothing new.[53] However, few have considered the 'beyond IP' approach as applied to evolving forms of art – often times, the argument

---

49 Ibid.
50 Ibid. at 166.
51 Ibid. at 166.
52 Ibid. at 168.
53 The literature I consider 'beyond IP' takes many forms. It could, for example, consider government regulation or incentives (i.e., prizes), open source data and other forms of public domain (i.e., 'non-IP' means of governing knowledge production). To name just a few examples, see Amy Kapczynski, The Cost of Price: Why and How to Get Beyond Intellectual Property Internalism, 59 UCLA L. Rev. 970 (2012); Jessica Litman, The Public Domain, 39 Emory L.J. 965 (2003); Yochai Benkler, Freedom in the Commons: Towards a Political Economy of Information, 52 Duke L.J. 1245 (2003).

is centred on, instead, adapting IP law to accommodate art.[54] In this section, I consider alternatives to land art protection that do not hinge on the medium-specific requirements of copyright law. These approaches, I argue, are more fitting for the expanded field of certain forms of art production, as they are also superior in that they do not seek either common law or statutory schemes that inevitably evolve more slowly than the art forms they would address. Rather, real property and tort law principles work for land art.

4.1 But First, Why Shouldn't Land Art Be Considered 'Architecture' for Copyright Purposes?

The decisions in *Phillips* and *Kelley*, while they have been roundly criticized by the art law community, seem to reach the most logical conclusion regarding the vexed relationship between the new expanded sculpture and moral rights. After all, work that invites modification or alteration from viewers themselves – in the form of footprints on the 'Spiral Jetty', for example, or a plucked flower from Chapman's 'Wildflower Works' – cannot then claim the right to be free from modification or alteration. Similarly, land art's communitarian function suggests at least some need for a balancing of rights between public interest and artistic control – both *Phillips* and *Kelley*, after all, involved highly public works in city parks. Like architecture, which serves a public function as a monument or tourist attraction, it would seem at least reasonable that the public remain free to photograph large works of copyrighted land art. All of which might lead one to suggest – why not just definitively categorize land art as architecture? We would hence solve both the moral rights problem (as architectural works are not protected under VARA) and allow for public engagement with these highly visible works in the form of a pictorial representation exception.[55] Moreover, land artists have themselves already acknowledged the architectural influences in their work, so the analogue seems a natural one.[56]

And yet some would argue that the right to control how, for example, the 'Spiral Jetty' appears in photographs and films is as important or more

---

[54] See, e.g., Amy Adler, Fair Use and the Future of Art, 91 N.Y.U. L. Rev. 559 (2016); Megan Carpenter and Steven Hetcher, Function over Form: Bringing the Fixation Requirement into the Modern Era, 82 Fordham L. Rev. 2221 (2014).

[55] 17 U.S.C. § 120 (2006).

[56] See Krauss, supra note 10, at 287 (detailing artists' explorations of architecture, drawing, and site to form a hybrid she terms 'marked sites', which includes art others have termed site-specific art).

important than the jetty itself – after all, most of us only know the jetty as a 'neatly framed . . . striking still photographic image that is endlessly reproduced as the preferred symbol, elegant and concise, for "earth art"'.[57] Yet the question in this instance necessarily becomes a case-by-case one: a widely circulated, unauthorized image of the jetty might forever tarnish its reputation, but a relatively unknown photograph published by a tourist on his blog might not affect the jetty's public image at all.

This is where I suggest that artists working with the new sculpture in the expanded field similarly expand their legal remedies beyond that of copyright law to land art's natural analogue: real property law and the doctrines of trespass and nuisance. The former goes straight to the heart of ownership rights, and the latter to rights of the public.

4.2 Trespass as One Form of Creative Control

The doctrine of trespass is probably the more straightforward of the two. It acknowledges that owners have a 'bundle' of rights to their property, including the right to exclude others from entering and the right to place limitations on what one may do upon entering. Using a mix of property and contract law, then, owners of land art (or really, the owner of the site on which the land art sits – oftentimes they are aligned, as the Dia Art Foundation and Walter de Maria are in the example below) may appropriately place conditions on those entering – requiring, for example, any photography to be solely for personal use, or banning photography all together. Such is precisely the method that Dia uses with Walter de Maria's 'Lightning Field', which does not permit visitors to simply drop in, look around, and snap photos. Rather, access to and from the 'Lightning Field' is tightly controlled by Dia, which ferries visitors to the site only upon a two-night stay.[58] Further, they are explicit about retaining the right to control the way 'Lightning Field' appears in photos.[59] This makes sense for works like de Maria's, as the poles of Lightning Field undoubtedly look best when struck with lightning and lit up against a dark night sky. Dia, therefore, does not need a copyright claim to ban any unauthorized photos of 'Lightning Field' – they can simply assert a breach of contract action, as every visitor to the site has explicitly agreed not to publish, or take, any such photos.

However, Dia's approach to 'Lightning Field' does not transfer over as well to, for instance, the 'Spiral Jetty' context, as the sheer mass of it

[57] Reynolds, supra note 2, at 73.
[58] Geoff Dyer, Poles Apart, The New Yorker (18 April 2011) at 62, 65.
[59] Ibid.

makes the object visible from the air. Thus, even if Dia could easily limit what visitors do on the ground, they cannot contract with those chance passersby who spot the jetty from above and snap a few photos of it. The same applies to other mass-scale earthworks visible (and really, only viewable as whole) from above, like Michael Heizer's 'Effigy Tumuli', a massive compilation of earth mounds made to resemble creatures when viewed from the air.[60]

Furthermore, as noted above, trespass works only when the interests of the artist and the land owner are perfectly aligned. It should be every land artist's hope that if he does not himself own the land, then at least whoever does will act in the best interest of the work, neither destroying it nor altering it for the worse, and working to ensure that any unauthorized, unfavorable reproduction in film or photographs is halted. In a world in which we expect property owners to behave like rational actors, working diligently to increase property value and preserve what is valuable, this should not be an unrealistic expectation. Yet the current legal troubles of the jetty, in which the Dia Art Foundation (entrusted with the care of the jetty after Smithson passed) failed to renew its lease and has now (at least temporarily) lost control of the site to the State of Utah, leaves room to be pessimistic about the future of land art.[61]

In the absence of legal ownership of the land on which the art sits, artists hoping to preserve a certain degree of control over their works must look past the simpler gate-keeping methods of trespass and contract law. My proposal, then, turns to an unlikely source: tort law and applying the principles of nuisance to the unauthorized aerial photography/film context.[62]

---

60 Michael Heizer, Effigy Tumuli, Double Negative, http://doublenegative.tarasen.net/effigy_tumuli.html (last visited 21 September 2012).

61 Glen Warhol, Control of Iconic Sculpture Spiral Jetty in Dispute, The Salt Lake Tribune, 23 June 2011, http://www.sltrib.com/sltrib/news/51970092-78/dia-jetty-lease-spiral.html.csp.

62 This chapter focuses on the reproduction and distribution rights of copyright (17 U.S.C. Section 106(1) and 17 U.S.C. Section 106(3) (2006)) because these are the two rights land artists will be most concerned with. Consider the other exclusive rights a land artist could assert if he held a copyright: the right to prepare derivative works and the right to display his work. However, land art's invitation to the public to interact with it renders a prohibition on derivative works impractical and likely directly inapposite to the artist's intentions, while its immovable state means that it can only be displayed in the one place it was built – the intended site.

4.3 Nuisance Law: Balancing Public Interest with Private Right

Nuisance as a real property injury is a relatively new development in Anglo-American law, dating to the beginning of the 20th century.[63] Unlike trespass, which is a hard-and-fast rule with bright-line applicability (any unauthorized trespass upon a land is de facto wrongful),[64] nuisance is a balancing test conducted by judges that weighs the competing interests of both parties.[65] And unlike trespass, no physical intrusion is necessary – intangible intrusions, like dust particles, light, air, and noise – can constitute an actionable invasion.[66] Courts have defined a nuisance as 'anything which interferes with the rights of a citizen, either in person, property, the enjoyment of his property, or his comfort . . . A condition is a nuisance when it clearly appears that enjoyment of property is materially lessened'.[67] Courts will balance the landowner's interest in the enjoyment of his property with the social value of the activity alleged to cause the harm.[68]

Activities like air pollution emitting from a nearby factory and onto homeowners' land or noxious gas issuing from a smelting plant that substantially damaged plaintiffs' hedges and shrubs have been enjoined.[69] Further, the plaintiff is entitled to monetary damages stemming from 'the depreciation in the market value of the property injured' (a 'diminution of value' standard).[70] I argue, then, that a nuisance cause of action is just as apt in the land art context, where publication of photographs or films that misrepresent, distort, or otherwise present works like the jetty unfavorably, detracting from its overall appeal, could cause irreparable

---

63 See *Hendricks v. Stalnaker*, 380 S.E.2d 198 (Sup. Ct. Va 1989) (discussing early cases dating from the late 1800s through the mid-1900s applying nuisance law).

64 See *Jacque v. Steenberg Homes, Inc.*, 563 N.W.2d 154, 159 (Sup. Ct. Wis. 1997) (awarding punitive damages in the absence of actual harm due to 'the loss of the individual's right to exclude others from his or her property').

65 Restatement (Second) of Torts § 826 (1979) (noting that 'the gravity of the harm[]' must 'outweigh[] the utility of the actor's conduct' to be considered a nuisance).

66 See, e.g., *Medford v. Levy*, 8 S.E. 302 (W. Va. 1888) (cooking odors as nuisance); *St. Helen's Smelting Co. v. Tipping*, 11 H.L.C. 642, 11 Eng. Rep. 1483 (House of Lords 1865) (noxious vapors constitute a nuisance); *Boomer v. Atlantic Cement Co., Inc.*, 257 N.E.2d 870 (Ct. App. N.Y. 1970) (dirt, smoke, and vibration emanating from nearby plant actionable as nuisance).

67 Hendricks, 380 S.E. 2d at 200.

68 See, Restatement (Second) of Torts § 826 (1979).

69 See, Greenberg, supra note 44.

70 10 A.L.R.2d 669 (1950).

damage to the value of the jetty itself and therefore constitutes a nuisance that must be enjoined.

While almost all nuisance cases up to date have balanced a homeowner's right to the enjoyment of his property against another party's right to carry on some arguably harmful activity, a land art owner's right to the enjoyment of his property is fundamentally different from that of a resident. Specifically, the value of his property lies in its public presentation – it is mostly through the dissemination and mass circulation of images or film that favorable public regard is garnered to a remote work of land art otherwise only accessible via pilgrimage. The owner of a work of land art, unlike a homeowner, does not derive enjoyment from living peaceably on the property, which provided the justifications for categorizing noise, pollution, and noxious gases as nuisances. Rather, land art owners derive enjoyment solely via the good reputation (translating directly to higher market value) his work enjoys.[71] For the same reasons that a loud plant operation must be enjoined if it interferes with a homeowner's right to quiet enjoyment, so the mass-scale circulation of unfavorable unauthorized photographs must be prohibited to prevent the depreciation in value of the art owner's property.

Using nuisance in lieu of copyright essentially weighs the same balancing interests that copyright law is concerned with. That is, copyright is not an absolute monopoly. The fair use doctrine, which allows the appropriation of a copyrighted work in limited instances, is, like nuisance law, a balancing test that weighs the interest of the copyright holder against the interest of the defendant.[72] The four-factor fair use test weighs the purpose and character of the infringing use (whether it is transformative – that is, adding 'new aesthetics, new insights and understandings' to the original work)[73] with the nature of the copyrighted work (fictional works get more copyright protection than nonfictional works),[74] the amount of the work taken, and the effect of the infringing use on the market or value of the copyrighted work.[75]

---

[71] There has been much literature written on the effect an artist's reputation has on the value (translated via market price) of his work. See, e.g., Henry Hansmann and Marina Santilli's article, Authors' and Artists' Moral Rights: A Comparative Legal and Economic Analysis, 26 J. Legal Stud. 95, 104–5 (1997); Gladys Engel Lang and Kurt Lang, Recognition and Renown: The Survival of Artistic Reputation, 94 Am. J. Soc. 79, 105 (1988); Tang, supra note 9, at 234–5.

[72] 17 U.S.C. § 107 (2006).

[73] Pierre N. Leval, Toward a Fair Use Standard, 103 Harv. L. Rev. 1105, 1111 (1990).

[74] *American Geophysical Union v. Texaco Inc.*, 60 F.3d 913, 925 (2d Cir. 1994).

[75] 17 U.S.C. § 107 (2006).

A nuisance balancing test applied in the art context would take much of the same considerations into play. Specifically, the 'nuisance' – in other words unauthorized representation in the form of a photo or film – must be harmful in some way, and, as in fair use, severely affect the market or value of the work (the 'diminution of value' standard) without adding any tangible benefits or transformative value. Such a requirement recognizes that there is a broad public interest in art that copyright law itself acknowledges and serves – the public interest in being enriched, in appropriating knowledge, and in building on the works of others in order to create more works. By balancing the rights of the real property owner and the social value of the activity said to constitute a nuisance, nuisance law strangely mimics fair use's balancing test of the rights of the copyright holder against the social value of the new, infringing work.

We can also see this concern with balancing public good and private rights in the unfavorable treatment *Kelley* and *Phillips* accorded site-specific art. Underlying both courts' reluctance to allow the artists' complete control of these works, installed and displayed on public grounds, could be the fact that these works were precisely that – public art, sponsored and owned by the city and created with public funds. If copyright law was already deeply concerned with the appropriate balancing of a rights holder's monopoly power and the public interest in art's progress (specifically, the Constitution's Copyright Clause allows Congress '[t] o promote the Progress of Science [broadly meaning knowledge] . . . by securing for limited Times to Authors . . . the exclusive Right to their . . . Writings'),[76] then allowing one artist to prohibit the public from either touching a sculpture or else taking a photograph as a keepsake seems to take the right too far. Land art, in bringing art out of the sacred space of the museum and into life, should likewise welcome the contingencies and human interaction, modification, recreation, or personal memorialization that life thrusts upon it.

At last, it may seem obvious but nonetheless bears pointing out that nuisance, unlike fair use, does not require that the new use be valuable at all. We might analogize nuisance law's requirement that any interference with an owner's property be substantial to the doctrine of de minimis use in copyright law. That is –trivial uses (like reproducing an image to hang on one's refrigerator)[77] are not considered copyright infringement. Likewise,

---

[76] U.S. Const. art. I, § 8, cl. 8.

[77] *Ringgold v. Black Entertainment Television, Inc.*, 126 F.3d 70, 74 (2d. Cir. 1997).

images taken of Smithson's jetty from a personal trip and posted on one's travel blog, for example, are likely so de minimis as to constitute a non-nuisance. In that instance, an artist cannot argue that such a reproduction has materially interfered with his rights in his work.

5. CONCLUSION

Some great works of land art will no doubt become as iconic as the natural wonders that draw visitors from the world over each year. Smithson's man-made jetty thus could easily rival the Grand Canyon in the years to come. But the future of the jetty – including its status as a copyrightable artwork – is hard to predict.

The continued development of these works in the expanded field must similarly expand beyond the modernism-inflected, medium-specific, self-contained ethos of copyright law. To be like life, to be in the everyday, to expand the realm of possibility, the new sculpture must necessarily appropriate legal remedies once thought of as more fitting for homeowners than artists, for real, grounded property rather than intellectual property. Similarly, courts should remain open to these untraditional remedies as one way of addressing the non-fixed, land/art hybrid of the new expanded medium. These remedies, of course, are imperfect: as with Chapman's wildflowers and Phillips' stoneworks, the artwork owner's interests may be opposed to the artist's interests, and in many best-case scenarios artists must rely on property owners with aligned interests, like the Dia Art Foundation, for a work's conservation and preservation. But then again, life itself, which earthworks take as their base, is imperfect. The desire for absolute control over a work of one's own, like the desire to leave one's mark on the world – etched in larger-scale format, permanent for now, and hopefully for all time – is a natural, albeit megalomaniacal, human urge. Land art's inevitable impermanence lends it mortality's tragedy, along with a strange sense of otherworldly potency and urgency.

I recently journeyed across the American southwest, on a mission to check certain of these works off of the great bucket list of life – the jetty, the 'Lightening Field', Nancy Holt's 'Sun Tunnels' in Utah, and, finally, Michael Heizer's 'Double Negative' in Nevada. On the day we went to 'Double Negative', it had been pouring, and the 'road' leading to the artwork was all but washed out. After twenty minutes of driving through brush, we finally got out of the car and peered over the edge of the work – two man-made canyons carved into opposite sides of a mesa. It was beautiful, despite the fact that time had already started to take its toll on the work – the 'canyons' were only about half as large as they used to be, as the sedimentary rock had been crumbling for years. There was not a single other person there. The downpour had subsided into a light mist. Above us, the sky was gray, and all around us, the land was barren, arid, shades of brown and dust red. The work was remarkable. But it was only until we clambered to the edge of the mesa and looked out at the lush green of the valley below that the Wordsworthian sublime, of man at the very top of that summit, began to overtake me. Heizer may have made the canyons, but he didn't make that river running through the valley, which stained all the brown and rust with an impenetrable green, the green of life. At the end of the day, the art had been bested – by life itself. Or perhaps, after all, that was the point. Where did the work end, and the rest of life begin?

2. Copyright and conceptual art

Shane Burke

1. INTRODUCTION

The first issue one is confronted with when conducting any study pertaining to conceptual art relates to definition and temporality. It has previously been stated that '[c]onceptual art's factions have frequently been at odds, usually over definitions and often after the fact'.[1] Conceptual Art is frequently cited as a movement that was active between the years of 1966 and 1972,[2] yet alternately, a recent Tate Britain exhibition defined the pertinent time frame in terms of the UK as between the years 1964 and 1979.[3] Notwithstanding these time frames, it has also been stated that the term conceptual art encompasses a range of practices that may be traced from the work of Marcel Duchamp into the practice of some contemporary artists,[4] the latter group sometimes being termed Neo-Conceptualists. It may also be argued that conceptual art strategies still have a more general 'effect on many artists working today'.[5] Accordingly, historian, curator, practitioner and critic Charles Harrison speaks in terms of 'a continuing generic conceptual art as a long-term outcome of the historically specific Conceptual Art movement'.[6]

If one was seeking to characterize a core tenet in relation to the nature of the authorship inherent in conceptual art strategies, it could be argued that 'conceptual art wasn't about art that had a concept, but about

---

[1] Frazer Ward, Some Relations between Conceptual and Performance Art, 56 Art J. 36 (1997).

[2] Lucy R Lippard, *Six Years: The Dematerialization of the Art Object* (University of California Press, 1973).

[3] Tate, Conceptual Art in Britain 1964–1979, http://www.tate.org.uk/whats-on/tate-britain/exhibition/conceptual-art-britain-1964-1979/conceptual-art-britain-room-guide (last accessed 3 November 2017).

[4] Peter Goldie, Conceptual Art and Knowledge. In Peter Goldie and Elizabeth Schellekens, eds., *Philosophy and Conceptual Art* (OUP, 2007) 158.

[5] Ibid.

[6] Peter Goldie, Art and Language, Emergency Conditionals. In Peter Goldie and Elizabeth Schellekens, eds., *Philosophy and Conceptual Art* (OUP, 2007) 259.

interrogating the concept of art'.[7] Central to such an outlook was a process of dematerialization of the traditional art object. It must be stated however, that such an approach 'was not a literal challenge to the art object, but to the 'doctrine of "objecthood"'.[8] As Stapleton has argued 'dematerialisation was grounded on the possibility of the artist retreating into the production of intellectual properties and its cognate forms of creative labour'.[9] So therefore, it may be argued that it was the type of the creative labour involved that was paramount to conceptual art strategies rather than a per se rejection of the art object. This dematerialization in practice may be said to involve the privileging of the idea over the execution, the linguistic over the visual and the performative over the static. Copyright law and conceptual art therefore stand in apparent contradistinction to one another, the former professing not to protect ideas[10] but rather sufficiently developed expression while the latter decries expression as relatively unimportant while privileging conception. This however is just the prima facie position. On closer examination, there are synergies and overlaps that lead to an irregular and unpredictable landscape in terms of what qualifies as protectable subject matter. Conceptual art does in a sense concern idea-focused artistic expression but nonetheless produces a range of materials that may or may not be protected by copyright law depending on their final realized form. While Article 2(1) of the Berne Convention provides a non-exhaustive list of protected artistic works, stating that protectable works 'shall include every production in the literary, scientific and artistic domain, whatever may be the mode or form of its expression', in practice copyright law tends to be less than accommodating when it comes to many forms of contemporary artistic expression, with conceptual art frequently on the margins in terms of what is considered protectable subject matter, a US Court for example stating, 'the law must have some limits; not all conceptual art may be copyrighted'.[11]

---

7 Jens Hoffmann, Deputy Director of Exhibitions and Programs at the Jewish Museum, cited in Isaac Kaplan, Artsy, If You Don't Understand Conceptual Art, It's Not Your Fault (31 March 2016), available at https://www.artsy.net/article/artsy-editorial-if-you-don-t-understand-conceptual-art-it-s-not-your-fault (last accessed 3 November 2017).

8 Jaime Stapleton, Art, Intellectual Property and the Knowledge Economy (PhD Thesis, Goldsmiths, University of London, 2002) 93.

9 Ibid. at 92.

10 It was stated in *SAS Institute v. World Programming Ltd* [2013] EWCA Civ 1482 [20] that, 'It is a cliché of copyright law that copyright does not protect ideas: it protects the expression of ideas. But the utility of the cliché depends on how ideas are defined'.

11 35 F 3d 290, 304 (7th Cir 2011) citing *Feist v. Rural*, 499 US 340 (1991).

Barron has averred that 'copyright law assumes its objects to be much the same kind of entity that modernist criticism across all the arts assumes its objects to be: stable fixed, closed, self-contained, and autonomous of its context and audience'.[12] She argues that in the UK as copyright is a property right it, in concert with Modernist art theory, employs a taxonomic approach which requires that '[a]ll the arts can be confined within a closed list of mutually expressive genres, each with an essential character'.[13] Rahmatian has also stated of copyright law that in certain instances:

> the determinative factors as to what constitutes an artwork in law are in fact not abstract legal notions of "work" or "originality" but the value judgments of art theory and aesthetics which the law endows with normative force when deciding on "work" criteria contrary to its professed neutrality.[14]

It will be argued that a medium-specific approach is implicit in the judgments of courts in a number of jurisdictions irrespective of whether they adopt a closed-list approach to subject matter or not. With a focus on the use of language as a 'signature' style of conceptual art, this chapter, through an examination of copyright protection for text-based instructional works, will seek to highlight the fact that copyright law sometimes struggles to accommodate the traversing of media boundaries, in this case between the textual and the visual. It will explore whether the artistic authorship inherent in text instructions may be protected through a broad analysis of the scope of protection of literary works, how such text may be protected as an artistic work, and, lastly, it will examine how a number of court decisions in the UK, US and Germany on copyright protection for recipes and literary characters highlight the fact that such expression may struggle to gain copyright protection due to judicial aesthetic biases.

2. LANGUAGE- AND TEXT-BASED INSTRUCTIONAL WORKS

The rise of the linguistic as a means of artistic expression can be ascribed to a number of factors and can be set against the backdrop of the general project of dematerialization. The use of language may be seen as a method of spanning any gap that may be perceived to exist between the viewer and

---

12 Anne Barron, Copyright Law and the Claims of Art, 4 Intell. Prop. Q. 368, 370 (2002).

13 Ibid. at 372.

14 Andreas Rahmatian, *Copyright and Creativity: The Making of Property Rights in Creative Works* (Edward Elgar Publishing, 2011) 189.

the work of art. It is evidence of an aspiration to engage cognitively more directly with the viewer, of a desire to generate ideas for the purposes of theorizing the meaning of art, and can be seen as part of a broad political project to help dematerialize the art market and also to aid communication with a wider audience.[15] It may also be contended that the validity of the use of language in the production of art was further strengthened when viewed in conjunction with the contemporary cultural context and what has been called the 'linguistic turn'. This concept was taken from the name of a prominent anthology edited by Richard Rorty published in 1967. In this work, Rorty opined that 'philosophical problems are problems which may be solved (or dissolved) either by reforming language, or by understanding more about the language we presently use'.[16] It may be noted also that the English conceptual art group Art and Language founded in 1968 were seen as pioneers in this area, with their practice questioning 'the critical assumptions of mainstream modern art practice and criticism'.[17] A more recent manifestation of the use of language in instruction art can be seen in the 'do it' exhibitions and the subsequent publication 'do it; the compendium'.[18] This project, curated by Hans Ulrich Obrist, showcases art as instruction with a large array of contributors, from Marina Abramović, Tracey Emin, Ai Wei Wei to Carl Andre, to Tino Sehgal and David Lynch. The instructions in question here may contain graphic elements but most are primarily or solely text based.

Sol LeWitt, one of the architects of the Conceptual Art movement, also utilized language as part of a more systemic approach to the production of art. LeWitt, especially in his earlier works, often used language as the means of communicating his ideas to others who were responsible for physically executing the final visual piece of art. These text instructions were often contained within the certificate of authenticity which was sold with the work. LeWitt maintained distance in his work from 'a single authorial presence whilst maintaining a certain degree of control; like defining the rules of a game without predetermining the state of play, or the eventual outcome.'[19] He once remarked that '[t]he serial artist does not

15 Tony Godfrey, *Conceptual Art* (Phaidon Press, 1998) 163.

16 Richard Rorty, *The Linguistic Turn: Recent Essays in Philosophical Method* (University of Chicago Press, 1992) 3.

17 Tate, Art and Language, http://www.tate.org.uk/learn/online-resources/glossary/a/art-and-language (last accessed 3 November 2017).

18 See generally, Independent Curators International, http://curatorsintl.org/special-projects/do-it (last accessed 3 November 2017).

19 Anna Lovatt, Seriality and Systematic Thought in Drawing c 1966–1976 (PhD Thesis Courtauld Institute 2005) 67.

attempt to produce a beautiful or mysterious object but functions merely as a clerk cataloguing the results of his premise'.[20] The instructions may vary in detail and specificity depending on the work in question.

There are a number of contributions in this more systemic form of artistic production that may be relevant in terms of the subsistence of copyright, the text instructions, the supervision by the artist of the execution of the work and the actions of the draftsperson who actually executes the work, all of these 'inputs' potentially giving rise to a claim of authorship or joint authorship. While the issues surrounding joint authorship have been dealt with very effectively elsewhere,[21] this chapter will instead focus on whether copyright law adequately protects the artistic authorship inherent in the text instructions in and of themselves and how this highlights the fact that the law, in the context of both infringement and subsistence analysis, struggles to deal with situations where there is a blurring of the line between the textual and the visual.

3. THE UNITED KINGDOM

Under UK copyright law, where a closed-list approach to categorization of subject matter exists, there is no subsistence of copyright unless the expression fits into one or more of the given categories of works.[22] The classification of an expression as a particular type 'work' has implications for the type of the protection offered to it under copyright law. For example, the making of a three-dimensional reproduction of an artistic work from a two-dimensional drawing may constitute infringement.[23] It is the case however that, despite the fact that reproduction is prohibited in 'any material form',[24] the making of a work that is disclosed through a description or set of instructions contained in a literary work, has generally not been considered an infringement, given both the principle that copyright law does not protect ideas and also the fact that any ensuing visual work

---

[20] Simon Wilson, *Tate Gallery: An Illustrated Companion* 263 (Tate Gallery, 1991).

[21] See Lionel Bently and Laura Biron, Discontinuities between Legal Conceptions of Authorship and Social Practices: What, If Anything, is to be Done? In Mireille van Eechoud, ed., *The Work of Authorship* (Amsterdam University Press, 2014) 237.

[22] Pumfrey J in *Sandman v. Panasonic UK Ltd* [1998] FSR 651, 658; citing the example of 'an E E Cummings poem about a cat in the shape of a cat' as that of a work in which both literary and artistic copyright may subsist.

[23] Copyright Designs and Patents Act 1988, s 17(3) (hereinafter CDPA).

[24] CDPA s 17(2).

would normally not be seen as sufficiently similar to the underlying literary instructions so as to constitute an infringement.[25] A central argument thus being that there must be sufficient objective similarity between a copyright work and an alleged infringing work, which must in turn 'in some real sense represent the claimant's'.[26] The realization of an artist's work through the use of literary instructions has been held as an example where this is clearly not the case.[27] Despite a previous judicial linking of artistic works with visual significance in the context of derivative works,[28] Copinger has noted that the definitions of drawings and diagrams have recently been broadly interpreted, leading to a situation where these categories may now include works which, despite involving a small level of visually significant elements, are in essence 'little more than written instructions for creating an article of visual appeal'.[29] The 16th edition of Copinger disagrees with such an approach, contending that works which can primarily be read rather than appreciated visually should only be classed as literary works, stating that '[i]f it were otherwise, purely written instructions describing how to make an artistic work could be protected as an artistic work, which is clearly not the correct position'.[30] Reflecting such an approach, the Supreme Court of Victoria, in *Cuisenaire v. Reed*, held that the copyright in a book which described a system of teaching was not infringed by the making of 'rods' that demonstrated that method. Page J famously stating that to do otherwise would lead to a situation where: 'everybody who made a rabbit pie in accordance with the recipe in Mrs. Beeton's Cookery Book would infringe the literary copyright in that book'.[31]

---

25 *Hollinrake v. Truswell*, [1894] 3 Ch 420, 427 (Lindsey LJ), 428 (Davey LJ).

26 Kevin Garnett et al., *Copinger and Skone James on Copyright*, Vol. 1, § 7-12 (17th edn, Sweet & Maxwell, 2016), citing *Purefoy Engineering Co Ltd v. Sykes Boxall & Co Ltd* (1955) 72 RPC 89, 99.

27 Ibid. at § 7-12.

28 *Interlego v. Tyco Industries* [1988] RPC 343, 371.

29 See Kevin Garnett et al., supra note 26 at § 3-112, citing *Lerose Ltd v. Hawick Jersey International Ltd* [1973] FSR 15; *Vermaat v. Boncrest Ltd* [2001] FDR 43.

30 Kevin Garnett et al., *Copinger and Skone James on Copyright*, Vol. 1, § 3-57 (16th edn, Sweet & Maxwell, 2011).

31 [1963] VR 719, 735. In relation to the same facts, this reasoning was supported by a ruling of the Canadian Supreme Court in *Cuisenaire v. South West Imports Limited* [1969] SCR 208. See also *J & S Davis v. Wright Health Group* [1988] RPC 403, 414, where Whitford J also observed that you do not infringe copyright in a recipe by making a cake. See also *Brigid Foley Ltd v. Elliot* [1982] RPC 433, where knitting instructions were held not to be infringed by the making of the garment.

Similarly, in *Plix Products Limited v. Frank Michael Winstone (Merchants)*, the High Court of New Zealand held that:

> a written verbal description of an artistic work, however precise and explicit, is not an infringement of any copyright subsisting in that work. That, of course, is because the two media are so completely different that one can never, in a real sense, be a medium in which it is possible to reproduce the other – just as a painting cannot be played on a gramophone record.[32]

It is interesting to note however that while courts have taken such a medium-specific approach to issues of infringement in the context of literary works, they have also been willing to accept that in cases of indirect copying, where the link between the first work and the copy is either a written or verbal description, that description may nevertheless be sufficient to enable a copy of the artistic work to be made.[33] In other words, a written or verbal description may provide the causal link in a case of indirect copying where this description allows for the reproduction of a substantial part of a work.

Notwithstanding the foregoing, in the UK, a number of decisions have sought to classify text instructions as artistic works in and of themselves.[34] In *Abraham Moon v. Thornber and Ors*,[35] a Patents County Court decision, it was held that a ticket stamp, a set of instructions for the production of fabric comprising two pages of words, letters and numbers, constituted an artistic work on the grounds that an experienced fabric designer looking at the ticket would find that it had 'real visual significance', counsel having argued that such a person 'could look at a weaving ticket and visualise what the fabric looks like'.[36] Judge Birss QC (as he was then), although acknowledging that counsel had argued that terms like 'graphic work' should be given their ordinary meaning, stated 'I do not think it is doing

---

[32] *Plix Products Limited v. Frank Michael Winstone (Merchants)* [1986] FSR 63, 89–90. On this argument generally see Staniforth Ricketson, The Law of Intellectual Property (Law Book Company, 1984) 179–80.

[33] See Kevin Garnett et al., supra note 26 at § 7-20, citing *Plix Products Limited v. Frank Michael Winstone (Merchants)* [1986] FSR 608, 615 (NZ CA) (written) and *Ultra Marketing (UK) Ltd v. Universal Components Ltd* [2004] EWHC 468 (Ch) [15] (verbal).

[34] See e.g., *Lerose Ltd v. Hawick Jersey International Ltd* [1973] FSR 15, where it was held that, under the 1956 Act, knitting instructions consisting of a point pattern including hieroglyphics, figures and words was an artistic work which rendered an article made by the defendants to the pattern an infringement.

[35] [2012] EWPCC 37.

[36] [2012] EWPCC 37 [102].

violence to the language to regard the ticket as a graphic work. It may be of an unusual sort but it is a record of a visual image'.[37]

While it is unclear whether or how closely the decision will be followed, the Abraham Moon judgment provides for a potentially broader interpretation of what may constitute an artistic work and hence a more inclusive scope of protection that may include some conceptual artworks. But it must also be acknowledged that the decision has its limitations, and whether it could lead to the protection of the artistic authorship inherent in LeWitt's instructions is a moot point given their predominantly artistic as opposed to technical character, and also the fact that there may be varying levels of indeterminacy that may or may not be allowed for in relation to the execution of LeWitt's works, which may mitigate against the possibility of a definitive 'reading' by a sufficiently skilled person. It remains to be seen also how the law in this area will develop given the impact of EU harmonization. Counsel for the claimant in Abraham Moon argued that in light of recent CJEU decisions[38] that intellectual creations should be protected regardless of any distinction between literary and artistic works, while the court declined to follow the invitation.[39] It remains to be seen if UK law will develop in this vein, particularly so now with Brexit looming.

4. THE UNITED STATES

The US engages more expressly with the idea-expression principle in that it has placed it on a statutory footing.[40] In *Baker v. Selden* it was held that 'the mere copyright of Selden's book did not confer upon him the exclusive right to make and use account-books, ruled and arranged as designated by him and described and illustrated in said book'.[41] The Court further stated:

> To give to the author of the book an exclusive property in the art described therein, when no examination of its novelty has ever been officially made,

---

37 [2012] EWPCC 37 [107].

38 See e.g., *Infopaq International A/S v. Danske Dagblades Forening*, Case C-5/08 [2009] ECR I-6569 (ECJ, Fourth Chamber); *Eva-Maria Painer v. Standard Verlags GmbH and Others*, Case C-145/10 [2012] ECDR (6) 89 (ECJ, Third Chamber); *Football Dataco Ltd and Ors v. Yahoo! UK Ltd and Ors*, Case C-604/10 [2012] 2 CMLR (24) 703 (ECJ, Third Chamber).

39 [2012] EWPCC 37 [98]. For a more positive view on such arguments see *SAS Institute Inc. v. World Programming Ltd* [2013] EWHC 69 (Ch) [27].

40 It is enshrined in section 102(b) of the Copyright Act 1976 Act.

41 101 US 99, 107.

> would be a surprise and a fraud upon the public. That is the province of letters patent, not of copyright.[42]

However, looking at the artistic authorship inherent in the instructions of LeWitt, it is clear that the authorship in question here very much relates to artistic expression rather than with more functional concerns and therefore, while the instructions may be an element in a more systemic form of artistic production, they are not subject matter that patent law would be concerned with. The issues surrounding copyright protection for recipes provide a further illustration of the difficulties in this area.

In a recent First Circuit decision, the principle that 'recipes are functional directions to achieve a result and therefore not copyrightable' was approved of.[43] But is such a blanket assertion justified? Buccafusco argues while the fact that recipes will necessarily be dictated by functional concerns and therefore will lack the required originality may be true of 'well-established dishes like apple pie or coq au vin', it is clearly not the case as regards more inventive contemporary forms of cuisine.[44] Similarly, Goldman has argued that in the case of recipes, the combination of 'the collectivity of the list of ingredients and directions that make up an original recipe arguably defeat these limiting principles'.[45] It would appear therefore that a rigid application of the idea-expression dichotomy could lead to undesirable results where the creativity and originality inherent in the instructions may not be protected without a focused assessment in order to determine whether the expression in question was in fact dictated by function.

It must also be acknowledged that a broader form of protection would be available should the recipes be regarded as a fixation of a work of authorship rather than simply literary works. Such an approach would also limit the impact of the idea-expression principle by virtue of the fact that the instruction/fixation would now no longer be different from any other form of fixation; for example, sheet music is not deemed inappropriate fixation for a musical work simply because it is instructional in nature.[46]

---

42 Ibid., 102.

43 *Lorenzana v. S Am Restaurants Corp*, 799 F 3d 31, 35 (1st Cir 2015).

44 Christopher J Buccafusco, On the Legal Consequences of Sauces: Should Thomas Keller's Recipes be Per Se Copyrightable?, 24 Cardozo Arts Ent Law J 1121, 1130 (2007).

45 Michael Goldman, Cooking and Copyright: When Chefs and Restaurateurs Should Receive Copyright Protection for Recipes and Aspects of Their Professional Repertoires, 23 Seton Hall J Sports & Ent L 153, 176–177 (2013).

46 Ibid., 172.

However, the prospect of the protection of text instructions against reproduction of a resultant visual realization or the classification of them as a fixation of an artistic work would, similar to the position in the UK, suffer from the fact that courts in the US tend to privilege visual over written representation despite the fact that reproduction may extend to copying in different media.[47] An examination of judicial decisions in the context of the protection of characters, or what the courts have termed in some instances 'word portraits', tends to support a conclusion that despite no per se bar to protection there is a judicial aesthetic bias against text towards visual reproductions.

UK courts have been reluctant to grant such copyright protection to literary characters.[48] The Whitford Committee has previously expressed its reservations, stating 'there would be very real difficulty in defining what a "character" is'.[49] In the US however, it has been established that characters may be protected by copyright when they are contained as part of a literary work.[50] It has nonetheless been the case that literary characters have traditionally received a relatively low level of copyright protection.[51] While copyright has been held to protect only the 'author's means of expression' of a character,[52] it is however true that it has more regularly been held to be infringed in cases where a three-dimensional figure has been created based on two-dimensional cartoons.[53] The position in relation to the two- or three-dimensional reproductions of literary characters has proved somewhat more complicated. In *Walt Disney Productions v. Airpirates*,[54] it was stated:

---

[47] *Midway Mfg Co v. Bandai-America Inc*, 546 F Supp 125, 139 (DNJ 1982) ('It is clear that there can be substantial similarity and copyright infringement between works in different media'); *D C Comics Inc v. Unlimited Monkey Business Inc*, 598 F Supp 110, 117 (ND Ga 1984) (fictional characters held to be infringed by singing telegrams).

[48] *Kelly v. Cinema Houses Limited* [1928–35] MacG CC 362; *Tyburn Productions Ltd v. Conan Doyle* Ch 75 (1991).

[49] Copyright and Designs Law: Report of the Committee to Consider the Law on Copyright and Designs, HMSO (Cmnd 6732), 229 [909] (1977).

[50] *Burroughs v. MGM Inc*, 519 F Supp 388, 391 (SDNY 1981); *Nichols v. Universal Pictures Corp*, 45 F 2d 119, 122 (2d Cir 1930).

[51] See, e.g., *Rokeach v. Avco Embassy Pictures*, 3 Med L Rptr (BNA) 1774, 1779 (SDNY 1978).

[52] *Rokeach v. Avco Embassy Pictures*, 3 Med L Rptr (BNA) 1774, 1779 (SDNY 1978).

[53] *King Features Syndicate v. Fleischer*, 299 F 533 (2d Cir 1924); *Geisel v. Poynter Prods Inc*, 295 F Supp 331 (SDNY 1968); *Fleischer Studios Inc v. Freundlich*, 5 F Supp 808 (SDNY 1934).

[54] 581 F2d 751, 755 (1978 9th Cir).

> Although it is difficult to delineate distinctively a literary character, when the author can add a visual image, the difficulty is reduced; while many literary characters may embody little more than an unprotected idea, a comic book character, which has physical as well as conceptual qualities, is more likely to contain some unique elements of expression which are copyrightable.

It would appear therefore that once a literary character is more than a mere idea and has been sufficiently delineated that copyright protection is possible. US law does not draw a fundamental distinction between the literary and the visual in this regard and merely focuses on the delineation issue. It has been argued however that, despite this lack of a per se prohibition of copyright for characters, the test for delineation makes it extremely difficult for literary characters to gain protection. For example, in *Warner Bros v. ABC*,[55] it was stated that because of the existence of visual depictions, 'Superman is more readily protectable than he would be had he been merely a word portrait'. It has also been stated of this judgment that it 'makes clear that word portraits are protectable, if less "readily" so'.[56]

It has been argued that 'copyright incorporates aesthetic valuation implicitly through the law's "distinctive delineation" standard',[57] and as a result 'copyright law thus contravenes, at least in theory, its own aesthetic nondiscrimination principle' as outlined in the *Bleistein v. Donaldson Lithographing Co.*[58] judgment.[59] The aforementioned test for delineation arguably makes fundamental aesthetic assumptions based on the nature of the medium in question.[60] The Seventh Circuit judgment in *Gaiman v. McFarlane*, a case concerning protection for comic book characters, evidences such an analysis, Judge Posner stating, 'The reason is the difference between literary and graphic expression. The description of a character in prose leaves much to the imagination, even when the description is detailed'.[61] He continued, '[a] reader of unillustrated fiction completes the work in his mind; the reader of a comic

---

55 530 F Supp 1187, 1193 (SDNY 1982).

56 *JD Salinger v. Colting*, defendants-appellants, 2009 WL 6865323 (CA 2), 23; 2009 WL 6865323 (CA 2) (appellate brief). For examples of where protection has been granted to literary characters see *Bach v. Forever Living Products US Inc*, 473 F Supp 2d 1110, 1118 (WD Wash 2007); *Salinger v. Colting*, 607 F 3d 68, 73 (2d Cir 2010).

57 Zahr K Said, Fixing Copyright in Characters: Literary Perspectives on a Legal Problem, 35 Cardozo L Rev 769, 806 (2014).

58 188 US 239, 251 (1903).

59 Supra note 57 at 805.

60 Supra note 57 at 805.

61 360 F 3d 644, 660 (7th Cir 2004).

book or the viewer of a movie is passive'.[62] These passages illustrate an analysis which is based on an assumption that 'cast[s] the reader as passive when viewing images or television and active when reading texts, which . . . require a higher level of semiotic participation'.[63] While it has been stated that it is reasonable to argue that there is a level of reader involvement in the creation or actualization of meaning within literary text, Reader Response theorists have also stated that this is equally true in the case of visual works.[64] Such arguments have also been advanced in relation to the act of listening, where that act assumes the role of active participation; such works are structured around a 'post cognitive' interpretation, in essence creating work to be completed in the mind of the viewer/listener.[65] Nonetheless, Leaffer has also concurred with a Posner-type approach, stating that '[p]ictorial characters present less theoretical difficulties between literary and graphic expression. The description of a character in prose leaves much more to the imagination, whereas the impact of a graphic character is immediate, concrete and unmediated'.[66]

Tushnet has argued that pictures are treated by the courts as either 'transparent' or 'opaque'.[67] In the former case, interpretation of visual images is deemed unnecessary; in the latter, interpretation through discussion and analysis via language is deemed as impossible and therefore not worthy of attempting. Judge Posner in *Gaiman* was clearly opting for this 'transparent' view of images. This mistrust of language as a means of interpreting art may also extend to the court doubting whether visual works are capable of descriptions through language. Kasunic cites the distinction between visual- and language-based expression, defining the latter as discursive in its articulation, in other words, it presents its constituent parts successively rather than simultaneously, so that while visual 'reading' may be completed in one act of visualization, with literature consistent elements are presented sequentially. He contends that 'the law has not squarely addressed how the courts should assess

---

[62] Ibid.

[63] Supra note 57 at 807.

[64] Zahr Said, Embedded Advertising and the Venture Consumer, 89 N. C. L. Rev 99, 124 (2010).

[65] Brandon LaBelle, *Background Noise: Perspectives on Sound Art* (Bloomsbury, 2010) xiii.

[66] Marshall A Leaffer, Understanding Copyright Law (6th edn, Carolina Academic Press, 2014) 92.

[67] Rebecca Tushnet, Worth a Thousand Words: The Images of Copyright, 125 Harv. L. Rev. 683, 684 (2012).

the meaning of non-discursive expression'[68] and speculates as to whether non-discursive expression can truly ever accurately be transformed into words, asking: 'Is such a translation anything more than a subjective interpretation? Does non-discursive expression have one meaning? Does it have a "meaning" at all, or is meaning purely a discursive concept?'[69] He concludes, 'words can never fully translate non-discursive expression. Its meaning is intrinsically bound to its non-discursive form which words can only attempt to describe',[70] adding, '[a]n idea that contains too many minute yet closely related parts, too many relations within relations, cannot be "projected" into discursive forms; it is too subtle for speech'.[71]

Such a logic may also be seen as underpinning a number of judgments in other jurisdictions. For example, the decision of the German Bundesgerichtshof in *Re 'Pippi Longstocking' Character* also demonstrates a logic that is implicitly informed by the discursive/non-discursive distinction. In a case concerning whether two-dimensional advertising images infringed Astrid Lindgren's literary character Pippi Longstocking, the Court first stated: 'a literary description also can create in the mind's eye of the reader an equally clear "picture" of a character in a book'.[72] It then quickly qualifies this assertion with the following passage:

> It must be borne in mind that, with the means of language, precisely the formative character qualities of a fictitious person can be portrayed in a much more differentiated way than with the means of fine art. Consequently, a detailed description of character qualities can readily compensate for a representation of the external appearance of a person which can be conveyed by means of language to only a limited extent.[73]

The Bundesgerichtshof's scepticism about the limitations of language in conveying a visual picture is assuaged only by the fact that the text also details character qualities which are seen as a mitigating fact. This approach was further reinforced when the Court stated:

---

[68] Rob Kasunic, The Problem of Meaning in Non-Discursive Expression, 57 Copyright Soc'y USA 399, 400 (2010).

[69] Ibid.

[70] Supra note 68 at 406–7.

[71] Supra note 68 at 399.

[72] (Case I ZR 52/12) Bundesgerichtshof [BGH] [German Federal Supreme Court] July 17, 2013, [27] European Commercial Cases (E.C.C.) 300, 306, (2014); *Re 'Pippi Longstocking' Character* (Infringement of a Copyright User's Rights)

[73] Ibid.

> Separate protection for a fictitious character therefore depends on the creator endowing the character with an unmistakable personality by means of a combination of distinctive character qualities and particular external features. The test for that purpose must be stringent. A mere description of the external form of a character or his or her appearance will not as a rule be sufficient.[74]

Yet paradoxically, in asserting that there would not be infringement on the basis that only a number of external characteristics were used in the allegedly infringing visual works, the Court did acknowledge that the character was recognizable through those allegedly infringing images.

Similarly, it may be argued that the Canadian decision of *Anne of Green Gables Licensing Authority Inc v. Avonlea Traditions Inc*,[75] was also implicitly decided on a similar basis. This case, decided using a conceptually comparable approach to the US delineation analysis, concerned whether two- or three-dimensional representations of the literary character Anne of Green Gables constituted an infringement of the copyright in that character, in other words 'does copyright protect a verbal portrait?'[76] In answer to that question, Wilson J in the Ontario Supreme Court stated:

> [T]he character 'Anne' and the situations in which she finds herself in the book are clearly delineated, distinctive, thorough, and complete. The defendant has reproduced or translated these descriptions in her two- and three-dimensional representations of 'Anne' and various other characters and situations from the text of the book. The literary work Anne of Green Gables contains a detailed verbal portrait that captures Anne's physical image and her personal qualities, a portrait which I conclude is protected by copyright. I conclude that . . . the copyright in the book Anne of Green Gables extended to the two- and three-dimensional images based on descriptions of characters and situations found in the book.[77]

While there are no explicit statements about the limitations of literary descriptions in terms of visual representations thereof, the judgment, like that of the German Court, appears not to conceive of protection based on physical image alone but rather includes personal qualities as a necessary part of any verbal portrait.

On the basis of the foregoing, it may be stated that even where a seemingly progressive sufficient delineation analysis is undertaken that this may still be undermined by a judicial scepticism that a visual image may

---

[74] Ibid.

[75] *Anne of Green Gables Licensing Auth. Inc. v. Avonlea Traditions Inc.*, No. 95-CU-89192, 4 C.P.R. (4th) 289 (O.S.C.J. March 10, 2000).

[76] Ibid. at 94.

[77] Ibid. at 121.

only be described by means of text to a limited extent. As Said has argued, any distinctions between the visual and literary subject matter in terms of copyrightability, whether maintained for reasons of providing a more transparent subject matter for evidentiary ease or simply just providing evidence of an inherent aesthetic preference, should be jettisoned as they are unjustifiable given the professed role of copyright in the promotion of the useful art and the sciences.[78] With conceptual art in general, and the works of Sol LeWitt in particular, it is arguable that such an approach is untenable given the fact that many text instructions would be eminently recognizable from a visual representation thereof, given that they are minimal in nature and in that sense are not similar to more traditional or figurative painting.

5. CONCLUSIONS

This chapter has sought to highlight the fact that irrespective of the existence of criteria for protection, such as that of being a 'work' or having 'originality', and regardless of whether a closed-list approach to copyright protection is adopted, there is a judicial tendency to employ an aesthetic bias when confronted with a visual realization of a textual instruction or character description. It may be argued that any distinction between discursive and non-discursive expression is irrelevant given the nature of conceptual art and its conscious attempt to blur the line between traditional media. It may also be suggested that the 'elevation of "creativity" as an organizing principle' both under EU law[79] and in the US may lead to a more inclusive regime of copyright protection, but this assertion needs to be critically engaged with.

Since the CJEU ruling in *Infopaq International A/S v. Danske Dagblades Forening*[80] there has been effective harmonization of the originality standard across all works of authorship,[81] with the standard of 'author's own intellectual creation' replacing the previous 'skill, labour and/or judgement' originality requirement in the UK. A number of CJEU

---

[78] Zahr K Said, Fixing Copyright in Characters: Literary Perspectives on a Legal Problem, 35 Cardozo L Rev 769, 812 (2013).

[79] Jonathan Griffiths, Dematerialization, Pragmatism and the European Copyright Revolution, 33 Oxford J Legal Stud. 767, 782 (2013).

[80] Case C-5/08, *Infopaq Int'l A/S v. Danske Dagblades Forening*, 2009 E.C.R. I-06569.

[81] See also Case C-393/09, *Bezpecnostni softwarova asociace v. Ministerstvo kultury*, 2010 E.C.R I-13971.

decisions have developed the boundaries of this new standard by stating that it should reflect the 'personal touch' of the author and the making of 'free and creative choices'.[82] An aspect of 'creative freedom'[83] has also been cited as a desired characteristic. As a result of this elevation of creativity as the central criterion for protection, it has also been argued that the UK fixation requirement,[84] as well as the requirement for visual significance[85] may no longer be justifiable. It has also been suggested that the closed-list approach and therefore the UK concept of the 'work' may also now be untenable.[86] Barron has previously posited that the 'work' has been 'the key site of intersection between copyright, art, and art theory, because it is the manner in which copyright law defines the work, rather than the manner in which it defines the author, that exposes copyright's aesthetic prejudices (such as they are) most vividly'.[87] While a more open approach to subject matter could potentially mitigate such prejudices and solve some of the potential issues highlighted above in relation to the idea expression principle, it still remains to be seen whether it will allow for the protection for more conceptual art of the type in focus here.

In the US, a similar shift in the originality standard has occurred with a move from the 'sweat of the brow' approach to a new threshold of 'a modicum of creativity' as per the *Feist*[88] judgment. It has been posited that 'most "artistic" works will be a mile above the threshold'.[89] It has also been opined that 'Feist established, arguably for the first time with this level of clarity, that creative choices were what gives a work its originality'.[90]

The CJEU has thus far 'not been willing to differentiate between different categories of "creation" or types of creativity';[91] it has also been stated that the CJEU may begin to 'distinguish between different forms

---

[82] Case C-145/10, *Painer v. Standard Verlags*, 2011 E.C.R. I-12533 [hereafter referred to as *Painer*]; Case C-604/10, *Football DataCo Ltd. v. Yahoo! UK Ltd.*, 2012 E.C.R. 115.

[83] Case C-403/08, *Football Ass'n Premier League, Ltd. v. QC Leisure*, 2011 E.C.R I-9121; Joined Cases C-403/08 and C-429/08 *Football Association Premier League Ltd and Others v. QC Leisure and Others and Karen Murphy v. Media Protection Services Ltd*, [2011] E.C.R. I-09083.

[84] Supra note 79 at 785.

[85] Supra note 79 at 788–9.

[86] *SAS Institute v. World Programming* [2013] EWHC 69 (Ch) [27] per Arnold J.

[87] Supra note 12 at 279.

[88] *Feist v. Rural*, 499 US 340 (1991).

[89] Daniel J Gervais, Feist Goes Global: A Comparative Analysis of the Notion of Originality in Copyright Law, 49 J. Copyright Soc'y U.S.A. 949, 955 (2002).

[90] See ibid. at 974.

[91] Lionel Bently and Brad Sherman, Intellectual Property Law 106 (4th edn, OUP, 2014).

of creativity as it jurisprudence matures'.[92] Eechoud has noted however that any prospect of building a harmonized notion of the work around the existing directives is complicated by 'the differences among Member States with respect to the (often controversial) copyright status of food design, perfumes, conversations, fashion shows, and conceptual art'.[93] The question therefore remains open as to how the notions of 'creative choice' and 'personal touch' will help to inform the development of jurisprudence going forward.

Gervais has put forward a working definition of the notion of 'creative choice' that would satisfy, inter alia, the *Feist* decision, stating:

> a creative choice is one made by the author that is not dictated by the function of the work, the method or technique used, or by applicable standards or relevant good practice. Conversely, purely arbitrary or insignificant selection is insufficient. A conscious, human choice must have been made, even though it may be irrational.[94]

This is potentially a promising standard in relation to some conceptual art practices as physical realisation of the work may be of less crucial importance[95] than may have previously been the case. Yet turning to the requirement of 'personal touch' as enunciated by the CJEU, such a standard arguably reinforces a deeply subjective notion of originality, which Nadia Walravens has in the past indicated seemed appropriate for figurative art but less so for many contemporary artistic practices.[96] It has also been argued that, as opposed to the position in relation to 'creative choices', given the requirement of 'personal touch', certain contributions

---

[92] Griffiths, supra note 79 at 789.

[93] Mireille van Eechoud, Along the Road to Uniformity – Diverse Readings of the Court of Justice Judgments on Copyright Work, 3 J. of Intell. Prop. Inf. Tech. & Elect. Comm. Law 60, 71 (2012). For examples of the range of works covered under French law see Tanya Aplin, Subject Matter. In Estelle Derclaye, ed., *Research Handbook on the Future of EU Copyright* (Edward Elgar Publishing, 2009) 58.

[94] See Gervais, supra note 89 at 976–7.

[95] Elizabeth F Judge and Daniel Gervais, Of Silos and Constellations: Comparing Notions of Originality in Copyright Law, 27 Cardozo Arts & Entertainment 375, 380 (2009), citing the French decision of Cour d'appel [CA] [regional court of appeal] Paris, 4e ch., April 28, 2000, R.I.D.A. 2000 where a test similar to that of 'creative choices' was applied in relation to authorship and collective works.

[96] Nadia Walravens, The Concept of Originality and Contemporary Art. In Daniel McClean and Karsten Schubert, eds., *Dear Images: Art, Copyright and Culture* (Ridinghouse, 2002) 172–3.

to collaborative works, particularly in relation to large-scale artistic projects, may find it harder to satisfy that criterion.[97] This may have implications for more systemic forms of artistic production such as those employed by LeWitt and it now remains to be seen whether the conceptual artist functioning as a clerk documenting the results of their premise will be regarded as having an appropriate contribution from the perspective of copyright law, and how developing jurisprudence will either mitigate or facilitate judicial aesthetic biases which currently may affect the protection of the artistic authorship inherent in textual instructions.

[97] Stef van Gompel, Creativity, Autonomy and Personal Touch: A Critical Appraisal of the CJEU's Originality Test for Copyright. In Mireille van Eechoud, ed., *The Work of Authorship* (Amsterdam University Press, 2014) 130.

3. Copyright in bio art

Jani McCutcheon*

1. INTRODUCTION

Bio art exquisitely demonstrates how developments in technology, creative curiosity and new modes of thinking about art have facilitated new methods of making art, including creations for which there is no presently existing taxonomy.

Bio art fuses art and science; biology, design and architecture. Copyright is very familiar with art, design and architecture, but science and copyright are relative strangers.

While the confluence of biology, science and art is fertile creative territory, it concurrently presents challenges which current models of copyright law struggle to resolve. This can result in a lack of copyright protection for bio art creations. Artists have conventionally been granted copyright protection for the work they produce. As bio art makes stronger claims to a legitimate artistic practice, it precipitates a reflection on whether it should stand on an equal footing with other artistic works protected under copyright law.

This chapter describes the broad spectrum of creations that might arguably fall within the contested definitions of bio art. It then explains the potential misalignment between bio art and copyright. Finally, it explores the practical and normative ramifications of this dissonance.

2. BIO ART DIVERSITY

The contested definitions of bio art are enthusiastically debated. Robert Mitchell identifies two broad subcategories of bio-art,[1] which we might

* The author is grateful to Oron Catts for his comments on an earlier draft of this chapter.

[1] Robert Mitchell, *Bioart and the Vitality of Media* (University of Washington Press 2010) 26. Suzanne Anker identifies three distinct, yet overlapping

usefully term 'dry' and 'wet' bio art. Some artists create conventional re-presentations of biotechnology,[2] or artistic portrayals of biomedical research using non-biotechnological media such as drawings, sculptures, video or photography.[3] These works are relatively uncontroversial in terms of copyright.

In contrast, vitalist, or 'wet' bio art is art usually generated in laboratories using a wide array of biotechnical media and tools, such as DNA, tissue, blood, bacteria, plants or other organisms, including higher organisms like mice, butterflies and rabbits. For example, in 1996 Oron Catts and Ionat Zurr established the Tissue Culture & Art Project, an artistic research project that interrogated the question of whether engineered living tissue could be a valid form of artistic expression.[4] This entailed an incursion into the biological laboratory to conduct an experiential investigation of 'the artistic potential of manipulating living matter'.[5] These 'wet' creations, which constitute, originate from, or incorporate living materials, present the greatest challenges for copyright. This chapter focuses on this sub-set of bio art.

Bio art has been referred to as a 'tangled cluster of contemporary art movements'.[6] This description is apposite, and highlights something perhaps unique about bio art's incongruence with copyright. As a mix of technological and scientific process, artistic concept, performance and biological engineering, many of the copyright challenges experienced by bio art are shared by other discrete genres of contemporary art such as land art, conceptual art, performance art, and even computer-generated art. The point of distinction is that bio art frequently faces all of these challenges at once.

---

sub-categories in the genre. See Suzanne Anker, The Beginnings and the Ends of Bio Art, 34:3 Artlink (2014) at 14.

[2] Oron Catts, Bioart Special, Waag Society, Institute for Art, Science and Technology (2012), https://waag.org/sites/waag/files/media/publicaties/bioart_special.pdf at 3.

[3] Thomas Margoni, Bioart Special, Waag Society, Institute for Art, Science and Technology (2012), https://waag.org/sites/waag/files/media/publicaties/bioart_special.pdf at 7.

[4] About the TC&A Project, The Tissue Culture and Art Project, http://lab.anhb.uwa.edu.au/tca/about/.

[5] Oron Catts, Shining (broken) Mirrors is not Gadgeteering, Nitro: Non|Traditional Research Outcomes (July 6 2017), https://nitro.edu.au/articles/2017/7/6/shining-broken-mirrors-is-not-gadgeteering.

[6] Karel Stibral, George Gessert, Green Light: Toward an Art of Evolution, 1 Estetika: The Cent. Eur. J. of Aesthetics 127, 127 (2011).

3. COPYRIGHT ESSENTIALS

3.1 Copyright Subsistence Criteria

Across jurisdictions, copyright is heterogeneous. It is therefore impossible to provide a universal explanation of the essential criteria for copyright subsistence. However, with this caveat in mind, most jurisdictions require that copyright works are original, requiring some independent intellectual effort, authored by a human, and fixed in material form. Some jurisdictions also award copyright to less creative productions which don't require an author,[7] or productions which don't involve creative choices.[8] Many jurisdictions require the work or other production to fit within an exhaustive list of categories; others have more open-ended and inclusive protected classes, although this does not always guarantee that courts or commentators take an expansive approach to protectable subject matter.[9]

3.2 Categorised Protection

While Andy Warhol apparently once declared that 'art is whatever you can get away with', copyright law doesn't tend to work that way. The breadth of material caught by copyright stops well short of the astonishing breadth of material expressed as art. First, copyright protection for artistic works is not predicated on something being 'art', thus avoiding that endless, insoluble and metaphysical debate. Second, as mentioned, in most jurisdictions the artistic creation must slot within an existing category of protected subject matter. In Australia, for example, the only available 'boxes' are literary, artistic, dramatic and musical works, and films, sound recordings, broadcasts, published editions and performances.[10] While it is beyond the scope of this chapter to indicate how courts have interpreted each of these categories, that discussion would demonstrate a judicial

---

[7] See, for example, the 'subject matter other than works' protected under the Copyright Act 1968 (Cth) pt 4 (Austl), which include films and sound recordings.

[8] See, for example, the reduced term of protection of 25 years for 'ordinary' photographs under Urheberrechtsgesetz, [UrhG] [Copyright Act], September 9, 1965, BGBL. At 3037, § 72 (Ger.).

[9] For the situation under US law, see the discussion in Jani McCutcheon, Shape Shifters: Searching for the Copyright Work in Kinetic Living Art, 64 J. Copyright Soc'y USA 309 (2017).

[10] See Copyright Act 1968 (Cth) ss 32(1) for provisions relating to the nature of copyright in original works and pt IV div 2 for provisions relating to the nature of copyright in subject matter other than works (Austl).

tendency to adopt a narrow construction, particularly in common law countries.[11]

Bio art may be, or encompass, one or many of the protected categories of copyright material, or fall completely outside them. That bio art will often traverse a number of categories is problematic, because most copyright models require that the creation align with one category and don't recognise multi-media creations.

A number of examples of bio art demonstrate that it is frequently a *mélange* of conceptual work, performance, art object, science and installation. Cat Hope's 'Sound of Decay'[12] is described as a 'sound installation'. It comprises a decaying cane toad inside a desiccator, with microphones inserted into its body. Through a set of headphones, listeners can hear the 'animation' of decay. Artist Leslie Garcia flips this motif and records the sounds of life in her piece 'Pulsu(m) Plantae', which captures the sounds of a plant's growth.[13]

Guy Ben-Ary's piece 'CellF'[14] (pronounced 'self') constitutes a synthesiser made from the artist's own cells. Using a biopsy from his arm, Ben-Ary converted the skin cells into stem cells, and then into neural stem cells, ultimately creating a neural network of the stem cells grown on a multi-electrode array. This tiny 'brain' is connected to a modular analogue synthesiser, which is housed in a custom-built, trumpet-shaped mini-lab. When the 'brain' receives electrical stimulation from live musicians, it responds with electrical impulses, creating sounds through the synthesiser.[15] Ben-Ary dubs the result 'the world's first neural synthesizer'.[16] Stelarc's 'Ear on Arm' project (2006 ongoing) involves the surgical insertion into Stelarc's arm of a prosthetic ear.[17]

Often, bio artists work with cells, viruses, genes and bacteria. The

---

[11] See for example, Anne Barron, Copyright Law and the Claims of Art, 4 Intell. Prop. Q. 368 (2002).

[12] 'Sound of Decay', Cat Hope (2013), http://www.cathope.com/sounds-of-decay-2013.html.

[13] Steve Cahill, Pulsu(m) Plantae Project Gives Plants a Voice, MassEDMC (December 3, 2013), http://massedmc.com/pulsum-plantae-project-gives-plants-voice/.

[14] 'Cellf', Guy Ben-Ary (2014), http://guybenary.com/work/cellf/ (last accessed June 13, 2018).

[15] Penny Durham, Guy Ben-Ary's CellF: Playing with Neurons, The Australian (August 4, 2016), http://www.theaustralian.com.au/arts/music/guy-benarys-cellf-playing-with-neurons/news-story/db746384f6b5ca3f15c3c711a7002725.

[16] Anya Tchoupakov, Artist Makes Synthesizer from His Own Stem Cells, Vice Creators (November 25, 2015), https://creators.vice.com/en_au/article/kbnxkn/artist-makes-synthesizer-from-his-own-stem-cells.

[17] 'Ear on Arm', Stelarc, http://stelarc.org/?catID=20242 (last accessed June 13, 2018).

integral elements of the art thus exist at the microscopic level, and may never be seen, or only brought to light through the use of high-powered photographic equipment. For example, in Kathy High's 'Blood Wars' (2010 ongoing), she stages a contest between two individuals' white blood cells in a Petri-dish 'theatre'.[18] Eventually, one set of cells is vanquished by the other. High describes 'Blood Wars' as a 'performance and installation art-and-science project'.[19] A number of bio artists manipulate DNA and genomes, inserting text into them.[20] Similarly, in Verena Friedrich's 'Cellular Performance' (2011–2012), skin cells have been manipulated in the laboratory to form letters and words to lampoon the promises of the cosmetics and pharmaceutical industries.[21]

These examples illustrate the immediate difficulty in categorising bio art as a holistic entity. Is the entire installation comprising the arranged, composite parts of each of these works a 'sculpture'? This seems doubtful, with some case law questioning whether an assembly of disparate objects can be a 'sculpture' within copyright law.[22]

While some bio art creations might be regarded as a 'compilation' of material, they are not a compilation that copyright law is familiar with, which is usually limited to literary works.[23]

It is more likely that the individual pieces comprising the installation or compilation may have independent copyright, provided they are original. This may particularly apply to CellF's custom-built lab, which is most likely a sculpture, but, notably, this was not built by Ben-Ary (raising issues of multiple and joint authorship, discussed below). In the case of Stelarc's 'Ear on Arm', the ear model itself is possibly a sculpture, as are

18 'Blood Wars', Kathy High, http://kathyhigh.com/project-blood-wars.html (la st accessed June 13, 2018).

19 Ibid.

20 Examples include: 'Genesis', Eduardo Kac (1999), http://www.ekac.org/geninfo2.html, in which Kac created a synthetic gene by translating a sentence from the biblical book of Genesis into Morse Code, converting the Morse Code into DNA, and then incorporating the Genesis gene into bacteria; and poet-bio artist Christian Bök's 'The Xenotext' project (2006 ongoing), where he used a 'chemical alphabet' to translate two lines of poetry into sequences of DNA and implanted this into the genome of an almost indestructible bacterium.

21 Verena Friedrich's Cellular Performance, LifeSpace Science Art Research Gallery, http://lifespace.dundee.ac.uk/artists/verena-friedrich (last accessed June 13, 2018).

22 *Creation Records Limited v News Group Newspapers Limited* [1997] EMLR 444 (Lloyd J) (UK).

23 For example, Copyright Act 1968 (Cth) s 10(1) (Austl.) provides that a '"literary work" includes . . . a table, or compilation, expressed in words, figures or symbols'.

the 'horns' inserted in bio artist Orlan's body modifications.[24] However, the holistic creation of an ear situated under the skin of an arm, or horns embedded under the skin of a forehead, are not within any known copyright category. Interestingly, in 2014 and 2016 Orlan sued Lady Gaga in France and the US respectively for alleged copyright infringement as a result of Gaga's utilisation of her 'look',[25] but it is perhaps significant that she relied on copyright in a sculpture 'Bump-Load', 2009,[26] and a still from a performance, 'Woman With Head' (1996), depicting Orlan's decapitated head on a table.[27] Her claims were not apparently based on her own living body 'sculpture'.

If the sounds made by Cat Hope's 'Sound of Decay', Leslie Garcia's 'Pulsu(m) Plantae', or Guy Ben-Ary's 'CellF' are recorded, then this creates a protectable sound recording. Possibly, the sounds produced by the decaying toad, the growing plants or the neural network might constitute a musical work, but who is the author? The dead toad? The plant? Ben-Ary's 'external brain'? The dominant view is that copyright needs a human author, an issue which is discussed further below.

Clearly, a number of photographs and films may also be made in these bio art pieces, which are often integral to our experience of this art, such as in Kathy High's 'Blood Wars' and Verena Friedrich's 'Cellular Performance', where in each case the biological activity under a microscope was filmed using time-lapse microscopy. These images result in protectable photographs and films. Finally, much of the process (and rationale) of bio art is recorded in documentary form, possibly comprising literary works.[28]

Arguably, installations and even the processes of bio art are all essentially performances and protectable as such. This may better describe bio

---

[24] Teresa Wiltz, The Operating Room is her Studio: French Performance Artist Orlan Undergoes Plastic Surgery to Express Artistic Vision, Chicago Tribune (December 4, 1998), http://articles.chicagotribune.com/1998-12-04/features/9812040380_1_plastic-surgeon-french-orlan.

[25] Brian Boucher, Orlan Must Pay Lady Gaga $22,000 After Losing Forgery Lawsuit, Artnet (July 26, 2016), https://news.artnet.com/art-world/orlan-loses-lawsuit-pay-lady-gaga-576311.

[26] Sculpture, Orlan, http://www.orlan.eu/works/sculpture/ (last accessed June 13, 2018).

[27] Claire Voon, Artist Orlan Tries Again to Sue Lady Gaga for Plagiarism, Hyperallergic (January 8, 2016), https://hyperallergic.com/266869/artist-orlan-tries-again-to-sue-lady-gaga-for-plagiarism/.

[28] See for example the long description of the entire process of Guy Ben-Ary's CellF project; cellF – The Process, Guy Ben-Ary (2014), http://guybenary.com/work/cellf/#The_Process (last accessed June 13, 2018).

art, which is often more a performance than a lasting artefact. In Jalila Essaïdi's project entitled '2.6g 329m/s', she reinforces in vitro human skin with spider silk produced by transgenic goats and worms. The material she produces is exceptionally strong, even resistant to bullets fired at reduced speeds.[29] How does one describe Essaïdi's endeavour? One commentator suggests that, '[l]ike other good art and biology projects, 2.6g 329m/s is a multi-façade act of cultural scrutiny, an artistic performance'.[30] However, labelling bio art a 'performance' is not a panacea for the categorical conundrum, because 'copyright has never fully conceptualized performance, and this has led to persistent confusion about what copyright protects'.[31] In particular, copyright law may require the performance to be of a recognised 'work',[32] so there is circularity of classification here. The reality is that, in many, if not most cases, there is just no obvious copyright 'home' for these biologically-based installations.

3.3 Bio Art as 'Project' and Process

It is also clear that most exhibited instantiations of bio art often follow a long process of investigation, experimentation, trial and study, which may be integral to the work. Bio art is fundamentally process-oriented. This process over time is often as important as any end object. Indeed, it is no accident that most bio art is situated in, and referred to as, a 'project'. The contrast with conventional art is striking. We don't think of Leonardo Da Vinci's 'Mona Lisa' as a 'project'. 'Projects' don't fit comfortably in

---

[29] Bulletproof Human Skin Made From Spider Silk, Fox News (August 17, 2011), http://www.foxnews.com/tech/2011/08/17/breakthrough-bulletproof-human-skin-made-from-spider-silk.html.

[30] Oron Catts, Bulletproof Skin vs. Utility Proof Art or When a Symbolic Gesture is Seen as Real Action. In Jalila Essaidi, ed., *Bulletproof Skin: Exploring Boundaries By Piercing Barriers* (2012) 31.

[31] Rebecca Tushnet, Performance Anxiety: Copyright Embodied and Disembodied, 60 J. Copyright Soc'y USA 209, 209 (2013).

[32] For example, under Copyright Act 1968 (Cth) s 84 (Aust), a performer's rights apply (only) to the following live performances:

- dramatic work, or part of such a work, including a performance given with the use of puppets;
- musical work or part of a musical work;
- dance;
- circus act or a variety act or any similar presentation or show; and an expression of folklore; or
- reading, recitation or delivery of a literary work or part of a literary work, or the recitation or delivery of an improvised literary work.

copyright boxes. Copyright focuses on 'works', or films, or sound recordings, but in doing so struggles to accommodate the process and practice integral to bio art and without which the works would not eventuate.

4. AUTHORSHIP

Under most copyright frameworks, a copyright work requires an 'author'. Authors remain generally undefined in copyright law, despite their fundamental importance to it.[33] They are frequently referred to as '[t]he person who brings the copyright work into existence in its material form'.[34] However, the author is most importantly the source of originality, another cornerstone of copyright subsistence.[35] Originality is also a nebulous concept in copyright law, but it has been described as the independent (ie uncopied) intellectual or creative effort of a human author[36] which is 'directed to the particular form of the expression'.[37] Indeed, originality is correlative to authorship.[38]

The requirement that a work is the result of human independent intellectual effort means that an author may be lacking from expression which

---

[33] Notably, the author is ordinarily the owner of the copyright.

[34] *IceTV v Nine Network Australia* ('*IceTV*') (2009) 239 CLR 458, 494 [98] (Gummow, Hayne and Heydon JJ) (Austl.), quoting Hugh Laddie, Peter Prescott and Mary Vitoria, The Modern Law of Copyright 243 [6.6] (1st edn, LexisNexis, 1980); *IceTV* (2009) 239 CLR, 474 494 [33] (French CJ, Crennan and Kiefel JJ) (Austl.): authors 'bring into existence' copyright works. See also *Donoghue v Allied Newspapers Ltd* [1938] 1 Ch 106 at 109 (Farwell J) (Eng.): 'the person who has clothed the idea in form'.

[35] Jane C Ginsburg, The Concept of Authorship in Comparative Copyright Law, 52 DePaul L. Rev. 1063, 1067 (2003) (noting how 'in most copyright/authors' rights jurisdictions, originality is the overarching standard of authorship').

[36] *IceTV* (2009) 239 CLR 458, 474 [33] (French CJ, Crennan and Kiefel JJ) (Austl.): 'originality means that the creation (ie the production) of the work required some independent intellectual effort'.

[37] *IceTV* (2009) 239 CLR 458, 481 [54] (Gummow, Hayne and Heydon JJ) (Austl.). French CJ, Crennan and Kiefel JJ also stated that 'there can be no doubt that copyright is given in respect of "the particular form of expression in which an author convey[s] ideas or information to the world"': *IceTV* (2009) 239 CLR 458, 471 [26], citing *Hollinrake v Truswell* [1894] 3 Ch 420, 424 (Lord Herschell LC) (Eng.).

[38] *Sands & McDougall Pty Ltd v Robinson* (1917) 23 CLR 49, 55, 57 (Isaacs J) (Austl.) See also *IceTV* (2009) 239 CLR 458, 474 [34] (French CJ, Crennan and Kiefel JJ) (Austl.): 'There has been a long held assumption in copyright law that "authorship" and "original work" are correlatives; the legislation does not impose double conditions'.

is substantially shaped by software[39] or nature.[40] Nature's impact on bio art is obvious. The critical question is whether the author shapes the form, or whether they devolve responsibility for that to nature. And of course there is a continuum of authorial intervention, with many creative outputs occupying a shadowy middle ground on that continuum.

Despite their heavy reliance on nature and existing biotechnology, many bio artists make strong authorial claims to their work. For example, Eduardo Kac's transgenic 'Natural History of the Enigma' (2003–2008) is a human/petunia hybrid created by introducing Kac's DNA into the plant, which Kac named Edunia.[41] He has claimed that 'nature didn't make it, I did'.[42] In contrast, George Gessert, who creates flower hybrids in the name of art,[43] considers himself more of an enabler than an author. He says '[p]lants, like ink spots, generate themselves. My job is to facilitate',[44] and '[t]he job of the artist is not so much to create, as to help what is latent in things manifest itself. In a way the job of the artist is to leave himself out'.[45]

In Marta de Menezes' 'nature?' (2000), she modified the wing patterns of butterflies by manipulating the caterpillar's wing site using microsurgical tools during its transition to butterfly pupae. This resulted in one normal wing, and one altered wing with a new design and colour element, such as new eyespots. De Menezes claims that the modified wings are 'designed by an artist',[46] and that her intervention creates a 'new pattern never seen before in nature ... simultaneously natural, but resulting

---

39 Jani McCutcheon, The Vanishing Author in Computer-Generated Works: A Critical Analysis of Recent Australian Case Law, 36(3) Melb. U. L. Rev. 917 (2013).

40 Jani McCutcheon, Natural Causes: When Author Meets Nature in Copyright Law and Art. Some Observations Inspired by Kelley v Chicago Park District, 86(2) U. Cin. L. Rev. (forthcoming 2018).

41 'Natural History of the Enigma', Eduardo Kac, http://www.ekac.org/nat.hist.enig.html.

42 Megan Gambino, Cracking the Code of the Human Genome: The Story of How an Artist Created a Genetic Hybrid of Himself and a Petunia, Smithsonian.com (February 22, 2013), http://www.smithsonianmag.com/science-nature/the-story-of-how-an-artist-created-a-genetic-hybrid-of-himself-and-a-petunia-25148544/.

43 Gorge Gessert, V2 Lab for the Unstable Media, http://v2.nl/archive/people/george-gessert (last accessed June 13, 2018).

44 'Bioart Through Evolution: George Gessert', Revolution Bioengineering, http://revolutionbio.co/bioart/bioart-through-evolution-george-gessert/ (last accesse d June 13, 2018).

45 Ibid.

46 'Nature', Marta de Menezes (2012), http://martademenezes.com/portfolio/projects/ (last accessed June 13, 2018).

from human intervention'.[47] There are complex authorship issues here. Is de Menezes responsible for the corruptions in the butterfly wings, or is nature? Causally, she precipitated the abnormality, but did she know how that would materially manifest, and does she need to, in order to be an author for the purposes of copyright? This is an unanswered question in copyright law. Australian law seems to require authorial responsibility for the 'particular form of expression'.[48] In other words, de Menezes may need to be responsible for more than just an abnormality; she might need to be responsible for the particular abnormality that eventuates.

Playing with living systems necessarily injects an element of spontaneity and accident which may disturb authorial vision. For example, the Tissue Culture and Art Project's 'Victimless Leather' is a miniature jacket-shaped polymer matrix seeded with immortalised cell lines.[49] When exhibited at The MoMA, it grew too fast and out of control, dropping an arm and shedding cells. At Toyko's Mori Art Museum, the jacket became infected by fungi.[50]

These more aleatory expressions can be usefully compared to more predetermined bio art. Despite unanticipated adventures, the Tissue Culture and Art Project engages with semi-living artefacts which 'are coerced to grow in predetermined shapes'.[51] Examples include 'Pig Wings' (2000–2002), where tissue culture was grown into a particular wing shape over a constructed scaffold, and their 'Extra Ear – ¼ Scale' project, where they grew a quarter-scale replica of Stelarc's ear using human cells.[52] In the same way, Diemut Strebe grew a replica of Vincent van Gogh's ear over a polymer-based scaffold.[53] Another example is bio fashion,[54] which uses organic material to form fashion outfits in a discernible and preconceived

---

47 Ibid.

48 *IceTV v. Nine Network Australia* (2009) 239 CLR 458, 481 [54] (French CJ, Crennan and Kiefel JJ) (Austl.).

49 Victimless Leather – A Prototype of Stitch-less Jacket Grown in a Technoscientific 'Body', The Tissue Culture and Art Project, http://www.tca.uwa.edu.au/vl/vl.html (last accessed June 13, 2018).

50 Jennifer Johung, Vital Maintenance: Tissue Culture & Art, Artlink (September 1, 2014), https://www.artlink.com.au/articles/4212/vital-maintenance-tissue-culture-26-art/ (last accessed June 13, 2018).

51 'Extra Ear – ¼ Scale', The Tissue Culture and Art Project, http://www.tca.uwa.edu.au/extra/extra_ear.html (last accessed June 13, 2018).

52 Ibid.

53 Carina Storrs, Art Imitates Life: Artist Bioengineers Replica of Van Gogh's Ear, CNN (November 13, 2015), http://edition.cnn.com/2015/11/13/health/van-gogh-ear-art-science/index.html.

54 See, for example, Donna Franklin's Fibre Reactive 2004 at Donna Franklin, Symbiotica, The University of Western Australia, http://www.symbiotica.uwa.edu.au/residents/franklin, and Gary Cass and Donna Franklin's Micro'be' project

form. There isn't anything inherently problematic about these structures, except perhaps their occasional tendency to go rogue, and their ephemerality as living tissue, an issue explored further below.

4.1 Joint Authorship

Issues of joint authorship and allocation of authorship and ownership between multiple contenders also frequently arise in bio art, because it is often a collaborative process. Consider the long list of collaborators in Guy Ben-Ary's CellF, including synthesiser designer, lab creator, musicians and scientists, the plant biologists working with Kac to produce the Edunia, de Menezes' butterfly collaborators,[55] the phlebotomists who drew the blood in Kathy High's 'Blood Wars', and the 'who's who' of engineers and scientists Diemut Strebe worked with to create the replica of van Gogh's ear.[56]

Many multi-party bio art projects would struggle to comply with the frequently stringent and narrow definition of a work of joint authorship in copyright law. For example, the Australian and UK definitions of a work of joint authorship require that all participants be (1) authors; (2) collaborators; and (3) that their contributions be inseparable from the contributions of the other author or authors.[57]

While there may be intense collaboration among bio art project teams, their contributions are frequently distinctly separable. This results in protection being parsed out to disparate individuals for discrete expressions falling within copyright's protectable categories.

Another requirement of this definition is that each collaborator must work with the others as authors. Whether each collaborator in a bio art project is also an author is contestable. In many cases they will be involved because the artist lacks the technical skills to conduct the scientific work, not because

---

garments at Donna Franklin and Gary Cass, Micro'be' Fermented Fashion, Bioalloy, http://www.medinart.eu/works/bioalloy/ (last accessed June 13, 2018).

55 Who seem always to be specifically acknowledged in descriptions of her work. See, for example her website statement, 'This project was developed at the laboratory of Professor Paul Brakefield, University of Leiden, Holland, with the scientists: A Monteiro, M Bax, K Koops, R Kooi and P Brakefield' in 'Nature', Marta de Menezes (2012), http://martademenezes.com/portfolio/projects/ (last accessed June 13, 2018).

56 Storrs, supra note 53.

57 Copyright, Designs and Patents Act 1988, c. 48 § 10(1) (UK) defines a work of joint authorship as 'a work produced by the collaboration of two or more authors in which the contribution of each author is not distinct from that of the other author or authors'. The definition of a work of joint authorship in s 10 of the Copyright Act 1968 (Cth) (Austl.) is virtually identical.

they are contributing to the 'authorship' of the work. At the very least, the collaborative nature of bio art blurs authorship and ownership of it.

5. ORIGINALITY

5.1 Original Expression

The broad concept of originality in copyright doctrine was explained above. Significantly, copyright demands originality of expression, not ideas. This means that originality cannot be supplied by merely overlaying existing, unoriginal expression with subjective, artistic purpose, unless that process generated new expression.

This poses problems for a number of bio art creations, such as Kac's 'Edunia'. Despite Kac's ideational originality, the difficulty is that the plant looks like any other petunia. It lacks expressive originality. The same might be said of Stelarc's 'Ear on Arm'. The idea of inserting an ear shape under the skin of a forearm is certainly novel,[58] but is the ear shape itself original? It probably meets the low threshold of originality due to the reality that there are numerous ways to craft the unique configurations of an ear. But at best this would prevent second comers from copying the particular configuration of the Stelarc ear. It would not prevent others implanting ears under their skin. In contrast to Stelarc's ear, Diemut Strebe's replica of Vincent van Gogh's ear may not be original because it apparently was made to closely approximate van Gogh's ear, using a photograph of his ear as a model. Similar problems might also beset Stelarc's ear, if it was based on a direct mould of his own.

5.2 Originality, Context and Authorial Intention

It is only if copyright recognises a notion of 'contextual' or 'purposive' originality, where unoriginal expression derives originality through its contextual placement, or the subjective purpose of the creator, that Stelarc's ear installation may be considered an original work. Such a concept was hinted at by Mann J in *Lucasfilm Ltd v Ainsworth*[59] where he stated:

---

[58] Although it was clearly inspired by Vacanti's 'ear mouse', as discussed in Karl Kruszelnicki, Mouse with Human Ear, ABC Science (June 2, 2006), http://www.abc.net.au/science/articles/2006/06/02/1644154.htm.

[59] *Lucasfilm Ltd v Ainsworth* [2008] EWHC (Ch) 1878 [118] (Mann J) (Eng.).

> A pile of bricks, temporarily on display at the Tate Modern for 2 weeks, is plainly capable of being a sculpture. The identical pile of bricks dumped at the end of my driveway for 2 weeks preparatory to a building project is equally plainly not. One asks why there is that difference, and the answer lies, in my view, in having regard to its purpose. One is created by the hand of an artist, for artistic purposes, and the other is created by a builder, for building purposes.[60]

Relying on this logic, we may argue that the ear/arm fusion could constitute an original installation, based on Stelarc's subjective artistic purpose and the ear's incongruous placement in his arm.[61] In contrast, a surgeon affixing a prosthetic ear to an earless child would not create an original artwork. And if uncoupled from his arm, Stelarc's ear sculpture may again become unoriginal.

Bio artists are a bit schizophrenic, a hybrid of scientist, technologist and artistic visionary. Because of this, it may be very difficult to discern the purpose of their creative practice and isolate the scientific from the artistic. For example, Kathy High apparently pursues genuine scientific objectives, claiming that 'Blood Wars will ultimately give a better understanding of the processes of blood cell division and cell membrane fluid exchange, the immune system and the mythologies of inheritance and blood pathologies'.[62] In other cases, the claim to 'art' and the subjective purpose of the artist may be the only thing to distinguish the art from pure science. For example, Eduardo Kac's 'Edunia' was built on the existing biotechnological practice of inserting human genes into plants, using them like factories to manufacture drugs on a large scale or to produce human antibodies to diagnose diseases. The plant biologist who inserted a copy of Kac's gene fragment into the petunia acknowledged the utilisation of existing scientific protocols, but argued: '[b]ut you don't have plants that have been made to explore ideas . . . Eduardo came to this with an artistic vision'.[63] This may be the case, but whether this vision is even relevant in the context of copyright subsistence is questionable.

---

60 Ibid.

61 Nevertheless, as argued above, this doesn't resolve the problem of categorisation that would beset most putative copyright owners, at least in common law jurisdictions.

62 Katy High, 'Blood Wars', Dublin Science Gallery, https://dublin.sciencegallery.com/visceral/blood-wars/. It is arguable, however, that this claim is part of High's 'performance', since her experiment does not apparently comply with conventional scientific conventions such as using a control or a double blind protocol.

63 Gambino, supra note 42.

6. FIXATION, EPHEMERALITY, CHANGEABILITY

6.1 Bio Art and Kinesis

Most, although not all, copyright systems require that a work be fixed in material form. Bio art generates two major issues in the context of the fixation doctrine: ephemerality of the art, and changeability of its form.

Because of its reliance on living systems or semi-living material, most bio art is ephemeral. The growth activity of cells, bacteria, moulds, plants and viruses is transient, and the disintegration of cane toads is generally inevitable. Most bio artists accept that their work will disintegrate, or at least fundamentally change. Often that is the point of the work. For example, Marta de Menezes' 'Decon: Deconstruction, Decontamination, Decomposition' (2007) replicates Piet Mondriaan's geometric paintings as a support medium for Pseudomonas putida bacteria, which progressively denudes the colours in the piece until it is colourless.[64] The 'life supports' for the Tissue Culture and Art Project's exhibited 'semi-living' works are inevitably switched off and a symbolic 'funeral' may be held.[65]

However, most bio art would satisfy copyright's fixation doctrine, notwithstanding diverse definitions of fixation.[66] Most of the semi-living forms would be crystallised long enough to be observable and stored in material form. Few bio art projects are so fleetingly transitory that they are imperceptible or cannot be recorded or stored for a sufficient period. The more germane (and difficult) issue is identification of the work – or works – at issue and their categorisation within a recognised class.

As noted above, documentation also serves a pivotal role in bio art, capturing the ephemerality of the biology in video, notes, photography, schematics or microscopy. Especially for bio artists working at the microscopic

---

64 'Decon: Deconstruction, Decontamination, Decomposition', Marta de Menezes (2012), http://martademenezes.com/portfolio/decon-deconstruction-decontamination-decomposition/.

65 Johung, supra note 50, describing the 'killing ritual at the end of each exhibition'.

66 The Copyright Act of 1976, 17 U.S.C § 101 (2012) provides that '[a] work is "fixed" in a tangible medium of expression when its embodiment in a copy . . . is sufficiently permanent or stable to permit it to be perceived, reproduced, or otherwise communicated for a period of more than transitory duration' (defining 'fixed'). Copyright Act 1968 (Cth) s 10(1) (Austl.) defines 'material form' in relation to a work as including 'any form (whether visible or not) of storage of the work . . . (whether or not the work . . . can be reproduced)'. The Copyright, Patents and Designs Act 1988, c. 48 § 3(2) (UK) requires a work to be 'recorded, in writing or otherwise'.

level, these are the necessary proxies through which a work is known. However, as noted above, the composite parts of bio art projects and installations also potentially multiply the possible authors, because there may be a distinction between the art and the photo (or film) of the art, and the authors of each. Thus if the biological subject matter of the photograph or film is not a copyright work (which may often be the case, as this chapter has argued), then the artist must rely on the copyright in the photograph or film as a means of controlling the bio art depicted in it. This requires them to be the owner of the copyright in the film or photograph, which may not always be the case if they didn't take the photograph or arrange to make the film. In any event, the photograph or film is not in truth the work. Its flat static depiction does not reflect the work's true essential, biological character. Ben-Ary also recorded aspects of CellF on video, and acknowledges that '[w]hen we talk about art and science, there's a lot of people who think [the video] is it . . . [they say] "Put it in a gallery because it's beautiful"'.[67] But Ben-Ary claims 'the work we're doing is not about beauty. It's part of it, but for me this beautiful movie is just part of the process'.[68]

The dynamism of the living systems and the processes associated with bio art also complicates fixation, making it difficult to identify the work. This vitality sometimes results in an ever-changing form, raising the question – when is the work? At what point on its kinetic pathway do we identify the work? Or are there in fact several separate works formed along a temporal continuum? If so, when does each one crystallise into a discrete work? Or rather, is the entire continuum of shifting forms the work, and if so, how do we categorise it?

7. IDEA-EXPRESSION: BIO ART AS CONCEPT

The 'idea-expression dichotomy' is a copyright doctrine that preserves ideas in the public domain. The result is that a reproduction of the detailed expression of a work may infringe copyright, but the reproduction of a general idea underlying that work will not. There is a clear policy rationale for this. First, it is impossible to police a legal monopoly over broadly described ideas, and second, it would unreasonably restrict the creative freedom of subsequent authors to use those ideas.

---

[67] Penny Durham, Guy Ben-Ary's CellF: Playing with Neurons, The Australian (August 4, 2016), http://www.theaustralian.com.au/arts/music/guy-benarys-cellf-playing-with-neurons/news-story/db746384f6b5ca3f15c3c711a7002725.

[68] Ibid.

Bio art is often fundamentally conceptual and process-focused. It might aptly be termed conceptual art utilising living forms, with the artistic personality imbued not in the material form, but rather in the idea. Under the idea-expression doctrine, the protection of abstract concepts is anathema to copyright, which looks for things, form and expression and assumes a stable object created by an author or joint authors. The familiar policy concerns resonate with bio art. Because bio art is so process-driven, if the ideas underlying that process are protected, it could risk monopolising a bio-artistic 'territory'. For that reason, the idea-expression dichotomy would prevent Marta de Menezes from monopolising the idea of modifying butterfly wings. At best, she would have copyright in the particular modifications she achieved.

Indeed, the 'art' in the bio art is frequently difficult to see, even if the idea is discernible. Take Kac's 'Edunia'. According to Kac, he selected a gene responsible for the identification of foreign bodies, and used a 'promoter' to ensure that his genes were expressed only in the plant's veins.[69] However, it appears that the visual legacy of this manipulation cannot be attributed to Kac or his scientific collaborators. The plant biologist working with Kac explains that they 'decided to start with a petunia that already has red veins'.[70] The intervention in 'Edunia' is the insertion of Kac's gene. However, this cannot be visually detected. The result is that to the naked eye 'Edunia' 'looks like a normal petunia. The trans-gene has had no effect on the plant's appearance'.[71] As Mitchell explains, with these types of bio art projects, 'the object of mimesis here is really the invisibility of the genetic revolution, its inaccessibility to representation'.[72] The result is that 'one is basically taking it on faith that the work exists and is doing what it is reported to be doing'.[73] This underscores bio art's preoccupation with 'the idea of the work and the critical debate that surrounds it as much as its material realization'.[74]

---

[69] Kac, supra note 41.

[70] Camille LeFevre, Of Transgenic Petunias and Glow-in-the-Dark Bunnies, MN Artists (May 13, 2009), http://www.mnartists.org/article/transgenic-petunias-and-glow-dark-bunnies.

[71] Kac, supra note 41.

[72] William JT Mitchell, The Work of Art in the Age of Biocybernetic Reproduction, 10(3) Modernism/Modernity 494, 495 (2003).

[73] Ibid.

[74] Ibid.

8. WHO CARES?

The above discussion has canvassed the dissonance that can exist between copyright doctrine and many examples from the broad and eclectic discipline of bio art. This means a significant proportion of bio art is likely to be unprotected by copyright. But this sparks an important question: Do bio artists need or want copyright protection?

Much of the scholarship critiquing copyright's inability to accommodate contemporary art either overtly or tacitly claims that the lack of copyright is worrisome. It is argued that '[c]opyright is valuable in that it protects artists allowing them to live in society and to persevere in the socially desirable task of creating';[75] and that '[g]ranting an artwork intellectual property protection is crucial for the artist to maintain both economic and cultural standing'.[76] Finally, '[s]ince the statutory framework makes classification of a work crucial to the rights that are reserved to the copyright owner, it is important for the postmodern artist to know where a work is located in the protective scheme'.[77]

The implication is that copyright protection for borderline art is laudable, necessary and warranted, but is wrongly denied. This clearly valorises copyright and suggests bio artists would be impoverished by the lack of it. Many commentators are at least puzzled by the discriminatory exclusion of contemporary art from the copyright reward, or lament it: 'copyright should not seek to discern socially worthy art from art that is not, lest a potentially great artist go underappreciated'.[78] This overlooks the question of whether copyright is necessary to affirm the 'socially worthy' status of contemporary art, including bio art.

Many of these responses proceed on the incentive theory, which regards copyright as an instrumental right necessary to encourage creation.[79] This raises the issue of the suspect link between copyright and the creation of

---

[75] Elettra Bietti, Defining Art for Copyright: Against the Museum without Walls, 16 UCL Juris. Rev. 131, 140 (2010).

[76] Lily Ericsson, Creative Quandary: The State of Copyrightability for Organic Works of Art, 23 Seton Hall J. Sports Ent. L. 359, 361 (2013).

[77] Steven Shonack, Postmodern Piracy: How Copyright Law Constrains Contemporary Art, 14 Loy. L.A. Eny. L. Rev. 281, 293 (1994).

[78] Megan Carpenter and Steven Hetcher, Function Over Form: Bringing the Fixation Requirement into the Modern Era, 82 Fordham L. Rev. 2221, 2226 (2014).

[79] See generally William M Landes and Richard A Posner, An Economic Analysis of Copyright Law, 18 J. Legal Stud. 325 (1989).

art.[80] As has been argued, '[w]e have built an entire system of copyright protection on the untested assumption that more remuneration motivates artists to create more and better art. Every aesthetic theory within the legal culture must be careful as it treads across this wide expanse of quicksand'.[81] Others have criticised the 'unexamined assumption that copyright law must adapt itself to new modes of authorship'.[82] The lament for copyright protection for post-modern art conveniently and ironically overlooks the existence of that art, which was created in the (apparent) absence of copyright protection.

A separate question, which needs more extensive empirical investigation, is whether copyright is relevant to contemporary artists whose work sits at copyright's borders. Bio artists, for example, are not taking to the streets in protest at a lack of copyright. They tend to be iconoclastic, post-modern and anti-establishment. Many may wilfully use copyright works in their practice and willingly share them. They may have a profound disinterest in copyright, even a distrust of it, or may outright reject it. They may rely on norms of professional practice which militate against unacknowledged appropriation of others' bio-artistic domains. Agitation for copyright reform on the basis that bio artists may need copyright or moral rights to protect their work may smack of paternalism.

Artistic works are significantly different to other copyright works. Literary and musical works traditionally generated revenue through the commercial dissemination of copies. Conventionally, artworks are not designed to be reproduced en masse. Their value lies in their uniqueness and rarity; their lack of dissemination through copies. This applies equally to much bio art, particularly the more fragile 'semi-living' constructs made from tissues and cells, which must be coaxed to 'life' and carefully maintained in sterile conditions. These are not easily replicable, unlike some other examples of conceptual art which can be

---

80 Copyright scholarship has consistently critiqued the persuasiveness of the incentive theory when explaining creativity. See, eg, Diane L Zimmerman, Copyrights as Incentives: Did We Just Imagine That?, 12 Theoretical Inquiries in Law 29 (2011); Jeanne C Fromer, Expressive Incentives in Intellectual Property, 98 Va. L. Rev. 1745 (2012); Adam Moore, Intellectual Property, Innovation, and Social Progress: The Case Against Incentive Based Arguments, 26 Hamline L. Rev. 601 (2003); and Jeffrey L Harrison, Rationalizing the Allocative/Distributive Relationship in Copyright, 32 Hofstra L. Rev. 853, 868 (2004) ('Common and observable market phenomena suggest that a great many works are not dependent on copyright').

81 Marci A Hamilton, Four Questions about Art, 13 Cardozo Arts Ent. L. J. 119, 122 (1994).

82 David Saunders, *Authorship and Copyright* (Routledge, 1992) 227–9.

simultaneously or progressively installed in a number of different institutions, sometimes under the remote control of the artist. However, some more inherently reproducible bio art may be different, such as Eduardo Kac's, Edunias', Gessert's plant hybrids and Edward Steichen's giant delphiniums, first exhibited in 1936.[83] Apparently Steichen dreamed of selling his delphinium seeds for 25 cents a packet.[84] Apart from the claim to practice the plant breeding as an art form, there is little objectively differentiating bio artists like Kac, Gessert and Steichen from conventional plant breeders, who may seek protection under plant breeders rights legislation or even patents, and broadly commercialise their plant varieties.

And yet there is no evidence suggesting that bio artists routinely generate income through these kinds of sources, or by selling artistic artefacts. Most, if not all, of the examples of bio art discussed in this chapter exist outside of the contemporary art market and the commercial gallery spaces. Bio artists raise funding through various sources for the creation and perhaps exhibition of their work, and rarely sell or distribute it through conventional channels (although photographs of the work may be licensed).[85] In this paradigm, copyright is apparently peripheral, if not irrelevant. This might change if and when these works become commodities circulating in the art market.

Rather, the ability to control others' reproductions and harmful treatments would lie at the heart of the copyright incentive for fine artists, including bio artists. Moral rights are predicated on copyright, and are designed to protect the personality interests of artists, guarding against derogatory treatment of the art, and non-attribution and false attribution of authorship. So perhaps if copyright is needed, it is probably needed primarily as a means of securing moral rights.

9. IS COPYRIGHT FOR BIO ART JUSTIFIED?

Bio art can be unnerving, and generates issues of policy, philosophy, theory and ethics that copyright has never traditionally grappled with.

---

[83] Apparently these hybrids were the first exhibited as art, and perhaps one of the earliest examples of bio art. George Gessert, *Green Light: Toward an Art of Evolution* (MIT Press, 2010) 48.

[84] George Gessert, *Green Light: Toward an Art of Evolution* (MIT Press, 2010) 48.

[85] For example, photographs of the Tissue Culture and Art Project's 'Pig Wings, 2000–2002' were licensed by Viscopy for reproduction in 34:3 ArtLinks (2014) at 69.

Copyright resists the admission of 'undesirable' inclusions to the copyright framework through doctrines like the idea-expression dichotomy, originality, authorship and material form, and through categorisation. Under current copyright doctrine, there is no separate basis for denying copyright to bio art because it is unsettling, or because it uses biotechnical materials like DNA, genes, tissue, blood, bacteria, viruses or even higher organisms like mice, butterflies and rabbits, or because it is 'weird' or even incomprehensible.

If there is potential for copyright to subsist in bio art, despite the hurdles outlined in this chapter, do we need some more effective exclusionary mechanism? If bio art was given copyright protection, what harms may ensue? Copyright is a potent property right which can exclude others from expression for a very long time, and exceptions to infringement are often uncertain and limited in scope and application, so it is worth interrogating this question.

We might have some legitimate concerns if new biological material of general public benefit is protectable through copyright. This is ordinarily the domain of patent law, which imposes a more rigorous innovation test and limits the monopoly to 20 years. Bio artists and scientists shadow each other, so their work can produce real, tangible, and potentially commercially valuable scientific outcomes. For example, bio art can stimulate development in bio surveillance technology through practices like animated tattoos, DNA chips, and human barcodes, and translate to the bio economy engaged in policing 'biological threats'.[86] Bio artist Adam Zaretsky observed increased production of antibiotics after playing Engelbert Humperdinck's Greatest Hits for 48 hours to a petri dish of E. coli bacteria.[87] The next stage of Diemut Strebe's work was to involve cloning the mitochondrial DNA of Vincent van Gogh's sister's descendant and inserting it into cells. This was 'absolutely the reason'[88] her scientific collaborators became involved, because 'no research group [had] ever convincingly managed to deliver cloned mitochondrial DNA into cells'[89] and success in doing so 'would give researchers a new platform to study mitochondrial diseases'.[90] There is also value in the data generated from the

---

[86] Elizabeth Abergel and Jamie Magnusson, The Art of (Bio)Surveillance: Bioart and the Financialization of Life Systems, 30–31 TOPIA: Canadian J. Cultural Stud. 237 (2013/2014), http://topia.journals.yorku.ca/index.php/topia/article/view/38430 (last accessed June 13, 2018).

[87] Steve Nadis, Science for Art's Sake, 407 Nature 668 (October 12, 2000).

[88] Storrs, supra note 53.

[89] Ibid.

[90] Ibid.

experiments undertaken by bio artists, such as DNA strain combinations and variations.[91]

But copyright has sufficient inbuilt safeguards to prevent a potentially disquieting patent-like monopoly. Copyright protects a particular form of expression. It doesn't protect the tools or media used to create that expression, and it doesn't protect the ideas underlying that expression. What matters is whether the use of that biological material results in a recognised copyright work. The fact that some new genetically engineered biological material was used to create that work will not confer protection on the biological material itself, but only the expression achieved through the utilisation of that biological material. If the material is only used as a tool, there will be sufficient safeguards.

Similarly, copyright is unlikely to extend to the particular chemical components of compounds such as Jalila Essaïdi's spider silk, or the particular functional characteristics of bio surveillance technology, or protect the idea of accelerated bacterial growth as a response to music. Nor does copyright generally subsist in raw data. Thus in the unlikely event that copyright extends to bio art, protection is likely to be relatively thin, covering the expression produced by the biological material, rather than the material itself.

10. CONCLUSION

This chapter has explored copyright's application to bio art through a number of examples of this perplexing, fascinating, innovative and provocative artistic practice. While the spectrum of bio art is vast, the chapter demonstrates a potentially profound disconnect between copyright and a significant proportion of bio art. This results primarily from the difficulty in categorising bio art as an artistic work within the largely rigid taxonomy of most copyright systems. But the fundamentally natural force of bio art also problematises authorship and originality, and it too often strays into the unprotectable realm of ideas. For now, at least, copyright and much bio art are likely to remain relative strangers.

---

[91] Margoni, supra note 3.

4. Street art, graffiti and copyright

Enrico Bonadio*

1. INTRODUCTION

This chapter examines to what extent street and graffiti art can be protected by copyright law. Although the focus will be on US law and to a lesser extent UK law,[1] a few references will also be made to how these forms of art are regulated in other jurisdictions.

Graffiti refers to a technique of painting names and letters on various urban surfaces, such as tube and railway trains as well as walls. The graffiti movement first started in Philadelphia and New York City in the late 1960s and early 1970s, evolved into a proper form of art through the 1970s, and then spread to other cities and countries in the world. The term 'street art' is instead used to define more elaborate forms of art, which have both evolved from the early graffiti movement by focusing on images rather than letters and developed independently by building upon, and mixing with, established artistic movements such as muralism. Street artists may nowadays use a variety of tools to paint and draw: not only spray cans, but also traditional instruments such as brushes, rollers and palettes as well as marker pens, chalks and charcoal. Ways of placing art in the street may also take forms different from painting such as attaching mosaic tiles, stickers and posters on walls as well as abandoning artworks.

Street artworks are often produced illegally, namely without the authorization of the owner of the property where they are placed, which can expose artists to serious legal consequences including jail time. Yet, as opposed to the early days of these artistic movements (especially graffiti) when most works were done illegally, many street and graffiti artists nowadays produce their pieces legally. Also, city authorities, private

---

* The initial research which eventually led to this chapter was carried out at CUNY Law School in New York City in the autumn of 2016.

[1] I have provided an expanded legal analysis of these issues under UK law in Copyright Protection of Street Art and Graffiti under UK Law, 2 Intellectual Property Quarterly 188–220 (2017).

property owners and businesses often make available spaces where street and graffiti art is tolerated and artworks change regularly.

Communities' interest in preserving street artworks is also growing, especially when they are created by famous artists, whose pieces are often protected by Perspex covers. The expanding interest in street art and graffiti is confirmed by the high number of magazines and websites dedicated to these forms of art; and by the many street art tours offered in several cities around the world to admire pieces which – it is often noted – end up beautifying and revamping local neighbourhoods.

Street art has also become a hot commodity. Due to their increasing popularity, works by renowned street artists such as Banksy are often offered for sale in galleries and reproduced (with or without their authorization) on products as diverse as posters, clothes, bags, mugs and home furnishing as well as on canvases and prints sold in popular street markets as well as the Internet. They are also increasingly commissioned and used as a backdrop in TV series, commercial ads and music videos or to promote products.

This graffiti and street art boom has attracted attention from advertisers and marketing gurus in their pursuit of new commercial ideas and the attempt to give their products an 'aura' of street credibility; indeed, cases of unauthorized exploitation of urban creativity are increasing alarmingly, especially in certain industries, including fashion,[2] fast food[3] and entertainment.[4]

In the following paragraphs I will expand on whether copyright law is capable of protecting street and graffiti artworks, also taking into account the (often) conflicting rights of the owners of the property upon which the works are placed. There have been thus far few reported decisions, especially by US courts, on copyright and moral rights protection of these forms of art. This is probably due to the fact that disputes in this field are usually settled out of court. After all, this seems to be the rule

---

[2] See for example Donald Robbins, *Noah Daar v. Oakley Inc.*, Doc. No. 2:18-cv-05116-PA-KS; *H&M v. Jason Williams a/k/a Revok*, Doc. No. 1:18-cv-01490; *Jason Williams et al. v. Roberto Cavalli, S.p.A. et al.*, Doc. No. 2:14-cv-06659 (C.D. Cal. August 25, 2014), Court Doc.; *Joseph Tierney v. Moschino S.p.A. et al.*, Doc. No. 2:15-cv-05900 (C.D. Cal. August 5, 2015), Court Doc.; *Anasagasti v. American Eagle Outfitters, Inc.*, No. 1:14-cv-05618-ALC (S.D.N.Y. July 23, 2014).

[3] See amongst other cases, *Jade Berreau v. McDonald's Corporation et al.*, No. 2:16-cv-07394 (Central District of California).

[4] See for example *Franco Fasoli et al. v. Voltage Pictures LLC et al.*, No. 2:15-cv-00889 (C.D. Cal. February 6, 2015); *Hayuk v. Sony*, No. 1:14-cv-06659 (S.D.N.Y. August 19, 2014).

in most cases regarding art.[5] This chapter, in particular, will investigate to what extent certain copyright law provisions and case law apply to various types of art placed in urban environments, particularly paintings on walls and other urban surfaces. It will focus on the following selected aspects: (i) subject matter and requirements for protection; (ii) moral rights, with a particular focus on the integrity right; and (iii) illegal artworks.

2. SUBJECT MATTER AND REQUIREMENTS FOR PROTECTION

Most forms of art placed in the streets such as murals, stickers, posters, mosaics and abandoned objects certainly fall within the category of artistic works under most national laws.[6] They are therefore within the scope of copyrightable subject matter.

Many street artworks are created relatively quickly. Take stencil art, which is a technique for reproducing designs and patterns by passing spray-paint or ink over holes cut in cardboard or metal onto the surface to be decorated: it allows the preliminary creative activity to be carried out in studio and the actual work to be subsequently placed on the street in a relatively quick way, which also minimizes the risk of possible arrest if the piece is created illegally. Banksy himself noted that the reason why at some point in his career he started focusing on stencils was to drastically reduce the time required for placing pieces on the street and thus reduce the risk of being noticed by the police.[7] The fact that some street artworks may be rapidly created and placed in the chosen location (sometimes it is just a matter of a few seconds) does not stand in the way of their copyrightability. Indeed, in most jurisdictions the time taken to create the work is not relevant for copyright purposes.

---

[5] See Molly Torsen Stech, *Artists' Rights: A Guide to Copyright, Moral Rights and Other Legal Issues in the Visual Art Sphere* (Institute of Art & Law, 2015) xi.

[6] See for instance 17 U.S. Code § 102(a)(5) (providing that copyrightable works of authorship include 'pictorial, graphic, and sculptural works'); Section 4 of the Copyright, Designs and Patents Act (CDPA) 1988 (stating that the category of artistic works includes graphic works, photographs, sculptures or collages, irrespective of artistic quality. It also specifies that the category of 'graphic works' includes paintings, drawings and other visual artworks).

[7] Banksy, *Wall and Piece* (Century 2005) 13.

2.1 Fixation

Most street and graffiti artworks are ephemeral as they disappear relatively promptly. They may be removed by the owner of the property upon which they are placed; they may be painted over by local councils that want to keep the neighbourhood clean; or they may end up being vandalized. Artworks may also deteriorate and eventually vanish just because of the passing of time and the effect of the elements.

Could the transient nature of these forms of art be an obstacle that prevents their copyright protection? One may answer in the affirmative by remembering that works should be fixated in a medium to be considered copyrightable. In the US, for example, copyright subsists in works of authorship 'fixed in any tangible medium of expression',[8] this provision being in line with the Berne Convention, which allows states to provide that works shall not be protected 'unless they have been fixed in some material form'.[9] Yet, this objection would be rather weak, as such a fixation requirement only demands that the work is 'sufficiently permanent or stable' to allow it 'to be perceived, reproduced, or otherwise communicated for a period of more than transitory duration'. Although street and graffiti artworks may not last for a long time, their temporary existence appears to be sufficient to satisfy this requirement.[10]

A similar point could be made under UK law. Although in *Merchandising Corporation of America v. Harpbond*[11] it was held that the facial make-up of the pop artist Adam Ant did not constitute a painting for the purpose of copyright ('if the marks are taken off the face there cannot be a painting. . . . a painting without a surface is not a painting'), subsequent case law seems to have made the transient or permanent fixation of the work on a surface a less important issue. In *Metix Ltd v. G.H. Maughan*[12] for example it was suggested that an ice sculpture, although obviously not permanent,

---

[8] See 17 US Code § 102; see also Section 3(2) of the UK Copyright, Designs and Patents Act 1988, which provides that literary, dramatic and musical works must have been fixed in a tangible form before copyright can subsist.

[9] Section 2(2) of the Berne Convention for the Protection of Literary and Artistic Works, September 9, 1886, as revised at Paris on July 24, 1971, and amended in 1979, S. Treaty Doc. No. 99-27 (1986) [hereinafter 'Berne Convention'].

[10] See also Chapter 6 'Copyright in Culinary Presentations', authored by Cathay Smith (making a similar point in relation to food creations, and noting that despite contrary case law such creations are fixed for a period more than a transitory duration when they are first made).

[11] *Merchandising Corp. of America Inc. & ors v. Harpbond Ltd & ors* [1983] FSR 32 (CA).

[12] *Metix (UK) Ltd v. G.H. Maughan (Plastics)* [1997] FSR 718.

should in principle be protected as a three-dimensional work made by an artist's hand.[13] What *Metix* entails is that not just a permanent materialization, but also an ephemeral fixation, of the tangible medium would suffice for copyright subsistence purposes. And what has been suggested in this case may easily apply to most street art and graffiti scenarios, even more so in light of the recent judicial developments at European Union (EU) level.[14] Indeed, under EU law and in particular in light of the *Infopaq* decision,[15] anything that constitutes an intellectual creation should be protected by copyright.

But even if it is eventually accepted that a permanent fixation is required to attract copyright, it could still be argued that taking photographs of street art and graffiti (which is common practice, within these communities, to 'preserve' the relevant artworks) could help in meeting this requirement.[16]

2.2 Originality

All copyrightable works must be original.[17] There is no doubt that many, if not most, street and graffiti pieces satisfy the originality requirement, whether we apply the originality test under US law (as affirmed in the seminal case *Feist v. Rural*),[18] the UK 'skill, labour and judgment'

---

[13] See also Lionel Bently and Brad Sherman, *Intellectual Property Law*, 4th edn (OUP, 2014) 92.

[14] It is not unlikely that even after the UK leaves the EU in March 2019, several pieces of EU legislation, including copyright-related directives, and CJEU decisions may still be applicable and relevant in the UK (directly or indirectly) for several years.

[15] Case C-5/08, *Infopaq Int'l A/S v. Danske Dagblades Forening*, 2009 E.C.R I-6569).

[16] Marta Iljadica, *Copyright Beyond Law: Regulating Creativity in the Graffiti Subculture* (Bloomsbury, 2016) 103. For a contrary opinion on whether photographed works satisfy the fixation requirement, see Megan M. Carpenter and Steven Hetcher, Function Over Form: Bringing the Fixation Requirement into the Modern Era, 82 Fordham L. Rev. 2221 (2014) (noting that the photographed work remains unfixed for copyright purposes and that any 'fixation of the piece . . . is merely representational of the piece rather than the piece itself').

[17] See 17 U.S. Code § 102, which provides that 'Copyright protection subsists . . . in original works of authorship'. Similarly, Section 1(1)(a) of the UK Copyright, Designs and Patents Act 1988 states that 'Copyright is a property right which subsists . . . in the following descriptions of work – (a) *original* literary, dramatic, musical or artistic works . . .' (emphasis added).

[18] In *Feist Publications, Inc. v. Rural Telephone Service Co.*, 499 U.S. 340 (1991) the US Supreme Court explained that the requirement of originality is not particularly stringent and is comprised of two elements: that the work be

Source: Photograph by Serk.

Figure 4.1 Tag by Serk (New York City, 2003)

standard,[19] or other originality requirements, such as the EU 'intellectual creation' test.[20] A quick look at the many pieces painted on our cities' walls is enough to confirm that.

While most street and graffiti artworks are certainly original, one may note that the so-called tags and throw-ups lack a sufficient level of originality, and in general are too trivial to attract copyright protection.

What are tags? Tags are street signatures painted or drawn by graffiti writers on walls and other urban surfaces, and may be executed in condensed calligraphic form (see Figure 4.1). They represent the writer's chosen name, and epitomize a strong desire to be recognized and appreciated within the subculture.

In graffiti jargon the term throw-up – which also represents the artist's name – refers to a one colour outline and one layer of fill-colour, painted in bubble style letters: it is basically an evolution of the tag (see Figure 4.2).

As mentioned, one may have doubts about the originality of tags and throw-ups, and point to the regulations of the US Copyright Office which clarify that 'words . . . such as names . . . lettering' are examples of works not protected by copyright and that applications for registration of such

independently created by the author (as opposed to copied from other works) and that it possesses at least some minimal degree of creativity.

19 See for example *Ladbroke v. William Hill* [1964] 1 All E.R. 465, 469 (Eng.).

20 See also Marta Iljadica, Graffiti and the Moral Right of Integrity, 3 Intell. Prop. Q. 266, 268 (2015).

Source: Photograph by Serk.

Figure 4.2 Throw-up by Serk (New York City, 2005)

works cannot be entertained.[21] And indeed the graffiti-style words 'Skate Key' designed by the famous Bronx-based writer Tracy 168 were denied registration by the copyright office.[22]

What if tags or throw-ups are made in a highly stylish way? Indeed, many writers develop and perfect over the years their own lettering style: a style which derives from countless hours of perfecting the image,

[21] US Copyright Office 37 C.F.R.

[22] As pointed out by Judge Mukasey in *Tracy v. Skate Key, Inc.*, 697 F. Supp. 748 (S.D.N.Y. 1988).

even if the final image may appear to members of the general public less than perfect.[23] Even tags and throw-ups which to an untrained eye and those outside the graffiti subculture happen to seem banal, meaningless and always similar may be considered sufficiently original instead.[24] As has been correctly noted, the assessment of their originality should be carried out by taking into account a specialized audience, namely the graffiti community, whose members are able to distinguish and appreciate differences and peculiarities, which people outside the scene cannot do.[25]

Also, the support on which tags and throw-ups are inserted and viewed (for example, a rubbish bin) may influence our assessment of their originality. Instinctively (and perhaps wrongly) one may be more inclined to recognize the same graffiti signature as being artistic if one sees it on a canvas or a print. Yet, such assessment would be biased as it would be based on a prejudice, namely that only certain objects can be the support of artistic outputs. That does not seem correct, also taking into account that the medium on which works are placed is not, and should not be, conclusive for the purposes of copyright subsistence.

But what is copyright judges' stance when it comes to the originality of graffiti-based lettering? An interesting US case has dealt with this issue, namely *Reece v. Mark Ecko Unlimited*.[26] The artist Reece sued a videogame producer over the use of a stylish version of the word 'Dip' in a graffiti-inspired game. The work was created in 1972 by Reece and also registered with the US copyright office (which seems to confirm that stylish lettering may be considered sufficiently original and thus copyrightable).[27] Judge Debra Freeman confirmed that '"DIP" was

---

[23] Mark Halsey and Alison Young, 'Our Desires Are Ungovernable': Writing Graffiti in Urban Space, 10 Theoretical Criminology 275, 294 (2006).

[24] After all, legal actions to react against the misappropriation of graffiti lettering are increasingly being considered: see for example the recent suit brought by the partner of the deceased graffiti artist Dash Snow against the fast food chain McDonald that had reproduced the artist's throw-up on the walls of hundreds of their restaurants (*Jade Berreau v. McDonald's Corporation*, complaint filed on 3 October 2016 at the US District Court Central District of California, Western Division). The legal action, however, was not successful on procedural grounds.

[25] Iljadica, supra note 16, at 152 (also proposing a comparison between graffiti and engineering drawings: the decision on whether the latter are to be considered original or not – Iljadica stresses – should depend on their significance to the category of persons to whom such drawings are addressed, namely engineers, as confirmed by the UK case *Billhofer Maschinenfabrik GmbH v. T H Dixon & Co Ltd* [1990] FSR 105).

[26] *Reece v. Mark Ecko Unlimited*, 2011 U.S. Dist. Lexis 102199 (2011), fn. 1.

[27] Reg. No. VAu001035632/2010-06-01.

Reece's "tag'" as a graffiti artist; in other words, that he used it as his personal identifying mark. Also, even though the judge did not find infringement in light of some differences between the tag and the image used in the videogame, she nonetheless made some interesting points that could be relied on to support the argument that stylish tags and throw-ups can be considered as sufficiently original and therefore copyrightable. Judge Freeman in particular noted:

> the form of the lettering at issue is arguably central to the artistic expression of particular words. In this unusual context, aspects of the lettering shown in the Reece art, such as the degree of abstraction of the letters, the connectivity of the letters, the relative size of the letters . . . may reveal the artist's personal style and strongly influence the 'total concept and feel' of the subject works.
>
> . . . although the words 'Dip' and 'Dipism' are not themselves subject to copyright protection, the original manner in which Reece expressed these words is protectable.

UK courts have not decided copyright cases expressly dealing with graffiti lettering. They have instead dealt with lettering style in general, and in some cases found that the rendition was sufficiently original. For example, in *IPC Magazines Ltd v. MGN Ltd*[28] it was found that the claimant's stylized version of the word 'Woman' in white on a red background was arguably copyrightable. Also, in *Future Publishing v. Edge Interactive Media*[29] the judge expanded on the lettering style of the word 'Edge' and confirmed that the latter was artistic and original.[30]

What the above points and findings seem to confirm is that lettering renditions (including stylish tags and throw-ups) can be considered works of art. So why should we deny them copyright protection? After all, the concept of artistic works, and of art more generally, is a broad one. Much material depicted visually with a modicum of creativity, by whatever means, may be nowadays interpreted as artistic.[31]

---

28 *IPC Magazines Ltd v. MGN Ltd* [1998] FSR 431.

29 *Future Publishing Ltd v. Edge Interactive Media Inc.* [2011] EWHC 1489 (Ch.).

30 For a similar finding see *R Griggs Group v. Evans* [2003] EWHC 2914 (Ch), para 18 (holding that 'if an artist uses his skill and labour to draw a word or a phrase in a stylised way, as in the case of a logo, his drawing is capable of being an original work, protected by copyright law').

31 Catherine Colston and Jonathan Galloway, *Modern Intellectual Property Law* 305, 3rd edn (Routledge, 2010).

2.2.1 The 'typeface' comparison

The point could also be made that tags and throw-ups are comparable to typefaces,[32] namely sets of letters and numbers which are all in the same style and are used in printing. Yet, under US law typefaces are not copyrightable because – it is often noted – they perform the function of building blocks of language and their aesthetic element cannot be conceptually separated from their functional element, which in this case is to graphically represent the letters of the alphabet.

The non-copyrightability of typefaces in the US has been criticized, however. Typeface designers may indeed exercise significant creativity and come up with highly artistic letters; [33] and this can occur despite the constraints of the design process, which does not just dictate the function of creating letters, but also leaves room for a significant dose of artistic genius. Making copyright available for such non-functional aspects of the letter, therefore, does not seem heretical. That typefaces may have (copyrightable) artistic elements in addition to (uncopyrightable) functional features is confirmed by international law. The Vienna Agreement for the Protection of Typefaces and their International Deposit, signed in 1973 by 12 countries including UK, Germany, France, Italy and the Netherlands (but never entered into force) requires typefaces to be protected by either copyright or industrial design law, or a sui generis right; and it clarifies that what cannot be protected are just typefaces of a form dictated by purely technical requirements.[34] UK law also considers typefaces as copyrightable, as is indirectly confirmed by Sections 54 and 55 of the CDPA, which assume and take for granted such protection.[35]

As with typefaces, also letters painted by graffiti artists (so as to form their names) often have strong non-functional and artistic elements,

---

[32] See Nicole A. Grant, Outlawed: Finding a Home for Graffiti in Copyright Law, *Selected Works of Nicole A. Grant* (2012) 24, available at http://ssrn.com/abstract=2030514 (last visited June 20, 2018). On copyright protection of typefaces in general see Chapter 13 of this book, authored by Arul George Scaria and Mathews P. George.

[33] Jacqueline D. Lipton, To © or Not to ©? Copyright and Innovation in the Digital Typeface Industry, 43 U.C. Davis L. Rev. 143, 155 (2009).

[34] Article 2(i)(c) of the Vienna Agreement for the Protection of Type Faces and Their International Deposit and Protocol Concerning the Term of Protection, June 12, 1973, Records of the Vienna Diplomatic Conference on the Protection of Typefaces 1973, at 10 (W.I.P.O. 1980).

[35] Section 54 exempts from copyright infringement certain activities in relation to what is defined as 'an artistic work consisting of the design of a typeface'. Section 55 uses an identical wording.

which go beyond serving the neutral function of representing the letters themselves. Such artists develop and practise their style by individualizing and transforming the alphabet, and frequently making the letters and the entire word unreadable to people outside the subculture. As has been correctly pointed out, copyright should be made available when the graffiti lettering does not only consist of alphanumeric elements, but it is also elaborate, personalized and extravagant enough to be perceived and appreciated as an artistic endeavour.[36]

2.3 Single Words and Short Phrases

Issues of originality may also be raised by a specific category of street artworks, namely those which consist of just a written word or short phrases. Such works are popular with several street artists. Take for instance John Fekner, a street and multimedia artist, who in the late 1970s and early 1980s created hundreds of environmental, social, political and conceptual works consisting of stencilled words spray painted outdoors; or again Banksy who is known for stencilling sarcastic and anti-conformist phrases on walls and other surfaces.[37]

In the US these works are not copyrightable, presumably because they are not considered original enough. The regulations of the US Copyright Office Code are clear in this regard: 'The following are examples of works not subject to copyright and applications for registration of such works cannot be entertained: (a) Words and short phrases such as . . . titles, and slogans'.[38]

Not only doubts about a sufficient level of originality could be raised when it comes to street words or messages: for example, in the UK relevant case law on the *de minimis* rule[39] could also stand in the way of their

---

36 Grant, supra note 32, at 23.

37 Take Banksy's iconic phrase 'This is not a photo opportunity', stencilled on several places such as the Palace of Westminster in London, the Eifel Tower in Paris and the Sydney Opera House. See Banksy, supra note 7, at 122–3.

38 US Copyright Office 37 C.F.R.

39 The principle *de minimis non curat lex* means that the work is so insignificant and trivial that it does not deserve copyright protection, and in general that the law does not concern itself with trifles. For example, in *Synamide v. La Maison Kosmeo* (1928) 139 LT 265 it was held that to quote the slogan 'Beauty is a social necessity, not a luxury' was too small a matter on which to ground a copyright infringement action. Moreover, in *Francis Day & Hunter v. Fox* [1940] AC 112, it was held that the song title 'The Man Who Broke the Bank at Monte Carlo' was not sufficiently substantial for copyright purposes. After all, doubts about whether short sentences or words can be protected by copyright have also been expressed in

copyrightability. Yet, in light of the more recent *NLA v. Meltwater* case[40] (where the court found that newspapers' headlines could be considered as original literary works and thus copyrightable), it could now be argued that certain sentences can also attract copyright in the UK.

3. MORAL RIGHTS

I now proceed to analyse whether and to what extent moral rights, in particular the integrity right, can be invoked by street and graffiti artists.

3.1 Paternity Right

The paternity right allows authors to claim to be recognized as the creators of their works. It seems this right fits well into street and graffiti art scenarios, as within these subcultures there is a strong bond between the artist and her work: the former strongly identifies herself within the latter. The tag, the character, the sticker, the poster or the whole piece of art created and placed in the street often represent a projection of the artist herself. This bond is confirmed by the frequent inclusion of the artist's name next to, or in, the work.

The paternity right could be relied on by street artists and writers, for instance, to make a complaint against the publication of their works in a photographic book where their authorship is not acknowledged. Such unauthorized uses are increasingly common, with artists being more and more willing to come out and complain. This occurred for example in the US in 2007, when a group of street and graffiti artists objected to the publication of a photographic book on New York City murals (entitled *Tattooed Walls*) and a related exhibition, because pictures of their pieces had been included without their authorization, and many artworks were misidentified, misinterpreted or misattributed.[41] The case was subsequently settled out of court.

---

relation to traditional paintings on canvases: see Anne Barron, Copyright Law and the Claims of Art, 4 Intell. Prop. Q. 368, 388 (2002).

40 *Newspapers Licensing Agency v. Meltwater Holding* [2010] EWHC 3099 (Ch) [72]; *Newspapers Licensing Agency Ltd. v. Meltwater Holding BV* [2012] RPC 1 (CA).

41 See David Gonzalez, Walls of Art for Everyone, but Made by Not Just Anyone, *The New York Times*, June 4, 2007, http://www.nytimes.com/2007/06/04/nyregion/04citywide.html (last visited June 20, 2018).

3.2 Integrity Right

The integrity right allows authors and artists to oppose treatments of their works, particularly when they are prejudicial to their honour or reputation. The challenge here lies in verifying whether this right may be asserted by street and graffiti artists to oppose the destruction of their works or other unwelcome treatments, including their removal for the purposes of being exhibited and/or offered for sale in indoor environments such as galleries, museums or auction houses (so-called 'de-contextualization').

Two conflicting interests are at stake here: the interest of property owners in disposing of the tangible object they have become owner of, and the interest of artists in preserving their pieces. Who should the law protect more strongly? A broad interpretation of the integrity right, so as to give street artists the possibility to prevent the above treatments of their artworks, would strongly protect their interests. Yet, it would take away the power of private property owners to fully control their space.[42]

3.2.1 Destruction

Can street artists rely on the right of integrity to prevent the owner of the tangible object incorporating their piece (eg, a wall) from destroying or in anyway damaging it? Several artists seem indeed increasingly interested in their street works being preserved; and some of them have taken action against property owners with a view to either preventing the destruction or mutilation of their pieces, or asking for damages after such treatment has occurred. Most of these suits have been brought in the US.[43]

3.2.2 The *5Pointz* case

The recent decision of the US District Court for the Eastern District of New York in the so-called *5Pointz* case is of seminal importance, as for the first time graffiti artists have been protected under the Visual Artists Rights Act (VARA), the piece of legislation introduced in the US in 1990

---

[42] Jamie Griffin, Illegal Graffiti: a Copyright Perspective, (2010) 22 Intell. Prop. L. Bull. 123, 124 (2010). See also Stina Teilmann, Framing the Law: The Right of Integrity in Britain, 27 Eur. Intell. Prop. Rev. 19, 23 (2005) (noting that courts, faced with such a dilemma, may tend to favour physical ownership of property owners over incorporeal ownership of artists).

[43] In the UK instead there has been no reported legal action aimed at objecting to destruction of artworks placed in the street. This is also probably due to the fact that the right to prevent destruction of works is not expressly provided under the CDPA provision on integrity right.

to offer artists' moral rights, including the right to prevent destruction of their works.[44]

5Pointz was a thriving New York mural point, also known as the 'Mecca' of graffiti, which was knocked down in August 2014 by the property owner, Mr Wolkoff, to build luxury condos. Judge Frederic Block awarded a total of $6.7 million damages to 21 graffiti and street artists who had sued the owner of the complex, with the biggest award of $1.3 million being given to 5Pointz's *de facto* curator Meres One.[45] The year before the demolition the artists had tried to oppose the warehouse's destruction, but the court refused to issue an injunction to prevent the Mr Wolkoff from knocking down the site. The owner had been approving the art on his building for many years, and also profiting from it, as the value of the complex went up from $40 million to $200 million as soon as the variance and the green light to build the condos were obtained. Destroying 5Pointz – the judge interestingly stressed – permitted Wolkoff to realize that gain.

The judge accepted that because the 45 artworks had 'recognized stature' they must receive protection under VARA. That artworks must have 'recognized stature' to be protected against destruction is an explicit VARA requirement.[46] The judge found, in particular, that the 5Pointz artworks met the test affirmed in *Carter v. Helmsley-Spear Inc.*, the leading case that had clarified that this requirement is a sort of gate-keeping mechanism, being protection (against destruction) available just to those pieces that art experts, the art community or society in general considers as having stature.[47] But in the 5Pointz case Judge Block embraced a more generous interpretation of the *Carter* test. He indeed cited and endorsed *Martin v. City of Indianapolis* (which had affirmed that the *Carter* test 'may be more rigorous than Congress intended');[48] and recalled that in *Martin*, the requirement was satisfied by merely relying on newspaper and magazine articles and various letters, including a letter from an art gallery director.[49]

That said, Judge Block found that all the categories mentioned in *Carter* (art experts, the art community and society at large) had recog-

---

[44] Visual Artists Rights Act of 1990, Pub. L. No. 101-650, 104 Stat. 5130 (codified as amended in various sections of 17 U.S.C. (West Supp. 1992)).

[45] *Cohen et al. v. G&M REALTY L.P. et al.*, Case No. 13-CV-05612(FB) (RLM), decision of 12 February 2018.

[46] 17 U.S.C. § 106A(a)(3).

[47] *Carter v. Helmsley-Spear Inc.* (861 F Supp 303 (SDNY 1994) and 71 F.3D 77 (2nd Cir 1995)).

[48] 192 F.3d 608, 612 (7th Cir. 1999).

[49] See pp. 8–29 of the decision.

nised the stature of 5Pointz artworks. The artists had indeed shown their professional achievements, including evidence of the placement of their works in movies, TV, and the Internet, letters from art professors around the country, letters and emails from visitors to 5Pointz as well as media coverage and social media presence and followers. The judge also stressed that all the 45 artworks had received sufficient academic recognition, having been appreciated by university professors, art teachers and other art experts.[50]

Judge Block also carefully examined Mr Wolkoff's behaviour. The artworks – even those that could be easily removed as they had been placed on plywood panels – had been whitewashed by the property owner without giving artists the 90 days' notice required by VARA. And he did so – the judge stressed – being aware that the artists were pursuing a VARA-based legal action. Such a wilful disrespect of an important legal provision pushed the judge to award the artists the maximum amount of damages. Indeed, although he did not grant the injunction asked by the artist in 2013, the judge had warned Wolkoff that he would be exposed to potentially high damages if the artworks were finally considered of 'recognized stature'. But the property owner was apparently unconcerned by this warning, and proceeded to paint over all artworks displayed at 5Pointz without giving artists any notice. Such behaviour – the judge concluded – was not acceptable.[51]

The court took also into account that 5Pointz had become an attraction for New York City visitors, with busloads of tourists, school children and weddings constantly heading to the site. Also due to Meres One's brilliant curation for more than a decade, not only was the complex regularly painted by talented graffiti and street artists from all over the world, but it also frequently attracted movie producers and singers as well as advertisers interested in using the art as a backdrop.

The judge finally did not attach much importance to the fact that several artworks at 5Pointz were not meant to be permanent, an argument which had instead been relied on by the property owner to claim that the pieces could not be protected. The Court reminded the parties that VARA

[50] See the Appendix of Judge Block's decision of 13 June 2018, rejecting a motion by 5Pointz owner to set aside the Court's findings of fact and conclusions of law and grant a new trail.

[51] In the decision of 13 June 2018 (see supra fn. 50), Judge Block clarified: 'I found out at the trial that Wolkoff had misled me at the preliminary injunction hearing. . . . I was misled that the demolition of the buildings was imminent when there was not even an application for a demolition permit extant' (pp. 7 and 14).

protects both permanent and temporary art. This is an important point, especially – the judge added – when all that makes a work transient is the site owner's expressed intention to remove it.

3.2.3 Other US cases

The *5Pointz* case was not the first successful VARA case involving art placed on public spaces. In *Hanrahan v Ramirez*[52] a group of young artists had painted a mural carrying an anti-drug message on the side of a liquor store with the authorization of the property owner. After the store owner whitewashed half of the mural three years later, the artists took legal action. The court held in 1998 that the piece had recognized stature because it had won a national prize, received the local community's support and had been displayed in a government building as a photograph. The judge awarded the artists £48,000 as compensation and ordered the restoration of the mural.[53]

The VARA 90-day notice rule has been at the centre of several legal actions started by muralists against property owners, with the former complaining about the failure to serve such a notice. The following cases, most of which settled out-of-court, are noteworthy:

(i) In 1986 the artist Jesus Campusano painted a mural on a building in San Francisco, being assisted by three painters and his partner. After the building's owner covered it in 1998, his partner and children sued under VARA in the California Northern District Court. They claimed the owner had not notified them about the decision to cover the mural. The case was then settled for $200,000.[54]

(ii) In 2006 the mural 'Edward Ruscha', made by muralist Kent Twitchell over the course of nine years in Los Angeles, was painted over with no preliminary notice being given to the artist. The latter sued the US government, which owned the building, and 11 other

---

[52] *Hanrahan v. Ramirez*, No. 2:97-CV-7470, 1998.

[53] Another case where the recognized stature requirement has been debated was *Henderson v. Ziman*, No. 2:14-cv-03042-SJO-AS (E.D. Cal. April 21, 2014). In April 2014, the artist Victor Henderson started a legal action under VARA over the destruction of the mural 'Brooks Avenue Painting' which he had co-created in 1969. Henderson claimed that his artwork had major historical significance. While Henderson seemed to have strong evidence showing that his mural was of recognized stature, the artist voluntarily abandoned the case.

[54] See University of Missouri-St Louis hand-out from 'The Artist as Entrepreneur' lessons, published at http://www.umsl.edu/continuinged/artist/artist-lesson-VARA-ho2.htm (last accessed November 30, 2017).

defendants for damages under VARA in the California Central District Court.[55] Almost two years later the case settled for $1.1 million.

(iii) Muralist Dan Fontes painted an artwork on the side of an Oakland building in 1987. In 2015 he filed a suit in the California Northern District Court against the owners of the building after the tenant had whitewashed the mural without notifying the artist as required by VARA, and asked damages for $400,000.[56] The case was later settled.

(iv) In 2017 muralist Monte Thrasher sued several individuals and corporate defendants in the Central District of California, claiming that they had painted over his Six Heads mural, without the artist's permission, and substituted it with another mural. The case is about the alleged violation of Thrasher's integrity and paternity rights and at the time of writing is still pending.[57]

3.3 De-contextualization

What we have also witnessed in recent years is the 'surgical' removal of street artworks from walls or other urban surfaces and their resurfacing in galleries or other indoor venues for the purposes of being exhibited and offered for sale. Such removals obviously bother and frustrate street and graffiti artists as they are perceived as unacceptable attempts to not only distort the message conveyed by the original art; but also to profit from pieces which are placed in urban environments with the aim of being freely enjoyed by members of the public. Also, many of these (removed) street artworks are often site-specific, which entails that the pieces maintain their artistic meaning as long as they are kept in their original environment.

Yet, it appears domestic laws protecting the integrity right in the UK and US, as they stand now, cannot be invoked by street and graffiti artists

---

[55] *Kent Twitchell v. West Coast General Corp et al.*, 06-cv-04857. (C.A.C.D. 2006).

[56] *Fontes v. Autocom Networks, Inc.*, C 15-02044 CRB (N.D. Cal. 2015).

[57] *Thrasher v. Siegel*, No. 2:17-cv-03047 (C.D. Cal. April 24, 2017). Another (apparently still pending) case is the one started by muralist Katherine Craig, who sued in January 2016 a real estate developer, Princeton Enterprises, after the latter allegedly threatened to destroy her iconic watercolor work named 'The Illuminated Mural'. According to the complaint, the artist entered into an agreement with the previous owner, specifying that the artwork would remain on the building for at least 10 years (see http://www.crainsdetroit.com/article/20160105/NEWS/160109914/detroit-artist-sues-princeton-enterprises-over-bleeding-rainbow (last accessed June 20, 2018).

to oppose attempts to remove and de-contextualize their artworks. In the UK the CDPA wording seems to preclude the integrity right being invoked to prevent uses of the work in contexts different from those chosen by the artist.[58] Section 80(2)(a) of the CDPA, indeed, just targets addition to, deletion from, alteration and adaptation of works. The same goes under US law, as the relevant VARA provision only speaks about 'intentional distortion, mutilation, or other modification' of a work.[59] According to current US case law, not even site-specific street artworks could be saved from removals and de-contextualization. Indeed, in two seminal cases – *Phillips*[60] and *Kelley*[61] – courts have confirmed that VARA does not cover, and cannot be invoked to prevent the removal of, site-specific works.

The UK and US provisions regulating the moral right of integrity are thus at odds with Article 6-bis(1) of the Berne Convention, which does contain a general clause allowing artists also to object to any 'other derogatory action in relation to a work'. This part of the provision was added at the Brussels Conference in 1948 with the specific aim of covering uses different from the categories of modifications already mentioned in the provision.[62] It is believed that by adding such a general clause, this provision of the Berne Convention can be interpreted as permitting artists to object to their works being placed and exhibited in different contexts.[63] Some civil law jurisdictions have followed this approach. Case law from

---

[58] Gillian Davies and Kevin Garnett, *Moral Rights* (Sweet & Maxwell, 2010) 238–40; Gerald Dworkin, The Moral Right of the Author: Moral Rights and the Common Law Countries, 19 Colum.-Vla. J.L. & Arts 229, 245–63 (1995); W.R. Cornish, Moral Rights Under the 1988 Act, 11 Eur. Intell. Prop. Rev. 449, 450 (1989 Hugh Laddie, Peter Prescott and Mary Vitoria, *The Modern Law of Copyright and Designs* (4th edn, LexisNexis, 2011) § 13.18).

[59] 17 U.S. Code § 106A.

[60] *Phillips v. Pembroke Real Estate, Inc.*, 459 F.3d 128 (1st Cir. 2006). The ruling relied on the 'public presentation' exception of VARA (17 U.S. Code § 106(A)(c)(2)), which provides that 'the modification of a work of visual art which is the result of conservation, or of the public presentation, including lighting and placement, of the work is not a destruction, distortion, mutilation, or other modification'. On this case see also Chapter 1 'Copyright in the Expanded Field' by Xiyin Tang in this volume.

[61] *Kelley v. Chicago Park Dist.*, 635 F.3d 290 (7th Cir. 2011).

[62] Sam Ricketson and Jane Ginsburg, *International Copyright and Neighbouring Rights: The Berne Convention and Beyond* (2nd edn, OUP, 2006) (2 volumes) § 10-11. See also the first edition of this book: Sam Ricketson, *The Berne Convention for the Protection of Literary and Artistic Works: 1886–1986* (Sweet & Maxwell, 1987) 469–70; Irini A. Stamatoudi, Moral Rights of Authors in England: The Missing Emphasis on the Role of Creators, 4 Intell. Prop. Q. 478, 504–5 (1997).

[63] Ricketson, supra note 62, at 469–70; Cornish, supra note 58 at 450.

the Netherlands,[64] Greece,[65] Spain,[66] Switzerland[67] and Israel[68] confirms that in these countries the removal and de-contextualization of street artworks may in certain circumstances infringe the right of integrity and could thus be opposed.

4. CAN ILLEGAL STREET ART BE PROTECTED BY COPYRIGHT?

We have seen that creating artworks in the street without authorization from the owner of the property on which such works are placed can expose artists to serious legal consequences, including jail time. Although nowadays artists increasingly seek and obtain permission to place their pieces in urban environments, much street art and graffiti are still created illegally. Can such illegally produced art be protected by copyright? US and UK laws do not give a precise answer.[69]

---

[64] Even a minimum de-contextualization of an artwork was considered by a Dutch Court as infringing the integrity right: see Quentin Byrne-Sutton, The Owner of a Work of Visual Art and the Artist: Potential Conflicts of Interest. In Norman Palmer and Ewan McKendrick, eds., *Interests in Goods* (Informa Law, 1993) 281, 294 (mentioning a court decision finding that a large mural painting which had been specifically designed to hang in a particular building at a height of 90 centimetres could not be hung at the height of 2.20 metres instead).

[65] The integrity right has been held as infringed in Greece when an artwork was placed in a location which contradicted its purpose and spirit, or where the work was unjustifiably transferred to a venue different from its original one: see Counsel of State, Decision No. 1465/1954, Themis, 1955, 174; Court of Appeal, Dodoni, Decision No. 47/1956, 861. See also *Antonia-Belika Koubareli v. Evangelos Volotas* [2003] ECDR 19, a decision enabling an author to rely on her moral rights to prevent the publication of her work in an offensive environment on the Internet.

[66] In January 2013, the Spanish Supreme Court held that the relocation of a work by the Spanish sculptor Andrés Nagel violated the author's moral right. See Stech, supra note 5 at 153.

[67] Swiss case law has confirmed that a violation of the integrity right under Article 119 of the Swiss Copyright Act of 9 October 1992 may occur indirectly, namely when the work itself remains unchanged, but where the surroundings of the work are modified in a way that affects the author's interests. See Jaques de Werra, Treatment of Moral Rights in Other Jurisdictions – Switzerland, in Davies and Garnett, supra note 58 at 586–7.

[68] Israeli courts have been willing to accept claims against types of use that damaged the author because of the context in which the artwork was placed. See Gadi Oron, Treatment of Moral Rights in Other Jurisdictions – Israel, in Davies and Garnett, supra note 58 at 754.

[69] For an overview of how British courts have dealt with the issue of copyright protection of illegally created works as well as works showcasing immoral content

The issue was tangentially dealt with by a US court in *Villa v. Pearson Education*.[70] Hiram Villa, a Chicago-based street artist, took legal action against Brady Publishing for publishing a picture of his work in its strategy guide for a videogame without the artist's permission. The defendant asked to dismiss the complaint, arguing that the work in question could not be protected by copyright because the piece had been created illegally. The court denied the motion, yet it also noted that the claim that the work was not copyrightable due to its illicit origin would require investigating the circumstances under which the work was created (the case was then settled out of court). Such obiter led some commentators to argue that judges may value an argument based on the artwork's illegality as a defence to copyright infringement, or anyway as a factor which affects its copyrightability. This defence may be considered as a specific application of the so-called unclean hands doctrine, an equitable defence in which the defendant claims that the plaintiff should not obtain a remedy and profit when the latter has acted unethically or in bad faith, or has anyway carried out an illegal activity.

Such a defence is quite popular amongst defendants accused of appropriating and profiting from street artworks that have been created illegally. It has been raised, for example, by the fashion company Moschino in *Joseph Tierney v. Moschino*, a copyright infringement case started by the graffiti artist Rime as he claimed that various elements of his Detroit mural 'Vandal Eyes' had been copied onto a Moschino dress which was subsequently worn by the pop-star Katy Perry at a 2015 glamorous event.[71]

As mentioned, whether illegally produced artworks can be considered copyrightable, or whether any existing copyright could be enforced in court, is not entirely clear. In other words, this is still a grey area of the (copyright) law in several jurisdictions, including the US and UK. Having said that, in some street-art- and graffiti-related cases courts did not in fact bother to enquire whether the relevant work had been produced without the authorization of the property owner. In the already mentioned *Reece v. Marc Ecko Unlimited* case, for example, the court did not refer to any illegality-related issue when rejecting the copyright infringement case brought by Reece against the producer of a graffiti-inspired game which

---

in old and recent cases, see Bonadio, supra note 1, at 215–17. For an analysis of the copyrightability of illegal works in general, see Chapter 19 of this volume, authored by Eldar Haber.

[70] See *Villa v. Pearson Educ., Inc.*, 2003 WL 22922178 (N.D. Ill. December 9, 2003).

[71] *Joseph Tierney v. Moschino S.p.A. et al.*, Doc. No. 2:15-cv-05900 (C.D. Cal. August 5, 2015). I also mentioned this case supra at note 2.

had incorporated some bits of his graffiti art. Another relevant case is *Mager v. Brand New School*.[72] Stylish eyeball stickers had been placed by an artist named Damien Mager on many billboards of New York City. The stickers however also appeared on a TV commercial for a few seconds, without the authorization of the artist. Mager took action and asked for compensation, with the court accepting that the stickers could have been placed on the streets without authorization. Although damages were not awarded as the copyright had been registered after the alleged infringement occurred, the judge did not focus on, nor was interested in, any illegality-related aspect of the artworks when deciding the case.[73]

It is also worthwhile mentioning the UK case *Creative Foundation v. Dreamland*,[74] the first decision by a British judge to expressly consider the ownership of walls on which artworks are placed. The judge, Arnold J., held that a mural painted by Banksy and cut from the wall by a tenant was a chattel that belonged to the landlord. Yet, he also noted obiter that there is no doubt the copyright belongs to the famous British street artist. The judge stressed such a point despite recognizing that the artwork had been created without the prior knowledge or the consent of the leaseholder and tenant (although Creative Foundation, that in the meantime had acquired the rights to the piece, later impliedly approved the work). It thus seems that the judge was not concerned with whether the work had been created legally or not.

Let's go back to the unclean hands doctrine. What makes it particularly unsuitable to govern cases of the misappropriation of street and graffiti art is the lack of connection between the illegal act committed by the artist (for example, painting an unauthorized mural on a wall) and the merit of these disputes, namely the reproduction, adaptation and (often) communication and making available to the public of the work by third parties, frequently for commercial purposes. In simpler words, the illegal behaviour of the street or graffiti artist does not have a negative impact on the individual or organization which has misappropriated the illegally placed art (it instead negatively affects the owner of the property upon which the work is placed, which however is not party to the proceeding). After all, that the doctrine in question cannot be invoked in cases where the plaintiff's

---

[72] *Mager v. Brand New School*, 78 USPQ 2d 1389 (2004).

[73] See also Danwill D. Schwender, Does Copyright Law Protect Graffiti and Street Art? In Jeffrey Ian Ross, ed., *Routledge Handbook of Graffiti and Street Art* (Routledge, 2016) 456.

[74] *Creative Foundation v. Dreamland & Others* [2015] EWCH 2556 (Ch), 11 September 2015.

misconduct is not linked to the merit of the litigation between the parties has been confirmed by US courts in several cases.[75]

Another reason why making copyright available for unauthorized street and graffiti artworks seems to make perfect sense lies in the fact that here the illegal aspects do not even concern the content of the work[76] – rather they pertain to the processes of creation of the piece. In my opinion, the way (legal or illegal) art is created should not affect the analysis related to copyright subsistence and enforceability. The copyright system should be neutral towards,[77] and blind about,[78] the way eligible subject matter is produced. After all, this line of reasoning has been shared by a German court in a copyright-related case involving an artwork painted on the Berlin Wall: 'it is not in principle relevant . . . that the way in which it [the artwork] was produced is evidently unlawful – in this case by virtue of an act of damage to property subject to civil and criminal sanctions . . .'.[79]

This argument sounds logical. If I steal a pen which I then use to draw a wonderful piece of art, why should I be denied the right to enforce a

---

[75] For example, in *Mitchell Bros. Film Group v. Cinema Adult Theater*, 604 F.2d 852, 863 (1979), a case on enforcement of copyright over obscene pornographic content, the US Court of Appeals for the Fifth Circuit held that, when the plaintiff's wrongdoing does not relate to the defendant, the unclean hands doctrine cannot be relied on. See also *Keystone Driller Co. v. General Excavator Co.*, 290 U.S., 245 (1933) (holding that this doctrine can be invoked only where the wrongful acts 'in some measure affect the equitable relations between the parties in respect of something brought before the court for adjudication'); *NLRB v. Fickett-Brown Mfg. Co.*, 140 F.2d 883, 884 (CA5, 1944) (noting that the plaintiffs' alleged wrongful conduct has not changed the equitable relationship between plaintiffs and defendants and has not injured the defendants in any way); *Lawler v. Gilliam*, 569 F.2d 1283, 1294 (CA4, 1978) (holding that the unclean hands doctrine does not purport to search out or deal with the general moral attributes or standing of a litigant). See also the more recent case *Dream Games of Arizona, Inc. v. PC Onsite*, 561. F.3D 983 (9th Cir. 2009), where the court found that an electronic bingo game was still protected by copyright when it was distributed in violation of state gaming laws.

[76] It should also be remembered that in several jurisdictions, such as the US, not even the possibly illegal or obscene content of a work affects subsistence of copyright, or its enforceability (see again *Mitchell Bros. Film Group v. Cinema Adult Theater*, 604 F.2d 852, 863 (1979)).

[77] See Celia Lerman, Protecting Artistic Vandalism: Graffiti and Copyright Law, 2 NYU J. Intell. Prop. & Ent. L. 295, 316 (2013).

[78] Owen Morgan, Graffiti – Who Owns the Rights?, Univ. of Auckland Bus. Sch. Working Paper 5, 16, 21 (2006); available at SSRN: http://ssrn.com/abstract=929892 (last accessed November 30, 2017).

[79] *Re Pictures on the Berlin Wall* (Case I ZR 68/93) [1997] ECC 553.

valid copyright and tolerate that someone else copies and takes economic advantage of my work? The same could be arguably claimed where a person takes a picture of a minor without the parent's permission or where a paparazzo takes photographs of celebrities in private locations without authorization.[80] It is simply unfair to allow persons other than the artist to rely on the illegal nature of a street artwork to copy and exploit it for their own commercial purposes,[81] for example by using it in advertising messages or as a decoration element of fashion products. As has been noted, '[f]ailing to recognize a copyright in [illegal] graffiti could then have the effect of legitimizing its co-option but not its creation'.[82] This result would also be absurd as it would reward blatant imitations by individuals or corporations that have nothing to do with either the perpetrator of the illegal act (the artist) or the victim (the owner of the property).

One may however argue that making copyright and enforcement rights available to street and graffiti artists that have placed works without the authorization of the property owner would be useless and not really convenient for them. Many artists indeed would find it risky to come out and start a legal action against imitators as this would entail revealing their identity and being exposed to serious legal consequences. Yet, such risks would be lower, if not totally absent, where a copyright suit is brought against alleged infringers at some point after the street artwork has been illegally created, namely after the statute of limitation has expired: in these cases artists or their heirs may nurture a strong interest in bringing legal actions.[83]

---

[80] See Lerman, supra note 77, at 317, citing the US case *Mavrix Photo Inc. v. Allieiswired.com et al.*, No 10-7591, complaint filed, C.D. Cal. October 12, 2010 (where a photographer filed for copyright infringement on his picture of pregnant actress Penelope Cruz, which he had taken without permission). See also Danwill Schwender, Promotion of the Arts: An Argument for Limited Copyright Protection of Illegal Graffiti, 55 J. Copyright Soc'y U.S.A. 257, 274 (2008).

[81] See also Maribeth A. Smith, Tagging the Lanham Act: Protecting Graffiti Art from Willful Infringement 81 Brooklyn L. Rev. 813,834 (2016); Grant, supra note 32, 26.

[82] Jamison Davies, Art Crimes? Theoretical Perspectives on Copyright Protection for Illegally-Created Graffiti Art, 65 Me. L. Rev. 27, 51 (2013).

[83] For example, in the above mentioned *Jade Berreau v. McDonald's Corporation* (supra note 24), the partner of the deceased graffiti artist Dash Snow – who in the 90s had been illegally tagging his signature extensively in New York – took action against McDonald's for an alleged violation of the copyright over such work.

4.1 Illegal Street Artworks and Moral Rights

What about moral rights? Could street and graffiti artists rely on the rights of paternity and integrity in relation to works produced illegally? Again, this is a grey area of the law in the US and UK, as no express provisions within the copyright statutes deal with these issues.

As far as the paternity right is concerned, it could be argued that those who create artworks in the street illegally should be able to claim such a right. It would be (again) unfair to deny artists the right to be recognized as authors of their work just because they have been placed without the authorization of the property owner. Such denial would, once more, bring about an absurd result, namely that any member of the public would be able to present himself as the father of the artwork and possibly receive recognition and fame for a piece of art that actually was not his own creation.[84] Also, enforcing such right would bring no harm to the owner of the surface (for example, a wall) where the piece has been illegally placed.

The other delicate issue regards the destruction of illegally created street works, and the extent to which the artists who produced them are able to rely on the integrity right to prevent such destruction or anyway ask for compensation in case their pieces are unlawfully destroyed, mutilated or damaged. On the face of it, it appears fair not to give street and graffiti artists a chance to complain about such destruction. Property owners may understandably dislike works placed on their space without their consent, no matter how artistic they might be. Their decision to destroy or remove such pieces may thus look reasonable, and obviously justified on property rights' grounds.[85] If instead the work has been authorized or commissioned by the property owner, one may argue that such an authorization or commission may affect in some circumstances the property owners' ability to rely on a defence of reasonableness should they later decide to get rid of the surface incorporating the work.[86]

---

[84] Yet, it is likely that artists who illegally create street artworks want to remain anonymous to avoid criminal prosecution and therefore may not be interested in asserting the paternity right. See Grant, supra note 32, at 35. Yet, the interest in invoking this right – it should be stressed again – may arise after the statute of limitation has expired.

[85] See Griffin, supra note 42, at 124. As opposed to cases where unauthorised street artworks are misappropriated for commercial purposes (where artists' behaviour does not jeopardize the alleged infringer), the placing of illegally created works on the street here directly harms the other party of the dispute, namely property owners.

[86] Griffin, supra note 42, at 125.

The US case *Ron English et al. v. BFC & R. 11th Street LLC*[87] dealt with these issues. It involved a group of related artworks installed in a community garden. A redevelopment of the site was planned that would entail moving the sculptures but leaving the murals intact, although the new construction would obstruct the view of the murals. The artists claimed under VARA that the work was conceived as a unified environmental work and that therefore removing the sculpture would destroy the whole work. It was held that VARA was inapplicable as the works had been illegally created. The court held in particular that protecting these illegal pieces 'could effectively freeze redevelopments of vacant lots by placing artwork there without permission. Such a construction of the statute would be constitutionally troubling, would defy rationality and cannot be what Congress intended in passing VARA'. Similarly, in *Botello v. Shell,*[88] where a gas station developer bulldozed a mural in violation of the California Preservation Act (which provides legal protection for artists' moral rights), the court noted obiter that 'the statute does not apply to [illegal] graffiti, which . . . is hardly classifiable as "fine art", and which is the subject of several criminal laws'.

Yet, in the subsequent case *Pollara v. Seymour,*[89] where an unauthorized work installed in the Empire State Building plaza in New York City was removed from its frame and significantly damaged, the court clarified that the ruling in *Ron English* only applied to illegal works that could not be removed without destroying them, therefore suggesting that in the case of illegal works that can be removed without harm (and many street and graffiti artworks belong to this category) VARA does apply.[90] Although it was concluded that the artwork was for the purpose of advertisement and therefore could not be protected under VARA, the court added that 'there is no basis in the statute to find a general right to destroy works of art that are on property without the permission of the owner'. This point confirms that US courts have not taken a clear stance when it comes to interpreting the integrity right for illegally produced artworks.

---

87 *Ron English v. BFC & R East 11th St. LLC,* No. 97 Civ. 7446, 1997 WL 746444 (S.D.N.Y. December 3, 1997).

88 *Botello v. Shell Oil Co.*, 280 Cal. Rptr. 535 (Ct. A 1991).

89 *Pollara v. Seymour*, 206 F.Supp.2d 333 (N.D.N.Y. 2002), aff'd on different grounds, 344 F.3d 265 (2d Cir. 2003).

90 This would entail that street and graffiti artists that illegally place their pieces would be able to rely on VARA protection unless (1) the building owner has made a diligent, good faith but unsuccessful attempt at notification of the artist of his removal intent, or (2) the building owner did provide notice, but the artist either failed to remove the work or to pay for its removal within 90 days after receiving notice (see VARA, Section 113(d)(2)).

Finally, we should wonder whether street and graffiti artists can rely on the integrity right to oppose the removal of artworks illegally placed in the public environment, and their subsequent exhibition in an indoor setting (this is what I labelled earlier 'de-contextualization'). We have already seen that the US and UK provisions on the integrity right do not expressly allow artists to object to this treatment of their works, which casts doubts as to whether such a right could be invoked in these circumstances. Other jurisdictions – we have also seen – are in this regard less strict towards artists. As a general remark, I note that, while the property owners' decision to destroy illegal art placed in its own location may look reasonable,[91] their position appears less strong when they deliberately remove, or have someone remove, the artwork for the purpose of exhibiting and/or offering for sale the carved-out piece in an indoor location. After all, in these circumstances it could be argued that, if property owners do this, that means they also impliedly approve of and 'legalise' ex post the piece.[92]

5. CONCLUSION

Street art and graffiti are considered by many as the most important forms of artistic expression in present times. Often stimulated by a highly competitive and creativity-based environment, more and more artists use urban spaces as raw material and canvas for their works.

Obstacles may stand in the way of copyright and moral rights protection of (at least, some) street and graffiti art. As we have seen, provisions on copyrightable subject matter and requirements for protection could be interpreted by courts in a way which may close the (copyright) door to certain street artworks, including tags, throw-ups and street messages, which members of the general public as well as judges may not consider art. Such an outcome would probably be the result of a widespread lack of knowledge about the creative processes which characterize these artistic subcultures. As society's perception of these creative communities is however slowly and progressively changing and becoming more appreciative and tolerant, so could and should – I believe – our (copyright) laws.

Street art and graffiti deserve, and need, copyright and moral rights

---

[91] Yet, see again *Pollara v. Seymour*, where the US District Court for the Northern District of New York, which seems to suggest that in the case of illegal works that can be removed without harm (like many street and graffiti pieces) VARA does apply.

[92] See also Alison Young, *Street Art World* (Reaktion Books, 2016) 40, 41.

protection. These forms of art are indeed vulnerable to both corporate appropriation and physical destruction, mutilation or other kind of damages. The prospect of relying on copyright and moral rights laws may not be the main motivation that pushes artists to place works in the street, as several scholars that have developed the IP negative space theory would argue.[93] Yet, many street and graffiti artists are increasingly interested in protecting their pieces, as is confirmed by the growing number of legal actions or objections by such artists against who appropriates or damages their works.

Once it is accepted that street and graffiti art can and should be considered copyrightable, all artists within these artistic communities obviously need to accept the rules of the game. That means they may lose the copyright infringement cases they bring against alleged infringers. This happened for example in 2011 in *Seltzer v. Green Day, Inc.*,[94] where the artist took legal action against the pop band Green Day as the latter had incorporated in a video backdrop (used at live concerts) an adapted version of one of Seltzer's posters ('Scream Icon') placed on a Los Angeles wall. The Californian federal district and Circuit judges rejected the artist's claims as the use of the artwork was found to be 'transformative and not overly commercial' and therefore fair.

Street and graffiti artists may also be condemned for copyright infringement. This occurred, for instance, in *Morris v. Guetta* (Guetta is a famous street artist, also known as Mr Brainwash), where the judge ruled that seven of Guetta's works (including a mural), based on photographer Dennis Morris' iconic 1977 picture of Sid Vicious, deceased lead singer of punk band The Sex Pistols, were not protected by fair use as they were not transformative.[95] A similar fate would have probably awaited

---

[93] See Elizabeth L. Rosenblatt, A Theory of IP's Negative Space, 34 Colum. J.L. & Arts 317, 365 (2011); Kal Raustiala and Christopher Sprigman, The Piracy Paradox: Innovation and Intellectual Property in Fashion Design, 92 Va. L. Rev. 1687, 1778 (2006). For an application of the IP negative space theory to street art, see Cathay Y.N. Smith, Street Art: An Analysis Under U.S. Intellectual Property Law and Intellectual Property's 'Negative Space' Theory, 24 DePaul J. Art, Tech. & Intell. Prop. L. 259, 293 (2014).

[94] *Seltzer v. Green Day, Inc.*, 725 F.3d 1170, 1173–4 (9th Cir. 2013).

[95] *Dennis Morris v. Thierry Guetta, et al.*, No. LA CV12-00684. See also *Friedman v. Guetta*, No. CV 10-00014 DDP (C.D. Cal. May 27, 2011), a case started in 2011 by photographer Glen Friedman against Thierry Guetta. The Central District of California granted summary judgment for Friedman, finding that Guetta's work (in this case, not a street artwork, but a canvas) was substantially similar to Friedman's famous photograph of the rap group Run DMC, and that Guetta's use of the photograph could not be considered fair use.

the famous street artist Shepard Fairey if he had not reached a settlement agreement with Associated Press, which had sued him arguing that his use of Associated Press imagery depicting Barack Obama (that resulted in the iconic poster 'Hope' which came to represent the former US president's presidential campaign) did not constitute fair use.

All in all, making sure that street and graffiti artists are able to assert and enforce copyright and moral rights against whoever appropriates or damages their works does seem to make sense. Certainly, artists' moral rights need to be properly balanced with the rights of property owners, so that such owners do not have to accept passively an excessive limitation of rights over their tangible possessions. This is particularly true of property owners that follow the correct procedures when removing artworks from their spaces (this did not happen, for example, in the above commented *5Pointz* case). Several factors could be taken into account when striking a fair balance between these two rights, including the existence of any agreement between the artist and the property owner, the length of time the artwork has been allowed to stay, the advantages obtained by the property owner from the piece and the public interest served by the proposed new use of the property once the street artwork is removed.[96]

[96] See Timothy Marks, The Saga of 5Pointz: VARA's Deficiency in Protecting Notable Collections of Street Art, 35 Loy. L.A. Ent. L. Rev. 281, 310 (2015), at 313–14.

5. Copyright protection of tattoos

Yolanda M. King

1. INTRODUCTION

Body art – permanent makeup, piercings, and especially tattoos – has risen in popularity and visibility worldwide.[1] The media, and even the courts, have devoted the most attention to tattoos.[2] Once associated with motorcyclists, gang members, and others outside the mainstream, tattoos are now the adornment of everyone from professional athletes and performing artists to everyday men and women in the workplace, college students, and even teenagers.[3]

Traditionally, tattoos have not been viewed as a type of art subject to the laws of copyright in the United States. For example, because American society has categorized tattoo artists and tattoo bearers as people on the fringes of society, tattoos were not viewed as having artistic merit.[4] However, the increasing prevalence of tattooing in recent years raises the question of whether modern copyright law should be more adaptable to the protection of tattoos.[5] This chapter concludes that it is advisable,

---

1 See, Jordan S. Hatcher, Drawing in Permanent Ink: A Look at Copyright in Tattoos in the United States 3 (April 15, 2005) (unpublished manuscript), http://papers.ssrn.com/sol3/papers.cfm?abstract_id=815116 (stating, for example, that '[i]n the West, many associate tattoos with prisons, biker culture, or gang membership').

2 See, Mik Thobo-Carlsen, How Tattoos Went from Subculture to Pop Culture, The Huffington Post (October 27, 2014) http://www.huffingtonpost.com/mik-thobocarlsen/how-tattoos-went-from-sub_b_6053588.html.

3 See, Catherine L. Fisk, Privacy, Power, and Humiliation at Work: Re-Examining Appearance Regulation as an Invasion of Privacy, 66 La. L. Rev. 1111, 1142 (2006) ('The meaning of tattoos appears to be undergoing a significant transition in the contemporary United States. Once considered the favored adornment of sailors of the lowest rank, and later regarded the insignia of renegade motorcyclists and gang members, tattoos are now said to be widely accepted among the younger generation').

4 See, Yolanda M. King, The Challenges 'Facing' Copyright Protection for Tattoos, 92 Or. L. Rev. 129, 144–5 (2013).

5 While the practice of tattooing has gained greater popularity and acceptance in American society, the legal community has been slow to embrace this form of

and almost inevitable, that tattoos will be recognized as artistic works protected by U.S. copyright law.

1.1 Copyrightable Subject Matter

The Act identifies eight categories of works of authorship.[6] Each category of works is entitled to a bundle of exclusive rights under United States copyright law.[7] However, the exclusive rights of the copyright owner of a work of authorship depend on the classification of the work.[8] Therefore, the categorization of a work is integral to the determination of the exclusive rights of the owner of the copyright in that work.

A 'tattoo' is 'an indelible mark or figure fixed upon the body by insertion of pigment under the skin or by production of scars'.[9] Tattoo artists provide pre-designed tattoos that originate from pre-drawn images called 'flash' and/or custom work to their clients.[10] Tattoos would fall under the category of pictorial, graphic, and sculptural works ('PGS works'), which are defined in Section 101 of the Copyright Act as 'two-dimensional and three-dimensional works of fine, graphic, and applied art, photographs, prints and art reproductions, maps, globes, charts, diagrams, models, and technical drawings, including architectural plans'.[11]

---

art as a protectable work. Thomas F. Cotter and Angela M. Mirabole, Written on the Body, 10 UCLA Ent. L. Rev. 97, 102 (2003) ('[W]hen body art cases finally make their way into court, judges too will have to take the claims of body artists seriously because, as Justice Holmes once suggested, judges are not art critics').

[6] 17 U.S.C. §102(a) ('Works of authorship include the following categories: (1) literary works; (2) musical works, including any accompanying words; (3) dramatic works, including any accompanying music; (4) pantomimes and choreographic works; (5) pictorial, graphic, and sculptural works; (6) motion pictures and other audiovisual works; (7) sound recordings; and (8) architectural works').

[7] 17 U.S.C. §106.

[8] For example, the right of public performance is limited to 'literary, musical, dramatic, and choreographic works, pantomimes, and motion pictures and other audiovisual works' and the right of public display is limited to 'literary, musical, dramatic, and choreographic works, pantomimes, and pictorial, graphic, or sculptural works, including the individual images of a motion picture or other audiovisual work'. Ibid.

[9] Tattoo, Merriam-Webster: https://www.merriam-webster.com/dictionary/tattoo (last accessed October 8, 2017).

[10] Aaron Perzanowksi, Tattoos and IP Norms, 98 Minn. L. Rev. 511, 524 (2013).

[11] 17 U.S.C. §101.

2. USEFUL ARTICLE DOCTRINE

The second sentence of Section 101 for PGS works further defines and limits the scope of protection for PGS works. The definition provides that:

> Such works shall include works of artistic craftsmanship insofar as their form but not their mechanical or utilitarian aspects are concerned; the design of a useful article, as defined in this section, shall be considered a pictorial, graphic or sculptural work only if, and only to the extent that, such design incorporates pictorial, graphic, or sculptural features that can be identified from, and are capable of existing independently of, the utilitarian aspects of the article.[12]

Section 101 defines a useful article as a product 'with an intrinsic utilitarian function that is not merely to portray the appearance of the article.[13] Congress has provided examples of useful articles, such as automobiles, airplanes, women's clothing, electronics, and other industrial products.[14] The Supreme Court has held that an 'artistic feature' of a useful article may be copyrighted:

> If the feature (1) can be perceived as a two- or three-dimensional work of art separate from the useful article and (2) would qualify as a protectable pictorial, graphic, or sculptural work either on its own or in some other medium if imagined separately from the useful article.[15]

While the human body certainly performs a number of functions, it is highly unlikely that Congress would have considered or U.S. courts would characterize a person as a useful article. Even if one assumes that the human body is a 'useful article', a tattoo on the human body can be perceived as a two-dimensional work of art separate from the tattoo bearer's body. The skin of the human body is performing an artistic, not useful, purpose in serving as a canvas for the tattoo.

Furthermore, a tattoo would certainly qualify as a protectable PGS work in some other medium, such as the traditional medium of paper,

---

[12] Ibid.

[13] Ibid. Congress sought to narrow the scope of the definition of the former, broad regulations of the U.S. Copyright Office, which defined a useful article as 'one with utility as its "sole intrinsic function"'. *Kieselstein-Cord v. Accessories by Pearl, Inc.*, 632 F.2d 989, 993 (2d Cir. 1980).

[14] Copyright Act Congressional Bill, H.R. Rep. No. 94-1476, at 51 (1976), reprinted in 1976 U.S.C.C.A.N. 5659, 5664 [hereinafter Copyright Act Congressional Bill].

[15] *Star Athletica, L.L.C. v. Varsity Brands, Inc.*, 2017 WL 1066261.

if the tattoo artist created a line drawing of the tattoo first.[16] The mere transfer of a tattoo from one medium, a sketch on paper, to another medium, the human skin, should not eliminate the copyrightability of the work itself.

3. DO TATTOOS SATISFY THE REQUIREMENTS FOR FEDERAL COPYRIGHT PROTECTION IN THE UNITED STATES?

The Berne Convention, the oldest international copyright treaty, determines the minimum protections to be granted by signatory countries to the works of their authors.[17] Berne defines protected works as any production of '"literary and artistic works" . . . whatever may be the mode or form of its expression'.[18] However, signatory countries may determine that works or specified categories of works are ineligible for protection 'unless they have been fixed in some material form'.[19] While Berne does not generally require originality as a minimum standard for copyright protection, it sets this standard for works based upon underlying copyrighted works, providing that '[t]ranslations, adaptations, arrangements of music and other alterations of a literary or artistic work shall be protected as original works without prejudice to the copyright in the original work'.[20]

---

[16] Perzanowksi, supra note 10, at 526 (observing that tattoo artists 'occasionally ink an image freehand directly on a client's skin. But more often, they create a detailed line drawing of the tattoo design on paper').

[17] Summary of the Berne Convention for the Protection of Literary and Artistic Works, http://www.wipo.int/treaties/en/ip/berne/summary_berne.html ('The Berne Convention deals with the protection of works and the rights of their authors. It is based on three basic principles and contains a series of provisions determining the minimum protection to be granted . . .'). The United States of America acceded to the Berne Convention on November 16, 1998 (http://www.wipo.int/treaties/en/ShowResults.jsp?lang=en&treaty_id=15). The Berne Convention Implementation Act of 1988, which implemented the Berne Convention in the United States, was enacted on October 31, 1998. H.R. 4262 – 100th Congress: Berne Convention Implementation Act of 1988: https://www.govtrack.us/congress/bills/100/hr4262. The Berne Convention entered into force in the United States on March 1, 1989. Appendix K: https://www.copyright.gov/title17/92appk.pdf.

[18] The Berne Convention for the Protection of Literary and Artistic Works, Art. 2, Sec. 1, September 9, 1886, as last revised, Paris, (July 24, 1971): http://www.wipo.int/treaties/en/text.jsp?file_id=283698#P85_10661.

[19] Ibid. at Art. 2 Sec. 2.

[20] Ibid. at Art. 2 Sec. 3.

In the United States, the Copyright Act of 1976 provides that 'copyright protection subsists . . . in original works of authorship fixed in any tangible medium of expression'.[21] Originality is a requirement under the U.S. Constitution.[22] The U.S. Supreme Court has interpreted the Patent and Copyright Clause of the U.S. Constitution,[23] which grants Congress the power to enact copyright laws, to presuppose a degree of originality. The Supreme Court set forth the modern definition for originality in *Feist Publications, Inc. v. Rural Telephone Service Co.*: 'Original, as the term is used in copyright, means only that the work was independently created by the author (as opposed to copied from other works), and that it possesses at least some minimal degree of creativity'.[24] The Court not only addressed the low standard of originality, but also the breadth of works that should be captured by this standard. In the words of the Court, '[t]he vast majority of works make the grade quite easily, as they possess some creative spark, "no matter how crude, humble or obvious" it might be'.[25]

Therefore, the modern definition of originality is comprised of two requirements: independent creation of the work by the author and display of a minimal level of creativity.[26] Both of these requirements are quite modest. A copyrightable work need not be unique or novel;[27] all that is required is that the author contributes something 'recognizably his own'.[28]

Some tattoos meet the U.S. originality standard. For example, custom tattoos are independently created by the author, not copied from another source. Originality means 'little more than a prohibition of actual copying'.[29] Therefore, a tattoo with even a modicum of creativity likely would be copyrightable. Even a tattoo that added original expression to

---

21 17 U.S.C. §102(a).

22 *Feist Publ'ns, Inc. v. Rural Tel. Serv.*, 499 U.S. 340, 345 (1991).

23 Article I, Section 8, Clause 8 of the U.S. Constitution authorizes Congress to 'secure for limited Times to Authors . . . the exclusive Right to their respective Writings', U.S. Const. Art. I, §8, cl. 8.

24 Ibid. (citing M. Nimmer and D. Nimmer, *Copyright*, 1, §§2.01 [A], [B] (LexisNexis, 1990)) (the Court conceding that the 'requisite level of creativity is extremely low; even a slight amount will suffice').

25 Ibid.

26 Ibid.

27 *Alfred Bell & Co. v. Catalda Fine Arts, Inc.*, 191 F.2d 99, 102 (2d Cir. 1951); see also *L. Batlin & Son v. Snyder*, 536 F.2d 486, 490 (2d Cir. 1976) ('Originality is . . . distinguished from novelty; there must be independent creation, but it need not be invention in the sense of striking uniqueness, ingeniousness, or novelty').

28 Ibid. at 103.

29 Ibid.

another work (in the public domain or used with the authorization of the copyright owner) would meet this standard in most instances.

However, certain tattoos, even if independently created by the author, would not possess the requisite spark of creativity to meet the modest standard of originality. The Compendium of the U.S. Copyright Office Practices states that '[a] work that is "entirely typical," "garden-variety," or "devoid of even the slightest traces of creativity" does not satisfy the originality requirement'.[30] A tattoo of one's date of birth or name, with few (or perhaps without) artistic features, would not meet the requisite level of original expression by an author.[31] Nor is a tattoo of a basic geometric shape, such as a circle, oval, or star, copyrightable,[32] unless the selection and arrangement of those shapes possesses the minimum level of creativity.[33] Tattoos used to create permanent makeup, such as eyebrows, lip liner, or a freckle on a person's face, would also fail to meet the requisite level of originality for a copyrightable work.[34]

Unlike originality, the U.S. Copyright Act defines the fixation requirement. In order for a work to qualify for copyright protection in the United States, it must be 'fixed in any tangible medium of expression, now known or later developed, from which [the expression] can be perceived, reproduced, or otherwise communicated, either directly or with the aid of a machine or device'.[35]

Courts have derived two sub-requirements from this statutory definition:

---

[30] U.S. Copyright Office, Compendium of U.S. Copyright Office Practices, §308.2 (3d ed.) (2014) (citing *Feist*, 499 U.S. at 362).

[31] Certain categories of works are not copyrightable and authors cannot seek registration of such works with the U.S. Copyright Office. The categories relevant to this chapter are '[w]ords and short phrases such as names, titles, and slogans; familiar symbols or designs; mere variations of typographic ornamentation, lettering, or coloring . . .' and '[t]ypeface as typeface'. Preregistration and Registration of Claims to Copyright, 37 C.F.R. §202.1(a), (e).

[32] Ibid.; see also *Glasscraft Door I, L.P. v. Seybro Door & Weathership Co.*, 2009 WL 3460372, *3 (S.D. Tex. 2009). ('Individual shapes such as ovals or stars are not copyrightable, just as individual words are not copyrightable').

[33] Compilations are copyrightable so long as they meet the requirements under the statute. *Feist*, 499 U.S. at 356–8. Congress defines the term compilation as 'a work formed by the collection and assembly of preexisting materials or of data that are selected, coordinated, or arranged in such a way that the resulting work as a whole constitutes an original work of authorship'. 17 U.S.C. §101.

[34] See Randy Schueller, Tattoo. In Deirdre Blanchfield, ed., *How Products Are Made* (7th edn, Cengage Gale, 2011) 384, 386 (discussing the use of tattoos as permanent makeup for burned and disfigured victims).

[35] 17 U.S.C. §101.

the embodiment requirement and the duration requirement.[36] The first requirement is that the work must be embodied in a medium, i.e., placed in a medium such that it can be perceived, reproduced, etc., from that medium.[37] The second requirement is that the work must remain embodied in that medium for more than a transitory period of duration.[38]

As far as embodiment is concerned, neither the language of the fixation definition in Section 101 of the Act nor the legislative history of the Act places any limitations on the copyrightability of a work fixed in a particular type of medium.[39] The U.S. statute provides no further guidance regarding limitations on the protection of various media.[40] There is simply no mention of any requisite media of expression or requisites that must be met by any of the media of expression exemplified in the legislative history of the Copyright Act.[41]

The more intriguing question regarding the copyrightability of tattoos is whether the human body fits the definition of a medium.[42] The Copyright Act does not define the term.[43] In the United States, when a statutory term is undefined, one must look to the 'plain and ordinary meaning of the word.[44] In this context, the 'plain and ordinary' meaning of the term 'medium' is 'material or technical means of artistic expression'.[45] If this definition is applied to tattoos, then certainly, the human body is a material means of artistic expression.[46] Just as a canvas[47] serves as a surface for one type of PGS work – a painting – the skin serves as a surface for another type of PGS work, a tattoo. The human body provides a material means of embodying a work of authorship.

As far as duration is concerned, U.S. copyright law recognizes that

---

36 *Cartoon Network LP v. CSC Holdings, Inc.*, 536 F.3d 121, 129 (2d Cir. 2008) ('[N]o case law or other authority dissuades us from concluding that the definition of "fixed" imposes both an embodiment requirement and a duration requirement').

37 Ibid. at 127.

38 See ibid.

39 King, supra note 4, at 152–3.

40 Ibid. at 153.

41 Ibid.

42 Ibid.

43 Ibid.

44 Ibid. at 154.

45 Ibid.

46 Ibid.

47 A canvas is defined as 'a piece of cloth backed or framed as a surface for a painting'. Canvas, Merriam-Webster: https://www.merriam-webster.com/dictionary/canvas (last accessed October 8, 2017).

various physical substrates may be inadequate bases to secure protection.[48] One scholar provides examples of subject matter that would not qualify for protection, such as writing a literary work in the wet sand before the oncoming tide or inscribing a pictorial work in the frost of a windowpane.[49] While it is certainly true that a tattoo may be removed from the human body, it does not wash away with the ease of a wave lapping over the sand on a beach.[50] Tattoo scholars have referred to the practice of tattooing as 'one of many forms of irreversible body alteration'[51] and believe that 'the idea of the permanence of the tattoo is critical'.[52] Tattooing is 'designed to last forever'.[53]

Despite the historical and symbolic permanence of tattoos, advancements in laser skin resurfacing technology make them removable.[54] However, tattoo removal is not a brief, effortless process.[55] A tattoo is removable through a series of laser treatments over a period of months – a process that is expensive, painful, and not always successful.[56] It is hardly questionable that a tattoo is 'sufficiently permanent or stable to permit it to be perceived, reproduced, or otherwise communicated for a period of more than transitory duration.[57]

---

[48] King, supra note 4, at 153.

[49] Ibid. (citing D. Nimmer, *Nimmer on Copyright*, 1, §§1.01[B][2][a], 2.03 [B][2] (1965).

[50] Ibid.

[51] Jane Caplan, Introduction. In Jane Caplan, ed., *Written on the Body: The Tattoo in European and American History* (Princeton University Press, 2000) at xi (classifying tattooing as 'one of many forms of irreversible body alteration, including scarification, cicatrization, piercing and branding, and it is the [sic] probably the oldest and most widespread of these').

[52] Susan Benson, Inscriptions of the Self: Reflections on Tattooing and Piercing in Contemporary Euro-America. In *Written on the Body: The Tattoo in European and American History*, ibid., 240 ('This, in a culture so committed to constant change and innovation, is surely what lies at the heart of the fear of the tattoo, for it is precisely the permanence of the tattoo, its evocation of the private depths of the self upon the surfaces of the body and its non-negotiable relationship to that body, that reminds us of the fixed end-point of all this mutability and self-fashioning, our own death').

[53] Caplan, supra note 51.

[54] King, supra note 4, at 155 (citing Bernadine Healy, M.D., The Dangerous Art of the Tattoo, *U.S. News* (July 25, 2008), http://health.usnews.com/health-news/family-health/articles/2008/07/25/the-dangerous-art-of-the-tattoo.

[55] King, supra note 4, at 155.

[56] King, supra note 4, at 155 (citing Bernadine Healy, M.D., The Dangerous Art of the Tattoo, *U.S. News* (July 25, 2008), http://health.usnews.com/health-news/family-health/articles/2008/07/25/the-dangerous-art-of-the-tattoo.

[57] Ibid.

4. A CASE ON COPYRIGHTABILITY OF TATTOOS

The copyrightability of tattoos is an unsettled question that courts have not yet decided.[58] Thus far, only one United States federal judge has expressed thoughts on the copyrightability of tattoos.[59]

On April 28, 2011, tattoo artist S. Victor Whitmill sued Warner Bros. Entertainment Inc. for copyright infringement based upon the production company's unauthorized copying and creation of a derivative work of a tattoo, authored by Whitmill, on one actor's face onto the face of another actor in the comedic motion picture *The Hangover Part II*.[60] Whitmill further claimed that Warner Bros. infringed his copyright by the unauthorized copying, distribution, and public display of the tattoo in advertising and promotion of the motion picture.[61]

In Whitmill's complaint, he alleged that he 'created and applied an original and distinctive tattoo to the upper left side of the face of the former, world heavyweight champion boxer Michael Gerard 'Mike' Tyson' on February 10, 2003'.[62] Whitmill also alleged that on the same day, Tyson signed a release acknowledging 'that all artwork, sketches, and drawings related to [his] tattoo and any photographs of [his] tattoo' were the property of Whitmill, who was doing business as Paradox-Studio of Dermagraphics.[63]

---

[58] See, Christopher A. Harkins, Tattoos and Copyright Infringement: Celebrities, Marketers, and Businesses Beware of the Ink, 10 Lewis & Clark L. Rev. 313, 332 (2006) ('In years past, copyright protection was a non-issue or ignored because tattooists were either too reticent to sue a customer or too complacent to challenge tattoo lore, favoring any and all available exposure for their work').

[59] Meredith Hatic, Note, Who Owns Your Body Art? The Copyright and Constitutional Implications of Tattoos, 23 Fordham Intell. Prop. Media & Ent. L.J. 396, 423 (2013) ('Before Whitmill v. Warner Brothers in 2011, no court had ever directly addressed the question of whether [tattoos] copyrights [sic] are in fact copyrightable').

[60] Complaint at 6–7, ¶17, *Whitmill v. Warner Bros. Entm't Inc.*, 2011 WL 2038147 (E.D. Mo. April 28, 2011), dismissed, No. 4:11-CV-752 CDP (E.D. Mo. June 22, 2011) [hereinafter Whitmill Complaint] (involving a tattoo artist alleging infringement of copyright in a tattoo design).

[61] See, ibid. at 7, 18.

[62] Ibid.

[63] Ibid. Artwork, sketches, or drawings of the Maori-style tattoo design authored by Whitmill on Tyson's face probably did not exist prior to the tattooing of the design on Tyson because Whitmill thought that he might have created the design in the first instance on Tyson's face. Whitmill Transcript of Preliminary Injunction Hearing at 17–18 (first session, May 23, 2011), Whitmill, No. 4:11-CV-752 CPD (motion denied May 24, 2011) [hereinafter Whitmill Hearing Transcript]

Whitmill sought injunctive relief, actual damages and profits, and costs and attorney's fees.[64] He moved for a preliminary injunction to stop the advertising and release of *The Hangover II*, which was due to be released on May 26, 2011.[65]

While Judge Catherine D. Perry of the United States District Court for the Eastern District of Missouri did not analyze the requirements of copyright law, she stated that tattoos are copyrightable:

> Of course, tattoos can be copyrighted. I don't think there is any reasonable dispute about that. They are not copyrighting Mr. Tyson's face, or restricting Mr. Tyson's use of his own face, as the defendant argues, or saying that someone who has a tattoo can't remove the tattoo or change it, but the tattoo itself and the design itself can be copyrighted, and I think it's entirely consistent with the copyright law.[66]

Judge Perry expressed sympathy for Whitmill's claims, stating that he had a 'strong likelihood of prevailing on the merits for copyright infringement'.[67] However, she denied the motion for a preliminary injunction, concluding that Warner Bros. would 'lose millions' and the harm to the public interest outweighed the harm to Whitmill.[68] Shortly following the court's denial of the preliminary injunction, the parties dismissed the case.[69]

Currently, at the time of writing, there is a copyright infringement lawsuit pending in the United States District Court for the Southern District of New York.[70] On February 1, 2016, Solid Oak Sketches, the

---

(Whitmill testifying that he did not remember if they drew sketches, but he drew an original creation on Tyson's face during the consultation).

64 Whitmill Complaint, supra note 60, at 7–8, 22.

65 Whitmill Hearing Transcript, supra note 63, at 3 (May 24, 2011).

66 Ibid.

67 Ibid.

68 Ibid. at 6–8. In her analysis of the Eighth Circuit's four factors considered in the determination of whether a preliminary injunction should be granted, i.e., plaintiff's likelihood of success on the merits, whether plaintiff is threatened with irreparable harm, the balance of the equities, and the public interest, Judge Perry also concluded that Whitmill had shown irreparable harm and would be in danger of further irreparable harm if distribution of the motion picture was not enjoined because he had lost control of his work, which was exemplified by the use of his work in a convenience store's promotions. Ibid. at 3, 6.

69 Ibid.

70 Complaint and Demand for Jury Trial, *Solid Oak Sketches LLC v. Visual Concepts et al.*, No. 16-00724 (S.D.N.Y. February 1, 2016); Yolanda M. King, Tattoo Copyrights May Finally Get Their Day in Court, *Law360*, (April 28, 2016): https://www.law360.com/articles/788694/tattoo-copyrights-may-finally-get-their-day-in-court.

exclusive licensor of eight tattoos displayed on the bodies of five National Basketball Association ('NBA') players who appear in the NBA 2K16 video game, filed a complaint for copyright infringement against the makers of the video game.[71] The lawsuit alleges that the defendants infringe the copyrights in the tattoos by depicting the NBA players, with the tattoos inked on them, in the video game.[72]

On August 2, 2016, Judge Laura Taylor Swain granted the defendants' motion to dismiss Solid Oak's claim for statutory damages and attorneys' fees because the first of the alleged series of infringements occurred before the registration of the copyrights in the works.[73] However, Solid Oak may continue to pursue proving actual damages due to the alleged infringements.

On August 9, 2017, defendants filed a motion for judgment on the pleadings for an order to, among other things, dismiss Solid Oak's copyright infringement claim.[74] Defendants also counterclaimed for declaratory judgments of de minimis use and fair use.[75] In support of de minimis use, the defendants asserted that 'the Tattoos rarely appear in NBA 2K as they are only displayed when the players on whom they are inked are selected from the over 400 other players that are available'.[76] Regarding fair use, the defendants argued that (1) their use in the creative context of a video game is completely different than that for which the tattoos were originally created; (2) the creative nature of the works must be weighed against the fact that the works had already been published and defendants use them to accurately depict real-world subject matter; (3) their use is 'reasonable given that its purpose was to depict real life accurately, and using any less of the Tattoos would defeat that purpose'; and (4) because

---

[71] See Complaint and Demand for Jury Trial, *Solid Oak Sketches LLC v. Visual Concepts et al.*, supra note 70.

[72] Ibid.

[73] Judge Swain concluded that the tattoos first appeared in NBA 2K14, which was released in 2013, but the U.S. Copyright Office did not issue certificates of registration for the tattoos until 2015. *Solid Oak Sketches, LLC v. 2K Games, Inc.*, 2016 WL 4126543. Under the Copyright Act, a plaintiff must register its copyright prior to the alleged infringement in order to seek statutory damages and attorneys' fees. 17 U.S.C. §412.

[74] Defendant's Notice of Motion for Judgment on the Pleadings, *Solid Oak Sketches, LLC v. 2K Games, Inc.*, No. 1:16CV00724 (S.D.N.Y. 2017). Defendant's Memorandum in Support of Motion for Judgment on the Pleadings, *Solid Oak Sketches, LLC v. 2K Games, Inc.*, No. 1:16CV00724 (S.D.N.Y. 2017).

[75] Ibid.

[76] Defendant's Memorandum in Support of Motion for Judgment on the Pleadings, *Solid Oak Sketches, LLC v. 2K Games, Inc.*, No. 1:16CV00724 (S.D.N.Y. 2017).

Solid Oak is contractually prohibited from inking tattoos on other people, defendants cannot harm that market.[77] Defendants concluded that it is unreasonable to expect Solid Oaks' permission every time an NBA player commercializes his likeness, which would necessarily include his tattoos if it is a realistic depiction of the player, in the media.[78]

On September 12, 2017, plaintiff filed a response to defendants' motion. Plaintiff asserted that defendants' use of the tattoos is not de minimis use because a user of the video game could pick each NBA player bearing the tattoos at issue for hours of play, which would result in continuous observability of the tattoos with minimal difficulty. Regarding fair use, plaintiff argued that: (1) defendants' commercial use of the tattoos does not transform the works so they convey a new message; (2) the copyrighted tattoos are creative, expressive works entitled to greater protection and their published nature does not fully withdraw them from protection; (3) reproduction of the works in their entireties is not material for the alleged purpose of the game and the alleged minimal use begs the question as to why defendants did not alter the tattoos to avoid improper use; and (4) defendants' use diminishes the commercial value of the copyrighted tattoos' potential licensing markets.

On March 30, 2018, the court denied the defendants' motion for judgment on the pleadings as to de minimis use, concluding:

> [T]here is no objective perspective as to how the Defendants' video game is generally played, or to what extent certain game features can be or are actually utilized, that would allow this Court to make determinations about the choices and subsequent observations of the 'average lay observer,' or about the observability and prominence of the Tattoos. The Court is thus unable to conclude without the aid of extrinsic evidence that 'no reasonable jury, properly instructed, could find that the two works are substantially similar'.[79]

Regarding the fair use defense, the court determined that 'further evidence must be considered in connection with the fact-intensive question of the applicability of the fair use defense'.[80]

Since Solid Oak filed suit, tattoo artists have filed other lawsuits against

---

[77] Ibid.

[78] Ibid.

[79] Memorandum Opinion and Order, *Solid Oak Sketches, LLC v. 2k Games, Inc. et al*, No. 1:16-CV-00724 (S.D.N.Y. February 1, 2016).

[80] Ibid.

the same defendant in various jurisdictions. Those lawsuits are still pending as well.[81]

4.1 Who is the Author of a Tattoo?

Berne determines the minimum protections to be granted by signatory countries to the works of their authors. However, it leaves the interpretation of authorship to the signatories.

'Few laws tell us who is an author, or what authorship is'.[82] Professor Jane C. Ginsburg observed that '[i]t is easier to assert that authors are the initial beneficiaries of copyright/droit d'auteur than to determine what makes one an author'.[83]

In the United States, copyright law only protects the works of authors. The U.S. Constitution grants Congress the 'Power . . . to Promote the Progress of Science and the useful Arts, by securing for limited Times to Authors and Inventors the exclusive Right to their respective Writings and Discoveries'.[84] This constitutional clause gives Congress the power to enact federal copyright statutes.[85]

The Copyright Act of 1976 vests ownership of a copyrighted work in the author of the work.[86] However, the Act, as well as the preceding copyright statutes, do not define the term author. The meaning of authorship has been developed in the common law. The U.S. Supreme Court has defined an author as 'he to whom anything owes its origin; originator; maker; one who completes a work of science or literature'.[87] Lower courts have observed that an author must be more than one who contributes creativity or originality to a work.[88] An author is the one 'who superintended the whole work, the "master mind"'.[89]

---

[81] *See, e.g. Complaint, Hayden v. 2K Games, Inc. et al,* No. 1:17-CV-02635 (N.D. Ohio December 18, 2017) and *Complaint, Alexander v. Take-Two Interactive Software, Inc. et al.,* No. 3:18-cv-00966-MJR-DGW (S.D. Illinois April 17, 2018).

[82] Jane C. Ginsburg, The Concept of Authorship in Comparative Copyright Law, 52 DePaul L. Rev. 1063, 1069 (2003).

[83] Ibid. at 1066.

[84] U.S. Const., Art. 1, §8, cl. 8.

[85] Yolanda M. King, The Challenges 'Facing' Copyright Protection for Tattoos, 92 Or. L. Rev. 129 (2013).

[86] 17 U.S.C. §201(a) ('Copyright in a work protected under this title vests initially in the author or authors of the work. The authors of a joint work are co-owners of copyright in the work').

[87] *Burrow-Giles Lithographic Co. v. Sarony*, 111 U.S. 53, 58 (1884).

[88] See, *Aalmuhammed v. Lee*, 202 F. 3d 1227, 1233 (9th Cir. 2000).

[89] Ibid. (citing *Burrow-Giles Lithographic Co.*, 111 U.S. at 53, 61).

Under U.S. copyright law, the author of a tattoo would likely be the tattoo artist because he or she is the originator of the work in many instances. However, in some instances, the author, or the superintendent of the work, could be the tattoo artist's client. Alternatively, due to the complexities of the creative process between the tattoo artist and his or her client, both the artist and the client may be joint authors of the tattoo, and therefore, co-owners of the copyright in the work.[90] As co-owners of the copyright in a jointly authored work, each owner 'has the right to use or to license the work, subject to an accounting to the other co-owners for any profits'.[91]

In *Reed v. Nike, Inc.*, tattoo artist Matthew Reed filed a lawsuit for copyright infringement against athletic footwear and apparel company Nike, Inc., advertising agency Weiden + Kennedy, and former National Basketball Association player Rasheed Wallace, based on the unauthorized use of a tattoo on Wallace's arm, allegedly authored by Reed, in a commercial broadcast on television and over the internet.[92] The commercial included a close-up of the tattoo and featured the tattoo being re-created by computerized simulation with a voice-over from Wallace explaining the meaning of the tattoo.[93] In the complaint, Reed describes the process of creating the tattoo.[94] Reed took notes and made sketches of Wallace's description of his concept of 'incorporating an Egyptian themed family design of a king and a queen and three children with a stylized sun in the background', while relying on his experience and research of Egyptian artwork.[95] After another meeting with Wallace and additional suggestions by Wallace, Reed incorporated those changes into the final drawing and then applied the tattoo to Wallace's arm.[96]

---

90 A jointly authored work is defined in the 1976 Copyright Act as 'a work prepared by two or more authors with the intention that their contributions be merged into inseparable or interdependent parts of a unitary whole', 17 U.S.C. §101. In addition, the statute provides that '[t]he authors of a joint work are co-owners of copyright in the work', ibid. at §201(a).

91 *Erickson v. Trinity Theatre, Inc.*, 13 F.3d 1061, 1068 (7th Cir. 1994) (citing *Childress v. Taylor*, 945 F.2d 500, 505 (2d Cir. 1991); *Weinstein v. University of Illinois*, 811 F.2d 1091, 1095 (7th Cir. 1987); Melvin B. Nimmer and David Nimmer, *Nimmer on Copyright*, 1, §6.02, at 6–7 to 6–8 (Matthew Bender, 92nd Rev.)).

92 Complaint, 4–6, 14–16, *Reed v. Nike, Inc.*, No. 3:05-CV-00198 (D. Or. Feb. 10, 2005), dismissed (D. Or. October 19, 2005) [hereinafter Reed Complaint].

93 Harkins, supra note 58, at 317.

94 Reed Complaint, supra note 92, at 10, 12.

95 Ibid. at 10.

96 Ibid. at 12–13.

It is customary in the tattoo industry for the tattoo artist to collaborate with the client in the creation of the tattoo.[97] The collaboration between the tattoo artist and the tattoo bearer in Reed illustrates the fluidity and informality of the process of tattoo creation. In such circumstances, it can be challenging to determine who is the author – the superintendent or mastermind of the work?[98] Reed himself acknowledged the possibility that Wallace was a joint author of the tattoo, and therefore, a co-owner of the copyright in it.[99]

A further complication to the determination of authorship is the work made for hire doctrine.[100] Under U.S. law, the work made for hire doctrine is an exception to the general rule that copyright in a work vests initially

---

[97] Matthew Beasley, Note, Who Owns Your Skin: Intellectual Property Law and Norms Among Tattoo Artists, 85 S. Cal. L. Rev. 1137, 1162 (2012).

[98] Courts have considered two approaches to determining whether one's contributions to a work amount to authorship. See *Erickson*, 13 F.3d at 1069–71. The late Professor Melville B. Nimmer proposed a test that only required that the combined product of the joint contributions be copyrightable (the 'de minimis test'). Ibid. at 1069–70. Professor Paul Goldstein requires that each contribution be copyrightable (the 'copyrightability test'). Ibid. at 1070–71. The majority of courts have adopted the copyrightability test. However, the de minimis test may be more appropriate in evaluating joint authorship in the context of tattoo creation between a tattoo artist and her client. The de minimis test provides greater protection for the contributions of tattoo bearers, which promotes creativity, and better recognizes the flexibility and fluidity of tattoo creation.

[99] Reed filed a contributory copyright infringement claim against Wallace, alleging that Wallace incorrectly 'advised the other Defendants that he had exclusive ownership of all intellectual property rights in connection with the tattoo on his upper arm', Reed Complaint, supra note 92, at 28. However, Reed, in the alternative, filed a claim for an accounting against Wallace in the event Wallace was the co-owner of the copyright in the tattoo. Reed Complaint, supra note 92, at 37.

[100] The 1976 Copyright Act defines a work made for hire as: '(1) a work prepared by an employee within the scope of his or her employment; or (2) a work specially ordered or commissioned for use as a contribution to a collective work, as a part of a motion picture or other audiovisual work, as a translation, as a supplementary work, as a compilation, as an instructional text, as a test, as answer material for a test, or as an atlas, if the parties expressly agree in a written instrument signed by them that the work shall be considered a work made for hire. For the purpose of the foregoing sentence, a "supplementary work" is a work prepared for publication as a secondary adjunct to a work by another author for the purpose of introducing, concluding, illustrating, explaining, revising, commenting upon, or assisting in the use of the other work, such as forewords, afterwords, pictorial illustrations, maps, charts, tables, editorial notes, musical arrangements, answer material for tests, bibliographies, appendixes, and indexes, and an "instructional text" is a literary, pictorial, or graphic work prepared for publication and with the purpose of use in systematic instructional activities', 17 U.S.C. §101.

in the author of the work.[101] An employer or other person who commissions the preparation of a work is the author of the work, and therefore, owns the copyright in the work.[102]

In the context of authorship of tattoos, the traditional application of the work made for hire doctrine would be a circumstance in which a tattoo artist is employed by a tattoo parlor. If any tattoos are created within the scope of the artist's employment, those works likely would be authored by the employer, and therefore, the parlor would own the copyrights in those tattoos.[103]

However, the more intriguing question is whether the work made for hire doctrine applies to the tattoo artist–client relationship. Some scholars have concluded that it is unlikely that a tattoo could qualify as a work made for hire,[104] while others have reached the contrary conclusion.[105]

---

[101] Ibid. at §201(a).

[102] Ibid. at §201(b) ('In the case of a work made for hire, the employer or other person for whom the work was prepared is considered the author . . . and, unless the parties have expressly agreed otherwise in a written instrument signed by them, owns all of the rights comprised in the copyright').

[103] See, *Community for Creative Non-Violence v. Reid*, 490 U.S. 730, 751–2 (1989) (stating factors considered in determining whether a hired party is an employee under the common law of agency, including 'the skill required; the source of the instrumentalities and tools; the location of the work; the duration of the relationship between the parties; whether the hiring party has the right to assign additional projects to the hired party; the extent of the hired party's discretion over when and how long to work; the method of payment; the hired party's role in hiring and paying assistants; whether the work is part of the regular business of the hiring party; whether the hiring party is in business; the provision of employee benefits; and the tax treatment of the hired party'); see also *Aymes v. Bronelli*, 980 F.2d 857, 861 (2d Cir. 1992) (noting that 'the Reid test can be easily misapplied' because '[the Court] gave no direction concerning how the factors were to be weighed'. The court then listed factors that 'will almost always be relevant and should be given more weight in the analysis', including: '(1) the hiring party's right to control the manner and means of creation; (2) the skill required; (3) the provision of employee benefits; (4) the tax treatment of the hired party; and (5) whether the hiring party has the right to assign additional projects to the hired party').

[104] Harkins, supra note 58, at 324–5 (concluding that a 'tattoo would probably not constitute a work made for hire under either paragraph of section 101'. However, Harkins does not rule out the possibility that there could be a basis for asserting a work made for hire status for a tattoo in unusual circumstances.); Aaron Perzanowksi, Tattoos and IP Norms, 98 Minn. L. Rev. 511, 534 (2013) (concluding that '[c]ustom tattoos are almost certainly not works made for hire as defined by the Copyright Act').

[105] Cotter and Mirabole, supra note 5, at 105 (listing possibilities for a tattoo to qualify as a work made for hire, such as when a business employs a tattoo artist,

Absent a determination of an employer–employee relationship between the tattoo artist and his or her client, which is a highly improbable, a tattoo cannot qualify as a work made for hire.[106]

5. CONCLUSION

Tattoos fall within the categories of copyrightable subject matter and meet the requirements of copyrightability under U.S. federal copyright law. In addition, tattoo artists, and in some instances, their clients, meet the vague criteria of conception and control to be considered authors of tattoo artwork. Therefore, U.S. copyright law should be interpreted as protecting tattoos. With the growing visibility and prevalence of tattoos among many segments of society, and the liberal discussion of the protectability of tattoos by one U.S. federal judge already, it is quite likely that courts will extend copyright protection to this unconventional subject matter.

---

when a customer 'employs' a tattoo artist, and when a party specially orders or commissions a tattoo for use as a contribution to a collective work, arguably an additional tattoo placed on the body of an individual with other tattoos); see also David M. Cummings, Note, Creative Expression and the Human Rights Canvas: An Examination of Tattoos as a Copyrightable Art Form, 2013 U. Ill. L. Rev. 279, 314 (arguing that tattoos originally created on human flesh should be classified as works made for hire under the independent contractor provision of §101 because the tattoo recipient 'likely provides some general idea or vision . . . analogous to that of a hiring party commissioning an independent contractor to create a work').

[106] Tattoos do not fall within one of the enumerated categories of specially ordered or commissioned works set forth in the second part of the work made for hire definition, so they cannot qualify as this type of work made for hire.

6. Copyright in culinary presentations

Cathay Y. N. Smith

1. INTRODUCTION

The President's cake was a knock-off. After a turbulent election-season defined by personal attacks and character assassinations, the realization that the newly-elected President Trump's inaugural cake was an unauthorized exact duplicate of President Obama's inaugural cake was 'icing on the cake' for many people in the United States. Chef Duff Goldman, the original creator of President Obama's inaugural cake and star of the reality TV show *Ace of Cakes*, took to Twitter to vent his frustration. He posted a photo of the cake he created next to its copy (pictured below) with the caption, 'The cake on the left is the one I made for President Obama's inauguration 4 years ago. The one on the right is Trumps. I didn't make it' (Figure 6.1).[1]

Over six thousand people commented on Goldman's tweet, which was retweeted over 130,000 times, liked by over 280,000 people, and picked up by media outlets such as the Washington Post, Jezebel, Slate, Huffington Post, Fox News, CNN, USA Today, and People Magazine. According to media reports, Trump's staff members brought a photo of the cake Goldman created for President Obama's inauguration to Buttercream Bakeshop, a bakery in Washington D.C., and asked it to replicate the cake.[2] When the bakery asked whether it should use the photo as inspiration instead of creating an exact duplicate of the original cake, Trump's staff purportedly said that they 'want[ed] this exact cake'.[3] And the exact cake is what they got.

1 See Duff Goldman, Twitter (9:22 PM, January 20, 2017), https://twitter.com/duffgoldman/status/822675780341641216; both quote and image.

2 Amy B. Wang and Tim Carman, Trump's Inaugural Cake was Commissioned to Look Exactly like Obama's, Baker Says, *The Washington Post* (January 22, 2017), https://www.washingtonpost.com/news/food/wp/2017/01/21/trump-had-a-huge-luxurious-inauguration-cake-was-it-plagiarized/?utm_term=.9933d6123809.

3 Ibid.

Figure 6.1 Twitter post photo of original cake and inauguration cake

This is not the first time a chef has been caught ripping off another chef's creations, and it will not be the last time a chef copies or reuses another's culinary presentation.[4] The culinary industry has traditionally operated

[4] See, for instance, the publicized instance of Chef Robin Wickens of Interlude (Melbourne) copying culinary presentations created by Chefs Grant Achatz, Wylie

in a low-intellectual-property environment. This is the case not only because there are limited intellectual property protections available to chefs, but also because chefs have long adhered to and understood the culture of sharing and hospitality in the cuisine industry. Most chefs recognize that '[t]he world's culinary traditions are collective, cumulative inventions, a heritage created by hundreds of generations of cooks'.[5] For instance, when Christopher Buccafusco interviewed a number of celebrity chefs about borrowing each other's recipes, techniques, or even culinary presentations, many chefs expressed ease with the idea that their creations may be borrowed or used by other chefs,[6] and some chefs even noted that their own creations could have been inspired by other culinary creations of the past.[7] Nevertheless, this culture of sharing in the cuisine industry may be changing. In light of the recent U.S. Supreme Court decision, *Star Athletica, L.L.C. v. Varsity Brands, Inc.* (discussed more fully below), culinary presentations now may be eligible for copyright protection, giving chefs a colorable claim against each other for copying culinary creations. Additionally, because diners now have access to smart phone cameras and social media platforms, chefs' blatant copying of other chefs' culinary presentations, techniques, or recipes are much more easily detected, such as the incident with the Presidential Inaugural Cake recounted at the beginning of this chapter.

This chapter explores culinary presentations and copyright law. In the past, articles have explored intellectual property law in recipes, or surveyed intellectual property law in food plating, including trade dress,

---

Dufresne, and Joe Andres: Sincerest Form, eGForums: a service of the soc'y of culinary arts and letters (March 20, 2006, 12:44 AM), at http://forums.egullet.org/topic/84800-sincerest-form/.

[5] Ferran Adria, Heston Bulmenthal, Thomas Keller and Harold McGee, Statement on the New Cookery, *The Guardian* (December 9, 2006), https://www.theguardian.com/uk/2006/dec/10/foodanddrink.obsfoodmonthly.

[6] Christopher J. Buccafusco, On the Legal Consequences of Sauces: Should Thomas Keller's Recipes be Per Se Copyrightable?, 24 Cardozo Arts & Ent. L. J. 1121, 1152 (2007) (interviewing a number of high-profile chefs: Thomas Keller: 'We're in the hospitality industry . . . We're innately hospitable, so why wouldn't you want to share? . . . I share my restaurant [and] my food'; Van Aken: 'Most chefs are sharing and caring individuals that tolerate quite a bit' and 'I write cook books and teach classes so folks will use my recipes').

[7] Buccafusco, supra note 6, at 1152 (Thomas Keller acknowledging that chefs are inspired by past creations: 'Look at the [salmon] cornets for example. Where did it really come from? . . . Did I really invent it? Did I create it? Or was it an inspiration from an ice cream cone that I just looked at differently? . . . Do I have the right to say that this is mine and nobody else's? I don't know . . .').

copyright, and design patents.[8] This chapter's focus is narrower. It will specifically examine copyright protection of culinary presentations, which are the creative designs, plating, or presentations of food to be eaten, with an up-to-date analysis of this issue in light of the recent U.S. Supreme Court's decision on copyright of useful articles in *Star Athletica, L.L.C. v. Varsity Brands, Inc.*[9] To explore copyright in culinary presentations and illustrate the copyright hurdles for culinary presentations, this chapter uses examples of three different styles of culinary presentations as case studies: Goldman's presidential inaugural cake, Thomas Keller's famous Salmon Cornets, and a traditional bowl of Vietnamese Pho. These three culinary presentations are each described in more detail below. As the analyses below show, each of these culinary presentations face unique challenges under copyright law.

2. THE THREE CULINARY PRESENTATIONS

To illustrate copyright law's limitation on cuisine, this chapter analyzes three distinct categories of culinary presentations. The first is cake design. Gone are the simple cylindrical baked doughs covered in icing – cakes nowadays can be created to look like anything. There are cakes designed to look like fish tanks, cakes designed to look like buildings, cakes designed to look like animals, ships, flowers, footballs, and even characters from movies. Cake designs have become so creative and have so thoroughly captured the imagination of the public that there are more than a handful of reality television shows dedicated to cake design alone.[10] The cake design this chapter focuses on is Goldman's cake

---

[8] See, e.g., Malla Pollack, Intellectual Property Protection for the Creative Chef, or How to Copyright a Cake: A Modest Proposal, 12 Cardozo L. Rev. 1477 (1991); Buccafusco, supra note 6; J. Austin Broussard, An Intellectual Property Food Fight: Why Copyright Law Should Embrace Culinary Innovation, 10 Vand. J. Ent. & Tech. L. 691 (2008); Emmanuelle Fauchart and Eric von Hippel, Norms-based Intellectual Property Systems: The Case of French Chefs, MIT Sloan Research Paper No. 4576-06 (2006); Emily Cunningham, Protecting Cuisine Under the Rubric of Intellectual Property Law: Should the Law Play a Bigger Role in the Kitchen?, 9 J. High Tech. L. 21 (2009); Cathay Smith, Food Art: Protecting Food Presentation Under U.S. Intellectual Property Law, 14 J. Marshall Rev. Intell. Prop. L. 1 (2014).

[9] 137 S.Ct. 1002 (2017).

[10] See, e.g., *Ultimate Cake Off* on TLC, *Cake Boss* on TLC, *Cupcake Wars* on Food Network, *Ace of Cakes* on Food Network, *Amazing Wedding Cakes* on WE, *DC Cupcakes* on TLC, and *Last Cake Standing* on Food Network.

created for President Obama's inauguration (hereinafter, 'Goldman's Cake'). Goldman's Cake is designed with nine cylindrical tiers in a patriotic red, white, and blue theme. The bottom tier of the cake is the widest, and is decorated in red vertical stripes on a metallic background. The second tier is white and decorated with gold-leaf applique-like frosting, red, white, and blue bunting, and a circular seal of President of the United States. The third tier is thin, and created to look like a baby-blue cake stand emblazoned with silver stars. The cake has a thick fourth tier in sky-blue decorated with five silver coins engraved with military seals. Sitting on the fifth tier, which is another baby-blue, silver-star cake stand, is a thin white-silver sparkling tier, and a blue, unadorned, wider tier of cake. Two navy-blue tiers balance at the top of the cake adorned with sparkling silver five-point cut-out stars. These silver stars are secured on the cake by thin stems sticking out of the top two tiers to look like the stars are floating above the cake.

The second type of culinary presentation this chapter will analyze is artistic cuisine created by chefs for diners at celebrity or avant-garde restaurants.[11] These chefs design artistic dishes with elaborate color and texture combinations and creative layering and arrangement of ingredients.[12] These dishes are then prepared for diners with each ingredient placed on the plate with artistic precision and perfection. One of celebrity chef Thomas Keller's (The French Laundry, Per Se, and the Bouchan Franchise) most famous dishes is the salmon cornet (hereinafter, 'Keller's Salmon Cornets'). Keller's Salmon Cornets consist of a cone-shaped black sesame tuile filled with a pink scoop of salmon roe and sweet red onion crème fraiche, topped with salmon tartare.[13] The cones are typically served in a standing rack, where each cone is fitted – pointy-side down – into perfectly-shaped holes to allow the cones to remain free-standing for serving.[14] Keller's Salmon Cornets, or at least knock-offs of them, have become a ubiquitous appetizer served at weddings or corporate functions.

Finally, the third type of culinary presentation this chapter will consider is a traditional food preparation – specifically, a bowl of Vietnamese Pho noodles ('Pho'). Pho is traditionally served in a large cylindrical bowl, and consists of off-white rice noodles and thinly sliced meat and meat balls served in light brown broth. The bowl of noodles is typically served with a

---

[11] Traditionally defined by large plates with artistically styled miniscule amounts of food.

[12] Smith, supra note 8, at 5.

[13] See Smith, supra note 8, at 6–7.

[14] Ibid.

side plate of herbs and spices, such as basil leaves, bean sprouts, cilantro, and lime for diners to add to their noodles and broth. Although Pho originated in Vietnam, it has become quite popular outside of Vietnam, and is now served at trendy restaurants such as Pho Soho in London.

3. ARE CULINARY PRESENTATIONS PROTECTED UNDER COPYRIGHT LAW?

To determine whether copyright law would protect the presentation of Goldman's Cake, Keller's Salmon Cornets, or Pho, we must first look at what copyright law protects and what it means for a work to be protected by copyright. In the United States (U.S.), copyright law protects original works of authorship fixed in any tangible medium of expression.[15] Specifically, to be eligible for copyright, a work must be original, it must be a work of authorship, and it must be fixed. There is no requirement to register the work with the U.S. Copyright Office or to provide any notice of copyright (©) for protection. Once a piece of work is fixed, it is automatically protected under copyright law for the author's life plus 70 years. Even though it is easy for expressive works to qualify for copyright, copyright law does preclude certain works from protection. For instance, copyright law does not protect useful articles, which are 'object[s] having . . . intrinsic utilitarian function[s] that [are] not merely to portray the appearance of the article or to convey information', such as bicycle racks, clothing, furniture, or dinnerware.[16] In spite of this preclusion, copyright law may still protect the pictorial, graphic, or sculptural features on a useful article if those features may be 'identified separately from, and . . . exist independently of, the useful article'.[17] Once an author's work is protected under copyright law, that author may prevent anyone else from reproducing the work, preparing derivatives of the work, publicly distributing copies of the work, or displaying the work publicly. The sections below explore how U.S. copyright law treats culinary presentations, and specifically analyzes whether Goldman's Cake, Keller's Salmon Cornets, and Pho may qualify as copyrightable works.

---

[15] 17 U.S.C. § 102.

[16] Definition of Useful Articles, U.S. Copyright Office, https://www.copyright.gov/register/va-useful.html (last visited September 30, 2017).

[17] Ibid.

3.1 The Work of Authorship Requirement

As a preliminary matter, to be eligible for copyright protection, a work must be a work of authorship. Copyright law does not protect ideas or concepts, nor does it protect inventions. Examples of works of authorship include literary works, musical works, dramatic works, pantomimes and choreographic works, pictorial, graphic, and sculptural works, motion pictures and other audio visual works, sound recordings, and architectural works.[18] Culinary presentations, technically, would qualify as sculptural works, which embody three-dimensional works of fine, graphic, and applied art. The Copyright Act does not dictate the medium necessary to create a sculptural work. Therefore, a sculpture carved from stone is not more entitled to be a sculpture than one carved from zucchini. Similarly, a sculpture created with cake and frosting is no less a sculpture than one created with ceramic. Indeed, several scholars have argued that chefs do, in fact, create legitimate 'art' with their design of culinary presentations where those presentations incorporate 'patterns of harmonious or contrasting flavors, textures, colors and plating arrangements that are intended to stimulate [diners'] aesthetic sense', just like traditional art.[19] And like museum visitors who view traditional artworks, diners in turn enjoy their meals by 'act[ing] as art critics when they contemplate their dishes and appreciate them as visual and flavorful expressions of art'.[20] In fact, the U.S. Supreme Court has granted certiorari to hear a legal case on whether cake design is artistic expression, entitling a baker to claim that he has the freedom to create or not create custom cake designs for certain customers.[21] Regardless of whether culinary presentations could be considered art protected under the freedom of expression, culinary presentations have no problem meeting copyright law's work of authorship requirement as a sculptural work.

---

[18] 17 U.S.C. § 102.

[19] Broussard, supra note 8, at 718; see also Pollack, supra note 8; Elizabeth Telfer, Food as Art. In *Food for Thought: Philosophy and Food* (Routledge, 1996) 41–60; Jacquelyn Strycker, From Palate to Palette: Can Food be Art?, Createquity (January 7, 2013), http://createquity.com/2013/01/from-palate-to-palette-can-food-be-art/; Jason Farago, Chef Ferran Adria and the Problem of Calling Food Art, BBC Online (January 14, 2014) (The Drawing Center in Soho exhibited the work of Ferran Adria, chef of El Bulli).

[20] See Broussard, supra note 8, at 718.

[21] See Adam Liptak, Cake is His 'Art.' So Can He Deny One to a Gay Couple?, *New York Times* (September 16, 2017), https://www.nytimes.com/2017/09/16/us/supreme-court-baker-same-sex-marriage.html.

3.2 The Originality Requirement

A possible hurdle to protecting culinary presentations under copyright law is the originality requirement. Copyright law only protects original works of authorship, which means the work must be independently created and exhibit at least a modicum of creativity.[22] Even though the standard for originality is low, it does exist.[23]

Some culinary presentations may have difficulty meeting the originality standard because of the culinary industry's norms and ethos of sharing, re-using, and borrowing, which traditionally encouraged chefs to use other chefs' creations and even build upon them.[24] Many of the culinary creations chefs create today are based on or at least inspired by past creations. Because these culinary creations are often based on past creations or at least borrow elements from other chefs' creations, they may not be considered 'independently' created. For instance, when asked about his famous Salmon Cornets, Keller explained 'Look at the [salmon] cornets for example. Where did it really come from? . . . Did I really invent it? Did I create it? Or was it inspiration from an ice cream cone that I just looked at differently?'[25] Cake designs also borrow from past designs. For instance, the multi-tiered cylindrical cake has been around since medieval England, where small spiced buns of cake were stacked on top of each other, creating a towering pile of cake, challenging a new bride and groom to kiss over the tiers.[26]

The most obvious example of a culinary presentation that would likely fail to meet copyright's originality standard is the Pho. The current iteration of Pho – light brown broth served with off-white rice noodles, thinly sliced meat and meatballs, basil and garnishes – can be traced back to the late 1880s when the French colonized Vietnam.[27] Even though Pho has since evolved, its basic ingredients are typically retained.[28] Pho

---

[22] 17 U.S.C. § 102.

[23] *Feist Publications, Inc. v. Rural Telephone Service Co., Inc.*, 499 U.S. 340, 362 (1991).

[24] See Smith, supra note 8, at 23–4; Buccafusco, supra note 6, at 1151–6 (discussing the 'culture of sharing and non-legal norms' by chefs).

[25] Buccafusco, supra note 6, at 1152.

[26] Carol Wilson, Wedding Cake: A Slice of History, 5 Gastronomica: The J. of Critical Food Studies 2 (May 5, 2005), https://gastronomica.org/2005/05/05/wedding-cake-a-slice-history/.

[27] The History and Evolution of Pho: A Hundred Years' Journey, LovingPho.com (April 8, 2009), http://www.lovingpho.com/pho-opinion-editorial/history-and-evolution-of-vietnamese-pho/.

[28] Ibid.

served at trendy Pho Soho, for instance, is described as 'an aromatic and delicious rice noodle soup served with a side plate of fresh herbs' in a bowl with thinly sliced meat and beef meatballs.[29] Chefs may attempt to personalize their own offerings of Pho, such as by including novel ingredients like sliced peppers, roasted garlic, or mushrooms, but these incremental changes to the underlying Pho presentation are not likely to be independently copyrightable or create new copyrightable derivatives. Indeed, one of the few cases in the U.S. to analyze culinary presentations and copyright law was *Kim Seng Company v. J&A Importers, Inc.*, which found a traditional bowl of Asian noodles to lack originality and the minimal creativity threshold for copyright eligibility.[30] In *Kim Seng Company*, the plaintiff hired an employee to arrange a bowl of Asian noodles for a photograph that was placed on the plaintiff's packaging materials. The employee created a bowl of noodles topped with egg rolls, grilled meat, and other garnishes.[31] To create the perfect presentation, the employee 'chose the foods out of thousands of possibilities, and directed their arrangement to be in a certain fashion out of infinite possibilities'.[32] When defendant, a competing Chinese-Vietnamese food supply company, used a photograph of an almost identical bowl of Asian noodles on the packaging of its own product, the plaintiff sued the defendant for infringing the plaintiff's copyright in the underlying bowl of food 'sculpture'.[33] The court in *Kim Seng Company* was not persuaded that the combination of a common bowl with the contents of a common Asian dish met the originality requirement of copyright.[34] In fact, the court in *Kim Seng Company* went further to explain that, 'regardless of which angle, quantity, or positioning of the various food items [the plaintiff] utilized, the unprotectable nature of the ingredients indicates a lack of originality'.[35]

Even if the individual elements in a culinary presentation – such as a common bowl, an ice cream cone, or a multi-tiered cake – may not satisfy the originality requirement for copyright, the original selection, arrangement, and combination of unprotectable elements could be

---

[29] Pho Soho Menu, http://www.phocafe.co.uk/wp-content/uploads/2017/06/Pho_Menu_Main_web.pdf (last visited September 26, 2017).

[30] 810 F. Supp.2d 1046, 1053-54 (C.D. Cal. 2011).

[31] Ibid. at 1050.

[32] Ibid. at 1053.

[33] Ibid.

[34] Ibid.

[35] Ibid. at 1053.

eligible for copyright protection as a compilation.[36] In other words, if a chef selects and arranges the ingredients or elements of her cuisine in an original and creative way, that original and creative arrangement may be eligible for copyright protection as a compilation, even if the chef cannot prevent others from using the underlying ingredients or elements in their dishes. For instance, although Keller's Salmon Cornets may have been inspired by a common ice cream cone, his creative selection of the tuile cone combined with a scoop of salmon roe and sweet red onion crème fraiche topped with salmon tartare arranged on a standing rack may make the Salmon Cornets original enough that their compilation constitutes an original work of authorship. Keller would not be able to prevent anyone else from offering salmon tartare served in a bowl with crackers to dip or spread individually on crackers or even served in circular tuile cups. He may, however, be able to prevent another chef from copying his unique and original compilation under copyright law. Similarly, individual features of Goldman's Cake – such as the cylindrical tiers, the floating stars, the colorful stripes or bunting – may not be protectable under copyright law because they do not meet the originality standard. Goldman's unique arrangement of those unprotectable features could, as a whole, make his cake eligible for copyright protection as a compilation. On the other hand, if the selection, coordination, and arrangement of a culinary presentation is entirely typical, then it would not satisfy the originality standard. For instance, Pho is traditionally defined by thin off-white noodles in light-brown broth, topped by thinly sliced meat, meat balls, and herbal garnishes. These ingredients are almost always served and displayed in the same way. Regardless of the position of the ingredients, or the angle at which each ingredient is placed, there are only so many ways to arrange the same ingredients in a bowl, so that none of those ways would likely be original enough to constitute an original work of authorship.[37]

---

[36] *Satava v. Lowry*, 323 F.3d 805, 811 (9th Cir. 2003); see also *Feist Publication, Inc.*, 499 U.S. at 362 (analyzing whether Rural selected, coordinated, or arranged uncopyrightable elements in an original way); *Belford v. Scribner*, 144 U.S. 488 (1892) (finding original compilations of unprotected recipes in *Common Sense in the Household: A Manual of Practical Housewifery*, by Marion Harland to be protected under copyright law).

[37] The 'merger doctrine' holds that 'if an idea and the expression of the idea are so tied together that the idea and its expression are one – there is only one conceivable way or a drastically limited number of ways to express and embody the idea in a work – then the expression of the idea is uncopyrightable because ideas may not be copyrighted'. Michael D. Murray, Copyright, Originality, and the End of the Scènes à Faire and Merger Doctrines for Visual Works, 58 Baylor

3.3 The Fixation Requirement

Copyright law requires a work to be fixed in a tangible medium to be eligible for protection.[38] Fixed in a tangible medium means that the work is 'sufficiently permanent or stable to permit it to be perceived, reproduced, or otherwise communicated for a period of more than transitory duration'.[39] The U.S. Copyright Act ('Act') does not dictate the medium in which the work must be fixed, in fact, 'any tangible medium of expression, now known or later developed', may satisfy the fixation requirement.[40] This broad allowance helps to 'avoid the artificial and largely unjustifiable distinctions . . . under which statutory copyrightability in certain cases [was] made to depend upon the form or medium in which the work is fixed'.[41]

One court has challenged food's ability to be considered fixed in a tangible medium. Specifically, in *Kim Seng Company v. J&A Importers*, the court found the bowl of Asian noodles, which the plaintiff's employee created with perishable food, to lack fixation for purposes of copyright law.[42] In coming to its decision, the court relied on the U.S. Court of Appeals for the Seventh Circuit's decision in *Kelley v. Chicago Park District*.[43] In *Kelley*, the Seventh Circuit held that Kelley's 'artistically arranged garden' did not meet copyright law's fixation requirement because it was 'not stable or permanent enough' to be a work of fixed authorship.[44] Analogizing the living garden in *Kelley* to the perishable bowl of food in *Kim Seng Company*, the court stated, '[l]ike a garden, which is "inherently changeable," a bowl of perishable food will, by its terms, ultimately perish. Indeed, if the fact that the Wildflower Works garden reviving itself each year was not sufficient to establish its fixed nature, a bowl of food which,

---

L. Rev. 779 (2006); but see *Kim Seng Co.*, 810 F.Supp.2d at 1055 (acknowledging that the employee who selected and arranged the ingredients in the original bowl of food sculpture may have 'demonstrate[d] sufficient creativity to surpass the low threshold established in *Feist*', but denying copyright protection to the bowl-of-food sculpture because of failure to meet the originality and fixation requirements).

38 17 U.S.C. § 102.

39 17 U.S.C. § 101.

40 2 *Patry on Copyright* § 3:22.

41 Ibid. (quoting *Matthew Bender & Co., Inc. v. West Pub. Co.*, 158 F.3d 693, 703 n. 18; H.R. Rep. No. 94-1476, at 52 (1976), reprinted in 1976 U.S.C.C.A.N. 5659, 5665).

42 *Kim Seng Co.*, 810 F. Supp.2d at 1053–5.

43 Ibid. at 1054; *Kelley v. Chicago Park District*, 635 F.3d 290 (7th Cir. 2011).

44 *Kim Seng Co.*, 810 F. Supp.2d. at 1054.

once it spoils is gone forever, cannot be considered 'fixed' for the purpose of [the Act]'.[45] This seemed to create a new standard for copyright law's fixation requirement – that in order to be fixed, the work must not be perishable. This new standard, however, is not in the Act, nor has it been applied by many other courts. In fact, the *Kim Seng Company* court's requirement that the work not be perishable seems to directly contradict the Act's definition of 'fixed' in a tangible medium, which merely requires that the work be 'sufficiently permanent or stable . . . for a period of more than transitory duration'.[46] The Act does not require the work to be permanent, or sufficiently permanent for the duration of the copyright, or even sufficiently stable for a full day (or two days if refrigerated). The *Kim Seng Company* court's decision also seems to carve out one type of medium – perishable materials – from being considered a 'tangible medium' for copyright eligibility. This also contradicts the Act's broad acceptance of 'any tangible medium of expression, now known or later developed, which can be perceived, reproduced, or otherwise communicated' to satisfy fixation.[47] Not surprisingly, scholars have criticized the *Kim Seng Company* decision. For instance, Zahr Said criticizes the court's requirement for permanent fixation because, contrary to the court's reasoning, 'the fixation requirement does not require permanence, or even that a fixed work last very long'.[48] Megan Carpenter finds the *Kim Seng Company* decision and court's reasoning inapt because 'copyright protection does not degrade in conjunction with the degradation of its subject works'; in other words, even where a work is later destroyed, it is still subject to copyright protection and does not fall into the public domain merely because the physical embodiment is destroyed.[49]

On the other hand, there are complications with finding works to be fixed for copyright purposes when they are designed to be destroyed. For instance, artist Felix Gonzalez-Torres's work, *Untitled (Portrait of Ross in L.A.)*, which has been exhibited in The Art Institute of Chicago and most recently at the Met Breuer, consists of a 175 pound pile of candy on the ground, from which museum visitors are encouraged to take pieces.[50] Torres's work symbolizes the process by which the artist's late partner

---

[45] Ibid. at 1054.
[46] 17 USC § 101.
[47] 17 USC § 102.
[48] Zahr K. Said, Copyright's Illogical Exclusion of Conceptual Art, 39 Colum. J.L. & Arts 335, 344 (2016).
[49] Megan Carpenter, If It's Broke, Fix It: Fixing Fixation, 39 Colum. J.L. & Arts, 355, 360 (2016).
[50] Stephanie Eckardt, The New Met Breuer Wants You to Take Candies,

slowly faded away and died from complications from AIDS.[51] Torres's work could, theoretically, be considered one large culinary presentation. But is Torres's work fixed in a tangible medium? If so, when is it fixed? When the pile of candy is first displayed in a museum? When the first museum patron takes a piece? When the pile of candy is half-way gone? All of the above? Culinary presentations, which are created to be eaten or are destined to perish, suffer the same fixation questions. Nevertheless, in spite of the contrary case law on the issue, most culinary presentations – unless the presentation incorporates an unstable or impermanent performance – are fixed for at least a period more than transitory duration when they are first created, and would technically be considered fixed under current copyright standards.

3.4 Useful Articles Exception and Separability: Pre-*Star Atheltica, L.L.C. v. Varsity Brands, Inc.*

The most difficult hurdle to protect culinary presentations under copyright law in the past has been the useful articles exception. Copyright law in the U.S. does not protect 'useful articles', which are articles 'having an intrinsic utilitarian function that is not merely to portray the appearance of the article or to convey information'.[52] Even though chefs may create art through their culinary presentations, the food they serve ultimately serves a utilitarian purpose. Some authors have claimed that this utilitarian purpose is caloric intake.[53] This chapter, however, takes a broader view of the utilitarian purpose of culinary creations – which is to be edible and eaten. In either case, whether the utilitarian purpose of culinary presentations is to be edible and eaten or purely for caloric intake, there is no dispute that culinary presentations are considered useful articles.

Useful articles themselves are not protected by copyright law, but pictorial, graphic, or sculptural features ('design features') that can be 'identified separately from, and that can exist independently of, the utilitarian article' may be eligible for copyright protection.[54] For instance, in the U.S. Supreme Court's decision *Mazer v. Stein*, the plaintiff designed

---

Not Photos, *W Magazine* (March 13, 2016), https://www.wmagazine.com/story/felix-gonzalez-torres-candy-the-met-breuer.

[51] Ibid.

[52] 17 USC § 101.

[53] Buccafusco, supra note 6, at 1139 (finding food's utilitarian purpose to fulfill the 'basic need to provide calories').

[54] U.S. Copyright Office, supra note 16.

and created sculptures of dancing figures out of clay to use as table lamp bases. The plaintiff used these sculptures as molds to create ceramic table lamp bases by adding a lamp shade, electric wiring, and sockets to the sculptures.[55] The plaintiff sold the sculptures as lamp bases and separately and independently as sculptures.[56] The Court found the dancing figure sculptures had the ability to be separable from their utilitarian function, and the separable sculptures individually met all of the copyright requirements. Therefore, because the sculptures were protected by copyright law, the plaintiff in that case was able to enjoin a competitor from creating lamps with the same sculptural base.

For a long time, the test to determine separability had been one of the oldest circuit splits in copyright law. Courts struggled to articulate clear doctrines to determine when a useful article may have design features that are separable from the underlying useful article and, therefore, copyrightable. For instance, courts in the past held that there were two ways of determining whether design features are separable – physical separability and conceptual separability. If a pictorial, graphic, or sculptural feature may be physically separated from the underlying utilitarian article, and that separated design feature meets all of the copyright requirements, it is eligible for copyright protection if it leaves the underlying utilitarian aspects of the article intact. The classic example of physical separability is a decorative hood ornament on a car. Cars are useful articles. A decorative hood ornament, which is a sculptural feature, may be physically removed from the car without jeopardizing the underlying utilitarian aspects of the car. If that separated hood ornament independently meets the standards of copyright, it may be protected under copyright law.

Even if the design features of a useful article could not be physically separated from the underlying utilitarian article, if they could be conceptually separated, those features were still eligible for copyright protection. The test to determine conceptual separability, however, was confusing. Not only were there significant splits between different jurisdictions on how to determine when design features may be conceptually separable from an underlying useful article, sometimes courts in the same jurisdiction would apply different standards. For instance, the U.S. Court of Appeals for the Second Circuit alone articulated three different tests to determine conceptual separability: one examined whether design features on a useful article were primary and utilitarian features subsidiary; another examined

---

55 *Mazer v. Stein*, 347 U.S. 201, 202 (1954).
56 Ibid. at 203.

whether the article 'stimulated in the mind of the beholder a concept that is separate from the concept evoked by its utilitarian function'; and a third case applied Robert Denicola's test, which examined whether the design features on the useful article were animated by functional considerations.[57] Other jurisdictions applied their own tests, including examining whether the useful article 'would still be marketable to some significant segment of the community simply because of its aesthetic qualities', or whether the design features of a useful article 'can stand alone as a work of art traditionally conceived'.[58]

Under the tests recounted above, design features in culinary presentations would not likely be physically or conceptually separable from their underlying utilitarian purpose. As a preliminary matter, culinary presentations do not typically have design features that are physically separable from the underlying useful article. Even if certain design features may be physically separable, they may not be independently copyrightable, nor would protection of those separated features protect the culinary presentation as a whole – which would be the goal for chefs or creators of culinary presentations. For instance, in a bowl of Pho, all of the ingredients in the culinary presentation are created and chosen for their edible qualities, and none are physically separable from that utilitarian function. The only aesthetic feature of Pho that may be physically separated is the bowl containing the ingredients, but the bowl itself would also be a useful article not subject to protection.[59] Even if the bowl itself may be protected under copyright law, that would not achieve the goal of protecting the culinary presentation. Similarly, all of the features that make up Keller's Salmon Cornets are also useful articles that are not separable from their underlying utility – which is to be edible and eaten. Even the inedible rack on which the salmon cornets are served is a useful article created to hold the cornets upright. Finally, certain aspects of Goldman's Cake may be physically separable from the underlying edible cake, such as the inedible floating cut-out stars pinned on the cake. Nevertheless, the silver-colored

---

57 *Kieselstein-Cord v. Accessories by Pearl, Inc.*, 632 F.2d 989, 993 (2d Cir. 1980); *Carol Barnhart Inc. v. Econ. Cover Corp.*, 773 F.2d 411, 422 (2d Cir. 1985); *Brandir Int'l, Inc. v. Cascade Pac. Lumber Co.*, 834 F.2d 1142, 1143 (2d Cir. 1987).

58 1 *Nimmer on Copyright* § 2.08[B][3], at 2-101 (2004); 1 Paul Goldstein, *Copyright* § 2.5.3, at 2:67 (1996).

59 *Star Athletica, LLC v. Varsity Brands, Inc.*, 137 S. Ct. 1002, 1010 (2017) ('Of course, to qualify as a pictorial, graphic, or sculptural work on its own, the feature cannot itself be a useful article or "[a]n article that is normally a part of a useful article" (which is itself considered a useful article)').

five-pointed stars would not be independently copyrightable because they lack originality.

Similarly, culinary presentations would not meet any of the conceptual separability tests. They would not meet any of the Second Circuit's tests: the design features on culinary presentations are likely subsidiary to the utilitarian features – the taste and edibleness of the ingredients; culinary presentations typically stimulate the utilitarian concepts of food and calories to consumers; and culinary presentations are necessarily animated by functional considerations – such as the flavors and tastes of the combination of ingredients.[60] They would not meet the other jurisdictions' tests either: if they are not edible, culinary presentations would not generally be 'marketable to some significant segment of the community'; and the design features of culinary presentations could not 'stand alone as a work of art traditionally conceived'. Indeed, the court in *Kim Seng Company* agreed that, at least in regards to a bowl of traditional Asian noodles, the sculptural features of the culinary presentation 'cannot be conceptually separated from their utilitarian function, which is to be eaten'.[61]

3.5 Useful Articles and Separability: Post-*Start Athletica, L.L.C v. Varsity Brands, Inc.*

Lower courts carried on with these various tests and standards to determine separability until March 22, 2017 when the U.S. Supreme Court issued its opinion in *Star Athletica, L.L.C. v. Varsity Brands, Inc.* The Court's opinion in *Star Athletica, L.L.C.* attempted 'to resolve widespread disagreement over the proper test' to determine separability by setting forth a two-part test for determining when design features of useful articles may be protected under copyright.[62] In its opinion, the Court first rejected the traditional distinction between 'physical' and 'conceptual' separability because, according to the Court, 'separability does not require

---

[60] Some commentators would disagree with this analysis, including Buccafusco, supra note 6, at 1138 (arguing that 'in many dishes, the caloric content of the food is secondary to the chef's creative expression. The primary purpose of the chef creating the dish or the diner consuming it may have very little to do with filling the belly' and that 'consumers are readily able to consider a dish's taste and expressiveness separately from any nutrition that they may derive from it'); Naomi Straus, Trade Dress Protection for Cuisine: Monetizing Creativity in a Low-IP Industry, 60 UCLA L. Rev. 182, 212 (2012) ('courts should easily be able to conceptually separate plating from the functional content of the dish, as plating does not reveal or affect (significantly) the flavors or caloric content of the dish').

[61] *Kim Seng Company*, 810 F.Supp.2d at 1053.

[62] *Star Athletica, L.L.C. v. Varsity Brands, Inc.*, 137 S.Ct. 1002, 1007 (2017).

the underlying useful article to remain'.[63] In other words, the Court rejected the requirement that in order to find separability, once the design features of an article are separated from the underlying useful article, the underlying article must retain its utility. Instead, the Court articulated a two-prong test to determine when a design feature incorporated into a useful article may be eligible for copyright protection:

> (1) if the [design] feature can be perceived as a two- or three-dimensional work of art separate from the useful article and (2) if the [design] feature would qualify as a protectable pictorial, graphic, or sculptural work – either on its own or fixed in some other tangible medium of expression – if it were imagined separately from the useful article into which it is incorporated.[64]

In *Star Athletica, L.L.C.*, the designer and marketer of cheerleading uniforms, Varsity Brands, acquired copyright registrations for certain geometric designs appearing on its cheerleading uniforms, including combinations and arrangements of chevrons, curves, stripes, angles, diagonals, coloring, and shapes.[65] A competitor, Start Athletica, copied five of Varsity Brands's designs on its own cheerleading uniforms and Varsity Brands sued for copyright infringement.[66] The Court found cheerleading uniforms to be useful articles. However, because the combinations and arrangement of the design features on the uniforms had pictorial and graphic qualities and, according to the Court, those features could be eligible for copyright protection if they were applied to another medium, the Court found the combinations and arrangements of the designs to be separately protected by copyright law. The Court's new test has not yet been applied to culinary presentations, but its application could result in an increase in protection of culinary presentations under copyright law, albeit on a case-by-case basis.

Specifically, to satisfy the Court's first prong of its test, a 'decision maker need only be able to look at the useful article and spot some two- or three-dimensional element that appears to have pictorial, graphic, or sculptural qualities'.[67] Take, for example, Keller's Salmon Cornets. Could a decision maker, looking at Keller's Salmon Cornets, spot some three-dimensional element that appears to have sculptural qualities? Certainly. To claim that there are no sculptural qualities in Keller's Salmon Cornets would likely run afoul of Justice Holmes's famous warning, that 'it would be a

---

63 Ibid. at 1014.
64 Ibid. at 1007.
65 Ibid.
66 Ibid.
67 Ibid. 1010.

dangerous undertaking for persons trained only to the law to constitute themselves final judges of the worth of [artistic merit]'.[68] This first prong of the Court's test, as the Court itself acknowledged, is 'not onerous' and is easy to meet.[69] In fact, it is so easily met that some argue that it does not create any hurdles for a creator to overcome. For instance, using Marcel Duchamp's 'readymades' art series as an example, in which Duchamp exhibited mass-produced, functional objects (such as an everyday wood and galvanized-iron shovel) as art, Justice Breyer argued in his dissenting opinion in *Star Athletica, L.L.C.* that 'virtually any industrial design can be thought of separately as a "work of art"', and there is nothing 'in the world that, viewed through an esthetic lens, cannot be seen as a good, bad, or indifferent work of art'.[70]

The second prong of the Court's test is more difficult to satisfy: that '[t]he decision maker must determine that the separately identified [design] feature has the capacity to exist apart from the utilitarian aspects of the article'.[71] According to the Court, in order to meet this element, the separately identified design feature must be eligible for copyright as a pictorial, graphic, or sculptural work on its own or when fixed in some other tangible medium.[72] In *Star Athletica, L.L.C.*, because the combination and arrangements of the designs on Varsity Brands's cheerleading uniforms could be eligible for copyright as pictorial or graphic works if they were applied to another medium, such as the cover of a notebook, the Court found those design features to be protected. While the test seems easy to apply to design features on cheerleading uniforms, attempting to apply the Court's logic and analysis in *Star Athletica, L.L.C.* to culinary presentations is a more difficult task. First, the Court's test seemed narrowly targeted to two-dimensional designs applied on the surface of useful articles, and did not necessarily help a decision-maker determine when three-dimensional sculptures could have separable design features. Of the three case studies examined in this chapter, the culinary presentation most likely to meet the Court's second-prong test by exhibiting design features that are separately eligible for copyright are the design features on Goldman's Cake. Even though the underlying stacked cylindrical

---

68 *Bleistein v. Donaldson Lithographing Co.*, 188 U.S. 239, 251 (1903).

69 *Star Athletica, L.L.C.*, 137 S.Ct. at 1010.

70 Ibid. at 1033–4; In Advance of the Broken Arm, MoMA Learning, https://www.moma.org/learn/moma_learning/marcel-duchamp-in-advance-of-the-broken-arm-august-1964-fourth-version-after-lost-original-of-november-1915 (last visited September 25, 2017).

71 *Star Athletica, L.L.C.*, 137 S.Ct. at 1010.

72 Ibid. at 1012.

layered cake in Goldman's Cake may not be protectable by copyright law because it is useful, the designs Goldman created and applied on the cake technically could be separable and protected, because those designs – if applied to another non-useful medium – could be eligible for copyright as pictorial or graphic works. For instance, if I were to decorate a cake by creating on it an impressionist painting out of edible frosting, the painting could, arguably, be eligible for copyright as a pictorial work on its own or if created on another medium such as a canvas. Indeed, cake artists argued in a recent brief that:

> a sufficiently original artistic design or structure on a cake would . . . be eligible for copyright protection as an artistic work if applied to a different medium of expression. . . . The fact that copyright law protects items like the uniform chevrons in Star Athletica as 'artistic works,' combined with the existence of copyrights on cake designs at a massive scale, demonstrate that cake-making is an artistic and expressive – rather than a purely mechanical or utilitarian – endeavor.[73]

There are two problems with that analysis. First, the designs on the face of the cake would necessarily be created with edible ingredients, such as frosting, so that the entire cake can be edible. Designs created with edible frosting serve both aesthetic and utilitarian purposes. Even though the painting would certainly be protected if it were created with oil or acrylic or even crayon, because the painting is created with edible frosting – the purpose of which is to be eaten – it may be considered a useful article itself requiring its own separability analysis. The Court specifically stated in *Star Athletica, L.L.C.* that, 'to qualify as a pictorial, graphic, or sculptural work on its own, the feature cannot itself be a useful article'.[74] Furthermore, the Court rejected the argument that one could claim a copyright in a useful article 'merely by creating a replica of that article in some other medium – for example, a cardboard model of a car'.[75] Therefore, according to the Court's logic, even if the impressionist painting could be recreated with oil or acrylic, the fact that it was originally created with edible frosting might technically prevent it from being protected under copyright law. Second, even if edible design features on a cake could overcome their own utility, those design features on the cake may still fail to meet copyright law's originality requirement.

---

[73] Brief for Cake Artists as Amici Curiae in Support of Neither Party, in *Masterpiece Cakeshop, Ltd. v. Colorado Civil Rights Commission*, available at 2017 WL 4004524, at *38–9 (2017).

[74] *Star Athletica, L.L.C.*, 137 S.Ct. at 1010.

[75] Ibid. at 1010.

Indeed, the specific design features on the Goldman Cake – stripes, bands of color, stars, coin-shapes, and bunting – have been used on many other forms of designs, even other cakes, and are not likely to meet copyright's originality requirement. Even if their color combinations and design arrangements may be creative enough to qualify as an original compilation, the combination of red, white, and blue, stars, stripes, and bunting may be considered *scènes à faire* for a patriotic-themed cake.[76] Similarly, the presidential seal, and the seals of the army, navy, marines, air force, and coast guard that also appear on Goldman's Cake are not independently created and, in fact, are designs that likely belong to the federal government.[77] Applying the second prong of the Court's separability test would be even more difficult for Keller's Salmon Cornets or traditional cuisine preparations, such as Pho. The Court in *Star Athletica, L.L.C.* did not discuss or provide any helpful examples on how to separate design features from useful articles that are, themselves, three-dimensional sculptural works. And, as discussed above, the Court explicitly rejected the replica approach to finding works copyrightable. It would likely be difficult for the Court to find any design features in Keller's Salmon Cornets or Pho to be separable from their underlying utilitarian articles.

In spite of these hurdles, however, since the Court issued its decision in *Star Athletica, L.L.C.*, the cuisine industry has renewed its interests in the possibility of protecting culinary presentations under copyright law, and lower courts may use the Court's reasoning and analysis in *Star Athletica, L.L.C.* to justify granting broader copyright protection to previously unprotected works, including culinary presentations.

4. SHOULD CULINARY PRESENTATIONS BE PROTECTED BY COPYRIGHT LAW?

The culinary industry has traditionally operated in a low-intellectual-property environment. It has long been established that recipes and cooking techniques are not subject matters protected by copyright law.

---

76 Under the doctrine of *scènes à faire*, 'courts will not protect a copyrighted work from infringement if the expression embodied in the work necessarily flows from a commonplace idea', *Ets-Hokin v. Skyy Spirits, Inc.*, 225 F.3d 1068, 1082 (9th Cir. 2000).

77 See, e.g., Baker Donelson, A Tale of Two Cakes: Can Copyright Law Protect this Cake Design?, Baker Donelson IP Watch (January 23, 2017), http://www.jdsupra.com/legalnews/a-tale-of-two-cakes-can-copyright-law-49760/.

Based on the discussion above, culinary presentations also face hurdles to copyright eligibility. Some may find this result inequitable. Many commentators and foodies might argue that chefs who create artistic culinary presentations are artists, and that their works transcend mere hospitality and should be contemplated as legitimate art worthy of intellectual property protection.[78] Chefs certainly spend time and effort coming up with new recipes and creative ways to present their cuisines to diners. It is indisputable that food creations are often just as creative, and sometimes even more creative today, than other works that are easily protected by copyright law such as photographs, paintings, or sculptures.

On the other hand, lowering the hurdles to allow culinary presentations to be protected by copyright law may harm – rather than help – the cuisine industry, chefs, and creators of food. In spite of not being able to protect recipes, techniques, or culinary presentations under copyright law, the cuisine industry has flourished. Projected sales for the restaurant industry in the U.S. are estimated to be $799 billion in 2017, representing 4% of the country's entire GDP.[79] New restaurants continue to open and chefs continue to create new, artistic (and edible) cuisines. As many scholars have explored in the recent past, and as indicative in the cuisine industry, intellectual property – as a formal legal entitlement – may not motivate or incentivize the creation of new artistic work.[80] If the cuisine industry is functioning and thriving in a low-intellectual-property environment, why provide additional copyrights where such rights not only do not motivate new creation but are also inherently exclusionary?[81] Furthermore, inserting ownership and exclusivity into the cuisine industry might even undermine the industry's traditional norms and ethos of sharing, and thereby decrease creative output.[82] As Buccafusco, Emmanuelle Fauchart, and Eric von Hippel have explored, chefs tend to embrace the hospitality and sharing ethos of the industry and sometimes even encourage others to copy

---

[78] See Smith, supra note 8, at 5; Jason Farago, Chef Ferran Adria and the Problem of Calling Food Art, BBC Online (Jan. 14, 2014), http://www.bbc.com/culture/story/20140114-can-food-be-art.

[79] Economic Engine, National Restaurant Association, http://www.restaurant.org/Industry-Impact/Employing-America/Economic-Engine (last visited September 25, 2017).

[80] See, e.g., Cathay Y.N. Smith, Street Art: An Analysis Under U.S. Intellectual Property Law and Intellectual Property's 'Negative Space' Theory, 24 DePaul J. Art, Tech. & Intell. Prop. L. 259, 292 (2014) (and sources cited therein).

[81] See, e.g., Kal Raustiala and Christopher Sprigman, Response. The Piracy Paradox Revisited, 61 Stan. L. Rev. 1201, 1225 (2009).

[82] See, e.g., Buccafusco, supra note 6, at 1156; Cunningham, supra note 8, at 38.

their works.[83] Copyright's exclusivity could undermine an industry built upon this ability to share, use, and reuse prior creations. For instance, if culinary cuisines can now be protected by copyright law, this could create uncertainty in the cuisine industry where chefs may not know whether or not another chef would one day enforce her copyright in her culinary presentations. This uncertainty could cause chefs to hesitate about using prior works and may slow the innovation of new cuisine creations. Chefs may also be less willing to share their recipes or creations with others in order to maintain exclusivity. Furthermore, instead of being motivated to innovate and create new techniques or cuisines in order to stay competitive, chefs may instead rely on their copyright-imposed exclusivity. The U.S. Supreme Court has acknowledged that copying is not always bad, and that it can have beneficial effects to society.[84] At least in the culinary industry, this certainly seems true, and the traditional lack of copyright protection for culinary presentations has likely fostered more sharing, innovation, and creativity in the industry, which ultimately benefited society.

5. CONCLUSION

There are many challenges to protecting culinary presentations under copyright law. These challenges include whether culinary presentations may qualify as sculptural works of authorship, whether presentations are independently created, and whether food may be fixed in a tangible medium. The most significant hurdle to the protection of culinary presentations under copyright law has traditionally been copyright law's exclusion of useful articles. However, the U.S. Supreme Court's recent decision in *Star Athletica, L.L.C.* changed the standard for determining when design features of a useful article may be eligible for copyright protection, and its effect on the culinary industry remains to be seen. It is possible that, after this decision, we might be seeing a change in the legal environment as well as sharing attitudes and norms in the cuisine industry and other industries that have traditionally operated in a low-intellectual-property environment.

---

83 Buccafusco, supra note 6, at 1156; Emmanuelle Fauchart and Eric von Hippel, Norms-Based Intellectual Property Systems: The Case of French Chefs (MIT Sloan School of Management, Working Paper No. 4576-06), https://papers.ssrn.com/sol3/papers.cfm?abstract_id=881781.

84 *TrafFix Devices*, 532 U.S. 23, 29–30 (2001).

PART II

Music and culture

7. Protecting traditional music under copyright (and choosing not to enforce it)

Luke McDonagh

1. INTRODUCTION

This chapter examines how music is defined and protected under copyright law, with a focus on traditional music, and outlines the reasons why, even if their music is protectable, traditional musicians might decide not to enforce copyright in their works. The particular jurisdiction I focus on is the UK, but the underlying principles of the chapter are relevant to copyright elsewhere, for example in the EU, the US, Ireland, Canada, Australia and New Zealand. This chapter complements the work of Mazziotti in this volume.[1]

The next part of this chapter outlines what traditional music is and why it poses a challenge to copyright. Section 3 explores how the musical work is defined within UK case law as well as in international conventions. Section 4 examines how musicians who add originality to the arrangement of a recorded tune or song may be entitled to copyright over that arrangement as a musical work in its own right.[2] Following on from this, the chapter argues that it is perfectly possible to protect traditional music under copyright (and to enforce it) – but there are cogent reasons why this might not be in the best interests of the process of music-making.[3] Moreover, while there is space for some use of innovative licensing, inspired by FOSS and Creative Commons, it should not be viewed as a panacea; instead, acknowledgement and maintenance of the social norms

---

[1] See Giuseppe Mazziotti, Chapter 8.

[2] Luke McDonagh, Is Creative Use of Musical Works without a Licence Acceptable under Copyright?, 43 Int'l Rev. Intell. Prop. & Competition L. (IIC) 401, 401–426 (2012).

[3] Olufunmilayo Arewa, From J.C. Bach to Hip Hop: Musical Borrowing, Copyright and Cultural Context, 84 N. C. L. Rev. 547 (2006).

within traditional music communities is of vital importance, rather than dramatic legal changes.

2. WHAT IS 'TRADITIONAL' MUSIC AND HOW IS IT RELEVANT TO COPYRIGHT LAW?

What we call 'traditional' or 'folk' music is generally thought of as old, ancient and pre-modern.[4] The terms 'folk' and 'traditional' music are sometimes used interchangeably, though it has been argued that the term 'traditional music' is preferable.[5] In this chapter I do not make a firm distinction between the two terms, but 'traditional' music will be used as much as possible, although 'folk' music will be used where related academic literature explicitly uses the term.

Historically, traditional music – whether European folk music,[6] African-American blues and jazz music,[7] or indigenous music in South America,[8] Asia[9] and Africa[10] – has been dependent on a process of person-to-person transmission, rather than a written, documented form such as in Western classical music. Today, however, many performers of traditional music use some form of documentation to aid their learning, usually recordings or transcriptions.[11] This illustrates one of the most important elements of traditional music: although music follows established patterns it is not set in stone; it is in a constant state of redefinition. Furthermore, although an established body of public domain

---

[4] Lillis O'Laoire, Fieldwork in Common Places: An Ethnographer's Experiences in Tory Island, 12 Ethnomusicology Forum 113 (2003).

[5] James Porter, *Traditional Music of Britain and Ireland: A Select Bibliography and Research Guide* (Garland, 1989).

[6] Jan Lind, *A History of European Folk Music* (University of Rochester Press, 1997).

[7] Tony Whyton, *Jazz Icons: Heroes, Myths and the Jazz Tradition* (CUP, 2010); see also Olufunmilayo Arewa, Copyright on Catfish Row: Musical Borrowing, Porgy and Bess, and Unfair Use, 37 Rutgers L. J. 277, 278–290 (2006).

[8] Malena Kuss, Performing Beliefs: Indigenous Peoples of South American, Central America and Mexico. In *Music in Latin America and the Caribbean: An Encyclopedic History* (University of Texas Press, 2010).

[9] William P. Malm, *Music Cultures of the Pacific, the Near East, and Asia* (Prentice Hall, 1996).

[10] J.H. Kwabena Nketia, *The Music of Africa* (W.W. Norton & Company, 1974).

[11] Jason Toynbee, Copyright, the Work and Phonographic Orality in Music, 15 Soc. and Legal Stud. 77, 79–99 (2006).

tunes/songs exists in many forms of traditional music, the creation of new compositions and arrangements plays a key role in maintaining the 'living tradition'.[12]

My focus in this chapter is on forms of secular traditional music in Europe and North America, especially Irish instrumental dance music (e.g. jigs and reels) and African-American blues and jazz music (e.g. the 12-bar blues progression).[13] What marks out these traditional forms of music from popular or 'pop industry' music is the style of composition and performance. Traditional songs and melodies tend to follow accepted patterns, often in line with specific dance steps. Jigs are usually comprised of two or three musical parts, played repeatedly in 6/8 or 9/8 time, while reels are played in 4/4.[14] 12-bar blues songs are often played in 4/4 or 3/4 and tend to feature a minor-based scale played over major chords.[15] New compositions that enter the tradition usually feature familiar rhythms and musical phrases. Where authorship is observable it is often viewed along a 'chain' of musicians who have added their own creativity and variations to an existing tune. If a musician performs or records an evocative variation of an old Irish reel it can lead to the musician's name becoming attached to the tune (e.g. 'Joe Cooley's Reel' – associated with the playing of the Irish accordion player Joe Cooley); similarly, a blues or jazz musician's adaptation of an existing song can become famous, and perhaps even definitive, within the tradition (e.g. 'Mississippi' John Hurt's version of 'Frankie and Albert').[16]

What makes traditional music an interesting topic for scholars of intellectual property is that the authorship and performance of traditional music does not correspond to the commonly accepted copyright

---

[12] McDonagh, supra note 2.

[13] Byron Dueck, Jazz Endings, Aesthetic Discourse, and Musical Publics, 33 Black Music Research Journal 91, 91–115 (2013).

[14] Brendán Breathnach, *Folk Music and Dances of Ireland* (Ossian, 1993) 57–61.

[15] Peter Spitzer, *Jazz Theory Handbook* (Mel Bay Publications, 2001) 63.

[16] Luke McDonagh, Does the Law of Copyright in the UK and Ireland Conflict with the Practices of Irish Traditional Musicians? A Study of the Impact of Law on a Traditional Music Network, 151–156 (PhD Thesis, School of Law, Queen Mary University, 2011), Roberta Freund Schwartz, Early Blues: The First Stars of Blues Guitar by Jas Obrecht (review), 55 American Studies 91, 91–92 (2016), and Bjorn Heile, Who Wrote Duke Ellington's Music? Authorship and Collective Creativity in 'Mood Indigo'. In Andreas Rahmatian, ed., *Concepts of Music and Copyright: How Much Music Perceives Itself and How Copyright Perceives Music* (Edward Elgar Publishing, 2015) 123, 123–141.

model within the music industry. Within that model, pop composers are usually easily identifiable as authors and owners of copyright works, whereas, in accordance with industry practice, performers have a different role; although they are entitled to performers' rights, they are not viewed as authors or owners of the underlying musical compositions.[17]

Traditional music is different. For one thing, in the case of older tunes and songs within a body of traditional music – many of which are in the public domain or are orphan works – it is often impossible to identify a single author or joint authors. Moreover, such songs and tunes can exist in several regional and local variants that have arisen from the creative arrangement of tunes by skilled performers (such as the Joe Cooley and Mississippi John Hurt examples cited above). Yet, new compositions by living authors can become accepted as part of the tradition over time. Thus, traditional music brings up many questions for copyright lawyers. I do not aim to answer all of them in a single chapter. What I intend to explore is whether traditional music can be protected under copyright; and, if it can be protected, what options do musicians have should they wish, or not, to enforce their copyrights? Finally, I consider the fact that the process of traditional music-making has parallels with the experiences of creators across a diverse range of non-conventional spaces, including free and open source software (FOSS), comedy, magic and graffiti.[18]

One important limitation of this chapter is that I do not intend to deal with the complex questions concerning the rights of indigenous or tribal communities in Asia, Africa and Latin America to traditional music that forms part of sacred traditions or religious ceremonies.[19] Nonetheless, the two are not mutually exclusive. Some of the problems faced by members

---

17 Richard Arnold, *Performers' Rights* (5th edn, Sweet & Maxwell, 2015).

18 Luke McDonagh, FOSS and Alternative Licensing in the United Kingdom: Assessing the Dual Importance of Contract Law and Copyright Law. In A. Metzger, ed., *Free and Open Source Software (FOSS) and Other Alternative Licensing Models: A Comparative Analysis of the Main Legal Issues* (Springer, 2016) 461, 461–476; Dotan Oliar and Christopher Sprigman, There's No Free Laugh (Anymore): The Emergence of Intellectual Property Norms and the Transformation of Stand-Up Comedy, 94 Va. L. Rev. 1787 (2008); Jacob Loshin, Secrets Revealed: Protecting Magicians' Intellectual Property without Law. In Christine A. Corcos, ed., *Law and Magic: A Collection of Essays* (Carolina Academic Press, 2010) 123, 140; and Marta Iljadica, *Copyright Beyond Law: Regulating Creating in the Graffiti Subculture* (Bloomsbury, 2016).

19 Karen Ralls-MacLeod and Graham Harvey, *Indigenous Religious Musics* (Routledge, 2000).

of traditional or indigenous communities are similar to the challenges faced by e.g. Irish traditional musicians or software programmers, as both make objections to individualist notions of copyright.[20] These objections are founded upon a belief that intellectual property law fails to adequately facilitate the processes of cultural innovation within their environment.[21] Given that this chapter evaluates these objections the analysis provided has relevance beyond its immediate focus.

3. DEFINING AND PROTECTING THE MUSICAL WORK UNDER COPYRIGHT

It is not the intention of this chapter to explore the history of copyright in great detail; expansive studies on this subject have been undertaken elsewhere.[22] However, a number of relevant points are worth noting regarding the beginning of the relationship between modern copyright law and music. Although the first 'letters patent' were issued to music publishers in England during the late 16th century,[23] at the time of the enactment of the Statute of Anne in 1710[24] music was 'not thought to be protected'.[25] This changed towards the latter half of the 1700s, when there was a 'shift in judicial understanding of the possible objects of property' to encompass

---

20 Kathy Bowery, Alternative Intellectual Property? Indigenous Protocols, Copyleft and New Juridifications of Customary Practices, 6 Macquarie Law Journal 65, 89–95 (2006).

21 Ibid.

22 See for example John Feather, The Book Trade in Politics: The Making of the Copyright Act of 1710, 8 Publishing History 19 (1980); Trevor Ross, Copyright and the Invention of Tradition, 26 Eighteenth-Century Studies 1 (1992); Mark Rose, *Authors and Owners: The Invention of Copyright* (Harvard University Press, 1993); Brad Sherman and Lionel Bently, *The Making of Modern Intellectual Property Law* (CUP, 1999); and Catherine Seville, *The Internationalisation of Copyright Law: Books, Buccaneers and the Black Flag* (CUP, 2006). See also, Justin Hughes, The Philosophy of Intellectual Property, 77 Georgetown Law Journal 287 (1988) and Jeremy Waldron, From Authors to Copiers, 68 Chicago-Kent Law Review 841 (1988).

23 Michael Carroll, Whose Music is it Anyway? How We Came to View Musical Expression as a Form of Property, 72 University of Cincinnati Law Review 1405, 1463 (2004).

24 Statute of Anne, 1710, http://avalon.law.yale.edu/18th_century/anne_1710.asp (last accessed November 24, 2017).

25 David Hunter, Musical Copyright in Britain to 1800, 67 Music and Letters 269, 274 (1986).

music.[26] In 1777, the seminal decision in *Bach v Longman*[27] established that musical compositions were covered by the Statute of Anne.[28]

Three conditions necessitated the application of copyright protection to music in England during the 18th century:[29] (i) the existence of printing technology; (ii) a general acceptance of the concept of 'intellectual property' in some form at a governmental level; and (iii) a rapidly expanding market for sheet music (and, consequently, a black market in 'unauthorised publication of musical works').[30] On this, Deazley notes that it was the economic rights of publishers that were of paramount importance to the legislature when the Statute of Anne was enacted – not the rights of authors or composers.[31]

Whatever the intent of legislators, the ability to enforce copyright over musical works had a tangible effect on the attitudes and practices of major classical composers, who began to take a more authorial view of their compositions during the 18th and 19th centuries.[32] Interestingly, due to the emphasis on oral learning/transmission and a general lack of commerciality, traditional music was relatively unaffected by this – the collective remained key, and individual musicians were rarely credited as authors/composers.[33] Sheet music was only of tangential interest in the traditional context, mainly to upper-class collectors of the music.[34] Thus, in the classical field, where music was commercially viable, composers increasingly saw themselves as authors; whereas the shift towards individual or 'Romantic' authorship did not occur in traditional music until the commercial 'folk boom' in the 1960s; and even then, collective

---

[26] Anne Barron, Copyright Law's Musical Work, 15 Social and Legal Studies 101, 106 (2006).

[27] *Bach v Longman* 98 ER [1777] 1274.

[28] John Small, J.C. Bach Goes to Law, 126 The Musical Times 526 (1985).

[29] David Hunter, Musical Copyright in Britain to 1800, 67 Music and Letters 269, 270–274 (1986).

[30] Ibid.

[31] Ronan Deazley, *On the Origin of the Right to Copy* (Bloomsbury, 2004) xix.

[32] Frederic M. Scherer, The Emergence of Musical Copyright in Europe from 1709 to 1850, 5 Review of Economic Research on Copyright Issues 3, 11 (2008).

[33] An exception in Irish traditional music is the harper Turlough O'Carolan, who was active in the late 17th and early 18th centuries and who blended elements of the folk tradition with contemporary classical themes –see Harry White, Carolan and the Dislocation of Music in Ireland, 4 Eighteenth-Century Ireland 55, 55–64 (1989).

[34] For a full account of this period see generally Matthew Gelbart, *The Invention of 'Folk Music' and 'Art Music': Emerging Categories from Ossian to Wagner* (CUP, 2007).

ideals remained strong in many areas of traditional music, and persist to this day.

3.1 International Conventions

In addition to UK common law and legislative developments, the development of music copyright in the centuries since *Bach v Longman* has been shaped by enactments at the international level.[35] Key here is the Berne Convention of 1886, which has been adopted by much of the global community and which provides an international framework for copyright in relation to musical works.[36] Many of its standards also form part of the 1994 TRIPS agreement.[37]

Berne protects the 'musical composition with or without words' and 'dramatico-musical works', but does not define 'music' or 'musical composition'.[38] Indeed, there are surprisingly few definitions of 'musical work' in national and international copyright law. In 1994 TRIPS merely adopted the terms of the Berne Convention and did not provide a further definition of musical work.[39] The WIPO World Copyright Treaty of 1996[40] does not attempt to define it. Similarly, the relevant US legislation does

---

[35] See Berne Convention for the Protection of Literary and Artistic Works, September 9, 1886, S Treaty Doc. 99-27, 1161 U.N.T.S. 3 (amended September 28, 1979) (hereinafter Berne Convention).

[36] Universal Copyright Convention (1952). This was enacted to provide international protection standards for countries that were unwilling to accept certain terms of the Berne Convention. Today the Universal Copyright Convention is less relevant due to the requirement that countries accede to the TRIPS agreement for WTO membership. Universal Copyright Convention, September 6, 1952, revised at Paris July 24, 1971, 25 U.S.T. 1341.

[37] Agreement on Trade-Related Aspects of Intellectual Property Rights, art. 27(2), April 15, 1994, Marrakesh Agreement Establishing the World Trade Organization, Annex 1C, 1869 U.N.T.S. 299 (hereinafter: TRIPS). However, the Berne Convention only covers literary and artistic works. The later 'Rome Convention' was enacted to provide protection for other rights such as rights over sound recording and performers' rights. See International Convention for the Protection of Performers, Producers of Phonograms and Broadcasting Organisations, October 26, 1961, 496 U.N.T.S. 43 (hereinafter Rome Convention).

[38] Article 2(1) Berne Convention, *op. cit.*

[39] Similarly, under TRIPS Articles 1–21 of the Berne Convention are adopted with no expansion of the definition of 'musical work', http://www.wto.org/english/docs_e/legal_e/27-trips_04_e.htm#1 (last accessed November 24, 2017).

[40] WIPO Copyright Treaty (1996) (hereafter referred to as WCT). This exists in compliance with Article 20 of the Berne Convention and it complies with the Berne definition of 'literary and artistic works'. The WCT was enacted primarily to address the issues surrounding copyright and digital technology, http://www.wipo.

not provide a definition.[41] What this indicates is that legislative bodies, both national and international, accept that the terms 'music' and 'musical work' are inherently difficult to define (and/or that it is not necessary to define the terms strictly for copyright purposes).[42] Interestingly, as explored in this chapter, the UK courts have, over time, developed a rich and useful definition.

3.2 Defining the Musical Work – UK Case Law

The Imperial Copyright Act of 1911 brought the Berne Convention standards into UK law (as well as to the rest of the Empire, becoming the foundational copyright statute for nations as diverse as India, Israel and Ireland).[43] It was followed by the Copyright Act of 1956.[44] Neither Act, however, provided a definition of 'musical work'. This changed in 1988 when the Copyright, Designs and Patents Act (CDPA) repealed the 1956 Act. Under the CDPA, a musical work is described as 'a work consisting of music, exclusive of any words or action intended to be sung, spoken or performed with the music'.[45] Nevertheless, the statute does not define what amounts to a piece of 'music'.[46]

---

int/treaties/en/ip/wct/trtdocs_wo033.html#P51_3806 (last accessed November 24, 2017).

[41] United States Copyright Act 1976 s 102(a)(2), http://www.law.cornell.edu/copyright/copyright.act.chapt1a.html#17usc102 (last accessed November 24, 2017).

[42] An important shift occurred when the Berne Convention was revised in 1908: it was decided that copyright should arise automatically, in other words that there should be no need for registration (though the US retained a registration system until the 1970s). Article 4, Convention for the Protection of Literary and Artistic Works (1908) (hereafter referred to as Berlin Act), http://en.wikisource.org/wiki/Convention_for_the_Protection_of_Literary_and_Artistic_Works_(Berlin_Act,_1908) (last accessed November 24, 2017). It was also agreed that the minimum term should be 50 years after the life of the author.

[43] Copyright Act, 1911 (hereafter referred to as CA 1911), http://www.opsi.gov.uk/RevisedStatutes/Acts/ukpga/1911/cukpga_19110046_en_1 (last accessed November 24, 2017).

[44] Copyright Act, 1956 (hereafter referred to as CA 1956), http://www.opsi.gov.uk/acts/acts1956/pdf/ukpga_19560074_en.pdf (last accessed November 24, 2017).

[45] CDPA s 3(1).

[46] Cheng Lim Saw, Protecting the Sound of Silence in 4'33" – A Timely Revisit of Basic Principles in Copyright Law, 27 European Intellectual Property Review 467, 469 (2005) and David Seymour, This is the Piece that Everyone Here has Come to Experience: The Challenges to Copyright of John Cage's 4'33", 33 Legal Studies 532, 537–548 (2013).

One advantage of the lack of a definition of 'music' within the CDPA is that it gives the courts leeway to take a broad interpretation of what is encompassed by a musical work.[47] The Court of Appeal decision in *Sawkins v Hyperion*[48] is the most recent, authoritative decision on the nature of the musical work in the UK. In *Sawkins*, the claimant successfully argued that he owned the copyright in performing editions that he had prepared of public domain works originally composed by Michel-Richard Lalande during the early modern period. Mummery L.J. stated that 'the essence of music is combining sounds for listening to'.[49] He further remarked:

> Music is not the same as mere noise. The sound of music is intended to produce effects of some kind on the listener's emotions and intellect. The sounds may be produced by an organised performance on instruments played from a musical score, though that is not essential for the existence of the music or of copyright in it . . . There is no reason why, for example, a recording of a person's spontaneous singing, whistling or humming or improvisations of sounds by a group of people with or without musical instruments should not be regarded as 'music' for copyright purposes.[50]

Clearly, this notion of music is not limited to harmony or melody. Mummery L.J. further stated that it would be incorrect to 'single out the notes as uniquely significant for copyright purposes and to proceed to deny copyright to the other elements that make some contribution to the sound of the music when performed, such as performing indications, tempo and performance practice indicators'.[51] Therefore, in the UK the musical work can be described as broad and flexible: it can encompass not only notes of music, but also other elements of musical practice and performance.[52]

---

47 Andreas Rahmatian, Music and Creativity as Perceived by Copyright Law, 3 Intellectual Property Quarterly 267, 268 (2005).

48 *Sawkins v. Hyperion Records Ltd* [2005] EWCA Civ 565; [2005] 1 WLR 3281.

49 Ibid. at para. 53.

50 Ibid.

51 Ibid. at paras 55–56.

52 A rare – though logical – limitation to this can be observed from the case of *Coffey v Warner* where it was held that a musical work cannot exist where it consists of mere 'extractions' from another work. Thus, to exist as a musical work in itself, a smaller work must be separable from a larger work – *Coffey v Warner/Chappell Music* [2005] FSR (34) 747. See also, Giuseppe Mazziotti, this volume, who points out that live performances/arrangements that are not fixed will not be protected. It is worth mentioning, however, that such unfixed arrangements are highly unlikely to be infringed, as potential infringers would have nothing to work with other than the original, unfixed, ephemeral performance.

The UK courts' open-minded view of music contrasts with the perspective of copyright taken by some scholars, who believe that the law imposes a necessarily 'Romantic' or 19th-century classical definition traced to the rise of individual authorship.[53] For example, Goehr describes copyright's musical work as 'a self-sufficiently formed unity, expressive in its synthesised form and content of a genius's idea'.[54] Yet, as Bently remarks, 'Mummery L.J.'s conception of the musical work' in *Sawkins* 'seems miles away from the image of the completed, notated score awaiting conversion by musical automatons – performers – into sounds appreciated by reverent, sedentary, passive audiences'.[55] Therefore, rather than viewing it as unduly formalist, it is more accurate to regard the musical work under UK copyright as a fluid concept, influenced by a number of aesthetic and abstract notions of musical work present in 'Romantic' and classical musicological literature, but not bound by them.[56] The reason this is a positive thing is that it is possible to envisage cases, e.g. involving 'avant-garde' music,[57] where the provision of a strict definition of musical work might end up creating problems for judges, who may be unable to fit an avant-garde work within a formalist definition.[58] By contrast, the UK courts' broad notion of musical work gives the courts room to manoeuvre.

In light of this, if there *are* challenges that arise from the application of copyright to traditional music under UK law, it can be said with confidence that that *these difficulties do not arise due to copyright's definition of musical work*. Nonetheless, as discussed further below, complexities can arise in relation to cases involving the distinction between two different types of musical work – the 'composition' and the 'arrangement' – as well as the twin notions of authorship and joint authorship.

---

53 Lionel Bently, Authorship of Popular Music in UK Copyright Law, 12 Information, Communication and Society 179, 184 (2009).

54 Lyndia Goeher, *The Imaginary Museum of Musical Works: An Essay in the Philosophy of Music* (OUP, 1992) 242.

55 Bently, supra note 53.

56 Anne Barron, Copyright Law's Musical Work, 15 Social and Legal Studies 101, 106–107 and 123–124 (2006).

57 Sam Ricketson and Jane Ginsburg, *International Copyright and Neighbouring Rights: The Berne Convention and Beyond* (2nd edn, OUP, 2005) 426.

58 Justine Pila, Copyright and its Categories of Original Works, 30 Oxford Journal of Legal Studies 229, 236–242 (2010).

4. THE DISTINCTION BETWEEN THE COMPOSITION AND THE ARRANGEMENT UNDER COPYRIGHT

Of particular importance to discussions of traditional music and copyright is the distinction between the copyright in the musical composition and the copyright in a subsequent 'arrangement' of that composition. Under UK law a separate copyright can subsist in an arrangement of a composition as long as the arrangement is sufficiently original[59] and the requisite originality comes from the arranger.[60] The owner of the original copyright in the composition is not the owner of the new arrangement copyright, which vests in the arranger.[61] In other words, copyright can recognise rights in multiple, original arrangements of the same composition.

Key is the notion of originality underpinning this idea. Although the Berne Convention does not expressly state that there is a requirement of 'originality',[62] there is 'a clear indication' that the notion of intellectual creation is 'implicit in the conception of a literary or artistic work'.[63] The same logic applies to a musical work. Crucially, whether one considers the traditional UK standard of 'skill, labour and judgment' or the CJEU standard of 'intellectual creation' it is clear that the threshold is not a burdensome one in the musical context: music performance and composition are inherently creative and thus have little problem satisfying this test.[64]

---

59 *Austin v Columbia* [1917–1923] MacG CC 398. See also *Robertson v Lewis* [1976] RPC 169; *Lover v Davidson* (1856) 1 CBNS 182, which involved accompaniment to an old air; and *Wood v Boosey* (1868) LR 3 QB 223, which involved an operatic pianoforte score.

60 *Redwood Music Ltd v Chappell & Co Ltd* [1982] RPC 109 (QBD). However, a straightforward 'cover' of a work may lack sufficient originality – see *Aston Barrett v Universal Island Rec. Ltd* [2006] EWHC 1009 (Ch).

61 This would usually be the case unless an alternative has been agreed between the two parties.

62 Ricketson and Ginsburg, supra note 57 402–403.

63 Article 2(5) Berne Convention, *op. cit.* See also Ricketson and Ginsburg, supra note 57 402–403, noting that the preparatory documents for the Brussels Revision Conference appear to acknowledge that the expression 'literary and artistic works' encompassed a notion of 'intellectual creation'.

64 *Infopaq International A/S v Danske Dagblades Forening* (C-5/08) [2009] ECR I-6569 (ECJ (4th Chamber)); [2009] ECDR 16 259; *Newspaper Licensing Agency v Meltwater* [2010] EWHC 3099 (Ch); and *SAS Institute Inc. v World Programming Ltd* [2013] EWHC 69 (Ch); [2013] RPC 17. See also Luke McDonagh, Headlines and Hyperlinks: UK Copyright Law Post-*Infopaq*, 1 Queen Mary Journal of Intellectual Property 184, 185–187 (2011); Jonathan

Virtually any creative contribution to an existing tune or song will be sufficient for the arrangement to be protected. However, the effective use of this new arrangement copyright would be subject to licensing requirements because the copyright in the new arrangement does not replace or nullify the copyright in the underlying work. This means that an arranger of a copyright work must obtain a licence from the owner of the underlying copyright work in order to release the new arrangement because the right to make 'adaptations' is one of the rights of the copyright owner.[65] The only exception to this is if the new arrangement is based on a public domain work; if this is the case then the arranger will own the copyright in the arrangement, with no requirement to obtain a licence (though copyright only covers the original elements of the new arrangement – the underlying work remains in the public domain).

The conflict between these rights – the right of adaptation and the automatic protection of new original arrangements – has long been acknowledged. At the time of Berne 'musical arrangements' were not protected as original works;[66] they were merely viewed as potentially 'unauthorised indirect appropriations of works' i.e. as examples of infringement.[67] However, following the Berlin Revision of the Berne Convention, arrangements were given protection under Article 2(3) as original works, 'without prejudice to the copyright in the original work'.[68] This debate over the terms of Berne underlines the fact that a new arrangement of a copyright composition can simultaneously be an 'original' work in its own right, and also an 'infringing' work with respect to the underlying copyright in the composition (unless it is properly licensed).

4.1 Authorship and Joint Authorship of Musical Arrangements

As noted earlier, an arranger can be the author of, and can own copyright in, an original arrangement of a public domain song/tune, though not the song/tune itself, which remains in the public domain. In theory, the arrangement copyright includes the right to object to 'sound-a-like'

Griffiths, Dematerialisation, Pragmatism and the European Copyright Revolution, 33 Oxford Journal of Legal Studies 767, 788–790 (2013); Jonathan Griffiths, Constitutionalising or Harmonising? The Court of Justice, the Right to Property and European Copyright Law, 38 European Law Review 65, 69–78 (2013).

[65] CDPA s 21.

[66] Ricketson and Ginsburg, supra note 57.

[67] Ricketson and Ginsburg, supra note 57 at 435.

[68] Article 2(3) Berne Convention, *op. cit.*

records which mimic the particular arrangement. However, such cases can be difficult to prove.[69]

There have been a number of UK music copyright cases where a particular arrangement of a composition has been the subject of a legal dispute. For example, in both *Godfrey v Lees*[70] and *Beckingham v Hodgens*[71] the disputes centred on the authorship of the specific copyright arrangements of existing songs. At the same time, courts have sometimes found it difficult to clarify the distinction between the underlying work and an arrangement, especially in joint authorship cases involving music. The nature of music performance as a group activity can lead to confusion over who is an author, and thus a copyright owner, and who is a mere performer, and thus not entitled to copyright protection.

In *Hadley v Kemp*,[72] a number of the members of the group Spandau Ballet took a copyright case against their fellow band member and principal songwriter, Gary Kemp. During the early 1980s, Kemp had written the lyrics, chords and basic melody to the song 'True', one of the group's biggest hits, and one of the works cited in the case. The key dispute over 'True' concerned its famous saxophone solo, which was performed on the recording by Steve Norman, a band member. The solo was significant – it lasted for 16 bars, approximately 9% of the song. Moreover, the melody of the solo was devised and improvised in the studio by Norman over the chords that Kemp presented to him. Despite Norman's creativity, the court took the view that Kemp was the 'author' and Norman was a mere 'performer'. Thus, it was held that the creation of the solo was not a 'significant and original contribution' to the work. Norman was left without a share in the copyright work.

As Arnold states, 'in assessing claims to co-authorship of musical works, the vital first step is for the court correctly to identify the work the subject of the claim to copyright and to distinguish it from any antecedent work'.[73] One of the reasons the *Hadley* judgment is problematic is that it is unclear as to whether the court considered the musical work, as composed and recorded in 'demo' form by Kemp, to be the same 'work' as the eventual version of 'True', as recorded by the entire band, or whether the eventual

---

69 *CBS Records Australia Ltd v Gross* (1989) 15 IPR 385 at 393. Further, see *ZYX Music GmbH v King* [1997] 2 All ER 129.

70 *Godfrey v Lees* [1995] EMLR 307 (Ch D).

71 *Beckingham v Hodgens* [2002] EWHC 2143 (Ch); [2002] EMLR 45.

72 *Hadley v Kemp* [1999] EMLR 589 (Ch D).

73 Richard Arnold, Reflections on 'The Triumph of Music': Copyrights and Performers' Rights in Music, 2 Intellectual Property Quarterly 153, 159 (2010).

band recording was an original 'arrangement' of Kemp's composition.[74] In addition, on the author/performer distinction, the onerous 'significant and original' requirement placed on the saxophonist went far beyond the ordinary standard of originality under copyright – another reason to doubt the validity of Park J.'s decision.[75]

Thankfully, the more recent case of *Fisher v Brooker*[76] – a case with similar facts to *Hadley* – provided clarity regarding how to assess musical arrangements in the context of joint authorship. In *Fisher*, authorship of the famous 1960s song 'A Whiter Shade of Pale' was disputed. Gary Brooker had always been credited with the copyright in the musical work because he wrote the chords and melody of the song. According to Brooker, he recorded this as a bare 'demo', and presented it in the studio to the other band members, who then performed on the final recorded and released work – 'A Whiter Shade of Pale'. The song is famous for its organ instrumental sections, which were not part of the song at the time the demo was recorded, but were instead improvised by band member Matthew Fisher during the performance and recording process. Therefore, as in *Hadley*, the instrumental sections in question were created by a band member (Fisher) in response and counterpoint to a chord structure devised by the main songwriter of the group (Brooker).[77]

Unfortunately, it its analysis the High Court in *Fisher* did not properly disentangle the composition, as represented by the demo, from the arrangement, as represented by the final recorded work. Thankfully, this was rectified by the Court of Appeal, which maintained a clear distinction between the demo and the final arrangement, and ultimately upheld the claim of Fisher that he ought to be viewed as a joint author and owner of the eventual recorded and arranged work known as 'A Whiter Shade of Pale'.[78]

The principle that *Fisher* highlights is that 'it will often be the case that

---

74 Ibid.

75 Luke McDonagh, Rearranging the Roles of the Performer and the Composer in the Music Industry – the Potential Significance of Fisher v Brooker, 1 Intellectual Property Quarterly 64, 66–70 (2012).

76 *Fisher v Brooker* [2006] EWHC 3239 (Ch); [2007] EMLR 9, on appeal to the Court of Appeal in *Fisher v Brooker* [2008] EWCA Civ 287, and on appeal to House of Lords in *Fisher v Brooker* [2009] UKHL 41; [2009] 1 WLR 1764.

77 *Fisher v Brooker* [2007] EMLR 9 at para. 36, noting that both the initial song, as apparently first presented to the band members in demo form by Brooker, and the organ solo devised by Fisher, were adapted to some extent from separate musical pieces originally composed by Bach, in other words works which reside in the public domain.

78 *Fisher v Brooker* [2008] Bus LR 1123, Mummery L.J. at para. 34.

a recorded piece of music created through performance is sufficiently original over any antecedent musical work to attract copyright'.[79] Again, this is not to say that the antecedent work will no longer have copyright protection; in fact, both works will have copyright protection, though as noted earlier, the owner of the subsequent or 'derivative' work will usually have to pay a licence fee for the use of the underlying, antecedent work. Thus, in *Fisher* the eventual famous version of the song, which Fisher contributed to and which was released as a hit single, would have required a licence for the use of Brooker's original musical work, as presented in the demo recording. To resolve this, the Court of Appeal held that various implied licences had been granted by the parties to each other over the prior 40-year period, and that these licences were only revoked once litigation between the parties commenced.[80] This further underscores the significance of licensing to cases involving musical arrangements.

5. PROTECTING TRADITIONAL MUSIC UNDER COPYRIGHT — CAN THE LAW AID MUSICIANS TO FACILITATE THE TRADITIONAL PROCESS?

What cases such as *Sawkins* and *Fisher* demonstrate is that (i) the definitions of music and musical work under the CDPA are broad and encompassing; (ii) arrangements can be protected as separate musical works to the underlying work; and (iii) the low threshold of originality will often be satisfied by the creative act of the performer(s) under the terms of authorship/joint authorship.

These three points give credence to the argument that *there is nothing about traditional music of itself that makes it unprotectable under copyright law*. Original arrangements of traditional works – whether public domain works or those in copyright – are clearly protected by copyright. However, this has consequences: where there are multiple arrangements of an original tune/song which is still under copyright, the law says that licences are required from the copyright owner of that original tune/song. Furthermore, if a later arrangement featured original elements of an earlier arrangement, a licence would need to be granted by the earlier arranger. Given how important arrangements of songs/tunes are to the

[79] Richard Arnold, Reflections on 'The Triumph of Music': Copyrights and Performers' Rights in Music, 2 Intellectual Property Quarterly 153, 163 (2010).

[80] Supra note 78 at paras 101–102.

practices of traditional music in a wide range of contexts – including US blues[81] and jazz,[82] North American fiddle music,[83] and Irish traditional music[84] – this issue of licensing cannot be ignored. If all arrangers of traditional music attempted to enforce their arrangements of traditional songs and tunes against others, it would severely affect the processes of traditional music, which rely on free sharing of tunes, variations and, effectively, arrangements.[85]

An empirical study I carried out with Irish traditional musicians during 2007–2011 demonstrates that one of the reasons why traditional music is shared so freely is the existence of social norms of sharing and reciprocity.[86] These norms are valued highly. At the same time, many Irish traditional musicians do record albums and perform in commercial settings, so some knowledge of copyright is necessary. The empirical data I gathered from 2007–2011 show that although musicians might seek to enforce their rights over e.g. a composition or arrangement if used in a film or TV programme, musicians rarely do so when their compositions/arrangements are recorded on albums, or are performed at gigs/festivals, by their fellow musicians. The work of Toynbee and Okpaluba shows that similar traits can be observed in other musical contexts.[87]

Even if the norms are paramount, given that their compositions and

---

[81] Candace G. Hines, Black Musical Traditions and Copyright Law: Historical Tensions, 10 Mich. J. Race & L. 463, 463–475 (2004). See generally Giles Oakley, *The Devil's Music: A History of the Blues* (BBC Books, 1976).

[82] Paul F. Berliner, *Thinking in Jazz: The Infinite Art of Improvisation* (University of Chicago Press, 2009) 61–85. See generally Ingrid Monson, *Saying Something: Jazz Improvisation and Interaction* (University of Chicago Press, 2009).

[83] Earl Spielman, The Fiddling Traditions of Cape Breton and Texas: A Study in Parallels and Contrasts, 8 Anuario Interamericano de Investigacion Musical 39, 41–48 (1972).

[84] McDonagh, supra note 16.

[85] Musicians may incorporate variations they have heard from other musicians into their own arrangements and live improvisations. See generally Marie McCarthy, *Passing It On: The Transmission of Music in Irish Culture* (Cork University Press, 1999).

[86] McDonagh, supra note 16. As detailed in the thesis, this study involved an anonymous online survey of Irish traditional musicians and face-to-face interviews with a representative sample of traditional musicians, so these findings are based on two different streams of data, underpinned by a rigorous empirical methodology.

[87] Jason Toynbee, Reggae Open Source: How the Absence of Copyright Enabled the Emergence of Popular Music in Jamaica. In Lionel Bently, Jennifer Davis and Jane Ginsburg, eds, *Copyright and Piracy* (CUP, 2010) 357, 357–373, and Johnson Okpaluba, 'Free-Riding on the Riddim'? Open Source, Copyright

arrangements are protected by copyright law, is there anything further that musicians working in a traditional idiom can do to balance their rights with those of the wider community or network? One plausible way in which composers and arrangers could try to ensure the traditional modes of sharing and reciprocity continue is to *license their works openly* using an alternative licensing system.[88]

5.1 How Alternative Licensing Systems Work

As Kelty states, alternative licensing systems 'rely on the existence of intellectual property to create and maintain the "commons" . . . even as they occupy a position of challenge or resistance to the dominant forms of intellectual property'.[89] Hence, alternative licensing systems do not attempt to break away completely from intellectual property, but instead they attempt to bend IP so that it can be tailored to suit individual or collective creators. Alternative licences, such as the 'free and open-source software' (FOSS) licences, have proven very successful, with prominent examples including Unix/Linux[90] and Google's Android operating system.[91] Perhaps the most influential licence is the GNU General Public Licence, which allows the creators of FOSS to release their software with 'open-source' code that can be improved by subsequent programmers working along the chain.[92]

It is notable that the FOSS movement thrives because it is a based upon a kind of *community ethos*. Indeed, social norms of sharing, responsibility and reciprocity are crucial to the success of community-based creativity within FOSS.[93] FOSS licences facilitate shared creativity – something that, like traditional music, could be described as a *chain of authorship*.

The most prominent attempt to bring the ethos of FOSS to other

---

Law and Reggae Music in Jamaica. In Lionel Bently, Jennifer Davis and Jane Ginsburg, eds, *Copyright and Piracy* (CUP, 2010) 374, 375–386.

[88] There is the possibility, since 2014, under UK law of arguing for a fair dealing defence on the basis of quotation within CDPA s 30. However, no such musical case has, as yet, arisen in UK, law and a full theoretical analysis of that possibility is beyond the scope of this chapter.

[89] Christopher Kelty, Punt to Culture, 77 Anthropological Quarterly 547 (2004).

[90] See Ubuntu, Ubuntu Enterprise Summit http://www.ubuntu.com/ (last accessed November 24, 2017).

[91] See Microsoft + Open Source, www.microsoft.com/opensource (last accessed November 24, 2017).

[92] See GNU Operating System, What is Free Software?, http://www.gnu.org/philosophy/free-sw.html (last accessed November 24, 2017).

[93] Guido Westkamp, The Limits of Open Source: Lawful User Rights,

cultural fields is the Creative Commons (CC) licence, a type of alternative licence for a wide range of creative works, including music, film and literature. CC is founded upon the principle of using licences to enable creators to claim 'some rights reserved' rather than 'all rights reserved'.[94] Today CC is the most widely known and widely used type of alternative licence for artistic works.[95]

Under a CC licence, copyright in the work typically remains with the author, but the author can choose one of the CC licences in order to regulate further uses of the work by other artists or users.[96] CC licences are provided in three forms: first in legal language; secondly in clear, readable language; and thirdly, as 'machine-readable' content. The terms of the CC licence are for the initial author to choose. For instance, it is possible for an author to retain only the attribution right, and to allow all – even commercial – uses of the work.[97] In contrast, it is possible to restrict all rights except non-commercial distribution. Therefore, the core aspects of a CC licence are the terms covering attribution, non-commercial reproduction and derivative use. In addition, the licences entrench the idea of 'share-alike'.[98] This share-alike notion envisages that 'derivative works' – akin to works created under the UK adaptation right, and thus relevant to arrangements – can be created using the licensed work, as long as these derivative works are themselves licensed under the same CC terms.

Thus far, several courts, including those in the US and the Netherlands, have accepted such alternate licences as being legally valid.[99] Nonetheless, CC licences still pose some difficulties. As noted earlier, under CC licences works are protected by the underlying copyright law, but are licensed contractually under a set of terms chosen by the licensor. In this context,

---

Exhaustion and Co-Existence with Copyright Law, 1 Intellectual Property Quarterly 14, 57 (2008).

[94] See Creative Commons: https://creativecommons.org/ (last accessed November 24, 2017).

[95] There are others such as the Artistic Licence 2.0 – see Open Source Initiative, Artistic Licence 2.0 http://www.opensource.org/licenses/artistic-license-2.0.php (last accessed November 24, 2017).

[96] Joëlle Farchy, Are Free Licences Suitable for Cultural Works?, 31 European Intellectual Property Review 255, 257 (2009).

[97] See Creative Commons, About the Licenses: http://creativecommons.org/about/licenses/meet-the-licences (last accessed November 24, 2017).

[98] Ibid.

[99] See for example, the US case of *Jacobsen v Katser*, 535 F. 3d 1373 Fd. Cir. (2008) and the Dutch case of *Curry v Audax*, Case no. 334492/KG 06-176 SR (March, 2006).

the question of what each term – for example, 'commercial use' – means is crucial. To take this one example, CC defines 'commercial use' as use exercised 'in any manner that is primarily intended for, or directed toward, commercial advantage or private monetary compensation'.[100] Yet, different jurisdictions may interpret and define this term in their own ways, and the boundary between 'commercial' and 'non-commercial' may not always be clear.[101] Moreover, jurisdictions may disagree over principles of copyright law itself – for instance, UK judges typically take a different interpretation of the moral right of integrity when compared with the courts in France.[102] These factors mean that legal uniformity is far from certain. To try to ensure better compatibility with domestic laws some 'local' versions of CC licences have been developed, but uncertainty remains.[103]

Furthermore, the terms of collective licensing vary from territory to territory. For instance, in France, authors generally do not 'fragment' their body of work for 'independent management of the parts'.[104] The same is largely true in the UK and Ireland with respect to PRS for Music and IMRO.[105] This reduces flexibility – if the author wishes to release one work under a CC licence, he or she may be 'thwarted by his or her status as a member of a collecting society'.[106] Even if authors do succeed in disaggregating the licences for their works, this has the potential disadvantage of increasing complexity for users. Indeed, the large number of different CC licences that are available can make it difficult for users to comprehend precisely which uses are acceptable and legal.[107] To try to assuage concerns

---

[100] Creative Commons, Attribution-NonCommercial 3.0 Unported, S. 4.b http://creativecommons.org/licenses/by-nc/3.0/legalcode (last accessed November 24, 2017).

[101] Adrienne K. Goss, Codifying a Commons: Copyright, Copyleft, and the Creative Commons Project, 82 Chicago-Kent Law Review 963, 964 (2007).

[102] Luke McDonagh, Plays, Performances and Power Struggles – Examining Copyright's 'Integrity' in the Field of Theatre, 77 Modern Law Review 533, 541–562 (2014).

[103] Catharina Maracke, Creative Commons International: The International License Porting Project – Origins, Experiences, and Challenges, 1 Journal of Intellectual Property, Information Technology and Electronic Commerce Law 4, 6–7 (2010).

[104] Joëlle Farchy, Are Free Licences Suitable for Cultural Works?, 31 European Intellectual Property Review 255, 257 (2009).

[105] PRS for Music: https://www.prsformusic.com/ (last accessed Nov. 24, 2017) and Irish Music Rights Organization: https://www.imro.ie/ (last accessed November 24, 2017).

[106] Farchy, supra note 104.

[107] Farchy, supra note 104.

over this, CC has created a database of works/licences to make this task easier for users. Nonetheless, in some circumstances a potential user may have to contact the copyright holder directly, which may be inconvenient and burdensome.[108] Thus, while the CC licence system is a 'worthy' attempt at creating a universal alternative licence for creative works, it is not without its own difficulties.[109]

6. WHY ALTERNATIVE LICENSING IS NOT A PANACEA – FORMALITY AND INFORMALITY

There is little doubt that CC licences can apply to musical works – some of the most visible successes of CC licensed works include works by popular musicians such as Radiohead[110] and Nine Inch Nails.[111] Moreover, given the centrality of *reciprocity*, the idea of utilizing an alternative licensing system like CC is potentially attractive in the traditional music context because traditional composers/arrangers are unlikely to want to avail of all the rights associated with copyright, and CC gives composers/arrangers the choice to license their works as they choose.[112]

For example, a composer could, via CC, stipulate that his/her composition should be freely usable by other musicians in non-commercial contexts while retaining attribution rights and rights over commercial uses. An arranger could do the same. On the face of it, this certainly has the potential to prove useful in the context of traditional music. Nevertheless, there are reasons to be sceptical about the overall benefit of utilizing such licences in traditional music contexts.

For one thing, the challenge of defining the key terms of CC licences –

---

[108] Creative Commons, Search: http://search.creativecommons.org/ (last accessed November 24, 2017).

[109] Niva Elkin-Koren, Exploring Creative Commons: A Sceptical View of a Worthy Pursuit. In Lucie Guibault and P. Bernt Hugenholtz, eds, *The Future of the Public Domain* (Kluwer Law, 2006) 325, 337–339.

[110] See Greg Sandoval, Radiohead's Web Venture Spooks Wall Street, CNET (December 17, 2017) http://news.cnet.com/8301-10784_3-9812275-7.html (last accessed November 24, 2017).

[111] See Greg Sandoval, Trent Reznor: Radiohead's 'In Rainbows' promotion was 'insincere', CNET (Mar. 15, 2018) http://news.cnet.com/8301-10784_3-9894376-7.html (last accessed November 24, 2017).

[112] Genevieve P. Rosloff, 'Some Rights Reserved': Finding the Space Between All Rights Reserved and the Public Domain, 33 Columbia Journal of Law & the Arts 37, 68–80 (2009).

in other words 'commercial use' – would not be any easier to overcome in the case of traditional music.[113] If a CC licence for a traditional song only allows 'free, non-commercial use' this could end up creating complications since, as noted earlier, many traditional musicians do in fact release recordings on LP, CD or mp3, and these 'uses' may fall into the 'commercial use' bracket.[114] As a result, a musician who records an arrangement of another musician's CC-licensed composition may breach this term.

There is a more fundamental reason to be sceptical of CC in the context of traditional music. As noted earlier, it is the *informal* social norms of sharing and reciprocity that are key to the processes of Irish traditional music.[115] A similar trait can be observed in other traditional contexts.[116] CC, by contrast, could have the effect of entrenching the primacy of individual property rights, and thus, *formalising* the system of free sharing that occurs within traditional music networks of, for example, Irish or North American fiddle players or jazz trumpet players. Here it is worth recalling the old adage that a 'leaky' copyright system works best.[117] To some extent CC licences effectively 'fill in the gaps' left by the copyright system, something that may lead to the entrenching of individual property rights in areas where individual authors have generally ignored copyright. For example, in traditional contexts where there has been no financial incentive to enforce copyright, composers and arrangers have sometimes allowed, or even encouraged, their fellow musicians perform, re-arrange, and record their music freely, while enforcing rights against broadcasters and film and TV companies for soundtrack use. Such scenarios could be

---

[113] The most popular CC licence is issued for 'non-commercial use' with the author retaining attribution rights; see CreativeCommons.org: http://creativecommmons.org/weblog/entry/4216 (last accessed November 24, 2017).

[114] Although, the average traditional blues, jazz or Irish folk CD sells only a tiny fraction of what a commercially available 'pop' album sells, the traditional recording would arguably still be 'commercially available'.

[115] McDonagh, supra note 16.

[116] Jason Toynbee, Reggae Open Source: How the Absence of Copyright Enabled the Emergence of Popular Music in Jamaica. In Lionel Bently, Jennifer Davis and Jane Ginsburg, eds, *Copyright and Piracy* (CUP, 2010) 357, 357–373 and Johnson Okpaluba, 'Free-Riding on the Riddim'? Open Source, Copyright Law and Reggae Music in Jamaica. In Lionel Bently, Jennifer Davis and Jane Ginsburg, eds, *Copyright and Piracy* (CUP, 2010).

[117] Siva Vaidhyanathan, *Copyrights and Copywrongs: The Rise of Intellectual Property and How it Threatens Creativity* (NYU Press, 2001) 184, referring to the opinion of James Madison.

described as the 'leaky' copyright system working well.[118] By contrast, encouraging the use of a new formal system of licensing may have unforeseen negative effects – it may even reduce the flexibility of 'free sharing' within the network. In other words, the informal non-enforcement of copyright, which under traditional music is governed by flexible social norms, may become a formal rigid system via alternative licensing.[119] Rather than facilitating 'open' culture, CC could lead to greater cultural 'commodification' by encouraging small-scale creators, many of whom have largely ignored IP in the past, to claim some formal rights over their works.[120]

It is also worth noting that CC, for all its grand aspirations, has only been a modest and uneven success. The iconic alternative licensing success stories have been communally created software and online knowledge projects such as Wikipedia, not CC-licensed creative works. With this realisation in mind, one conclusion to draw is that the type of licensed 'peer production' envisaged by Benkler may be more easily applicable in the context of software and web-based knowledge outlets than in other cultural contexts.[121] Perhaps it is also significant that software is a recorded product, and therefore susceptible to 'wrapping up' in licence terms, whereas music is essentially intangible, and for this reason perhaps less amenable to a pure licensing solution. In any event, for the reasons given, while CC should not be entirely ruled out as potentially useful to traditional musicians, neither should it be embraced wholeheartedly. Musicians should approach it cautiously, and if they choose to avail themselves of it, carefully consider what uses they want their fellow musicians to make of their works.

---

118 Jessica D. Litman, Lawful Personal Use, 85 Texas Law Review 1871, 1912–1914 (2007).

119 Dotan Oliar and Christopher Sprigman, There's No Free Laugh (Anymore): The Emergence of Intellectual Property Norms and the Transformation of Stand-Up Comedy, 94 Va. L. Rev. 1787 (2008); Jacob Loshin, Secrets Revealed: Protecting Magicians' Intellectual Property without Law. In Christine A. Corcos, ed., *Law and Magic: A Collection of Essays* (Carolina Academic Press, 2010) 123, 140 and Marta Iljadica, *Copyright Beyond Law: Regulating Creating in the Graffiti Subculture* (Bloomsbury, 2016).

120 Niva Elkin-Koren, Exploring Creative Commons: A Sceptical View of a Worthy Pursuit. In Lucie Guibault and P. Bernt Hugenholtz, eds, *The Future of the Public Domain* 325, 334–339 (Kluwer Law, 2006).

121 Yochai Benkler, *The Wealth of Networks: How Social Production Transforms Markets and Freedoms* (Yale University Press, 2006) 59–91.

7. CONCLUSION: REFLECTING ON THE IMPORTANCE OF THE NORMS OF SHARING AND RECIPROCITY

This chapter has demonstrated that copyright can, and does, protect works of traditional music, whether new compositions or original arrangements of traditional works. However, since traditional music emerges in the context of free sharing of songs and tunes, I have posited a key question: what would happen if these copyrights were enforced? The answer is that it would disrupt the practices of traditional musicians. As I have shown, arrangements of songs/tunes are crucial to traditional music in a wide range of contexts – including US blues[122] and jazz,[123] North American fiddle music,[124] and Irish traditional music.[125] If all arrangers of traditional music attempted to enforce their arrangements of traditional songs/tunes against others, it would severely affect the process of traditional music, which relies on free sharing.[126]

At the same time, while the use of alternative licensing systems, such as CC, is an option for traditional musicians, it does not necessarily provide the optimum solution. The legal challenges of multi-jurisdictional licensing, and the problem of bringing formal legal concerns into an environment that has thrived on informal sharing, mean that CC is far from being a panacea. In fact, taking inspiration from the work of Elinor Ostrom, I argue that where a collective resource is being shared successfully – as it is in many traditional music contexts – the law should be akin to a bystander, rather than an active player.[127] Ultimately, for traditional music to continue to thrive, the informal norms of sharing and reciprocity are what must be encouraged – not the cold formality of the law.

---

[122] Candace G. Hines, Black Musical Traditions and Copyright Law: Historical Tensions, 10 Mich. J. Race & L. 463, 463–475 (2004). See generally Giles Oakley, *The Devil's Music: A History of the Blues* (BBC Books, 1976).

[123] Berliner, supra note 82. See generally Monson, supra note 82.

[124] Earl Spielman, The Fiddling Traditions of Cape Breton and Texas: A Study in Parallels and Contrasts, 8 Anuario Interamericano de Investigacion Musical 39, 41–48 (1972).

[125] McDonagh, supra note 16.

[126] Musicians may incorporate variations they have heard from other musicians into their own arrangements and live improvisations. See generally McCarthy, supra note 85.

[127] Elinor Ostrom, *Governing the Commons: The Evolution of Institutions for Collective Action* (CUP, 1990).

8. Music improvisation and copyright

Giuseppe Mazziotti*

1. INTRODUCTION

On January 24th, 1975, during a road tour of Europe with ECM record producer Manfred Eicher and sound engineer Martin Wieland, Keith Jarrett had been invited to perform at the Opera House in Cologne. The concert had been scheduled at an unusually very late hour, immediately after an opera performance. That concert would never have been the one music lovers know through the 'Köln Concert' album if a number of important things had not gone awry.[1] The concert was organized by a teenager who had arranged the rental of a Bösendorfer grand piano that, because of a number of misunderstandings with the opera house staff, was never delivered to the theatre stage. What Jarrett and his record label's team found when they arrived at the venue was a much smaller baby grand – a backstage piano used mostly for rehearsals. Having an unexpectedly bad instrument at his disposal – thin in the upper registers and weak in the bass register – Jarrett could have easily cancelled the live recording session that ECM had been planned in Cologne since the beginning of the European tour. The label team decided to make a record of the performance anyway, since they had brought and set all necessary equipment on stage.

In an interview with Don Heckman, Jarrett also recalled that, before that concert, he was incredibly tired after a long drive from Zurich and not having slept for 24 hours.[2] Anger and disappointment at the conditions in

* The author would like to thank Jane Ginsburg, Guido Zaccagnini, Enrico Bonadio and Nicola Lucchi for their review of a draft of this chapter and their suggestions and criticism. Errors and/or omissions are the sole responsibility of the author.

1 See Tyran Grillo, Keith Jarrett: The Köln Concert (ECM 1064/65), ECM Review, 24 February 2010.

2 See Don Heckman, The Making of Keith Jarrett's The Köln Concert, The GRAMMYs, 7 January 2014. As reported in the interview, Jarrett was also very upset ('in almost hell') for an unpleasant experience he had at an Italian restaurant

which the performance was taking place can be perceived especially at the very beginning of the performance, in the first track, when someone in the audience laughs after having heard Jarrett starting his improvised performance by imitating the bells the theatre used to play in the lobby to warn the public that a performance was about to start.[3] In the same interview, Jarrett recalled that both the enthusiastic audience he found at the theatre and the challenge of having to use a surprisingly bad instrument strongly motivated him to play in a different way.

While performing on that night he did not know the rolling and rhythmically repetitive figures he played with the left hand, to strengthen the weak bass notes, and his virtuosic phrasing by the right hand, confined to the middle part of the keyboard to avoid the thin upper registers, would be permanently fixed and become legendary. The record ECM produced after that adventurous musical journey is still the best-selling album in the history of jazz and, over the years, became a cult object for musicologists, music lovers and performers themselves.

This story exemplifies how improvisation practices can place a skilled performer in a position to create original and highly regarded pieces of music directly on stage without having to rely on fixed musical materials and on musical texts. This type of music work can easily be boosted by the intrinsic unpredictability and aural dimension of live performances and unexpected, adventurous conditions, under which a skilled musician might be encouraged to perform in response to 'spontaneous creative impulses',[4] as Jarrett did in Cologne.

As this chapter shows, composing a piece of musical work in an improvisatory manner, without having previously fixed the work into a musical text or another tangible medium openly challenges copyright laws that, for historical reasons, especially in common law systems, are based on the primacy of notated or fixed musical compositions. This chapter intends to clarify, from a broadly comparative perspective, whether and how current copyright systems, in spite of the emphasis placed by existing laws on notated or fixed works, can accommodate the interests of music authors resorting to improvisation practices. Improvisation gives

---

where he had his pre-concert dinner ('Everyone else was eating, I was the one who was going to play in an hour, and I still didn't have any food. And then when they finally brought the food, I was still hungry, because I wasn't happy with the food they served').

3 The introductory motif played by Jarrett at the beginning of the first track of the album, and imitating the bells in the theatre hall, is G-D-C-G-A.

4 See Bruno Nettl, Thoughts on Improvisation: A Comparative Approach, 60 Musical Quarterly, 3 (1974).

rise to an unconventional type of copyright in so far as composition and performance occur simultaneously, so that the improvisation itself is the composition being performed.[5]

The chapter is structured as follows. Section 2 explains briefly why, from an historical point of view, copyright systems ended up disfavouring extemporaneous authorship, in spite of an international legal framework which is strongly protective of authors' rights. Section 3 shows how the requirements of originality and fixation apply to improvised musical compositions and how such an unconventional kind of copyright work is protected in different copyright systems. Section 4 critically evaluates the impact of the exclusive rights of authors and music publishers on derivative works, taking jazz music as an example of performing art based on quotation, elaboration and transformation of pre-existing copyright works. Section 5, finally, while considering pros and cons of implementing copyright paradigms in the realm of improvisation, shows that a broader understanding of musical authorship, which also encompasses the performative aspects of composition, and a higher consideration for the autonomy of follow-on works, would ensure a more adequate incentive to and reward for creativity that relies on improvisation techniques.

2. NOTION OF COPYRIGHT 'MUSIC WORK'

Even though music is, by its very nature, a performing art, early copyright statutes in Europe started affording legal protection to composers and, even more so, to music publishers in relation to their musical texts. Legal protection of musical works developed – in both the common law and civil law traditions – as an extension of the protection originally reserved for books and 'writings of all kinds' to written musical compositions.[6] In

---

5 See Warren Shaw, Copyrighting Improvised Music, 32 Copyright L. Symp. 109, 111–112 (1982).

6 See Martin Kretschmer, Intellectual Property in Music: A Historical Analysis of Rhetoric and Institutional Practices, 6 Studies in Culture, Organizations and Societies, 197, 2007 (2000) (recalling that the English Statute of Anne did not protect music); Anne Barron, Copyright Law's Musical Work, 15 Social & Legal Studies, 101, 112–113 (2006) (emphasizing that the 1710 Statute of Anne referred to the object of copyright as the author's 'writing', 'book', 'printed book' or 'copy'). As pointed out by Jane Ginsburg, A Tale of Two Copyrights: Literary Property in Revolutionary France and America, 64 Tulane Law Review, 991, 1015 (1990), the first French copyright law (1791–1793) afforded rights to authors in 'writings of all kinds' but also to 'all productions of the beaux arts' (including the work of dramatists, whose exclusive right to authorize public performances aimed

particular, in 18th-century Britain, the renowned Statute of Anne (1710) did not afford protection to musical works, at least until 1777, when Johann Christian Bach, the youngest son of Johann Sebastian Bach, one the leaders of the London music scene at that time, brought a successful case against publisher Longman for the unauthorized publication of some of his works.[7] As pointed out by Barron, the ruling by Lord Mansfeld in *Bach v Longman* extended the protection afforded by the 1710 Act from 'books and other writings' to musical scores on the assumption that musical compositions were 'writings'.[8]

Since the only way to preserve music was through written notation, copyright in music works developed historically as a form of intellectual property being incorporated primarily into music texts and afforded to music 'writers' (in other words, composers and lyricists), who sold their rights to publishers in exchange for royalties. At that time, statutes and courts took into consideration the only aspect of musical creation that could be fixed onto a medium and be commercially exploited, namely, music scores.[9] From a historical point of view the codification of the modern copyright system coincided, in the second half of the 18th century, with the emergence of a proper and independent concert scene and the formation of a market for theatrical music (for example, operas) and vocal or instrumental recitals that enabled composers to present new pieces to the public and to earn from the performance of their own music more than they gained from music publishers in terms of royalties. The music publishing business was boosted by the emergence, since the 18th century, of paying audiences, especially in theatres, which attended public performances of musical works that, until then, had been confined to small and closed circles in religious ceremonies or aristocratic gatherings. Demand for music scores increased with the

at breaking the monopoly of La Comedie Française in the production of theatrical works: see at 1006–1008).

[7] See Anne Barron, Copyright Law's Musical Work, supra note 6, at 116–117; Martin Kretschmer, Intellectual Property in Music, supra note 6, at 209; see also Olufunmilayo Arewa, Creativity, Improvisation and Risk: Copyright and Musical Innovation, 86 Notre Dame Law Review 1829, 1836 (2011).

[8] See Anne Barron, Copyright Law's Musical Work, supra note 6, at 116–118. The author quotes an excerpt (118) from *Bach v Longman* [1774]: 624, where Lord Mansfield states: 'Music is a science; it may be written; and the mode of conveying the ideas is by signs and marks . . .'.

[9] See for instance Warren Shaw, Copyrighting Improvised Music, supra note 5, at 112, recalling that the US Congress added musical works to the statutory enumeration of copyrightable subject matter in 1831: cf. Act of February 3, 1831, ch. 16, 4 Stat. 436 (repealed in 1870).

emergence of bourgeois audiences with surplus income, time and interest in performing in their private homes what they could appreciate in public performances.[10]

It was only with the invention of the phonogram that the performing aspects of music, and in particular the ephemeral sounds of improvisation, which had played an essential function in the transmission and evolution of music across history, started being recordable onto tangible media.[11] To use Walter Benjamin's lexicon – at that time music performances lost their 'aura' and their eminently ephemeral character while being incorporated into reproducible and transmissible items that were distinct from original and authentic performances which took place at a given time and in a certain place before a present audience.[12]

The demand for musical works continued to expand with the advent of new technologies, enabling large-scale dissemination of music through phonograms and the evolution and diffusion of mass media. This technological evolution was accompanied by an expansion of the scope of music copyright, which aimed at enabling composers and music publishers to control (or at least be remunerated for) radio and TV broadcasts of their works. At the same time, sound recordings became objects of mass consumption and provided copyright owners with a source of additional revenues based on mechanical reproductions and communication to the public of musical compositions.[13]

Although in different ways, and with a different timing in continental-European and Anglo-American jurisdictions, the artistic work of music performers and the investments of record producers gained momentum through the introduction of an additional layer of legal protection specifically targeted at fixations of sounds by music performers and the production of phonograms by record producers. The rise of these new 'products' led to the codification in continental Europe of so-called 'neighbouring'

---

[10] See Anne Barron, Copyright Law's Musical Work, supra note 6, at 115.

[11] Recording technology became available following Thomas Edison's 1877 development of tinfoil phonograph technology: see Olufunmilayo Arewa, Creativity, Improvisation and Risk, supra note 7, at 1833, who recalls that this new technology was not immediately commercialized for music.

[12] Walter Benjamin, *The Work of Art in the Age of Mechanical Reproduction* (1936).

[13] From the late 19th century composers and lyricists had already started to form associations and collecting societies whose aim was essentially that of monitoring as well as licensing rights of public performance and, at a later stage, radio and TV broadcasting. In addition to that, at least in continental European countries, authors' collecting societies also licensed the rights of mechanical reproduction for the production and distribution of sound recordings.

rights for performers and phonogram producers,[14] and, in the United Kingdom, of an entrepreneurial form of copyright in sound recordings, whose author was defined by law as the 'maker' of a sound recording (in other words, the record producer).[15] These legislative interventions preluded the conclusion of the 1961 Rome Convention for the protection of performers, producers of phonograms and broadcasting organizations.[16] Remarkably, the United States was not a party to the Rome Convention and ended up enacting a form of federal copyright protection for the 'fixation of series of musical, spoken or other sounds' only in 1971.[17] Interestingly, similarly to the rights in sound recordings created in the UK and Ireland at an earlier stage, US law protects sound recordings that are original in the sense that they are not copied from a previous work. However, US copyright law relies on a different definition of 'authorship', under which music performers, recording engineers and record producers can be regarded as joint authors of the sound recording.[18] Moreover the

---

[14] For instance, Italy started conferring protection on performers and record producers in 1941, through the enactment of a new copyright act which contains a section on rights related to copyright ('diritti connessi': see Italian Copyright Act, Law No. 633 of April 22, 1941, Title II, Gazzetta Ufficiale della Repubblica Italiana [Gazz. Uff.], July 16, 1941, No. 166). France implemented the 1961 Rome Convention, with regard to performers, in 1985 (Loi no 85-660 du 3 juillet 1985 relative aux droits d'auteur et aux droits des artistes-interprètes, des producteurs de phonogrammes et de vidéogrammes et des entreprises de communication audiovisuelle, Journal Officiel de la République Française [J.O.] 4 July 1985, 7495).

[15] The United Kingdom introduced rights in sound recordings through the Copyright Act 1956 (cf. Sect 12) whereas Ireland enacted the same right in the Copyright Act 1963; significantly, neither the UK nor the Irish acts granted any statutory rights to music performers. As pointed out by Jane Ginsburg, The Concept of Authorship in Comparative Copyright Law, 52 De Paul L. Rev. 1063, 1070 (2003), in the case of sound recordings UK law raises confusion between two distinct aspects of copyright law by conflating authorship with the vesting of copyright ownership.

[16] See Rome Convention, International Convention for the Protection of Performers, Producers of Phonograms and Broadcasting Organizations, Oct. 26, 1961, 19 ILM 1492 [hereinafter Rome Convention], available at http://www.wipo.int/export/sites/www/treaties/en/ip/rome/pdf/trtdocs_wo024.pdf.

[17] Sound Recording Act of 1971, Pub. L. No. 92 140, 85 Stat. 391.

[18] See Sect 101 of the US Code, Title 17, which defines a joint work as 'a work prepared by two or more authors with the intention that their contributions be merged into inseparable or interdependent parts of a unitary whole'. Under US law recording engineers are regarded as co-authors of the sound recording in so far as they make original contributions to the recording. If, however, recording engineers are employees of the record producer, then the producer becomes the co-author. As regards copyright ownership, moreover, Section 101 provides that 'authors of a joint work are co-owners of copyright in the work'. However, it is

scope of the US copyright in sound recordings is narrower than its UK and Irish equivalents since it is deprived of the right of public performance, apart from digital audio transmissions.[19]

As a result of the diversified legal protection of sound recordings and performances, from the second half of the 20th century onwards the structure of music copyright has subsequently relied on distinct and autonomous layers of intellectual property. Such layers target – respectively – musical compositions, fixations of performances and sound recordings in a rigidly pre-determined manner. Following the logic and the spirit of the 1961 Rome Convention, the aim of rights of performers and record producers is not that of rewarding authorship and originality of musical creation, in the same way as the rights of authors do; rather, their function is to protect the value of the artistic work of performers and, even more so in Anglo-American copyright systems, the investments made by record producers from slavish copying. From the perspective of improvised music performances, the 1961 Rome Convention text, while restating the principle of the primacy of the rights of authors over the rights of performers and record producers, establishes a clear and restrictive link between 'performances' and 'literary or artistic works', keeping them firmly distinct, as if certain improvisations could not constitute per se original works created without having to relate to pre-existing, fixed works.[20] The definition of 'performers' was significantly broadened under the 1996 WIPO Performances and Phonograms Treaty (WPPT), where an express reference to (unfixed) 'expressions of folklore' suggests that protected performances, in the countries which are parties to the WPPT (which include the United States) should not necessarily consist of performances of pre-existing literary or artistic works.[21]

---

customary in the US recording industry that at least employee performers and contributors to the sound recording (such as recording engineers) conclude 'work made for hire' agreements assigning all of their rights to a record producer, who subsequently acquires all rights of commercial exploitation.

[19] The Digital Performance in Sound Recordings Act of 1995 amended Section 106 of the US Code, Title 17, to introduce an exception to the rule of 'no public performance rights' for sound recordings by providing an exclusive right to perform sound recordings publicly by means of digital transmissions. See Digital Performance in Sound Recordings Act of 1995, Pub. L. No. 104-39, 109 Stat. 336 (1995) (codified in scattered sections of 17 U.S.C.).

[20] See the 1961 Rome Convention, Article 3(a): '. . . actors, singers, musicians, dancers and other persons who act, sing, deliver, declaim, play in, or otherwise perform literary or artistic works'.

[21] See World Intellectual Property Organization: Performances and Phonograms Treaty, adopted December 20, 1996, 36 I.L.M. 76, Article 2(a).

3. ORIGINALITY, AUTHORSHIP AND FIXATION OF IMPROVISED MUSIC

As with all intellectual creations, improvised music works can be protected by copyright protection on condition that a work is 'original' and, in Anglo-American systems, fixed in a tangible medium.

3.1 Originality

'Originality' is an essential requirement for copyright to subsist in a given intellectual creation. In the absence of a unitary notion under international copyright law, originality is a concept through which copyright systems define, in significantly different ways, the main characteristics a work should possess in order to trigger copyright protection.[22] As explained by Jane Ginsburg, originality is synonymous with authorship, in so far as it identifies the type of contribution which makes an intellectual work original.[23] In setting out higher or lower standards of intellectual labour, the various copyright systems require either a minimum of personal creative activity (or a 'modicum of creativity', as the US Supreme Court held in *Feist v Rural Telephone*)[24] or, in traditional common law, 'sweat of the brow', in other words, a standard which allows for copyright protection in so far as works entail expenditure of time and resources.[25] In the last two decades the standard of originality has evolved in Europe as a result

---

[22] As pointed out by Jane Ginsburg and Edouard Treppoz, *International Copyright Law: U.S. and E.U. Perspectives* (Edward Elgar Publishing, 2015) 278, the Berne Convention does not provide for a minimum threshold of creativity; however, Article 2(3) refers to 'original works', whereas Article 2(5) refers to works which constitute 'intellectual creations' 'by reasons of the selection and arrangement of their contents'.

[23] See Jane Ginsburg, The Concept of Authorship in Comparative Copyright Law, supra note 15, at 16.

[24] See *Feist Publications, Inc v Rural Telephone Service Co., Inc*, 499 U.S. 340 (1991), where the US Supreme Court, holding that the requirement of originality (which consists of independent creation plus a modicum of creativity) is mandated by the US Constitution, excluded that mere aggregation of data could constitute original creations.

[25] See Jane Ginsburg, The Concept of Authorship in Comparative Copyright Law', supra note 15, at 14–16, who emphasizes that the gap between civil law and common law copyright systems was even wider, at least until recently, since countries such as France and Germany formerly endorsed an assessment of originality based on the higher or lower artistic or aesthetic merit of works (for example, the fact that photographs were regarded as 'machine-mediated' works made these works less worthy than other creations).

of the EU harmonization of national copyright laws and the extension of the standard of the 'author's own intellectual creation' to the generality of copyright works.[26] According to this standard, originality subsists whenever an author makes free or subjective choices and stamps the work with a personal touch.[27]

Originality, as such, does not raise specific challenges to the protection of improvised music works. For instance, there is no doubt about Jarrett's authorship of the three solo piano pieces he improvised at the Cologne opera theatre, especially if one considers that – unlike the vast majority of jazz works – his improvisation did not rely on any pre-existing materials or compositions and took advantage of the unexpected conditions which stimulated Jarrett's talent and creativity. That type of performance would also qualify as original work if Jarrett, in our example, had played his fully improvised pieces not alone but with a group, such as the trio he has formed for decades with Gary Peacock (bass) and Jack Dejohnette

---

[26] Initially, EU law harmonized copyright's standard of originality in a sector-specific fashion. Directive provisions laid down the concept of 'author's one intellectual creation' with regard to specific categories of works, such as computer programs, photographs and databases: see, respectively, Council Directive 91/250/EEC, on The Legal Protection of Computer Programs, 1991 O.J. (L 122) 42 (replaced by Directive 2009/24/EC, of the European Parliament and of the Council of April 23, 2009 on the Legal Protection of Computer Programs, 2009 O.J. (L 111) 16); Council Directive 93/98/EEC of 29 October 1993, Harmonizing the Term of Protection of Copyright and Certain Related Rights, Official J. of Eur. Cmty. No. L 290/9 (1993) (replaced by Directive 2006/116/EC of the European Parliament and of the Council of 12 December 2006 on the term of protection of copyright and certain related rights, 2006 O.J. (L 372) 12: cf. Art. 6 with regard to originality of photographs); and Directive 96/9/EC of the European Parliament and of the Council of 11 March 1996 on the legal protection of databases, Council Directive 96/9, 1996 O.J. (L 77) 20. From 2009 onwards, several judgments of the Court of Justice of the European Union have generalized the standard of originality: see, for instance, C-05/08 *Infopaq International A/S v Danske Dagblades Forening* [2009], which found that 11-word sentences can be 'suitable for conveying to the reader the originality of a publication such as a newspaper article, by communicating to that reader an element which is, in itself, the expression of the intellectual creation of the author of that article . . .'; and Joined cases C-403/08 *Football Association Premier League v QC Leisure* and C-429/08 *Karen Murphy v Media Protection Services Ltd* [2011], where the CJEU rejected the Premier League's claim of originality of football matches since the rules of football leave 'no room for creative freedom'.

[27] See C-145/10 *Painer v Standard Verlags* [2011], paras 89–92, where the CJEU found that portrait photographs of a young girl were capable of meeting the requirement of originality, so that the use of the photographs by magazines and newspapers without the photographer's authorization constituted copyright infringement.

(drums). The difference, in comparison to a piano solo performance, is that a group improvisation would entail joint authorship of the works, with a subsequent allocation of the rights in the music work in accordance with either the contribution of each group member to the whole work and/or the deal the band leader might have closed with his fellow musicians before undertaking the performance.[28]

Moreover, considering that music improvisation presupposes a condition of simultaneity of creation and performance, and given the aforementioned structure of music rights, fixation of improvised music can easily give rise to a dual legal protection subsisting in the (abstract) music work and, at the same time, in the performance fixed into a sound recording. In the example of Jarrett's trio, in the case of a live or studio recording, the improvisers would acquire also exclusive rights in the resulting work, on the grounds of the dual protection deriving from the application of both copyright and sound recording rights in the work and in the phonogram, respectively.

Determining authorship of improvised music is certainly more complex and difficult to discern when improvisation draws upon and borrows from pre-existing works, irrespectively of whether the earlier work is still under copyright or in the public domain.[29] If a pianist improvises her own variations on public domain works such as Domenico Scarlatti's keyboard sonatas at the time of performing (and recording) such works, her copyright as author of the improvised works and her rights as performer of Scarlatti's earlier works might prove to be inseparable because of the fusion of the main elements of the old and of the new works. Since rights in fixations and sound recordings are shaped primarily as neighbouring rights (or, in the UK and US, as copyrights in sound recordings), there is an objective risk of confusion between the subject matter of the improvised work – which is by definition not expressed through notation, at least initially – and the portions of Scarlatti's musical work which are just performed by the pianist, with no original contributions on her part. In the example of music improvisation drawing on public domain compositions, the risk is that the improviser's musical variations or additions might

---

[28] See Joseph D. Kim, Copyright: Jazz's Friend or Foe?, Law Practice Today, The Intellectual Property Issue, October 2013, who stresses how, from a practical perspective, at least in the US, a band leader might actually engage the services of the other band members in a written 'work made for hire' agreement, which would make the leader own both the composition and the recording as improvised by the group.

[29] The conflict between music improvisation and the copyright in pre-existing musical compositions used for the purpose of elaboration and transformation is taken into consideration in section 4 below.

be viewed, in a very reductive manner, not as original works themselves – because of the original contributions and elaborations made by the follow-on improviser – but as mere performances of earlier works such as Scarlatti's sonatas.

3.2 Fixation

What would happen if a solo piano improvisation such as that of the Köln Concert had remained unrecorded? Would such improvisation have any legal value or consequences?

According to Article 2(2) of the Berne Convention, fixation is not a mandatory requirement for subsistence of copyright.[30] This means that such a substantive limitation on copyrightable subject matter is a prerogative of national laws.

In Anglo-American copyright systems, unfixed works of any kind are expressly excluded from copyright's scope, irrespective of how original an improvised music work, a lecture or a theatrical improvisation might prove to be. From a copyright perspective, in countries such as the US, the UK and Ireland it would be as if Jarrett's performance had never taken place, due to the fact that fixation of musical work is a mandatory requirement.[31] Unfixed improvised music cannot be protected under UK or Irish copyright law since improvisatory musical creations are not 'recorded in writing or otherwise' or 'fixed in any tangible medium of expression . . . from which they can be perceived, reproduced, or otherwise communicated . . .'.

Continental-European copyright systems, instead, are less inclined to establish formal restrictions of copyright's subject matter, even though the creation and authorship of musical works are difficult or impossible to prove in the absence of a tangible medium. Countries like France and Italy closely follow the approach of the Berne Convention in not expressly requiring fixation in order to confer automatic protection in response to creation (and not fixation) of all original works.[32] As pointed out in the literature, in those specific circumstances where French copyright law

---

[30] Article 2(2) of the Berne Convention provides that it shall be a matter for national legislation 'to prescribe that works in general or any specified categories of works shall not be protected unless they have been fixed in some material form'.

[31] See US Code, Title 17, Sect 101 and Sect 102(a); UK 1988 Copyright Designs Patents Act, Sect 3(2); as for Ireland, the 2000 Copyright and Related Rights Act, Sect 18(1).

[32] This is the same policy goal the Berne Convention pursues by explicitly providing under Article 5(2) that the existence and exercise of copyright should not

exceptionally requires fixation (for example, with regard to choreographic works and circus acts: cf. L 112-2-4), such a requirement has been interpreted by French courts as a rule of proof rather than a statutory restriction of copyright's scope.[33]

However, one might legitimately wonder what kind of value a copyright that cannot be effectively proved might have: just a symbolic value? If, in the aforementioned example, Jarrett's fully improvised solo piano performances in Cologne had not been recorded, the pianist would not have been in a position to claim authorship in his improvised pieces unless he had made a transcription of his own pieces immediately after the concert or had authorized someone else to do so (as happened many years after the recording).[34] Jarrett would not have received, in addition to his performer fees, any additional compensation related to his rights as extemporaneous author of the three pieces. Moreover, the local authors' collecting society would not have found Jarrett's improvised piano pieces in any of the existing musical repertoires performing rights societies administer before public performances of (fixed) music works occur.[35]

4. COPYRIGHT AND ITS BASIC TENSION WITH DERIVATIVE WORKS: THE EXAMPLE OF JAZZ

As pointed out above, music copyright follows, mainly for historical reasons, a rigid allocation of rights and a subdivision of roles between authors and performers that clashes with the core business of music genres where the contribution of performance to musical authorship is as important as the content of written scores. As highlighted in the literature, it was somehow inevitable that the adoption under modern copyright law of a concept of music 'work' entirely based on the aesthetically and economically predominant conceptions of musical creativity in Europe

---

be made dependent on the fulfilment of any formality such as registration, deposit or filing of copies, payment of fees, etc.

[33] See Jane Ginsburg and Edouard Treppoz, *International Copyright Law: U.S. and E.U. Perspectives*, supra note 22, pp. 90–291.

[34] See Keith Jarrett, The Köln Concert, Original Transcription, Schott, Mainz (1991).

[35] As clarified below (see section 5), the subsistence of copyright in improvised music works, even in the absence of a recording – at least in those countries where fixation is not an indispensable requirement – might actually encourage a more fair allocation of fees stemming from commercial exploitation of the rights in public performances whenever pre-existing copyright works are used, especially in the domain of jazz music, for improvisatory purposes in the live music business.

as developed from the beginning of the 19th century onwards, ended up excluding antagonist styles and artistic movements from the benefits of copyright systems.[36] In certain musical contexts – for example, jazz, hip hop, punk, rap – improvisation plays such an essential function that the distinction between the author's and the performer's role might prove to be blurred and very difficult to draw.

The historical development and philosophy of jazz help to clarify what obstacles copyright poses to music improvisation, depending on whether improvised works incorporate or develop melodies, harmony or rhythms borrowed from earlier musical compositions. Even though improvisation practices have been relevant for performance across the modern history of music – from Renaissance and Baroque music to rap and hip hop – jazz is certainly the genre for which improvisation has played – structurally and aesthetically – the most relevant part.

What makes jazz so relevant to understanding the artistic value of improvisation is mainly the way this music historically developed from the late 19th century onwards. Jazz originated from a fusion of Afro-American popular music, which reached the United States at the time of the slave trade, with cultured and sophisticated traditions originally brought to America by European immigrants. Drawings from early 19th-century New Orleans show that black musicians gathering in Congo Square were playing percussion and stringed instruments which were identical to instruments used in indigenous African music.[37] Work songs that could be heard from slaves in the plantations of southern states such as Texas and Louisiana were characterized by different forms of ritualized vocalizing such as field hollers, prison work songs and street cries that openly disregarded the Western system of notation and scales. African music performance traditions could also be clearly identified in religious vocal pieces such as gospels and spirituals, in spite of the attempts of religious organizations to convert African Americans to more edifying examples of Western music. This way black people ended up transforming psalms and hymns into music where groups of singers performed variations on the melody and, at the same time, blended their voices through ornamentations resulting in 'great unified streams of tone'.[38] Afro-American and Latin-American rhythms and melodies also ended up influencing works

---

[36] See Arewa, Creativity, Improvisation and Risk, supra note 7, at 1839–1840; Anne Barron, Harmony or Dissonance? Copyright Concepts and Musical Practice, 15 Social & Legal Studies, 25, 33–37 (2006).

[37] See Ted Gioia, *The History of Jazz* (OUP, 1997) 3, available at http://www.nytimes.com/books/first/g/gioia-jazz.html.

[38] See Ted Gioia, ibid., quoting an expression by musicologist Alan Lomax.

by renowned composers and performers from the late 19th and the early 20th centuries, such as virtuosic pianist Louis Moreau Gottschalk (a New-Orleans-born child of a German Jew and a creole woman from Haiti) and Scott Joplin, a native of Texas renowned for his ragtime piano.

From a copyright-related perspective a crucial aspect triggered by the historical process of fusion and elaboration of different musical traditions is that jazz originates from an inextricable mix of folkloric and, yet, highly sophisticated instrumental and vocal practices which remained essentially unwritten and were transmitted only orally for decades. It was only with the advent of the first sound recordings that the eminently performance-based and improvised elements of what we call 'jazz' (from an expression used in brothels where legends such as that of phenomenal pianist and composer Jelly Roll Morton originated) became more available to wider audiences and easier to understand, perform and develop on a much bigger scale.

From the early 20th century jazz developed as an artistic form where performers combine music which is written down with music which is improvised. Jazz musicians interact with each other on the grounds of well-established musical structures, where a 'chorus' embodies a pre-existing composition which is used by an arranger or a band leader to create a 'head' theme (A) and a 'bridge' theme (B). The chorus is normally preceded by an introduction and follows, traditionally, a structure where the head and the bridge themes are alternated (for example, A-A-B-A). The chorus structure allows all performers in a group (starting with the lead instruments) to improvise their own solos on a rhythmic and harmonic base provided by other instruments (for example, bass, drums, piano) which loop the initial chord changes, with frequent alterations and substitutions of harmonies.[39]

This structure shows that jazz has developed largely as a form of appropriation art based on borrowing, elaborating and eventually transforming existing themes, harmonies and rhythms. A simple explanation of how improvisation has been conceived and organized in the context of jazz music shows why this artistic movement inevitably clashes with the exclusive rights to control derivative works in so far as the pre-existing composition is still protected by the copyright of composers and music publishers.

---

[39] As pointed out in the literature, the number of solos or choruses is not pre-determined, so that, if each member of a jazz group plays three choruses the underlying music work would essentially make up only 14% of the total work: see Note, Jazz Has Got Copyright Law and That Ain't Good, 118 Harvard Law Review, 1940, 1943–1944 (2005).

The musical compositions which have traditionally been used for jazz improvisation are catalogued and inserted into famous collections of songs, such as the 'Real Book' and the 'American Songbook', which are still regarded as works in progress. These pieces are commonly referred to as 'standards' and were composed mostly in the 1930s, '40s and '50s for films and musicals by musicians such as George Gershwin, Cole Porter, Irving Berlin, and many others.[40] Improvised music based on elaboration and transformation of copyright standards evidences that the exercise of copyright by owners of popular music repertoires ends up placing music improvisers in a precarious and disadvantaged position with regard to both live improvisations and improvisatory works fixed in studio recording sessions. As far as live music is concerned, improvisations are generally tolerated although not formally authorized by performing rights societies such as ASCAP and BMI, so that improvisers cannot acquire any rights in the derivative works they create at the time of their performance. With regard to studio recording sessions, improvisers need the express authorization of the owners of the copyright in the underlying musical composition whenever they use such a composition to create an improvisatory derivative work to be fixed in a phonogram.

The main obstacle to jazz music improvisation based upon pre-existing music materials comes from a copyright principle which emerged in the second half of the 19th century and was enshrined in the 1886 Berne Convention, according to which the author of a work protected by copyright must be granted the right to authorize (and object to) any adaptations, arrangements or alterations of her work.[41]

This means that, under international copyright law, derivative works entail permission by the copyright owner in the earlier work.[42] As pointed out above, copyright ends up protecting primarily authors and publishers (as assignees of the whole or a part of authors' rights) of notated works (for example, the Real Book songs) by granting them exclusive rights of repro-

---

[40] Standards are 'pieces that a professional musician may be expected to know': see Note, Jazz Has Got Copyright Law and That Ain't Good, ibid., at 1942.

[41] Cf Article 12 of the Berne Convention (Right of Adaptation, Arrangement and Other Alteration).

[42] The fact that such a right to control transformations and adaptations also has a moral right connotation under Article 6-*bis* of the Berne Convention amplifies the chilling effects on music improvisation, especially in countries where the rights to paternity and integrity of copyright works – which are actionable in cases where at stake is the reputation or honour of the author – are more effectively and generously enforced.

duction, public performance and – most importantly – the aforementioned right of adaptation and alteration of an original work. The application of this principle to an eminently performative art such as music – where copyright, as we have seen, subsists mainly in musical texts and the roles of authors and performers are rigidly pre-determined – greatly disfavours all music which is not (or not entirely) encompassed by notation and fixations occurring before performance of the work.

In light of the above-mentioned structure of rights and prerogatives in musical compositions, it is easy to understand why musical contexts and artistic movements where borrowing, modifying and transforming earlier creations have structural functions and an undisputed artistic value can be so greatly penalized.

If the principle of 'no derivative works without permission' were enforced tightly in the jazz music sphere, copyright would inevitably stifle (instead of encouraging) improvisation practices that jazz musicians have historically developed and resorted to. If, in live jazz performances, copyright were enforced consistently with its structure of rights, the vast majority of music performances would amount to large-scale infringements of copyright in compositions used by jazz musicians for their transformative purposes.

As things stand, the only way for music improvisers to escape the aforementioned bounds raised by copyright, especially in domains where creation of derivative works is pervasive, is by relying on the transformative character of their works. Transformative uses can be successfully excluded from music copyright's scope under the fair use doctrine, as the US Supreme Court did while considering that a rap version of Roy Orbison's 'Oh, Pretty Woman' by group Live Crew was transformative enough to constitute a parody of the original song.[43] In European copyright systems, instead, in the absence of a broadly construed exception to copyright such as 'fair use', transformative uses can be excused under exceptions such as parody and quotation, or, as explicitly provided in certain countries, whenever copyright law explicitly grants musical variations the status of autonomous copyright work.[44]

Whenever copyright exceptions do not apply or the autonomy of the

---

[43] *Campbell v Acuff-Rose Music, Inc.*, 510 U.S. 569 (1994). See Note, Jazz Has Got Copyright Law and That Ain't Good, supra note 39, at 1951.

[44] See, for instance, Article 2 of the Italian Copyright Act (Law n. 633/1941 and subsequent amendments), which reads as follows: 'In particular, protection includes . . . 2) musical works and compositions, with or without words, dramatic musical works and musical variations constituting, as such, original works . . .' (author's own translation).

improvised derivative work from the earlier work cannot be successfully invoked or relied on, copyright laws – as mandated by the Berne Convention – require permission for music improvisers to be entitled to adapt, elaborate or transform earlier copyright works. The impact of such a principle, however, is different depending on whether improvised works are fixed in sound recordings (also during a live performance) or remain unrecorded since performance occurs in live concerts just before a present audience (for example, in a theatre, concert hall or a live music club):

- As far as recorded music is concerned, an improvised piece drawing upon a pre-existing work acquires protection under copyright law as a derivative work in so far as it is original. If the earlier work is still protected by copyright, the improviser and/or her record producer will have to clear not only the right of mechanical reproduction but also the right of adaptation, which is normally exercised by individual composers and/or music publishers, unless a compulsory licence applies and authorizes adaptations. It should be considered that even in the United States, where a compulsory licensing scheme exists (since 1909) for mechanical rights in copyright compositions, the musician wishing to protect his or her original musical additions in a sound recording must seek permission from the copyright holder in the earlier/underlying composition in order to acquire a derivative work copyright in the additional materials.[45]
- As regards live performances, improvised music relying on the use of pre-existing copyright songs and other protected musical materials turns out to be de facto unrestricted since authors and music publishers do not expressly authorize, through their collecting societies, adaptation or transformation of standards. What copyright owners licence is just the right of public performance. This means that, when enjoying unlimited freedom on stage and using

[45] See Note, Jazz Has Got Copyright Law and That Ain't Good, supra note 39, at 1945. See Section 115 of the US Code, Title 17, which provides a compulsory licence to make and distribute a sound recording of a copyright musical work once a recording of such work has already been distributed to the public in the United States with the copyright holder's consent, subject to certain terms and conditions of use. Section 115 specifies that if the follow-on sound recording relies on a new arrangement of the musical composition, such arrangement '. . . shall not change the basic melody or the fundamental character of the work, and shall not be subject to protection as a derivative work . . . except with the express consent of the copyright owner'. This condition explains why the compulsory licensing scheme at issue does not cover highly transformative improvised works, for the recording of which the improviser shall clear the right of adaptation.

repertoires protected by copyright, jazz musicians do so at their own risk, and, at the same time, do not earn a cent from what collecting societies charge concert organizers and/or music venues for public performance of their members' compositions. Although enjoying de facto aesthetic freedom to develop pre-existing materials in their gigs, music improvisers are systematically deprived of any shares of the royalties stemming from the rights of public performance in their own original improvisations, and end up being treated – at least in the live music business – as mere performers of someone else's work. This means that, unless music jazz performers improvise on the grounds of their own tunes (which is increasingly often the case) the only beneficiaries of royalties from live concerts are the authors and publishers of jazz standards.

5. CAN COPYRIGHT GUARANTEE A BETTER TREATMENT OF MUSIC IMPROVISATION?

If copyright were expected to afford protection to all authors of original music works, irrespective of how authorship materializes, there would be no reason to exclude authors of improvised music from the realm of music copyright owners. If copyright systems sought to genuinely follow the broadly protective and formalities-free character of artistic property under the Berne Convention, barriers to a sufficiently broad legal recognition and financial reward of this form of musical authorship should not be raised or maintained, even though such barriers are authorized (as for the fixation requirement) or built (when it comes to derivative works) into the Berne Convention system.

In the relevant literature, it was argued that extending current copyright paradigms to the realm of music improvisation would be counterproductive, if one considers that the implementation of intellectual property rights would impact negatively on dynamics of improvisation practices, at least in certain contexts. It is pointed out that implementing copyright paradigms in artistic environments where creativity is community-based and authorship is primarily a collective, collaborative process would amount to commodifying relations among musicians who are not interested in claiming exclusive rights in their own musical work and in being credited for such work.[46]

---

[46] Cf Larisa Mann, If It Ain't Broke . . . Copyright's Fixation Requirement and Cultural Citizenship, 34 Columbia Journal of Law and the Arts 201 (2011),

Other commentators claimed, instead, that the current under-protection by copyright law of music genres based on improvisation practices has not allowed these genres to grow in terms of dissemination and popularity, and eventually hindered their creative development by widely discouraging vital reinterpretation of pre-existing music.[47] Jazz, for instance, accounts for a very low percentage of total record sales and no longer influences pop culture as it once did.[48]

As shown in the next paragraphs, in spite of persisting discrepancies at national level on significant fronts – for example, the notion of musical 'work' and the allocation of different rights to composers and performers; the scope of exceptions allowing certain transformative uses – copyright systems could ensure a better understanding of and provide a stronger support to improvised music. Firstly, authorship could rely on a broader understanding of what kinds of original contribution are copyrightable. Secondly, adaptations and elaborations of earlier music works, at least in certain musical sectors, could be regarded, on the grounds of their essentially transformative character, as exempted from the scope of the right to control the derivative works that improvisers create by modifying or quoting earlier works.

5.1 A Broader Notion of 'Work' Should Encompass Performative Aspects of Creation

As briefly recalled above, the requirement of originality under both European and US copyright laws sets out a low threshold of 'authorship'. The fact that the Court of Justice of the European Union, on the grounds of an EU-wide standard of originality based on the notion of the 'author's own intellectual creation', affirmed the copyrightability of 'works' such as 11-word extracts from protected newspapers or simple portrait photographs shows that copyright protects also minimum levels of creativity.[49] If, as a matter of consistency, at least in the EU, this broadly protective approach were applied also to musical authorship, one would need to replace the current narrow concept of 'work', based primarily on musical texts, with a broader concept encompassing also the aspects of composition which are not embodied in musical texts but are added at the time of

---

available at http://ssrn.com/abstract=1691266 (discussing whether the requirement of fixation should be abolished in the US).

[47] See Note, Jazz Has Got Copyright Law and That Ain't Good, supra note 39, at 1940–1941.

[48] Ibid. at 1940.

[49] See, respectively, notes 26 and 27.

performance and fixed on a tangible medium. This means that, at least in certain music genres where performers are expected to provide original contributions, authorship should logically be evaluated on the grounds of the medium embodying the final rendering of the music work, which is normally a phonogram.

At the end of the day, excluding oral expressions of music or 'non-visual' musical practices from the notion of 'work' is no longer easy to justify.[50] From the time when music publishing could be viewed as the most profitable copyright business, reality and technology have radically changed. The original commercial (and artistic) value of music sheets and scores has been preserved mostly in sectors where notation is still expected, as with music from the Romantic generation, from the beginning of the 19th century onwards, to encompass the act of musical creation in its entirety.[51] In these repertoires, the idea was (and still is) that the role of the performer remains subjected to that of the composer, so that performers are not expected to take any freedoms to alter and modify the musical text, which is therefore 'sacralized'.[52]

In the aforementioned domains it does make sense to preserve the current notion of music work and to continue to give priority to notation and sheets. To the contrary, for jazz and other music genres whose practices are at odds with the aforementioned narrow notion – since acts of creation and performance are intertwined – to prove authorship (which could often be joint authorship) musicians should be allowed to rely on the performative aspects embodied primarily in the sound recording version of a piece, which would define a broader notion of 'work' that corresponds to the reality of practice/performance.

The main point here is that the fact that recordings as such are protected through a neighbouring right in countries which have implemented the Rome Convention or a limited copyright in countries like the US should not hinder the qualification of extemporaneous music composers as *both* authors and performers of their music 'work'. The

---

50 See Anne Barron, Harmony or Dissonance?, supra note 36, at 48, and Jason Toynbee, Copyright, the Work and Phonographic Orality in Music, 15 Social & Legal Studies 77, 95 (2006) (discussing also, in a critical way, whether or not the notion of authorship and musical work should extend to aspects of a sound recording as performance style).

51 It should be recalled that even in the history of classical music there are avant-garde repertoires in which the notation leaves room for improvisation, and performers are requested to make their own extemporaneous creative choices while interpreting musical compositions by composers such as John Cage, Bruno Maderna, Sylvano Bussotti and Karl-Heinz Stockhausen.

52 See Arewa, Creativity, Improvisation and Risk, supra note 7, at 1842.

fact that fixation of an improvised performance triggers the acquisition of rights in the ensuing sound recording should also help evidence the acquisition of a related copyright in the musical composition created simultaneously.

In domains where performance naturally entails the addition of musical elements to the ones fixed through notation, an issue arises with regard to the version of a work which should be regarded as the definitive one, especially when such a work has to be compared with other creations because of plagiarism or infringement claims.[53] When jazz flutist James Newton claimed that the rap group Beastie Boys had infringed his copyright in a piece for solo flute ('Choir') by using, without his authorization, a sample from a record of the piece Newton had made for ECM, the Central District Court of California found against him on the assumption that the recording sample triggered the clearance of the sole sound recording right, which was licensed by ECM.[54] The Court (whose findings were upheld by the Court of Appeals for the Ninth Circuit)[55] distinguished between the right in the musical composition and the right in the sound recording, arguing that 'A musical composition protects an artist's music in written form . . . A musical composition's copyright protects the generic sound that would necessarily result from any performance of the piece'.[56] As pointed out by Toynbee, the main issue arising in this case was whether the author-performer could claim authorship in his piece as rendered in the sound recording, which contained a peculiar effect, known as 'multiphonics', produced by singing and playing a wind instrument at the same time.[57] This vocalization was not indicated specifically in the music text, which merely instructed the performer to sing into the flute and play simultaneously, without indicating the notes (C-D flat-C) that Newton improvised at the time of fixing his own piece into the ECM record. The Court dismissed the infringement claim, arguing that the additional elements coming from Newton's performance and fixation of sounds could not be regarded as protected by copyright since they were not notated in the score. The only right which could apply to the rap sampling was

---

[53] See *Newton v Diamond*, 204 F. Sup 2d 1244 (CD Cal. 2002). For a thoughtful analysis of the case see Toynbee, Copyright, the Work and Phonographic Orality in Music, supra note 50, at 91–92.

[54] See Toynbee, Copyright, the Work and Phonographic Orality in Music, supra note 50, at 91.

[55] *Newton v Diamond*, 349 F. 3d 591 (9th Circuit 2003).

[56] Ibid., at 249.

[57] See Toynbee, Copyright, the Work and Phonographic Orality in Music, supra note 50, at 91.

the (narrower) right in sound recordings owned, and duly licensed to the Beastie Boys, by ECM.[58]

As this chapter advocates, in proving originality and authorship of a certain music work, the performative aspects of musical composition should ideally be taken into consideration, and, if justified by a certain kind of musical authorship, prevail over the mere text-based version. In other words, the performed version of a music piece, as fixed in a phonogram, could legitimately be regarded as a more reliable form of fixation than a musical text. As pointed out above, the recorded versions of jazz, rap or hip hop pieces can easily contain additional elements which are improvised or added by performers only at the time of the recording production and can make the recording of a work remarkably different from what the composers notated on a music sheet.[59]

The advantages of a flexible notion of work could be appreciated also the other way around, in order to evaluate the subsistence of musical authorship whenever a music performance embodies additional original input incorporated into performing editions used for the fixation of a certain composition. Here an excellent example comes from the practice of ancient music, whose notation often leaves broad room for improvisation at the time of performance, giving musicians a varying degree of interpretative freedom. Notation in manuscripts from the 16th and 17th centuries is often poor, full of lacunae and/or indecipherable by musicians accustomed to modern notation. This means that performance of such a repertoire often requires the intervention of a musicologist having expertise in that repertoire and being able to fill the existing gaps in the score. As held by a UK court in a case concerning the production of a recording of Baroque music by the French composer Lalande, if the musicologist's work ends up filling significant gaps in the original manuscript and making the music works playable by adding musical lines, arrangements, bass line figures and ornamentations or variations, those original additions and materials can qualify as copyright works.[60] This entails that the record producer having fixed performances based on musical texts and on performing editions added by a musicologist would infringe the musicologist's copyright

58 Ibid., at 92.

59 As pointed out in this volume by Luke McDonagh, 'Protecting Traditional Music Under Copyright (and Choosing Not to Enforce It)', adopting a broad and flexible notion of 'musical work', which encompasses not only notation but also other elements of musical practice and performance, helps to apply copyright law also in the domain of traditional music, where musical arrangements of pre-existing tunes can qualify as original works and acquire copyright protection.

60 See *Hyperion Records v Sawkins* [2005] EWCA Civ 565.

if the record producer had not cleared the mechanical rights of the author of the performing editions.

If, as it has been suggested, the notion of music 'work' were no longer defined too narrowly and took account of the significant impact of performance on the creation of a work, it would be much easier for music improvisers to exercise their rights as 'authors' of original works in all sorts of exploitation of such works, from production of phonograms to the live performance business.

5.2 More Room for Free and Autonomous Transformative Music

Assessing whether a music improviser re-using someone else's musical work should be treated as author of an autonomous creation or, more reductively, as an author of a derivative work can be a hard task, legally speaking. Whether copyright in existing works should legitimately raise barriers and limits to musical adaptations and elaborations which characterize and define entire genres is a question that deserves, from a copyright policy perspective, a convincing answer, especially if one considers that a few of these genres have escaped the contexts and communities where they initially developed.[61]

For a matter of contractual fairness and an adequate balancing of interests, courts or bargaining parties should be placed in a position to predict the circumstances under which an improvised music work drawing upon (or at least initially quoting) an earlier copyright work would be regarded as a proper, autonomous work, and not as an elaboration or mere performance of earlier songs or compositions. For instance, a fair and balanced interpretation of the originality requirement could enable courts to understand to what extent Coltrane and his band fellows were indebted to Rodgers and Hammerstein in the making of their well-known version of 'My favourite things' and whether Coltrane's version constitutes a proper autonomous work or a derivative work.[62]

If one wished to foster improvisation drawing on pre-existing creations, which is typical of jazz music, extemporaneous composers would need to gain more freedom from legal ties to earlier works, as established by the copyright principle on derivative works. At the moment copyright law allows authors of copyright songs such as the

---

[61] See Anne Barron, Harmony or Dissonance, supra note 36, at 33, recalling how rap emerged out of the New York hip hop scene in the late 1970s, under the influence of dub music and DJs who had moved from Jamaica to New York in the early 1970s.

[62] See John Coltrane, My Favorite Things, Atlantic Records (1961).

Real Book or American Songbook standards to restrict or monetize widely transformative re-uses of their works made by music improvisers. In this regard, music copyright can be regarded as financially rewarding past composers while raising a barrier to follow-on musical creativity, especially after the 20-year extension of copyright's term of protection in Europe and in the United States in the mid- and late 1990s.[63] If obstacles to the making of derivative works were removed or lowered through one of the aforementioned legal tools – depending on their concrete availability and scope in a given copyright system – improvisers and their record producers would largely benefit from having to pay no fee for the clearance of rights in the repertoires they use, with subsequent, significant savings.

As advocated in the relevant literature, at least in certain musical contexts where earlier works are used as standardized elements of creation, such as the jazz music domain, copyright songs used as a starting point for musical improvisation could be treated as mere facts or ideas, whose elaboration and transformation should not entail permission and/or payment of any licence fees.[64] At the end of the day, if underlying musical compositions used by jazz performers constitute mainly raw materials, it is unfair (or widely exaggerated) to grant their copyright owners the right to control follow-on creativity.[65]

Alternatively, as pointed out above, one solution could be that of resorting to a broader interpretation of exceptions which shield highly transformative uses from infringement actions. Obviously, US courts would be better equipped than European courts considering the potential scope and flexibility of the fair use doctrine, which seeks to valorize the transformative character of follow-on work. However, the application of European copyright exceptions such as quotation and parody should not be underestimated a priori, especially at a time when the CJEU has embarked on a process of interpretation and harmonization of copyright exceptions, including parody, with EU-wide effects.[66]

---

[63] See, respectively, Directive 93/98/EEC harmonizing the term of protection of copyright and certain related rights and the 1998 Sonny Bono Copyright Term Extension Act in the US (Pub. L. No. 105-298, 112 Stat. 2827 (1998).

[64] See, Note, Jazz Has Got Copyright Law and That Ain't Good, supra note 39, at 1947ff.

[65] See, Note, Jazz Has Got Copyright Law and That Ain't Good, supra note 39, at 1948 (arguing that jazz standards are 'not intended to be the end product that reaches the listener' but simply '. . . the idea from which the predominantly improvisatory expression flows').

[66] See, with regard to parody, C-201/13 *Deckmyn v Helena Vandersteen and others* [2014].

Finally, when it comes to unrecorded music works improvised on stage, the ephemeral character of performance inevitably makes it hard or impossible, especially in the context of group improvisation, to discern and prove authorship from a legal point of view and to ascertain what has been created, by whom, and how much has been borrowed and transformed by the various performers.[67] However, at least in countries where fixation is not a mandatory requirement for subsistence of copyright, these improvisatory works would still qualify for copyright protection, at least on paper. An objective problem persists, instead, in common law jurisdictions, where an effective solution resulting in a fair qualification of musical authorship in the domain of improvised works would require a relaxation of the fixation requirement under copyright law statutes.

In these situations, if one wished to grant music improvisers freedom to use and elaborate someone else's work and to gain remuneration for their improvisations, it would not be unrealistic – also as a matter of local cultural policy – to encourage a change in the way authors and publishers of copyright works license, through their performing rights societies, their own works for live performances. If one wished to place authors at the centre of the copyright system, it would seem fair not only to recognize the aforementioned freedom, but also a right to remuneration justified by the original work music improvisers perform on stage on the grounds of earlier copyright works. If – to foster use of their works by music improvisers – performing rights societies licensed not only the authors' rights of public performance but also the rights of adaptation and transformation for purposes of improvisation, the concerned rights holders could contractually agree to reserve a fair share of public performance rights revenue to musicians having performed in specialized music venues where improvisation is an everyday business (jazz clubs, theatres or other music venues hosting improvised music).[68] This is actually the model followed by the French collecting society SACEM for the allocation of royalties stemming from rights of public performance in France and Luxembourg. This system of remuneration is based on the enrolment of SACEM members in a register of 'improvisers of jazz music', who are entitled to use, every time they perform copyright works publicly, a special form to be submitted to SACEM. To be eligible for such improvisation rights, these musicians are required to specify the title of each work, the performance and the name(s) of the improviser(s) in addition to the names of authors and composers of

---

[67] See, in particular, Mann, If It Ain't Broke, supra note 46.

[68] See Sacem F, http://sacem.fr (last accessed November 29, 2017) and Sacem Luxembourg, http://sacem.lu (last accessed November 29, 2017).

the underlying musical compositions. On the grounds of this information, SACEM regulations provide that two-thirds of the collected fees should be allocated to the owners of copyright in the original work and on-third to the improvisers whose names appear in the live music program.

6. CONCLUSION

As this chapter has advocated, there is no reason in a legislative environment which provides strong and broad protection to authorship, at both international and European level, to place authors of improvised music works in a disadvantaged position. This chapter has shown that it is mainly for historical reasons that the notion of musical 'work' ended up being defined on the sole grounds of notation, with the extension of copyright protection originally conceived for books and other literary works to musical writings. Music is, at the end of the day, an eminently performative art. In certain music genres composition inevitably occurs at the time of performance, in such a way that authorship and performance can easily coincide and be attributed to the same musician or group of musicians. This chapter has pointed out that, for improvised works which are recorded by using one of today's myriad digital recording technologies and devices, it is unreasonable to restrict the notion of 'work' to notated music.

As it has been said, such a 'sacralized' notion might still be acceptable for those works, especially in modern and avant-garde music repertoires, where texts are designed to entirely encompass the act of musical creation, in a way that performers are not asked or encouraged to create additional elements of a music piece at the time of performing it. In other music genres (this chapter has used jazz as an historically and aesthetically prominent example) where musical texts are no more than a source of guidance or raw materials for musical performance, instead a fair definition and finding of authorship should logically (and legally) be based on the final rendering of a musical work. In those genres it has been stressed that the creative elements and choices which meet the requirement of originality and, as a result, deserve copyright protection, can be much better appreciated and weighted, at least in their entirety, in a sound recording than in sheet music. In those genres where improvisation has a prominent role, a definition of authorship based on the final rendering of the work should not be hindered by the existence of compartmentalized categories of intellectual property rights, whose emergence was historically due to technological developments and to the need to foster and financially reward the work of individuals and/or entrepreneurs who contributed to the making and distribution of sound recordings without being, legally speaking, 'authors' of musical compositions.

It goes without saying that, as this chapter has sought to clarify, music improvisers shall be legally treated as holders – in a condition of simultaneity – of a plurality of exclusive rights granted under copyright laws to authors and, separately, to performers. This chapter has pointed out also that whenever original impromptu works draw upon someone else's work, as in the case of improvisations upon copyright jazz standards, a fair definition of musical authorship should make it possible for authors of widely transformative works or musical variations to acquire an autonomous copyright in their compositions without having to acquire permission from the composer and/or publisher of the work he or she transforms. As the chapter pointed out, such a non-derivative and autonomous character of improvised works could be found on the grounds of different legislative provisions (for example, parody and other copyright exceptions; the autonomous protection of musical variations) and doctrines (for example, 'fair use' in the US) that creators, content producers and, eventually, courts can dispose of in different copyright systems. Last but not least, it has been argued that, as regards unrecorded improvised performances in a genre like jazz, a relaxation or repeal of the mandatory requirement of fixation in jurisdictions such as the United States, the United Kingdom and Ireland would make it possible for collecting societies to strike a fair balance between the interests of composers and music publishers whose works are used for improvisatory purposes and the interests of music improvisers who use those works for their own impromptu creations in live music venues. As shown by the example given by SACEM in France and Luxembourg, there is nothing (apart from the above-mentioned fixation requirement in common law jurisdictions) which restricts parties from contractually agreeing on a fair split of the revenue stemming from the clearance of rights of public performance and of musical adaptation between the copyright holder in a pre-existing copyright composition and the author of an improvised work which draws upon such composition.

9. Original compilations of musical works: can DJ sets be protected by copyright?

Tom Iverson

1. INTRODUCTION

This chapter examines whether DJ sets and playlists, which are essentially compilations of sound recordings, can or should be protected by copyright, both in common law countries like the UK and the US, and civil law jurisdictions, such as Germany and France. The question, of course, must be assessed on a country-by-country basis using the relevant national law given the territorial nature of intellectual property rights such as copyright. But the underlying principles concerning whether DJ sets should be given copyright, or some other form of intellectual property protection, depends around the world upon the same points regarding creativity and the need to protect the creative output of individuals in order to encourage others to use their skills in a similar manner, thereby benefitting the whole of society through entertainment and cultural diversity. There is no question that authors, songwriters and directors should be given such protection, so why not DJs and others who create music compilations? Many world-famous DJs are paid large sums of money to put together the perfect combination of tracks, whether that's as a radio DJ broadcasting to millions, a nightclub superstar DJ spinning tracks for thousands of revellers, or a famous DJ working with a record label to create a multi-platinum-selling dance compilation.[1]

In terms of the requirements for copyright protection, DJ sets do not suffer from the classic copyright dichotomy of idea versus expression.[2] The

[1] In this chapter, the term 'DJ set' is intended to cover each of those various music compilations.

[2] In *(1) Ministry of Sound Group Ltd (2) Ministry of Sound Recordings Ltd v Spotify Ltd* (Claim No. HC13C03860) (an important case on the subsistence of copyright in music compilations, that is considered in the following paragraphs),

requirement for fixation[3] is also satisfied so long as the DJ set is recorded in some material form: a playlist created before or after the set, or a sound recording of a live set, for example. Fixation may be a problem, however, for a live DJ set with impromptu track choices which is neither written down nor recorded.

The crux of the issue, therefore, is most likely to be the requirement of originality, as is often the case in copyright matters. In this respect, anyone who has listened to a DJ in a nightclub, at a wedding or even a house party knows that there can be genuine artistic creativity involved in compiling and executing the set. (Simply standing at the decks and playing some popular tracks one after the other is rarely enough.) This art form is well recognised and has been for decades: big names in the world of dance music are paid big bucks to select tracks and mix them together for adoring fans, who pay significant sums to be entertained by the DJs, in a different way to the artists whose tracks are being played. So is this the sort of creativity that should be protected by copyright?

This chapter seeks to understand the law in the UK, where the issue has been considered most closely, with some comparisons with other common law and civil law jurisdictions around the world, and give consideration to what protections (if any) are currently in place for DJs and DJ sets.

2. CAN COPYRIGHT SUBSIST IN AN ORIGINAL MUSIC COMPILATION OR DJ SET?

2.1 Common Law Countries: UK

Unfortunately, the law in the UK is far from certain on this issue. Unlike many other jurisdictions, however, there have been legal proceedings brought and argued on the very closely related issue of whether there is copyright protection in a music compilation. The case was between two of the best-known names in music: Ministry of Sound and Spotify.

Ministry brought proceedings in the English High Court in September

---

Spotify argued that Ministry had substantially overstated the creative process behind creating its dance music compilation albums and that the concepts were simple and banal themes and that there is no copyright in a concept. The case settled prior to judgment, but it is submitted that this is a weak argument.

[3] Fixation is not a requirement for protection in all countries, but is in the UK.

2013 against Spotify.[4] Sadly for music and copyright lawyers, however, the case was settled some six months later in February 2014.[5]

Although there was no judgment given due to the relatively early settlement, the stances taken by both sides in their respective statements of case[6] make the opposing arguments as to whether there can ever be copyright in a music compilation or set list. Obviously, some of the arguments raised were fact-specific, but many are of general application. This chapter also considers what the outcome might have been had the case proceeded to trial and its application to copyright subsistence in DJ sets generally.

2.1.1 *Ministry of Sound v Spotify* (2013)

In 2013, Ministry was a well-known group operating mainly in the field of dance music. Its main source of income derived from sound recordings, including compilation albums, as well as its world-famous nightclub in Elephant & Castle, London, and certain other music-related operations. The nightclub opened in 1991, and Ministry released its first sound recordings in 1993. In the 20 years between 1993 and 2013, Ministry's compilation albums sold around 50 million copies worldwide, selling around 3 million copies of each of the 30 or so physical compilations and 50 or so digital compilations that Ministry released each year. The brand was hugely successful, having 45 albums reach number 1 in the UK Compilation Albums Chart. This popularity endured, and in the year before proceedings were issued, 11 of the UK Top 40 dance albums were Ministry compilations.[7] Ministry was arguably the biggest dance music brand in the world, had the planet's largest independent record label (since acquired by major label, Sony Music UK, in 2016), and its compilations had won numerous awards.

In 2013, Spotify's 21st-century business operated a music streaming service, providing content from a range of major and independent labels[8] with a repertoire of over 20 million tracks.[9] (In 2017, that had grown to

---

[4] *(1) Ministry of Sound Group Ltd (2) Ministry of Sound Recordings Ltd v Spotify Ltd* (Claim No. HC13C03860).

[5] See, for example, *The Guardian*, 'Spotify and Ministry of Sound Settle Music Playlists Copyright Lawsuit', Feb, 27, 2014, http://www.theguardian.com/technology/2014/feb/27/spotify-ministry-of-sound-lawsuit-settlement.

[6] Particulars of Claim dated 2 September 2013 prepared by Edmund Cullen QC [hereinafter Ministry Particulars] and Defence dated 15 October 2013 prepared by Henry Carr QC and Guy Hollingworth [hereinafter Ministry Defence].

[7] Ministry Particulars at para. 6.

[8] Ibid. at para. 11.

[9] Ibid. at para. 5.

around 30 million tracks.) Part of Spotify's service then (and now) allowed users to create 'playlists' of tracks and other users to search for and share those playlists. Users could name and arrange the playlists in any way that they wanted.

2.1.2 The claim

Ministry alleged that Spotify had infringed its copyright in certain music compilations by the carrying out, or authorising the carrying out, of certain acts in relation to the whole or a substantial part of those compilations without Ministry's consent, contrary to section 16 of the Copyright, Designs and Patents Act 1988,[10] namely copying the compilations or communicating them to the public. It stated that in numerous instances, Spotify users were compiling playlists that were the same, or substantially the same, as various Ministry compilations. Indeed, many of the playlists referred to Ministry compilations in their titles. The playlists were accessible to other users, who could then use the Spotify service to stream those tracks.

Ministry asserted that:

- Spotify had made copies of the whole or a substantial part of the track listings in certain Ministry compilations;
- Spotify had made those track listings available to the public by electronic transmission in such a way that members of the public could access them from a place and time individually chosen by them; and
- Spotify users had infringed Ministry's copyright in the compilations by copying them or communicating them to the public, and so: (i) by providing the service, Spotify had given users the means to do so; and (ii) by failing to take steps to prevent the creation of such playlists or to remove infringing playlists, Spotify was liable.

2.1.3 Ministry's position

Ministry claimed that copyright subsisted in each of its compilations as a database under sections 1, 3(d) and 3A(1) of the CDPA. This was on the basis that each compilation was an original copyright work (a database) because the selection and/or arrangement of its contents constituted its 'author's own intellectual creation', pursuant to section 3A(2) of the CPDA. Section 3A(1) states that a database is:

---

[10] Copyright, Designs and Patents Act of 1988, S.I. 1990/1496 (UK) [hereinafter CPDA].

a collection of independent works, data or other materials which –

- are arranged in a systematic or methodical way, and
- are individually accessible by electronic or other means.

Ministry explained that its compilations were usually created by its in-house team, consisting of music industry experts, DJs and individuals specifically recruited for their knowledge of music. The 15-strong team had decades of experience in creating compilation albums. They used not only their personal knowledge, but also sources such as DJ charts and airplay and sales statistics.

The Ministry team would start by putting together a concept for a new compilation album, which might derive from the music being played in clubs in the UK or abroad which had not yet reached the mainstream, or may tap into a trend in popular culture, or may be based on a historic back catalogue. Alternatively, Ministry would hire consultants from time to time to bring in new ideas to the compilations team. The individual members of the team working on a particular project were picked depending on the specific individual's area of expertise, age and musical tastes.

Once a concept had been determined, the team working on the project would compile a long-list of approximately 100 tracks based on their collective personal knowledge, by researching previous albums, music charts and radio playlists, while also considering the likelihood of being able to secure a licence for any particular track. Compiling such a long-list would typically take approximately two weeks, including making numerous revisions to ensure that the list matched the concept. The long-list would then be reduced to around 60 tracks, in part depending on the licences that could or could not be obtained.

The team members with the most relevant knowledge then created a final track list. Each CD or mix would have a different theme within the concept.[11] The team member with the best knowledge of the relevant repertoire arranged the compilation based on factors including exclusivity and familiarity of the tracks, balance of sound, artists and mix of each CD and the chronology of the selected tracks.[12] Individuals within the team revised and re-ordered the tracks as necessary to finalise the compilations. Revisions might be required during the mixing process or as a result of licensing issues. The final running order would be approved by Ministry's Senior Product Manager and the Head of Compilations.[13]

---

[11] Ministry Particulars at para. 23.
[12] Ibid. at para. 24.
[13] Ibid. at para. 25.

2.1.4 Spotify's position

In defending the claim, Spotify raised numerous points, the most significant of which was that copyright did not subsist in the particular compilations which were the subject of the claim or, indeed, in any music compilation at all.

2.1.5 No 'author's own intellectual creation'

Spotify argued that Ministry had substantially overstated the creative process.[14] Instead, the concepts were simple and banal themes generally relating to an established type of dance music, and, of course, there is no copyright in a concept. Further, many of the tracks were obvious 'standards' within a particular genre, were taken from pre-existing compilations, or were based on chart popularity, and, as such, did not reflect the authors' own intellectual creativity.[15] Reducing the long-list to the tracks to be included was primarily dictated by licensing and commercial decisions, Spotify argued, meaning that even if any authors' own intellectual creative choices had been made, these were not reflected in the ultimate compilations.

Spotify also asserted that the arrangement of tracks reflected banal and obvious choices, such as arranging tracks chronologically or putting popular tracks at the start and end of the running order.[16] To the extent that any authors' own intellectual creativity had been used in editing and mixing the tracks together, this had not be reproduced in the playlists (given that the Spotify playlists were created from individual tracks, not mixed together, and all of the individual tracks were properly licensed).

For these reasons, Spotify claimed that the process referred to by Ministry did not constitute the authors' own intellectual creation within the meaning of section 3A of the CDPA.[17] Alternatively, even if copyright could subsist in a compilation of recordings of musical performances, the authors' own intellectual creation in this case was insufficient to overcome the strong presumption in Recital 19 of the Database Directive.

2.1.6 A compilation of sound recordings is not covered by the Database Directive

Spotify argued that a compilation of sound recordings on an album is not covered by database copyright. Recitals 17 and 19 of the Database Directive state:

---

[14] Ibid. at para. 20.
[15] Ibid. at para. 22.
[16] Ibid. at para. 27.
[17] Ibid. at para. 29.

(17) Whereas the term database should be understood to include collections of works, whether literary, artistic, musical or other, or of other materials such as texts, sounds, images, numbers, facts, data; whereas it should cover collections of works, data or other independent materials which are systematically or methodically arranged and can be individually accessed; whereas this means that a recording or an audio visual, cinematographic, literary or musical work as such does not fall within the scope of this Directive; . . .
(19) Whereas, as a rule, the compilation of several recordings of musical performances on a CD does not come within the scope of this Directive, both because, as a compilation, it does not meet the conditions for copyright protection and because it does not represent a substantial enough investment to be eligible under the sui generis right.

The sui generis database right is, of course, a separate right from copyright in a database. It subsists if there has been a substantial investment in obtaining, verifying or presenting the contents of a database (Regulation 13, Copyright and Rights in Databases Regulations 1997). It should be noted that Ministry did not claim that Spotify had infringed the sui generis database right.

Spotify argued that Recitals 17 and 19 reflected a decision made by the legislature to exclude compilations of sound recordings from the scope of the Database Directive. The reasons for this were that phonograms are not protected as authors' rights at international level, but are instead given lesser protection as neighbouring or related rights. The Database Directive did not intend to give a term extension to the protection already afforded to sound recordings.[18] As such, giving database copyright protection to sound recordings, such as the Ministry compilations, on the basis that they constitute a compilation of other sound recordings, would be an unwarranted extension of the term of protection.[19] In the alternative, Spotify argued that the exceptional circumstances required to justify an extension of term did not exist in the circumstances. Additionally, Spotify argued that the Database Directive applies only to databases that are literary and artistic works within the scope of the Berne Convention, which it asserted a compilation of sound recordings will rarely, if ever, be.

2.1.7 What would the court have held had the case proceeded to trial?

Copyright in a database only subsists if it is 'original' in the sense that, by reason of the selection and/or arrangement of its contents, the database constitutes the 'author's own intellectual creation' (section 3A(2), CDPA).

---

[18] Ministry Defence, at para.16.

[19] Copyright in a database lasts for 70 years from the end of the calendar year in which the author of the database dies.

In other words, picking the tracks and the running order must be the compiler's own intellectual creation.

Recitals 15 and 16 to the Database Directive state:

> (15) Whereas the criteria used to determine whether a database should be protected by copyright should be defined to the fact that the selection or the arrangement of the contents of the database is the author's own intellectual creation; whereas such protection should cover the structure of the database;
> (16) Whereas no criterion other than originality in the sense of the author's intellectual creation should be applied to determine the eligibility of the database for copyright protection, and in particular no aesthetic or qualitative criteria should be applied;

So, eligibility for protection depends on whether it is the 'author's own intellectual creation', and that alone. As the CJEU stated in C-604/10 *Football DataCo v Yahoo! UK Ltd*:[20] '. . . the significant labour and skill required for setting up that database cannot as such justify such a protection if they do not express any originality in the selection or arrangement of the data which that database contains'.

So as long as the author's own intellectual creativity is in fact used in the selection and arrangement of the music compilation or playlist (a question of fact in each case), the only significant problem appears to lie with the general rule under Recital 19 to the Database Directive.

In some ways, it is hard to see that compilations of musical performances do not, as a rule, 'meet the conditions for copyright protection'. Recital 16 states that no criterion other than originality, in the sense of the author's own intellectual creation, should apply to determine eligibility for copyright protection, and *Football DataCo* states that the author's own intellectual creation must be applied to the selection or arrangement of the data within the database. This view appears to be shared by the authors of Laddie, Prescott and Vitoria[21] who state their belief that:

> . . . a person can research and put together a compilation of songs and create something which must be properly considered as constituting the author's own intellectual creation. Whether or not a compilation album amounts to the author's own intellectual creation will depend on all the facts. Some will qualify, others will not. It is submitted that it depends on what kind of work goes into the creation of the compilation. Notwithstanding the words of Recital 19 there

---

[20] See Case C-604/10, *Football DataCo Ltd. v Yahoo! UK Ltd.*, 2012 E.C.R. 115.

[21] Hugh Laddie et al., *The Modern Law of Copyright and Designs*, 32.32 (4th edn, LexisNexis, 2000).

> is no absolute rule that a compilation of recordings of musical performances cannot qualify for copyright protection.[22]

That said, even if that view is correct, there could be a strong public-policy argument against the subsistence of database copyright in music compilations, or, indeed, any multi-track sound recording (where a similar level of creativity might be involved). As noted in Spotify's defence, whether or not it is fair, sound recordings are treated differently to literary, artistic and musical copyright works and so the term of protection granted is generally for a fixed term from the end of the year of first release. Although it was extended in the UK in 2013[23] from 50 years to 70 years (put simply), this is significantly less than the usual 'life plus 70 years' awarded to author's works. There has been a great deal of debate over the term of protection for sound recordings for many years, and for a parallel form of extension to come into effect, as it were, 'by the back door', in the case of album track listings (and not just music compilations), would cause a major stir.

Some might see that as a just reward for the creativity that goes into track selection and arrangement in many cases, something often undertaken by individuals who do not benefit from copyright in the sound recording or in the music and lyrics. Perhaps the real problem is for the label investing in the process, which would welcome the back-up of a property right, rather than the slender advantage of being the first to market. At the same time, streaming services want their subscribers to be free to copy their favourite playlists, whatever the source. Understandably, it seemed that neither Ministry nor Spotify wanted to set an unhelpful legal precedent.

2.1.8 Application of the *Ministry v Spotify* arguments to DJ sets generally

In theory, any DJ set could constitute a collection of independent works (the tracks mixed together or simply played back-to-back) which are arranged in a methodical way (a way that the DJ believes 'fits' together well) and are individually accessible (section 3A(1) of the CDPA). And the selection and/or arrangement of the DJ set arguably at least constitute its 'author's own intellectual creation' (section 3A(2) of the CPDA), so long as the tracks are not simply played in alphabetical or chronological order, for example. Being an experienced DJ or not should make no difference. The literary work of a 10-year-old child is equally capable of copyright

---

22 It should be noted that the fourth edition of Laddie pre-dates the CJEU decision in *Football DataCo*.

23 Copyright and Duration of Rights in Performances Regulations 2013, s. 6(b).

protection as the work of a Booker-Prize-winning novelist, in theory at least. Selection and arrangement (and also mixing) are the critical aspects leading to the success of a DJ set and therefore must be considered the most valuable and worthy of protection.

DJs playing a live set are also less likely to be affected by the licensing issues sometimes faced by the Ministry of Sound team, so their 'intellectual creation' should not be inhibited in the way it was asserted by Spotify that it had been. Similarly, many DJ sets do not follow simple and banal themes, contain obvious 'standards' within a particular genre, or are based on chart popularity, for example – all arguments raised by Spotify. Aside from the selection of the tracks, a DJ is unlikely to arrange tracks in a set list chronologically, albeit the DJ might put popular tracks at the start and end of the running order.

Further, a DJ set may also use the author's own intellectual creativity in editing and mixing the tracks together – something which had not been reproduced in the playlists at issue in the *Ministry v Spotify* claim (given that the Spotify playlists were created from individual tracks, not edited or mixed together, and all of the individual tracks were properly licensed).

The same arguments apply, however, to DJ sets generally, as were raised by Spotify, that Recitals 17 and 19 of the Database Directive reflected a decision made by the legislature to exclude compilations of sound recordings from the scope of the Database Directive. Similarly, the same counter-arguments apply that Recital 16 states that no criterion other than originality (i.e. author's own intellectual creation) should apply to determine copyright protection.

2.1.9 Other (non-copyright) types of protection for DJ sets

There are also potentially various other types of intellectual property rights which might arguably give some protection to DJ sets, DJs and music compilations, in addition to, or instead of, copyright. Again, however, the law in the UK on these matters remains untested.

2.1.10 Sui generis database right

In *Ministry v Spotify*, the claimant chose not to run an argument of infringement of this alternative type of database right. On the facts, that was a sensible decision. In order for this related protection to apply, there must have been a substantial investment in obtaining, verifying or presenting the data.[24] The purpose of the legislation was to ensure protection for

---

24 Regulation 13(1) of the Copyright and Rights in Databases Regulations 1997.

those who have spent significant sums in creating databases,[25] but without using the creativity of the 'author's own intellectual creation' in order to make the database an 'original' work in the copyright sense.

2.1.11 Performers' rights

It is arguable that performers' rights might protect DJ sets, perhaps in a way that would not protect music compilations generally. In many countries, including in the UK and continental Europe, the US and Australia, performers of musical performances have rights against those who seek to make unauthorised, so-called 'bootleg', recordings of their performances.[26] This right subsists in addition to any copyright which exists in the underlying recordings (or which may exist in the music compilation in the form of the set list). Musical performers (along with actors) generally have the right to refuse consent to their performances being recorded, distributed or broadcast without their consent.

It is likely, however, that such protection would only apply (if at all) to live DJ sets where the DJ was doing more than simply playing a pre-existing track selection or playing songs back to back, without mixing the tracks together. It is the latter which would arguably constitute a 'performance' in the performers' rights sense, which would be worthy of protection. Under section 180(2)(b) of the CDPA a 'musical performance' is protected 'which is, or so far as it is, a live performance given by one or more individuals'. Musical performance is not itself defined within the CDPA, although 'musical work' is defined at section 3(1) as 'a work consisting of music, exclusive of any words or action intended to be sung, spoken or performed with the music'. Protection for the latter is given by section 180(2)(c) for 'a reading or recitation of a literary work', although it is hard to see how it could be argued to apply to the playing of the lyrical parts of a DJ set (as opposed to the musical elements) without a very wide interpretation of the word 'recite'. Again, the law in this area has not been tested, but it is certainly open to argument.

2.1.12 Moral rights

Although not relevant if copyright does not subsist in DJ sets, moral rights might also be a potential source of protection for DJs and other creators of music compilations, including the right of attribution, the right

---

25 See Recital 19 of the Directive 96/9/EC of the European Parliament and of the Council of 11 March 1996 on the Legal Protection of Databases, 1996 O.J. (L 77) 20 [hereinafter Database Directive].

26 In the UK, for example, performers' rights are protected by Part II of the Copyright, Designs and Patents Act 1988.

against false attribution and the right of integrity, albeit the former is the most likely to be useful. Performers' moral rights will also exist where a performance is protected by performers' rights.

2.1.13 Passing off and false endorsement

Aside from copyright and related rights, the copying of DJ sets and music compilations could even result in a passing off or false endorsement claim. For example, a superstar DJ might have a claim in circumstances where the entity reproducing his or her set list had deceived members of the buying public because they were led to believe the new work had been authorised by the superstar DJ, and such misrepresentation caused damage to his or her goodwill.[27] Again, however, this is not an area that has been tested in the courts.

2.2 Other Common Law Countries

The law in the biggest market for music in the world, the USA, has also yet to be tested on this particular issue. The starting point, however, is to look at section 101 of the Copyright Act, 1976.[28] A 'compilation', which is capable of being protected by copyright, is defined as a: 'work formed by the collection and assembling of pre-existing materials or of data that are selected, coordinated, or arranged in such a way that the resulting work as a whole constitutes an original work of authorship'. 'Collective works' are a particular type of 'compilation', in which a number of contributions, constituting separate and independent works in themselves, are assembled into a collective whole,[29] and it has been held that a sound recording album is a collective work of individual songs, for example.[30] Whether a DJ set falls within this section and is therefore capable of being protected by copyright has yet to be decided. But to be protected, a compilation or collective work must stem from some original effort that results in more than de minimis creativity.[31]

---

[27] See the leading English law false endorsement cases: *Irvine v Talksport Ltd* [2003] EWCA Civ 423; and *Robyn Rihanna Fenty v Arcadia Group Brands Limited* [2015] EWCA Civ 3.

[28] 17 U.S.C. § 101 (2010).

[29] Eric J. Schwartz and David Nimmer, United States. In Paul Edward Geller and Lionel Bentley, eds, *International Copyright Law and Practice* (LexisNexis, 2015) at 2[3][a].

[30] *King Records, Inc. v Bennett*, 438 F. Supp. 2d 812, 840–41 (M.D. Tenn. 2006).

[31] Eric J. Schwartz and David Nimmer, United States. In Paul Edward Geller and Lionel Bentley, eds, *International Copyright Law and Practice* (LexisNexis, 2015) at 2[3][b].

The leading related case, which must also be considered, on compilations of 'facts', is that of *Feist Publications, Inc. v Rural Telephone Service, Co.*[32] In that case, the highest court in the US, the Supreme Court, found that listings of names, towns and telephone numbers in a telephone directory were not capable of being protected by copyright. This was on the basis that such listings lacked the 'modicum of creativity necessary to transform mere selection into copyrightable expression'. The court was not prepared to grant copyright protection on the basis of 'sweat of the brow' alone, which was not the 'touchstone of copyright protection'.[33]

It must be strongly arguable at least, however, that when it comes to a DJ set, there is a strong possibility that it is a work formed by the collection of pre-existing materials (the underlying tracks) that are selected, coordinated or arranged in such a way that the resulting work as a whole constitutes an original work of authorship by virtue of there being creativity around the selection and arrangement. This might be harder to argue if, for example, a DJ set or music compilation simply contained each of The Beatles' number one singles in chronological order. But in the majority of instances, when it comes to a DJ set, at least, that is something which is unlikely to happen.

The problem, of course, is where to draw the line. At what point is there a transformation from mere selection and arrangement to selection and arrangement that is creative enough to warrant copyright protection?[34] Similar public policy concerns are likely to arise in the US as have been argued in the UK: giving copyright protection to individuals who select and arrange songs in a way which is sufficient to make original works could create an unmanageable amount of monopolies.[35] On the other hand, however, copyright is not a monopoly right which is similar to a patent. One can only rely upon it to bring infringement proceedings where there has also been access and copying. If a hundred different individuals all came up with the same DJ set or music playlist independently, none would have a claim against any of the others.

A similar public policy argument arises around whether society would benefit from millions of 'thin' copyrights which last for the life of the

---

32 *Feist Publications, Inc. v Rural Telephone Serv. Co., Inc.*, 499 U.S. 340 (1991).

33 Marc A. Fritzsche, Copyrightability of Music Compilations and Playlists: Original and Creative Works of Authorship?, 6 Pace. Intell. Prop. Sports & Ent. L.F. 258, 260 (2016).

34 Ibid. at p. 60.

35 Ibid. at p. 61.

author plus 70 years.[36] The reality, however, is that claimants would be unlikely to sue where no loss or damage could be proved,[37] and only genuine claims relating to, say, similar circumstances to the UK's *Ministry v Spotify* claim would be litigated.

Elsewhere, in Australia, 'literary works' are protected by copyright, and this includes a 'compilation' (under the Copyright Act 1968, s. 10). A compilation is a work formed by the selection, collection, co-ordination and arranging together of pre-existing materials, so that the work as a whole constitutes an original work of authorship.[38] There had been a similar case to the *Feist* case in the US – *Telstra Corporation Pty. Ltd. v Phone Directories Company Pty. Ltd.* [2010] FCAFC 149 – where the Full Federal Court refused to recognise copyright in telephone directories. Under Australian law, for copyright to subsist in a literary work, the work must be original (Copyright Act 1968, s. 32). The criteria were clarified in Telstra: while a lot of skill and labour may go into the preparatory steps of making a database, for originality to subsist, there must be an independent human intellectual effort to generate the material form of the work, and not an automated computerised process.[39]

Following the UK's *Ministry v Spotify* case, however, in December 2014, the Australian Copyright Council produced a guidance note[40] for DJs which stated: 'If you are creating playlists that have sufficient skill and effort in the selection or arrangement of the songs, these playlists may be protected by copyright as a compilation'. Sufficient 'skill and effort' – something which is more akin to the old originality test in English law (prior to the current, more European, 'author's own intellectual creation' test) – applies under Australian law. Arguably, it sets a lower threshold to be met, which a skilful DJ, who puts sufficient effort into the curation of his or her set, must be capable of meeting.

2.3 Civil Law Countries

The law on this issue remains undecided in another major music market – Germany. Its laws, like France, are contained in civil codes (as opposed to being developed through common law). In civil law countries the protec-

---

36 Ibid.

37 Albeit, statutory damages might be available if the work had been registered.

38 Brad Sherman, Australia. In Paul Edward Geller and Lionel Bentley, eds, *International Copyright Law and Practice* (LexisNexis, 2015) at 2[4][b].

39 Ibid. at 2[1][b][i].

40 See Australian Copyright Council, Information Sheet G089v4: Music: DJs 1 (December 2014) Australian.

tion offered to a creative 'author', for example, a songwriter, has always been greater than that extended to, for instance, those owning sound recording copyrights.

Under German law, copyright works are 'personal intellectual creations', with the author being the 'creator' of the work. Pursuant to Article 4 of the German Copyright Act of 9 September 1965 (as amended to comply with the Database Directive), 'Collections and Databases' are protected as follows:

> Collections of works, data or other independent elements which by reason of the selection or arrangement of the elements constitute the author's own intellectual creation (collections) are protected as independent works without prejudice to an existing copyright or related right in one of the individual elements.

It is safe to assume that a DJ set or music compilation would constitute a collection of works. The only requirement for protection is originality, which is met if, through the selection or arrangement, the author expresses his or her creative ability in an original manner by making free and creative choices.[41] The author's own intellectual creation requires a mental content to be shown which goes beyond the sum of its parts (e.g. the individual tracks), taking particular account of the overall impression of the work.[42] Examples of protected works include educational course materials, where the selection of the individual works, and their specific sequence, manifested a creative choice that went beyond the sum of the individual elements,[43] and a database of over 1,000 titles of 18th- and 19th-century German poems,[44] the titles being selected and arranged according to a specific methodology by a German academic. This would appear to be a reasonably low threshold for copyright protection.[45]

As the originality test is again broadly the 'author's own intellectual creation', the same arguments as to subsistence of copyright in music

41 Michael Gruenberger and Adolf Dietz, Germany. In Paul Edward Geller and Lionel Bentley, eds, *International Copyright Law and Practice* (LexisNexis 2015) at 2[3][b].

42 See Marc A. Fritzsche, Copyrightability of Music Compilations and Playlists, supra, note 33 at 268.

43 OLG (Court of Appeals) Frankfurt a.M., November 4, 2015 WRP 124.

44 BGH (Federal Court of Justice), May 2007 – Gedichttitelliste I (List of Poem Titles I), 2007 GRUR 685 (transl. in 39 IIC 985 (2008)).

45 Michael Gruenberger and Adolf Dietz, Germany. In Paul Edward Geller and Lionel Bentley, eds, *International Copyright Law and Practice* (LexisNexis 2015) at 2[3][b].

compilations (including DJ sets) must apply as were put forward in the UK's *Ministry v Spotify* case.

Copyright law in France is principally governed by the Law of Literary and Artistic Property of 11 March 1957. Article L 112-2 refers to 14 categories of work which are considered 'works of the mind' and which warrant protection by the 'author's right', as distinct from the (lesser) 'neighbouring right' protection given to performances, sound recordings, videograms and broadcasts, for example.

Pursuant to L 112-3, for an anthology or collection (including a database), 'author's right' protection is granted where: 'by reason of the selection or arrangement of their contents, they constitute "works of the mind". This is in line with the already mentioned Database Directive, according to which a database must be protected by the 'author's right' with regards to the 'creative' selection and arrangement of its contents (in contrast, the sui generis database right only gets 'neighbouring right' protection).

Traditionally, French law has equated copyright 'creativity' with a work needing to 'bear the mark of the personality of its author and confers on the created object a specific aspect'.[46] And historically, when considering compilations (as distinct from databases), the French court has protected a compilation where 'the choice, the order and the method under which the ensemble is presented incontestably constitute a work of the mind so as to give rise to a private right of property',[47] and has stressed that there must be 'an intellectual contribution of the author characterising an original creation'.[48] But, for example, tables of football matches 'did not reach the rank of intellectual creation, either by choice or disposition of material';[49] nor were an alphabetical listing of professionals, a graphic layout of annual tide tables, a map of France's wine regions or a system of classifying bank notes protectable.[50]

Again, it would seem that the same general arguments as presented in *Ministry v Spotify* in the UK would be likely to apply in France too.

---

[46] Cour Cass, 23 November 1983; A Lucas and H-J Lucas, Traité de la Propriété Littéraire et Artistique (1st edn, LexisNexis, 1994) para. 81.

[47] Cass. crim, 18 May 1938; Gz Pal 1938, 2, p 31: A Lucas and H-J Lucas, Traité de la Propriété Littéraire et Artistique (1st edn, LexisNexis, 1994) paras 81 and 90.

[48] 'Coprosa', Cass 1 civ, 2 May 1989: (1990) 143 RIDA 309.

[49] 'Henault', Cass crim, 2 June 1983: (1983) 117 RIDA 85.

[50] Andre Lucas and Pascal Kamina, France. In Paul Edward Geller and Lionel Bentley, eds, *International Copyright Law and Practice* (LexisNexis, 2015) at 2[3][b].

3. CONCLUSION

In the jurisdictions referred to in this chapter, which are many of the major music markets in the world, the law is far from settled on the question of whether DJ sets are protected by copyright or any related intellectual property rights. Until the question is fully considered by the courts, or legislative changes are made to introduce some certainty, this remains an area in flux.

At the highest level, the skill of the DJ cannot be in doubt: why else would major venues and events pay superstar DJs six-figure sums to work the decks to thousands of adoring fans? The same can be said of those DJs who are also paid six-figure sums to curate 20-, 40- or even 60-track dance compilations which can sell hundreds of thousands of copies. Should such creativity not be protected?

The counter argument, of course, remains that this sort of creativity should not be given very long-lasting and monopolistic protection. To prevent someone from copying the same DJ set or music compilation for many years is, it could be argued, not what the Statute of Anne or Berne Convention had in mind, when the underlying works already have various levels of protection. Copyright must protect only genuine creativity. But this issue is the crux of the question. Is putting together a DJ set or music compilation something that is genuinely creative and which therefore warrants very lengthy and wide-reaching protection (more so than for a sound recording)? The same concerns that were raised in the US in the Feist case surely cannot apply universally. Or should copyright protection depend on (in Europe) whether the work in question was the DJ's 'own intellectual creation', or (in the US) whether he or she has used the requisite 'modicum of creativity', or (in Australia) whether 'sufficient skill and effort' has been expended. Those who have been unfortunate enough to hear a terrible nightclub or wedding DJ will agree that not all DJ sets warrant protection; but, for example, the fact that labels like Ministry have others copying their dance compilation albums track for track suggests that there are others that do.

Some commentators have concerns that granting copyright protection to music compilations would have extensive ramifications: the potential for significant differences between jurisdictions depending upon the tests applied; millions of 'thin' copyrights coming into existence; and legal uncertainty around creativity.[51] Then again, copyright has always been an

---

[51] See Marc A. Fritzsche, Copyrightability of Music Compilations and Playlists, supra, note 33 at 271.

area where a fair degree of subjectivity ends up being used when it comes to determining creativity, with the inevitable consequence being a lack of certainty around what is protected and what is not.

Others argue that DJs have become the entertainment stars of our time and the DJ sets they create are like musical collages. In fact, the 'whole' that DJs create, in term of their sets, is far more than the sum of its parts. They have almost infinite possibilities in making their subjective choices in order to put together a set that creates a buzz on the dance floor or amongst listeners. This, it is argued, is a creative method of selection, arrangement and presentation of all of the different musical elements by the DJ. DJs are composers of their sets, as well as performing artists, and DJ sets and music compilations should be recognised as independent art forms.[52]

Copyright law has evolved over hundreds of years to embrace technological change as well as changes in cultural taste in order to encourage the creative arts and reward and protect those individuals and companies who do the most to advance those fields. Skilful radio DJs have been playing their skills for around a hundred years, and dance music DJs have been entertaining crowds from Ibiza to London warehouses for three decades. Perhaps it is time to recognise their talents through copyright law too.

---

[52] See Bjorn Schipper, The Art of DJing and Why DJ Sets should be Copyright Protected, available at https://www.djbroadcast.net/article/98962/the-art-of-djing-and-why-dj-sets-should-be-copyright-protected (last accessed July 20, 2018).

10. Copyright protection for modern comedic material

Trevor M. Gates*

1. INTRODUCTION

To some, a comedian entertains, provides relief from anxiety and distress, and creates a positive sense of well-being. To others, a comedian encourages them to think about issues otherwise not considered. A good joke can open up productive dialogues, and a successful comedian can reach people who would otherwise be unwilling to listen to more serious formats.

Humour has long been a part of our everyday life. We do not know when one homo sapiens first intentionally sought to make another laugh, but by the 6th century BCE the Greeks had popularized comedy as a form of theatre, and philosophers like Aristotle and Plato examined the role humor plays in society.[1] More recently, Charles Darwin looked for expressions of laughter in apes,[2] and Mark Twain dubbed humor mankind's greatest blessing.[3] Despite this, the use of the U.S. legal system among comedians and the application of intellectual property laws to protect comedic material is still in its infancy. Courts and comedians still struggle to solve one of comedy's most important problems – combating joke theft.

* Thank you to Mark Wittow, Michael Cohodes Payant, Enrico Bonadio, and Nicola Lucchi for their comments and suggestions. Thank you to my wife, KaLynn Gates, for her support and inspiration. The opinions in this chapter are solely my own.

[1] See, e.g., Carl Shaw, *Satyric Play: The Evolution of Greek Comedy and Satyr Drama* (OUP, 2014); Philosophy of Humor, Stanford University, https://plato.stanford.edu/entries/humor/ (last visited December 31, 2017); Mark Cartwright, Ancient Greek Comedy, http://www.ancient.eu/Greek_Comedy/ (last visited December 31, 2017).

[2] Charles Darwin, *The Expression of the Emotions in Man and Animals* (John Murray, 1872).

[3] Abhi Sharma, *The Great Book of Best Quotes of All Time* (Abhi Sharma, 2013) 88.

Copyright is the most relevant area of law that comedians could use to adequately protect their material. The U.S. Copyright Act[4] entitles the owner of a copyrighted work to certain benefits, such as the exclusive right to reproduce, distribute, and publicly perform or display a work, and to authorize others to do so.[5] Therefore, a comedian with a copyright in her joke should be able to prevent other comedians from using that joke without permission by asserting copyright ownership.[6] Yet, there have been very few reported copyright lawsuits stemming from joke theft, even though it is common to learn about public accusations of joke theft between comedians.

The absence of lawsuits may be due to a number of factors,[7] but the nature of comedic material in its current form seems to make reliance on copyright law to protect the rights of the author of comedic material an inaccessible, unpredictable, and burdensome option. From registering a copyright to successfully proving copyright infringement, comedians face steep challenges in relying on copyright to combat joke theft. This chapter examines these challenges and concludes that, while there are ways U.S. copyright law could be more flexible in its application to the comedy industry to help comedians enforce their rights in their material, substantially altering existing copyright laws to benefit the comedy industry is not the best solution. For example, the comedy industry appears to be thriving[8] and significantly altering U.S. copyright law may negatively disrupt the industry.

Before proposing any solutions, however, we must consider the goals of

---

4 Except where otherwise noted, this chapter focuses solely on U.S. copyright law and the U.S. Copyright Act (referred to herein as the 'Copyright Act').

5 See 17 U.S.C. §§ 106 & 201(d).

6 As the late comedian David Brenner stated: 'If we could protect our jokes, I'd be a retired billionaire in Europe somewhere – and what I just said is original'.

7 For a discussion of existing norms within the comedy industry, see Dotan Oliar and Christopher Sprigman, There's No Free Laugh (Anymore): The Emergence of Intellectual Property Norms and the Transformation of Stand-Up Comedy, 94 Va. L. Rev. 1787 (2008). Sometimes, comedians do not resort to legal recourse because existing norms within the industry lead to a more effective outcome. For example, allegations of joke theft can ruin a comedian's reputation and end a comedian's career, regardless of whether those allegations result in a lawsuit. However, these norms are unpredictable and do not provide uniform protection, as they may be dependent on a comedian's social status. See ibid. at 1825 ('Many of our interviewees stated that enforcement was relatively unlikely to succeed when the appropriator was a more popular comic than the originator . . . Also, intermediaries are less likely to enforce the norms or refuse to deal when the alleged thief enjoys public appeal').

8 Compare the comedy industry to, for example, other industries like music

U.S. copyright law. Any recommended solution should not conflict with the foundation upon which copyright law is based. The Copyright Act stems from Article I of the U.S. Constitution, which grants Congress the power to enact federal laws protecting works of authorship of all sorts. More specifically, the Constitution grants Congress the power to 'promote the [p]rogress of [s]cience and useful [a]rts'.[9] How best to promote this progress is subject to debate.[10]

Some argue the granting of strong protection over intellectual creations fosters creativity and promotes and incentivizes new works.[11] Under that assumption, a myriad of potential solutions would provide stronger protection for comedians. For example, Congress could substantially increase recoverable damages in lawsuits between comedians that are two times, five times, or even ten times the current amount, which may be as high as US$150,000 per infringing work. Congress could also significantly alter copyright law in an attempt to grant broader protection to a comedian's work, which may make the courts a more attractive option for some comedians. However, if this results in a system that does not promote

---

publishing that have more clearly suffered from the shifts from tangible media to downloads to streaming.

[9] U.S. Const. Art. I, § 8, cl. 8 (which states '[t]o promote the Progress of Science and useful Arts, by securing for limited Time to Authors and Inventors the exclusive Right to their respective Writings and Discoveries').

[10] See, e.g., William Fisher, Theories of Intellectual Property Law. In Stephen R. Munzer, ed., *New Essays in the Legal and Political Theory of Property* (CUP, 2001) 168 (Professor Fisher outlines the following four leading theories of intellectual property protection: utilitarian, labor, personhood, and social planning. For example, a utilitarian theory would interpret the U.S. Constitution's grant of power to Congress as one to 'maximize net social welfare', requiring lawmakers to find a balance between 'on one hand, the power of exclusive rights to stimulate the creation of inventions and works of art and, on the other, the partially offsetting tendency of such rights to curtail public enjoyment of those creations'. In contrast, the labor theory, which originates from the writings of John Locke, subscribes to the notion that a person who labors upon resources that are not owned has a natural property right to the 'fruits of his or her efforts').

[11] See Christopher A. Cotropia and James Gibson, The Upside of Intellectual Property's Downside, 57 UCLA L. Rev. 921, 922 (2010) (citing *Mazer v. Stein*, 347 U.S. 201, 219 (1954) which states that '[t]he economic philosophy behind the clause empowering Congress to grant patents and copyrights is the conviction that encouragement of individual effort by personal gain is the best way to advance public welfare through the talents of authors and inventors in "Science and useful Arts"').

the 'progress' of the industry, it would conflict with the foundation upon which copyright law is based.[12]

Perhaps this is why others argue that less copyright protection is needed. If the goal of copyright law is to promote progress, then prohibiting others from building on another's progress may be counterproductive.[13] For example, if one comedian tells a joke about a common event, should all others be prohibited from telling a similar joke, or from using that joke to develop additional material? In fact, copyright law in its current form is based on a system where new material may be protected for over a hundred years (life of author plus 70 years).[14] A solution must therefore find the appropriate balance between providing the right level of protection for comedians while continuing to promote the creation of new works.

A solution must also acknowledge the changing nature of the comedy industry and the varied and diverse nature of its participants and audiences. What is funny, entertaining, or an effective performance can fluctuate drastically over time and even across different cities and cultures within the same time period. [15] Notably, this chapter therefore focuses primarily on modern forms of comedy (stand-up comedy), which often take the form of a short, factual sentence followed by some commentary on that fact, whether that be about a sporting event such as the

---

[12] Some argue that we are currently seeing a version of this problem in the music industry, where some artists now avoid stating publicly by whom they may have been inspired when they create new music. See, Is the Threat of a Copyright Lawsuit Stifling Music?, BBC, http://www.bbc.com/news/entertainment-arts-40560477. During a popular copyright infringement lawsuit, singer and songwriter Michael Thicke admitted that he may have been 'inspired' by Marvin Gaye in creation of the song 'Blurred Lines'. Perhaps in part due to this testimony, Marvin Gaye's estate succeeded in winning a $7.3 million copyright verdict.

[13] See Planet Money podcast, NPR, #780: On Second Thought (arguing for less strict copyright laws because, among other things, they do not enhance creativity or incentivize new works, and they primarily benefit large corporations more than individual artists); cf. *Star Athletica, L.L.C. v. Varsity Brands, Inc.*, 137 S. Ct. 1002, 1035 (2017) (Breyer, J. dissenting) (arguing that Congress has chosen not to grant copyright protection to fashion designs in some respects because it would risk substantially disrupting a nearly $370 billion industry).

[14] See 17 U.S.C. § 302(a).

[15] For a historical look at how this (and the comedy industry generally) has changed over time, see, for example, CNN's History of Comedy series.

Super Bowl[16] or a political proposal such as an attempt to build a border wall.[17]

This chapter proposes that substantially altering copyright law to primarily benefit the comedy industry is not the best solution to helping comedians combat joke theft. Rather, we should consider ways to make it easier for comedians to use the legal system, if they choose to do so, while also considering solutions that do not rely on the legal system. To expedite a comedian's ability to seek legal relief, copyright infringement lawsuits should be allowed to proceed once a copyright applicant has filed a copyright registration for her material, rather than requiring an applicant to wait until the Copyright Office actually registers the work, which can delay by months a comedian's ability to rely on the courts. Additionally, courts faced with copyright infringement lawsuits that involve comedic material containing a factual statement could examine the relevant work as a whole and give greater weight to the comedian's choice and particular arrangement of the fact if the fact was selected or arranged in an original way. Separately, the establishment of an online licensing service could lower the transactional costs of exchanging comedic material and fund an alternative dispute resolution mechanism, such as an arbitration panel, that could provide comedians with a more efficient, less expensive alternative to relying on the courts. This service could also provide useful information for comedians to understand how best to leverage U.S. copyright law, such as when registering material with the Copyright Office may provide the most benefit.

Section 2 of this chapter provides a brief introduction to the statutory framework of the Copyright Act and examines requirements for protecting jokes under U.S. copyright law, as well as the difficulties of doing so. Section 3 examines how a U.S. court might currently treat a copyright infringement action between comedians involving comedic

---

[16] See *Kaseberg v. Conaco*, LLC, 260 F. Supp. 3d 1229, 1234 (S.D. Cal. 2017) ('Tom Brady said he wants to give his MVP truck to the man who won the game for the Patriots. So enjoy that truck, Pete Carroll'. Pete Carroll is the head coach of the Seattle Seahawks who lost to the New England Patriots in Super Bowl XLIX on February 1, 2015). Of note, the Copyright Office initially refused to register the 'Tom Brady Joke' but reconsidered upon request by Kaseberg and concluded the joke met the 'minimal degree of creativity' standard under *Feist*. See Second Request for Reconsideration for Refusal to Register A LITTLE BIT BAD, Blog Content, Brady Joke February 3, 2015; Correspondence ID: 1-25KMZ9Z; SR#: 1-2752094070 (July 17, 2017).

[17] See, Whose Joke Is It? Carlos Mencia? D.L. Hughley? George Lopez?, YouTube (February 19, 2007) http://www.youtube.com/watch?v=kPuu_VE7KOA (showing an example of an extremely similar joke about a border wall told by four different comedians over several months in 2007).

material. Section 4 proposes solutions to help comedians enforce their rights in their works.

2. STATUTORY BACKGROUND: BARRIERS TO COPYRIGHT PROTECTION

If specific requirements are met, the Copyright Act entitles the owner of a copyrighted work to certain benefits, such as the exclusive right to reproduce, distribute, and publicly perform a copyrighted work.[18] To garner protection under the Copyright Act, however, an author's work must be in a category of protectable copyright subject matter, it must be original,[19] and it must be fixed in a 'tangible medium of expression'.[20] Once those requirements are met, copyright protection in the United States exists automatically, although the work must be registered with the Copyright Office prior to filing any lawsuit for infringement.[21] Each of these requirements presents hurdles for comedians looking to rely on U.S. copyright law to enforce their rights in their material.

2.1 Subject Matter Restrictions and Factual Works

A valid copyright extends only to copyrightable subject matter.[22] While jokes can be copyrighted, the scope of copyright on jokes, like other creative works, is limited by subject matter restrictions. These limits on copyright protection often leave large portions of jokes unprotected and freely usable without exposure to copyright infringement liability. Ideas

---

[18] See 17 U.S.C. §§ 106 & 201(d).

[19] Jokes are considered under the ordinary originality standard generally applicable to other material. See Compendium II of Copyright Office Practices § 420.02 ('Jokes and other comedy routines may be registered if they contain at least a certain minimum amount of original expression in tangible form'); see also *Feist Publ'ns, Inc. v. Rural Tel. Serv. Co.*, 499 U.S. 340, 362 (1991) (noting that the particular work at issue lacked the 'modicum of creativity necessary to transform mere selection into copyrightable expression'). This is a low threshold and is not discussed in detail in this chapter.

[20] 17 U.S.C. § 102(a) (2016) ('Copyright protection subsists . . . in original works of authorship fixed in [a] tangible medium of expression').

[21] See 17 U.S.C. § 411.

[22] *Star Athletica, L.L.C. v. Varsity Brands, Inc.*, 137 S. Ct. 1002, 1008 (2017). Jokes could be written down and deemed a literary work under 17 U.S.C. § 102(a). The performance of a joke and subsequent fixation of that performance could also be protected by copyright.

are not copyrightable, nor are facts, independent of an author's creative expression of them.[23] This may prove problematic for many comedians because jokes are commonly created by taking a well-known fact and adding some twist to it that surprises the audience. A brief analysis of a joke reveals how little of it may be protectable under copyright law.

Consider a common format for stand-up jokes, consisting of a factual one-to-two-sentence introduction and a one-sentence punchline delivering the comedian's commentary on the fact, such as: The University of Alabama-Birmingham is shutting down its football program. To which the Oakland Raiders said: 'Wait, so you can do that?'[24] Currently, neither the fact that forms the basis for sentence one (that UAB shut down its football program) nor the general idea underlying sentence two (that Oakland may shut down its football program) may be protectable.[25] Rather, copyright law would protect, if at all, a comedian's particular expression of that joke.

Consider a similar joke: University of Alabama-Birmingham has decided to discontinue its football team. When they heard the news, New York Jets fans said: 'Wait can you do that?'[26] Both jokes point out that UAB discontinued its football team, which is fact, but each differs in its commentary on that fact – one comments on the Oakland Raiders football team and the other on the New York Jets football team and its fans. As this chapter examines later,[27] a court would analyze these jokes under the substantial similarity standard for copyright infringement, which would focus on the protectable parts of the works – that is, the commentary – which seems to differ, and not the facts or general ideas that form part of the jokes. After even this cursory analysis, it becomes evident how the actual elements of a joke that are protectable under copyright law can be

---

[23] See *Feist Publ'ns, Inc. v. Rural Tel. Serv. Co.*, 499 U.S. 340, 344–5 (1991); *Harper & Row Publishers, Inc. v. Nation Enters.*, 471 U.S. 539, 556 (1985); *Foxworthy v. Custom Tees, Inc.*, 879 F. Supp. 1200, 1218-19 (N.D. Ga. 1995); 17 U.S.C. § 102(b) (Copyright protection does not extend to any idea, procedure, process, system, method of operation, concept, principle, or discovery, regardless of the form in which it is described, explained, illustrated, or embodied in such work.); 37 C.F.R. § 202.1(d) ('Works consisting entirely of information that is common property containing no original authorship, such as, for example: Standard calendars, height and weight charts, tape measures and rulers, schedules of sporting events, and lists or tables taken from public documents or other common sources').

[24] *Kaseberg v. Conaco*, LLC, 260 F. Supp. 3d 1229, 1233 (S.D. Cal. 2017).

[25] 17 U.S.C. § 102(b).

[26] *Kaseberg*, 260 F. Supp. 3d at 1233.

[27] See infra pp. 2–14.

quite limited. Common jokes could be varied slightly and used by another without legal repercussions.

2.2 Fixing and Registering Jokes

The nature of the comedy industry sometimes makes fixing jokes a difficult burden for comedians. Even if a work is within a category of protectable copyright subject matter, copyright protection only subsists in works that are fixed in some form, such as in a recording or writing.[28] Original words spoken aloud can generally be copied by all if they have not been fixed appropriately. For many works, fixation may be commonplace. An author may naturally fix her completed story on paper. However, the nature of the comedy industry sometimes makes this requirement difficult for comedians to meet. It may require them to do something extra, such as write down a joke after it is performed.

Consider a comedian preparing for a Thursday night gig. She could write down her material before the performance, but it is likely she may come into that performance with two to three times as much material available to use, some of which may have been preplanned and other material that may vary depending on audience interaction, such as reading subtle cues or responding to hecklers. Afterwards, she may choose to fix those jokes, such as by recording the performance or writing down the finalized jokes, but copyright will not protect her material if she fails to fix it in some form. Also, a copyright on one version of a joke may not protect an altered, unfixed version of that same joke, so a comedian may need to fix different versions of a joke that is performed on different nights.

Similarly, comedians often lose out on copyright law benefits because they fail to register their jokes with the Copyright Office in a timely fashion, or at all. While fixation of a work is required to receive a copyright for that work, registration is not required, although it can provide significant benefits.[29] To be eligible to receive statutory damages up to $150,000 per work infringed in a copyright infringement lawsuit,[30] a work must typically

---

[28] See 17 U.S.C. §§ 101, 102(a); see also *Fritz v. Arthur D. Little, Inc.*, 944 F. Supp. 95, 100 (D. Mass. 1996); *MAI Sys. Corp. v. Peak Computer, Inc.*, 991 F.2d 511, 518–19 (9th Cir. 1993).

[29] See 17 U.S.C. § 411.

[30] The highest statutory damage penalty that is discretionary to the fact-finder is $150,000 under 17 U.S.C. § 504(c)(2) ('In a case where the copyright owner sustains the burden of proving, and the court finds, that infringement was committed willfully, the court in its discretion may increase the award of statutory damages

be registered within three months of publication or before the infringing conduct occurred.[31] Without statutory damages, a comedian must rely on actual damages and any additional profits of the infringer,[32] which in many instances may be too minimal to ever incentivize a comedian to file a lawsuit.[33] Further, for many comedians who perform with hundreds or thousands of jokes, the cost of registration may be prohibitive or a waste of time.[34] Registration of one joke, or even a compilation of jokes, may not provide any additional benefit for a comedian who later changes those jokes, or against an alleged joke stealer that uses a similar, unregistered version. Therefore, it is essential that comedians understand when it is advantageous to register their material.

Notably, registration is required prior to filing any lawsuit for copyright infringement, even though it is not required to receive a copyright.[35] Presently, there is a split in circuits regarding whether registration must be started or completed prior to filing a lawsuit.[36] Some courts find that

---

to a sum of not more than $150,000. In a case where the infringer sustains the burden of proving, and the court finds, that such infringer was not aware and had no reason to believe that his or her acts constituted an infringement of copyright, the court in its discretion may reduce the award of statutory damages to a sum of not less than $200').

[31] See 17 U.S.C. § 412. If a work is registered after the first act of infringement in a series of ongoing infringements, statutory damages may only be available for new, separate acts of infringements beginning after registration of the joke. See *Derek Andrew, Inc. v. Poof Apparel Corp.*, 528 F.3d 696, 700–1 (9th Cir. 2008) (also stating, in footnote four, that the Ninth Circuit is joining the other circuits that have addressed this issue, namely the Second, Fourth, and Fifth Circuits).

[32] See 17 U.S.C. § 504(a).

[33] For example, if a comedian delivered a 30-minute performance that contained 100 jokes, what remedies would be available if another comedian used only one joke out of 100?

[34] Depending on the application type, the fee is either $35 or $55 per online application. In addition to the application itself, the party seeking the copyright must deposit an electronic copy or hard copy of the work with the Copyright Office. United States Copyright Office, Circular 1, Copyright Basics 7 (2012), https://www.copyright.gov/circs/circ01.pdf. For some, the registration system may also seem complicated.

[35] 17 U.S.C. § 411(a) (stating that 'no civil action for infringement of the copyright in any United States work shall be instituted until preregistration or registration of the copyright claim has been made in accordance with this title').

[36] Compare *Apple Barrel Prods., Inc. v. Beard*, 730 F.2d 384, 386–7 (5th Cir. 1984) ('In order to bring suit for copyright infringement, it is not necessary to prove possession of a registration certificate. One need only prove payment of the required fee, deposit of the work in question and receipt by the

the registration requirement is met once a copyright owner files the copyright registration paperwork with the Copyright Office. Other courts find that the registration requirement is not met until the Copyright Office actually registers the work. The average copyright registration can take six to ten months to be granted,[37] so where a comedian is located can have a significant effect on the time it takes to sue. If a comedian must bring a lawsuit in a circuit that requires an issued registration before filing the lawsuit, she must wait for months or choose to pay an expedited processing fee.[38]

This creates a system where comedians may be faced with choosing whether to, on the one hand, meticulously fix their jokes and actively register them with the Copyright Office, potentially costing hundreds or thousands of dollars in registration fees, or, on the other hand, foregoing the opportunity for certain benefits such as statutory damages or being able to quickly seek immediate relief in court.

Copyright Office of a registration application') and *Cosmetic Ideas, Inc. v. IAC/InteractiveCorp*, 606 F.3d 612, 621 (9th Cir. 2010), (the 'application approach avoids delay without impairing the central goal of copyright registration' and further noting that 'in the worst-case scenario the registration approach could cause a party to lose its ability to sue' if the statute of limitations under 17 U.S.C. § 507 expired before the Copyright Office acted on the application) with *La Resolana Architects, PA v. Clay Realtors Angel Fire*, 416 F.3d 1195, 1200 (10th Cir. 2005) ('No language in the Act suggests that registration is accomplished by mere receipt of copyrightable material by the Copyright Office. Instead, the Register of Copyrights must affirmatively determine copyright protection is warranted . . . before registration occurs under the Act' (citation omitted)), abrogated by *Reed Elsevier, Inc. v. Muchnick*, 559 U.S. 154 (2010) and *Fourth Estate Pub. Benefit Corp. v. Wall-Street.com, LLC*, 856 F.3d 1338, 1341 (11th Cir. 2017) ('Filing an application does not amount to registration. . . . The Register of Copyrights [must] examine[] the material and determine[] whether the material deposited constitutes copyrightable subject matter' (citations omitted)).

37 See, I've Submitted My Application, Fee, and Copy of My Work to the Copyright Office. Now What?, United States Copyright Office, https://www.copyright.gov/help/faq/faq-what.html (last visited December 31, 2017).

38 United States Copyright Office, Circular 10, Special Handling 1 (2012), https://www.copyright.gov/circs/circ10.pdf ('To expedite the examination of a claim or document, you must pay a special handling fee in addition to the fee for filing an application to register a work or for recording a document. The special handling fee is $800 for registration . . .').

3. LEGAL LANDSCAPE: WITH A FOCUS ON *KASEBERG*

This section addresses how a U.S. court may currently treat a copyright infringement action between comedians involving comedic material, with a focus on a recent Ninth Circuit case: *Kaseberg v. Conaco*.[39]

3.1 *Kaseberg v. Conaco*

In 2017, Robert Kaseberg, a freelance and comedy writer for over twenty years, brought a copyright infringement lawsuit against late-night personality Conan O'Brien and others (collectively, the 'defendants') connected to Conan's late-night television show.[40] Kaseberg alleged the defendants had infringed the copyright in five jokes he had published on his blogs and on social media accounts in 2014 and 2015.[41] Each of the jokes generally consisted of a short, factual sentence followed by commentary, such as the UAB football joke discussed above.[42] According to Kaseberg, he began noticing Conan performing similar jokes on Conan's show shortly after Kaseberg posted them online.[43] He attempted to reach one of Conan's writers via Twitter.[44] He also phoned the writer and received no response.[45] Eventually, he filed suit.

To establish a claim of copyright infringement, a plaintiff must prove ownership of a valid copyright[46] and copying by the defendant of protected

[39] For other examples, see: *TCA Television Corp. v. McCollum*, 839 F.3d 168 (2d Cir. 2016), cert. denied, 137 S. Ct. 2175 (2017) (heirs of Bud Abbott and Lou Costello sued the producers of a theatrical production based on the 'Who's On First?' routine); *Castle Rock Entm't, Inc. v. Carol Pub. Grp., Inc.*, 150 F.3d 132 (2d Cir. 1998) (producers of a television program sued the author and publisher of a trivia book for use of material from the *Seinfeld* television series); *Foxworthy v. Custom Tees, Inc.*, 879 F. Supp. 1200 (N.D. Ga. 1995) (Comedian Jeff Foxworthy sued a T-shirt company alleging infringement of his phrase 'you might be a redneck'); *Smith v. Weinstein*, 578 F. Supp. 1297 (S.D.N.Y.), aff'd, 738 F.2d 419 (2d Cir. 1984) (a comedy writer sued a motion picture company based on the 1980 film, *Stir Crazy*).

[40] *Kaseberg v. Conaco, LLC*, 260 F. Supp. 3d 1229, 1232–3 (S.D. Cal. 2017).

[41] Ibid. at 1233.

[42] Supra p. 225. Another was the following airlines joke: Yesterday, a flight from Cleveland to New York took off with just two passengers. Yet somehow, they spent the whole flight fighting over the armrest. See *Kaseberg*, 260 F. Supp. 3d at 1233.

[43] *Kaseberg*, 260 F. Supp. 3d at 1233–4.

[44] Ibid.

[45] Ibid. at 1234.

[46] Ownership is typically established by way of a copyright registration. See 17 U.S.C. § 410(c).

elements of the work.[47] A plaintiff may prove copying by direct evidence but in most cases such evidence is unavailable,[48] so a plaintiff may show copying by establishing that the infringer had (1) access to the work and (2) that the two works are substantially similar.[49]

Proving access requires demonstrating that the defendant had a 'reasonable opportunity' to view the plaintiff's work,[50] such as by showing a particular chain of events between the plaintiff's work and the defendant's access to that work, or by showing that the plaintiff's work has been widely disseminated.[51] To find whether two works are substantially similar, a Ninth Circuit court conducts extrinsic and intrinsic analyses, except on summary judgment where it conducts only the extrinsic analysis.[52] The extrinsic analysis is said to be objective: the court looks to articulable similarities between the defendant's allegedly infringing work and protectable elements of the plaintiff's original work and uses 'analytic dissection' to determine the scope of copyright protection.[53] The intrinsic test, conversely, is the province of the jury.[54] It concerns 'whether the ordinary, reasonable person' would find 'the total concept and feel of the works' to be substantially similar.[55]

One of the primary points of contention analyzed in *Kaseberg* was the appropriate standard of evaluating the level of similarity between the works at issue. Because it was summary judgment, the court performed the extrinsic test, dissected the jokes, and found that some were comprised almost wholly of unprotectable elements. Those jokes were entitled to only 'thin' copyright protection.[56] The court found that when copyright protection is thin the appropriate standard for similarity may be 'virtual identity' of the protectable elements.[57]

---

47 *Feist Publ'ns, Inc. v. Rural Tel. Serv. Co.*, 499 U.S. 340, 361 (1991).

48 *Kaseberg*, 260 F. Supp. 3d at 1247 (citing *Shaw v. Lindheim*, 919 F.2d 1353, 1356 (9th Cir. 1990)).

49 *Kaseberg*, 260 F. Supp. 3d at 1247.

50 Ibid. at 1240 (citing *Three Boys Music Corp. v. Bolton*, 212 F.3d 477, 482 (9th Cir. 2000)).

51 Ibid.

52 Ibid. at 1243 (*citing Rice v. Fox Broad. Co.*, 330 F.3d 1170, 1174 (9th Cir. 2003)). This is the test in the Ninth Circuit. The test varies in other circuits. See, e.g., *Computer Assocs. Int'l, Inc. v. Altai, Inc.*, 982 F.2d 693 (2d Cir. 1992) (Second Circuit discussing substantial similarity between computer programs).

53 *Kaseberg*, 260 F. Supp. 3d at 1243–4.

54 See *Unicolors, Inc. v. Urban Outfitters, Inc.*, 853 F.3d 980, 985 (9th Cir. 2017).

55 Ibid. (citing *Pasillas v. McDonald's Corp.*, 927 F.2d 440, 442 (9th Cir. 1991)).

56 See *Kaseberg*, 260 F. Supp. 3d at 1243–5.

57 Ibid. at 1244 (citations omitted).

3.2 Application to the Comedy Industry

Kaseberg helps illustrate some of the difficulties comedians confront in enforcing their rights in their material. Comedians face unique challenges in proving access and substantial similarity, which are required elements of a copyright infringement lawsuit.

Demonstrating an alleged joke thief had access to a comedian's original joke is sometimes difficult, especially for less-heralded comedians. With the exception of those who have large tours or televised performances, many comedians do not broadcast their jokes to a particularly large audience. Because their jokes are not widely disseminated, a comedian would need to look to prove a chain of events linking the defendant to their joke, which may prove difficult.

Similarly, because large parts of jokes are often unprotectable, jokes that sound nearly identical may fall shy of the virtual identity standard. It is easy to imagine two jokes that share common words, phrases, or ideas, yet share no protectable elements. As examined above, after whittling down a joke into protectable and non-protectable elements – for example, by excluding those aspects of the joke that are factual[58] – it is likely that the protectable elements that remain will not be virtually identical. In addition, because the application of copyright analysis to jokes is still in its infancy, there is little precedent to suggest where the line for virtual identity will be drawn.[59] Moreover, even if the standard of substantial (or virtual) similarity is met, there are other foundational principles of copyright law that may apply to limit protection even further,[60]

---

[58] See *Feist Publications, Inc. v. Rural Tel. Serv. Co.*, 499 U.S. 340, 349 (1991) (stating that 'copyright in a factual compilation is thin').

[59] Because there are so few cases of legal action involving joke theft, the legal standards in cases of joke theft may remain amorphous, potentially contributing to a comedian's decision not to bring a lawsuit. Yet until comedians start to bring such claims, the standard will remain just that – amorphous.

[60] There are others. For example, if certain features of a joke are as a practical matter indispensable, or at least standard to a given field or type of work, they are not protected by copyright. See *Apple Comput., Inc. v. Microsoft Corp.*, 35 F.3d 1435, 1444 (9th Cir. 1994); *Landsberg v. Scrabble Crossword Game Players, Inc.*, 736 F.2d 485, 489 (9th Cir. 1984). Also, if a fact-finder concludes there is only a single way (or limited number of ways) to express a particular fact or punchline, it may not be copyrightable under the merger doctrine. See *Ets-Hokin v. Skyy Spirits, Inc.*, 225 F.3d 1068, 1082 (9th Cir. 2000).

such as the doctrine of independent creation,[61] de minimis copying,[62] or fair use.[63]

---

61 A finding of substantial similarity may be rebutted by proof that a comedian independently created the joke. See *Kaseberg*, 260 F. Supp. 3d at 1238–9 (finding that defendants had not infringed plaintiff's 'Delta Joke' because defendants created their joke before Kaseberg created his). The doctrine of independent creation seems especially relevant to jokes because it is not uncommon for multiple comedians to be inspired by the same real-world events and for those events to manifest themselves as very similar jokes. Millions of people may have a similar reaction to a political event. Likewise, similar jokes about universal experiences such as coming of age or navigating relationships do not necessarily indicate copying. For perhaps a more familiar example of independent creation, consider its application to the music industry. See, Ed Sheeran May Regret Photograph that Led to $20m Copyright Case, *The Guardian*, https://www.theguardian.com/music/2017/apr/11/ed-sheeran-20m-dollar-copyright-claim-matt-cardle-x-factor (last visited December 31, 2017) (Dennis Collopy, a senior lecturer at the University of Hertfordshire, stated that in music plagiarism cases it is important to acknowledge the narrow musical confines of what makes a hit song: 'We celebrate the fact that successful songs tend to fit into quite a narrow band of melody and structure. . . . When it comes to pop music, it's very hard to be completely original and successful at the same time'). On the other hand, certain jokes seem unlikely to be independently created. See, e.g., Stewart Lee, *How I Escaped My Certain Fate* (Faber & Faber, 2011) (comedian Stewart Lee describes in his book an example of alleged joke theft involving Irish comedian Michael Redmond telling a joke about standing in someone's yard. Michael is tall, has big bushy hair, a droopy mustache, and deep set eyes. He used to walk out on stage wearing a long brown trench coat and stand still in silence for about two minutes. Then, he would say to the audience: 'a lot of people say to me' . . . 'get out of my garden'. Stewart explained that Joe Pasquale, a successful English entertainer, performed this same joke in a 1995 Royal Variety Performance, sometime after Michael had performed it. Stewart called up Joe who, according to Stewart, could not recall how he came up with the joke or whether, for example, one of Stewart's writers came up with it. Stewart goes on to explain the relative unlikelihood this joke could have been independently created.).

62 The doctrine of de minimis copying would prevent a successful copyright infringement action if such a small amount of a protected joke is taken that substantial similarity cannot be said to exist between the original work and the alleged infringing work. See generally *Gordon v. Nextel Commc'ns & Mullen Advert.*, 345 F.3d 922 (6th Cir. 2003) (finding the advertising agency's copying of the artist's work was de minimis).

63 Fair use is a defense to copyright infringement that excuses the copying of copyrighted material if it is done for a limited, transformative purpose, such as to comment upon, parody, or criticize a copyrighted work. For a discussion of fair use as applied to humorous material, see Roger L. Zissu, Funny is Fair: The Case for According Increased Value to Humor in Copyright Fair Use Analysis. In Benjamin Kaplan et al., *An Unhurried View of Copyright Republished (and with Contributions from Friends)* (LexisNexis, 2005).

4. FINDING A BALANCED SOLUTION

Comedians do face challenges in combatting joke theft, but significantly altering existing copyright law to primarily benefit a single industry may prove difficult and inefficient. It may also substantially and negatively disrupt the comedy industry. A solution must find a balance between adequately protecting a comedian's works while continuing to promote the creation of new works.

4.1 Treatment of Facts

Where the nature of the fact is integral to the copyrightable expression, a court analyzing substantial similarity between two works in a copyright infringement lawsuit could examine the works as a whole, and the comedian's choice and particular arrangement of a fact could be given greater weight if the fact was selected or arranged in an original way. In *Kaseberg*, the court found jokes were only entitled to very thin copyright protection, partially because some of them consisted of facts. Under this standard, it is inevitable that many modern jokes, which are often based on short one-to-two-sentence monologues, will fail the substantial similarity infringement analysis. In analyzing comedic material, courts could consider whether a fact was selected or arranged in an original way. This analysis could be similar to how a court may examine whether compilations of material (including factual material) are entitled to copyright protection. Compilations may be eligible for copyright protection if the materials are selected, coordinated, or arranged in an original way.[64] Just as an original compilation of facts may be copyrightable, so should be a joke that contains a fact if the fact was selected or arranged in an original way.

This may be a difficult line to draw, but original authorship may be involved in a comedian deciding which fact to use and how to deliver the fact to create the best story, generate the best laugh, or effectively communicate the comedian's message. However, if a comedian uses a fact in a joke she should not have an automatic monopoly over that fact. If the use of the fact that makes up the joke is a purely mechanical task with no element of original selection or arrangement, copyright

---

[64] See 17 U.S.C. § 101 ('A "compilation" is a work formed by the collection and assembling of pre-existing materials or of data that are selected, coordinated, or arranged in such a way that the resulting work as a whole constitutes an original work of authorship . . .').

protection would not be available. But most jokes contain facts and often those are intentionally chosen and make up a significant portion of the joke. Without the fact, the copyrightable expression would likely not exist.[65]

4.2 Copyright Registration

Copyright registration should also be less of a bar to filing a copyright infringement lawsuit, and courts should consider resolving the circuit split to allow all copyright owners to file suit as soon as the owner files her paperwork, pays the fee, and deposits the work with the Copyright Office. As noted, some courts require copyright owners to wait until the Copyright Office takes action on an author's registration paperwork, which can take several months. Considering that most registrations are granted nearly every time, this hurdle is more of a formality that delays lawsuits in some circuits and prevents copyright owners from seeking immediate relief in court, or forces them to pay expedited fees.

4.3 Comedy Hub

Currently, there is no well-established process for licensing a joke from another comedian. It is often done orally and not written down. Creating a service that facilitates the licensing and sharing of comedic material could assist with this issue.[66] Through this service, or 'Comedy Hub', a work posted on the Comedy Hub, such as a joke, would be given a unique identifier that would allow a potential licensee to, via a click of the mouse, reach out to and connect with the computer of the owner of that work, similar to making an online purchase. The owner of the work could offer standard licenses for use of that material and coordinate the exchange of payment,[67] which would create a written record of the exchange. Fees from the licensing of material could also supplement certain dispute resolution procedures, such as a less expensive arbitrator. As part of the licensing process, comedians could

---

[65] There are other principles of copyright law that would limit the potential reach of this solution. For example, accused joke stealers could still plead independent creation. See supra pp. 224–8.

[66] A similar type of platform is in the process of being implemented in the United Kingdom, see http://www.copyrighthub.org/ (last visited December 31, 2017).

[67] The Comedy Hub could also take advantage of the use of specific types of currency, such as cryptocurrency, to help reduce fraud and theft.

agree to be bound by the arbitrator and not necessarily have to rely on the courts.

Additionally, the Comedy Hub could help comedians better understand copyright law and its benefits by acting as a user-friendly online repository of information about copyright law. For example, it could provide information on the registration of material and help comedians understand when it may be beneficial to register certain material, including individual jokes and compilations. The Comedy Hub could also encourage comedians to broadly disseminate works by providing a forum for facilitating the exchange of material. This would provide some practical benefits, but a comedian that chooses to actively publish her material through the Comedy Hub may also be able to more easily succeed in a copyright infringement lawsuit.[68]

5. CONCLUSION

While substantially overhauling existing U.S. copyright law to primarily benefit the comedy industry may not be the best solution to combatting joke theft, there are some tweaks that could increase protection for comedic material and help comedians more effectively enforce their rights in their material. Courts in copyright infringement lawsuits between comedians involving comedic material could consider giving greater weight to the choice and arrangement of a fact within a joke as chosen by a comedian, if the fact was selected or arranged in an original way. Registration of copyrightable works could become less of a bar to filing copyright infringement lawsuits if an author is allowed to file a lawsuit as soon as she takes the appropriate registration steps with the Copyright Office, rather than requiring the author to wait until the Copyright Office actually registers the work, which can delay by months the ability to rely on the courts.

Additionally, these changes could be supplemented by a Comedy Hub that leverages the exchange of material to create alternative dispute resolution options for comedians and lowers the transactional costs of licensing material. This Comedy Hub could also lay out useful information for comedians to understand how best to leverage existing copyright

---

[68] See *Kaseberg*, 260 F. Supp. 3d at 1240–3. Kaseberg had published his material online and reached out to the writers about his material, via Twitter and otherwise. While not dispositive, the court looked to this as one factor showing that the defendants had 'access' to Kaseberg's jokes.

laws, such as recognizing when registering or publicly disseminating material may provide the most benefit. These changes could provide some additional benefits for comedians looking to enforce their rights in their material without substantially disrupting existing copyright law or interfering with the current success of the comedy industry.

11. Now you own it, now you don't – or do you? Copyright and related rights in magic productions and performances

F. Jay Dougherty*

1. INTRODUCTION

Derek DelGaudio is a successful magician, who is noted for theater productions such as 'In and of Itself' that embed magic illusions in a more complex narrative, portraying character and evoking strong emotions from his audience. One review stated: 'In and Of Itself makes the case for the art of magic with the power of storytelling. Derek DelGaudio is not just performing magic tricks but presenting a moving deeply personal and compelling narrative, using magic as a metaphor to reach a deeper truth'.[1] It is clear that dramatic works are protected by copyright, but the creations and performances of magicians are among the types of expression at the edges of copyright and other intellectual property protection. There are many types of intellectual property law that may provide protection to some magic illusions and productions. This chapter will explore potential copyright protection for such creations and performances. The tricks and illusions of a magician, if novel, useful and non-obvious, are protectable

* The author would like to thank Ricky Jay for an inspiring interview and Ron Reitshtein (Loyola Law School,'09), Kevin Camisso (Loyola Law School,'17), and Alina Condurat (LLM, Loyola Law School,'16) for their research assistance. This chapter is a revised and updated version of F. Jay Dougherty, Now You Own It, Now You Don't: Copyright and Related Rights in Magic Productions and Performances, in Christine A. Corcos, ed., *Law and Magic: A Collection of Essays* (Carolina Academic Press, 2010) 101.

[1] Peter Sciretta, Frank Oz Directed a Magic Show and Its [sic] Unlike Anything You've Ever Experienced Before, Film Bloggin The Reel World (May 4, 2016), http://www.slashfilm.com/in-and-of-itself-magic-show/ (last accessed November 29, 2017).

by patent.[2] But magician inventors may be reluctant to obtain patents, because, as part of that process, the specifics of the invention must be disclosed to the public, including competing magicians. If not patented,[3] the secrets behind those tricks and illusions, having commercial value and being maintained as secrets, may receive protection as trade secrets.[4] The basic requirements for patent and trade secrets, along with their limiting doctrines, limit the value of those doctrines in protecting magicians. That, coupled with limits on potential copyright protection,[5] has led magicians to seek to protect exclusivity in their magic pieces through custom or, at times, through consensual self-regulation.[6]

By contrast, copyright protects original expressions of authorship, fixed in a tangible medium of expression. Various limiting doctrines in copyright also reduce its value in protecting magicians and their acts. Ideas,[7]

---

[2] See *Goldin v. Clarion Photoplays, Inc.*, 195 N.Y.S. 455 (N.Y. App. Div. 1923).

[3] See *Goldin v. R.J. Reynolds Tobacco Co.*, 22 F. Supp. 61, 65 (S.D.N.Y. 1938) (woman sawed in half illusion lost trade secret status by having been patented).

[4] But limitations on this type of protection may make it difficult to apply against those who reveal the secrets. See supra note 3; *Harrison v. SF Broad.*, (E.D. La. 1998) (not reported in F.Supp.).

[5] Not a great deal has been published specifically about legal protection for magic performances. In 2000, the World Alliance of Magicians, Inc. published *The Protection of Magicians' Secrets* (G. Weissenberger, ed.) This work briefly reviewed each of the types of intellectual property laws as it may apply to magic, including limitations on each type of protection. It also included some excellent articles concerning those subjects. The book is apparently not publicly available at the time of writing this article. The author is grateful to Walter Blaney and Abb Dickson, former presidents of the WAM, for providing him with a copy. For other articles on the subject, see Rick Lax, The Tricky Business of Innovation: Can You Patent a Magic Trick?, Wired (Jul. 8, 2013), https://www.wired.com/2013/07/the-tricky-business-of-innovation-can-you-patent-a-magic-trick/; Janna Brancolini, Abracadabra! – Why Copyright Protection for Magic Is Not Just an Illusion, 33 Loy. L.A. Ent. L. Rev. 103 (2013); Jared R. Sherlock, The Effects of Exposure on the Ecology of the Magic Industry: Preserving Magic in the Absence of Law, 6 Cybaris 1 (2015).

[6] See K. Seegrave, *Piracy in the Motion Picture Industry* (McFarland, 2003) 19–21 (describing disputes handled by the Joint Complaint Bureau of the Vaudeville Managers' Protective Association and the National Vaudeville Artists); Milbourne & Maurine Christopher, *The Illustrated History of Magic* (updated edn, Avalon, 2006) 269 (on Horace Goldin using that system with respect to the 'woman sawed in half' illusion). See also Jacob Loshin, Secrets Revealed: How Magicians Protect Intellectual Property Without Law, in Christine A. Corcos, ed., *Law and Magic: A Collection of Essays* (Carolina Academic Press, 2010) 123.

[7] See The Copyright Act of 1976, Pub. L. No. 94-553, 90 Stat. 2541 (1976); 17 U.S.C. § 102(a) (2007). Subsequent citations to the Copyright Act of 1976 will only refer to the codified provisions.

words and short phrases,[8] and 'scenès à faire'[9] are not protectable. 'Stock' characters, standard character types without original creative delineation, are treated similarly.[10] Copyright has limited application to literary and visual material that is utilitarian. Utilitarian works and useful methods, processes and articles are all excluded from copyright protection.

With all of those elements carved out of potential copyright protection, what, if anything, remains? This chapter will explore that question, primarily under U.S. law. It will discuss several types of copyrightable subject matter that might in some circumstances be found in a magician's performance. Then, it will discuss in some detail two judicial decisions that confronted claims for protection of magic productions. Finally, it will briefly discuss potential legal claims protecting the performance itself. Some other nations recognize a type of right related to copyright, known as 'performers' rights'. The chapter will conclude with a brief discussion of certain other legal theories that might apply to a magician's performance, namely, 'common law' copyright, unfair competition and the right of publicity.

2. BASIC U.S. COPYRIGHT CONCEPTS AND LIMITATIONS

U.S. copyright law fundamentals are addressed elsewhere in this book, so discussion here will be limited. Some minimal degree of creativity is implied by the word 'author'. Certain types of material are not viewed as sufficiently original or creative to merit copyright protection, even if not directly copied from another work. Hence, the basic building block components of various types of works are not viewed as copyrightable. For example, social dance steps would not be protected as choreography,[11] simple geometrical shapes would not be protected as pictorial works,[12] words and short phrases (and of course letters and typography) would not be protected as literary works. Important examples for the purposes of this

---

[8] 37 C.F.R. § 202.1 (2007).

[9] See, e.g., *Apple Computer, Inc. v. Microsoft Corp.*, 35 F.3d 1435, 1444 (9th Cir. 1994).

[10] See, *Nichols v. Universal Pictures Corp.*, 45 F.2d 119 (2d Cir. 1930); *Rice v. Fox Broad. Co.*, 330 F.3d 1170, 1176 (9th Cir. 2003); *Satava v. Lowry*, 323 F.3d 805, 810 (9th Cir. 2003); *See v. Durang*, 711 F.2d 141, 143 (9th Cir. 1983).

[11] H.R. Rep. No. 94-1476, at 54 (1976).

[12] See, e.g., *John Muller & Co., Inc. v. New York Arrows Soccer Team, Inc.*, 802 F.2d 989 (8th Cir. 1986).

chapter are words and short phrases (for example the title of a work, short magical incantations, or the name of a magician or character) and 'scènes à faire'; that is, scenes that are necessary or at least commonplace in a particular genre of expression. Similarly, basic 'tricks' or movements, for example, techniques for palming coins or other small objects, or distracting the audience, would not likely be viewed as original or creative enough to achieve copyright protection.

Included among the categories of potentially protectable works of authorship are literary works, dramatic works, pantomimes and choreographic works, pictorial, graphic and sculptural works and audiovisual works, such as motion pictures. Many of those types of works could be embodied in a magic production. Of course, there are numerous types of 'magic'.[13] Some magic involves sleight of hand. Examples of that would include card and coin tricks. Sometimes, sleight of hand is coupled with some type of prop or device, such as the classic 'cups and balls' magic. Such smaller-scale illusions are sometimes referred to as 'close magic'. Other magic involves complex devices, such as mirrors and mechanical mechanisms that can be used to create illusions. In either type of magic, the illusions could be simply enacted by an ordinarily dressed individual on a bare stage. Or they could be performed by a 'character' or incorporated into a dramatic work, perhaps with dialogue and costumes. Productions such as 'In and Of Itself' reflect that approach, and find their roots in historical magic productions. This chapter will consider below several aspects of such variations of magic that might achieve copyright protection.

2.1 'Now You Own It . . .': Potential Copyrightable Subject Matter in a Magic Performance

There are at least three broad areas of potential copyrightable subject matter to consider with respect to magic productions. First, this chapter will consider magical devices, props, masks and costumes. Second, magical tricks and illusions may be performed by a character and embedded in a larger work, such as a dramatic work, a work of choreography or pantomime, or a motion picture. Though copyright in such a work may

---

[13] For highly entertaining histories of modern magic, see generally Jim Steinmeyer, *Hiding the Elephant: How Magicians Developed the Impossible and Learned to Disappear* (Carroll & Graf, 2003); Milbourne & Maurine Christopher, *The Illustrated History of Magic* (Avalon, 2006) 269. See also, Grand Illusions: The Story of Magic (Paragon Productions/Image Entertainment, 2002) (12-part video series); Mysteries of Magic (Unapix Entertainment/TLC Video, 1998) (3-DVD program).

not extend to individual tricks or mechanisms, a substantially similar production could be an infringement of copyright.[14] Third, although many tricks and illusions are not original to the contemporary magician and indeed may be in the public domain, copyright law can provide protection for an original, minimally creative selection, ordering or arrangement of non-protectable elements, as a 'compilation'.

3. THE 'USEFUL ARTICLE' CONUNDRUM: PROPS, DEVICES, COSTUMES AND MASKS

Magic pieces occupy an interesting place in the murky boundaries between expression and function, copyright and patent. Devices used in creating illusions, such as the mirrored box permitting the impression of disappearance or a piece of glass used at a certain angle to create the illusion of a ghostly apparition, would likely be 'discoveries' protectable only under patent law,[15] not copyright. Putting aside issues of patent protection, which are beyond the scope of this chapter, could magical props be potentially protected as pictorial graphic or sculptural works? This is a surprisingly complex question. If one invents a new type of saw that can be used to cut a person in half, it would seem that the device would not be a subject of copyright, but if novel and non-obvious, would be a subject of patent. But what of a device that only appears to cut a person in half? Is such a device 'useful' or 'functional' when its only use is to entertain, to create an illusion?[16]

The copyright statute provides that the design of a 'useful article', i.e., an article with an 'intrinsic utilitarian function that is not merely to portray the appearance of the article or to convey information'[17] is not copyrightable unless it 'incorporates pictorial, graphic, or sculptural

[14] But in one case, a comedic sequence that incorporated some classic magic elements was found to be sufficiently important to a film that another film which copied the sequence was found to be an infringement. The second film was enjoined and damages were awarded. *Universal Pictures Co. v. Harold Lloyd Corp.*, 162 F.2d 354 (9th Cir. 1947). The court rejected arguments that the sequence was merely an unprotectable 'gag' or 'stage business'. Ibid. at 360.

[15] There are numerous patents for such devices. See, e.g., U.S. Patent No. 922,722 (filed May 25, 1909) (means for producing theatrical effects); U.S. Patent No. 1,785,347 (filed Sept. 24, 1926).

[16] Apparently, it would be sufficiently useful to obtain patent protection. See U.S. Patent No. 1,458,575 (filed September 9, 1921) (Horace Goldin's 'Illusion Device', used to create the illusion of sawing someone in half).

[17] 17 U.S.C. § 101 (2007) (definition, 'useful article').

features that can be identified separately from, and are capable of existing independently of, the utilitarian aspects of the article'.[18] Essentially, courts engage in a two-step analysis. First, is the item involved a 'useful article' under the Copyright Act? If so, the inquiry proceeds to consider whether or not the item includes separable pictorial, graphic or sculptural features. If so, those features are subject to copyright.

Of course, a device that portrays a saw capable of cutting someone in half but that doesn't actually do so is probably not utilitarian as a saw. Because a requirement for patentability is usefulness, the existence of patents for magic performance devices suggests that creating an illusion is itself a 'utilitarian function'.[19] But because an item is patentable does not necessarily preclude its having copyright protection for any appropriate subject matter.[20]

The primary usefulness or utilitarian function of a magic device or prop is to entertain by creating an illusion. That a work is 'useful' to entertain or amuse, or to evoke an emotion, does not constitute it as a 'useful article'.[21] Evoking emotion is an aesthetic function, not a 'utilitarian' one, at least for the purposes of copyright law. Otherwise, much art would not be copyrightable. For example, in *Gay Toys v. Buddy L. Corporation*, the district court denied copyrightability to a toy airplane, because it was 'useful and possesses utilitarian and functional characteristics in that it permits a child to dream and to let his or her imagination soar'.[22] The Sixth Circuit panel rejected that and reversed, noting that, under that reasoning, 'virtually any "pictorial, graphic, and sculptural work" would not be copyrightable as a "useful article." . . . The district court would have the "useful article" exception swallow the general rule, and its rationale is incorrect'.[23] An item that appears to be a useful article may not in fact be considered one. For example, in one case, an item that appeared to be a two-piece swimsuit

[18] 17 U.S.C. § 101 (2007) (definition, 'pictorial, graphic and sculptural works').

[19] See, e.g., U.S. Patent No. 1,458,575 (filed September 9, 1921) (claiming patent in a 'means for performing an illusion', specifically, 'a box whereby a person or object can be placed within and the container cut substantially in half, giving the effect to the audience of cutting the person or object in half').

[20] *Mazer v. Stein*, 347 U.S. 201 (1954); 37 C.F.R. § 202.10(a) (2007).

[21] *Masquerade Novelty, Inc. v. Unique Indus., Inc.*, 912 F.2d 663, 671 (3d Cir. 1990). See Stephen Langs, The Definitional Scope of an Intrinsic Utilitarian Function Under the 1976 Copyright Act: One Man's Use is Another Man's Art, 20 W. New Eng. L. Rev. 143 (1998) (arguing that if aesthetic, rather than functional, qualities are 'primary', an article should be viewed as a pictorial, graphic or sculptural work, rather than a useful article).

[22] *Gay Toys v. Buddy L. Corp.*, 522 F.Supp. 622, 625 (E.D. Mich. 1981).

[23] *Gay Toys v. Buddy L. Corp.*, 703 F.2d 970, 973 (6th Cir. 1983).

was found to require trial as to whether it was a swimsuit (ordinarily a 'useful article') or a work of art.[24] Hence, to be a 'useful article' a magic device must have some other 'utilitarian function', and it may be difficult to articulate exactly what that is, other than 'portraying the appearance' of something.

If a magic prop is determined to be a useful article, it still may include copyrightable artistic elements. Courts have struggled with how to analyze whether or not a useful article contains artistic features that are sufficiently 'separable' from the utilitarian aspects of the article. The fact that a magic device does include a component with a utilitarian function other than to create an appearance or illusion, might not eliminate its non-functional component's potential copyrightability as a work of applied art. In *Ted Arnold v. Silvercraft*,[25] a court found that a simulation of an antique telephone encasing a pencil sharpener was capable of physical separability from the useful component (the pencil sharpener), and was therefore a copyrightable work of art.

Garments are viewed as a type of useful article because of their function in covering the body and perhaps sheltering it from the environment. The Copyright Office will not register garments for copyright protection (unless the garment has pictorial or graphic material imposed on the garment, in which case that material may be protectable).

The U.S. Supreme Court has recently addressed the question of copyrightability of graphic elements embodied on a garment, finding that graphic designs on cheerleader uniforms could be protectable works, subject to other qualifications such as originality.[26] Some lawyers hoped that the Court would use this case to provide a definitive 'test' for conceptual separability. In a majority opinion by Justice Thomas, the Court approached the issue as purely a textual statutory interpretation matter, with little discussion of the various approaches applied by other lower courts. The Court rejected the physical/conceptual separability rubric, finding that separability is purely conceptual, and endorsing an approach that adds little clarity to the issue. The Court mentioned, but rejected, the 'designer judgment'[27] and 'likelihood of marketability' approaches to separability. Instead, the majority held that:

---

24 *Poe v. Missing Persons*, 745 F.2d 1238 (9th Cir. 1984).

25 *Ted Arnold, Ltd. v. Silvercraft Co.*, 259 F. Supp. 733 (S.D.N.Y. 1966).

26 *Star Athletica, L.L.C. v. Varsity Brands, Inc.*, 137 S. Ct. 1002 (2017).

27 *Star Athletica*, 137 S. Ct. at 1015 (rejecting this argument because not stated in the text of the statute; the court further held that the designer judgment approach would require the decision maker 'to consider evidence of the creator's design methods, purposes, and reasons', and the statute's text makes clear, that the

> [A]n artistic feature of the design of a useful article is eligible for copyright protection if the feature (1) can be perceived as a two- or three-dimensional work of art separate from the useful article and (2) would qualify as a protectable pictorial, graphic, or sculptural work either on its own or in some other medium if imagined separately from the useful article.[28]

It will continue to be difficult to assess copyrightability under this approach, as illustrated by the dissent of Justices Breyer and Kennedy, who endorsed the test, but reached a different conclusion as to the cheerleader uniforms at issue, because they could not perceive the designs on the uniforms as works of art separate from the useful article. In expressly rejecting the physical separability requirement, it would appear that designs on useful articles, such as the costumes of magicians and their assistants, and other artistic components of those costumes, will be more likely to receive protection than previously.

However, even before the *Star Athletica* decision, 'fanciful costumes,' for example Halloween costumes, were analyzed as 'useful articles' and registered if they contained separately identifiable pictorial or sculptural authorship.[29] It has been argued that theatrical costumes should not be treated as utilitarian articles, because their primary purpose is 'to convey information about the character'.[30] Indeed, at least a part of a fanciful

---

court's inquiry 'is limited to how the article and feature are perceived, not how or why they were designed').

28 *Star Athletica*, 137 S. Ct. at 1016.

29 Copyright Office Policy Decision, Registrability of Costume Designs, 56 Fed. Reg. 56530, 56531, 1991 WL 224879 (1991). See also *Nat'l Theme Prods., Inc. v. Beck*, 696 F.Supp. 1348 (S.D. Cal. 1988) (finding children's costumes copyrightable). Interestingly, one of the costumes in that case permitted the wearer to appear to be the well-known 'rabbit in a hat' trick. But see *Whimsicality, Inc. v. Rubie's Costumes Co, Inc.*, 721 F.Supp. 1566 (E.D.N.Y. 1989) (aff'd on other grounds) (finding Halloween costumes not copyrightable); *Whimsicality, Inc. v. Battat*, 27 F.Supp.2d 456 (S.D.N.Y. 1998) (same). The Whimsicality courts' decisions were arguably wrong. This is because they reasoned that enabling the wearer to masquerade was itself the utilitarian function. Of course, if that's the case, the artistic elements cannot be separated from that function. It is submitted that this is wrong for at least two reasons. First, to 'masquerade' is to portray the appearance of something, which, by definition, is not sufficient to constitute a 'useful article'. See *Masquerade Novelty, Inc. v. Unique Indus., Inc.* 912 F.2d 663, 664 (3d Cir. 1990) (animal nose masks not 'useful articles' for that reason). Second, 'masquerading', for anyone other than perhaps a criminal attempting to hide her identity, is arguably an aesthetic, rather than utilitarian, function.

30 Rebecca I. Foster, Protect the Bastard Child of the Arts: Copyright Protection for Theatrical Costumes, 22 Sw. U. L. Rev 431, 439 (1993).

theatrical costume has been found protected by copyright in the film including the character.[31] In that case, famous horror-film character, Freddy Kruger's, razor-fingered glove was found copyrightable and infringed by use in a music video without authorization.

Masks are not treated as 'useful articles,' since they 'generally portray their own appearance'.[32] Makeup design may also be copyrightable. In one case, makeup designs for the characters in the successful Broadway show 'Cats' were found to be copyrightable.[33] It seems unlikely that the *Star Athletica* decision would reduce copyright protection previously recognized in the context of costumes and masks.

It is possible that a magician's costume, if fanciful enough, could be treated more like a copyrightable costume than an uncopyrightable garment. If a mask, or even makeup design, is sufficiently original, it too, may be a copyrightable part of a magician's performance. Separable original pictorial, graphic or sculptural elements embodied on a magic device – such as a creative graphic design printed on the outside of a box for the 'woman sawed in half' illusion – would be copyrightable. Magical devices, as such, are not likely to be copyrightable, based on the usefulness of the means of creating an illusion, although there may be room to argue that, notwithstanding the patentability of such devices, their use is an aesthetic, entertaining one that should not be considered utilitarian for the purposes of copyright law.

4. PERFORMING ARTS: DRAMATIC WORKS, CHOREOGRAPHY AND PANTOMIME

4.1 Dramatic Works

Magic has long been presented as part of a play. For example, the famous 19th-century magicians, John Nevil Maskelyne and David Devant, had great success in that format[34] and modern magicians have also done so,

---

[31] *New Line Cinema Corp. v. Eastern Unlimited*, 17 U.S.P.Q. 1631 (E.D.N.Y., 1991).

[32] 56 F.R. 56530, 56531, 1991 WL 224879 (F.R., 1991). See also, Rebecca I. Foster, Protect the Bastard Child of the Arts: Copyright Protection for Theatrical Costumes, 22 Sw. U. L. Rev. 431, 450 (1993).

[33] *Carell v. Shubert Org., Inc.*, 104 F.Supp.2d 236 (S.D.N.Y. 2000).

[34] Jim Steinmeyer, *Hiding the Elephant: How Magicians Developed the Impossible and Learned to Disappear* (Carroll & Graf, 2003) 181–195. See Sam Sharpe, *The Magic Play* (Magic, 1976).

presenting plays on Broadway.[35] Jim Steinmeyer, noted magician and historian of magic, said:

> Ultimately, like any really successful storytelling, a magic show is an example of dictatorial control by the performer. The magician takes charge of every aspect of the story, pointing your head in certain directions at certain moments, emphasizing the importance of certain occurrences and minimizing others, hinting at every surprise, and imparting specific meanings. If the performance is successful, it is because the storytelling has been successful. If it stops being simply a puzzle and becomes magical, it is because the audience has succumbed to the magician's direction and interpretations.[36]

Magic 'pieces', illusions, and special effects have been part of dramatic performances since the earliest known dramas.[37] That dramatic work can be the subject of copyright protection.[38]

Moreover, the dramatic work need not be highly intellectual or sophisticated to secure protection. It is clear that courts are not to evaluate the aesthetic qualities of a work to determine its copyrightability, at least outside the narrowest limits. This was noted in some of the early decisions,[39] and confirmed by Justice Holmes for the Supreme Court.[40] A dramatic presentation would not have to be sophisticated or novel to receive copyright protection. For example, a sequence from a film 'short', which interestingly incorporated some common magic performance elements into a slapstick comedy routine, was found to be protectable.[41] The sequence involved the lead character inadvertently putting on a magician's coat at a social event. When the character dances with another character,

---

[35] See, e.g., *Ricky Jay and His 52 Assistants* (HBO 1996); Gordon Cox, 'In & Of Itself,' Produced by Neil Patrick Harris, Extends Off Broadway Run, *Variety* (May 8, 2017), https://variety.com/2017/legit/news/in-of-itself-extension-off-broadway-1202410796/.

[36] Steinmeyer supra note 34, at 180.

[37] See, Milbourne and Maurine Christopher, supra note 13, at 8–16.

[38] 17 U.S.C. § 102(3) (2007). There is no definition of 'dramatic works' in the statute. U.S. Copyright Office, Compendium of U.S. Copyright Office Practices, § 804.1 (3d ed. 2014), defines a dramatic work as 'a composition generally in prose or verse that portrays a story that is intended to be performed for an audience such as plays, musicals, or operas. Generally, a dramatic work represents the action as it occurs rather than simply narrating or describing the action. Some dramatic works include music'.

[39] See, e.g., *Henderson v. Tompkins*, 60 F. 758, 764 (C.C.D. Mass. 1894) ('topical song' sufficiently creative, and referring to the 'railroad scene' cases as exemplary).

[40] *Bleistein v. Donaldson Lithographing Co.*, 188 U.S. 239 (1903).

[41] *Universal Pictures Co. v. Harold Lloyd Corp.*, 162 F.2d 354 (9th Cir. 1947).

typical magic 'bits' occurred – doves, mice and eggs all emerged from the coat.[42] Although the setup seems to have been rather different in each of the films, those comedic but typical magic incidents were very similar. Though the defendant argued they were mere uncopyrightable 'gags' and 'stage business,' the finding of infringement was affirmed and damages awarded.

'Dramatic composition' (and the right to publicly perform such a work) was added to U.S. copyright law in 1856.[43] Whether or not a work was a 'dramatic composition' was important because the public performance right did not apply to literary works at that time. Over the next several decades there were a series of cases in which a plaintiff asserted copyright infringement in a dramatic composition, and courts had to decide whether or not certain types of performances qualified. This distinction is rendered somewhat less relevant now, because public performance rights are part of the copyright with respect to other types of works and because choreography and pantomime are protectable by copyright even if they are not 'dramatic works'. Still, because the legislative history for the 1976 Act did not define any of those categories of works, saying that they 'have fairly settled meanings',[44] earlier courts' discussions are relevant to deciding what should qualify as a dramatic work under current copyright law.

Several of the earliest cases involve the well-known 'railroad scene', in which a character is secured to a railroad track, a train is portrayed as approaching and the victim is rescued at the last moment by someone who must first dramatically free himself from some restraint.[45] This scene (or at least a version of the scene) was created by Augustin Daly as part of a popular play called 'Under the Gaslight', which Daly registered for copyright in 1867. Apparently, the railroad scene was quite sensational and an important part of the play's success, thus leading to imitation. A British author named Boucicault included a similar scene[46] in a play called 'After Dark' and Daly sued. The court found that the railroad scene was indeed a dramatic composition that was infringed, notwithstanding the absence of similar dialogue:

---

42 Ibid. at 358.

43 11 Stat. 158 (1857).

44 H.R. Rep. No. 94-1476, at 53 (1976).

45 See *Daly v. Palmer*, 6 F. Cas. 1132 (C.C. S.D.N.Y. 1868).

46 In Boucicault's version, the train was a subway. He changed the gender of the rescuer, the location from which the rescuer had to escape in order to rescue the person on the tracks, and the manner of that escape. *Daly*, 6 F. Cas. 1132, 1133–1134.

> Movement, gesture, and facial expression, which address the eye only, are as much a part of the dramatic composition as is the spoken language which addresses the ear only; and that part of the written composition which gives direction for the movement and gesture, is as much a part of the composition, and protected by the copyright, as is the language prescribed to be uttered by the characters. And this is entirely irrespective of the set of the stage, or of the machinery or mechanical appliances, or of what is called, in the language of the stage, scenery, or the work of the scene-painter.[47]

The court enjoined the public performance of Boucicault's play.

Copyright protection to a 'mechanical contrivance' was denied in another case, in which the plaintiff sued to stop the use in a competing play of a tank filled with water, which then flowed through a trough to simulate a river in a stage production.[48] The court found the allegedly dramatic incident (a character falling into the 'river') not to be original, and denied copyright protection to the device itself, saying:

> Such a mechanical contrivance, however, is not protected by a copyright of the play in which it is introduced. The decisions which extend the definition of dramatic composition so as to include situations and scenic effects, do not cover the mere mechanical instrumentalities by which such effects or situations are produced.[49]

One very early decision involved a dispute over a work the court described as a 'mere spectacle'.[50] Apparently there was very minimal 'scant and meaningless' dialogue, which the court characterized as 'a mere accessory to the action of the piece – a sort of verbal machinery tacked on to a succession of ballet and tableaux'. Essentially, this play seemed to be an early version of a burlesque show, involving 'the exhibition of women in novel dress or no dress, and in attractive attitudes or action . . . women lying about loose'. The court viewed such a 'spectacle' as not worthy of copyright protection as a dramatic composition.[51]

---

47 *Daly*, 6 F Cas. 1132, 1137.

48 *Serrana v. Jefferson*, 33 F. 347 (C.C. S.D.N.Y. 1888).

49 Ibid. at 348.

50 *Martinetti v. Maguire*, 16 F. Cas. 920 (C.C. Cal. 1867).

51 The court also found that, even if the work was a 'dramatic composition', its 'grossly indecent' character made it a work not 'suited for public representation', which was a requirement under the statute at the time. Although the statute did use that language, the judge's interpretation of that phrase to incorporate morality issues was unprecedented. More modern decisions do not interpret the Constitutional copyright power grant as precluding copyright for works viewed as indecent.

Another decision of that era illustrates that limit.[52] The plaintiff had created a 'stage dance', a series of 'tableaux' involving graceful movement and lighting that was described in literary form. The defendant had presented a very similar piece. Because there was no public performance right in literary works, the plaintiff argued that her work was a dramatic composition that would be infringed by an unauthorized performance. In denying her claim for an injunction, the judge said,

> It is essential to [a dramatic composition] that it should tell some story. The plot may be simple. It may be but the narrative or representation of a single transaction; but it must repeat or mimic some action, speech, emotion, passion, or character, real or imaginary. And when it does, it is the ideas thus expressed which become subject of copyright . . . The merely mechanical movements by which effects are produced on the stage are not subjects of copyright where they convey no ideas whose arrangement makes up a dramatic composition.[53]

On the other hand, the U.S. Supreme Court, in resolving an early motion picture copyright infringement case, confirmed that dramatic compositions do not require words, and can be infringed by technological means.[54] Justice Holmes said, 'drama may be achieved by action as well as by speech. Action can tell a story, display all the most vivid relations between men, and depict every kind of human emotion, without the aid of a word'.[55]

In examining works submitted for possible copyright registration, the Copyright Office uses a Compendium of Copyright Office Practices for guidance. Although it is not binding law, it does reflect the Office's view of the statute and case law, so it is informative to consider how the Compendium defines these types of works, left undefined in the statute or regulations. The current Compendium defines a 'dramatic work' as 'a composition generally in prose or verse that portrays a story that is intended to be performed for an audience such as plays, musicals, or operas. Generally, a dramatic work represents the action as it occurs rather than simply narrating or describing the action. Some dramatic works include music'.[56] According to the Compendium, the characteristic features of a dramatic work include 'plot, characters, dialog, and directions for performance, although each element is not necessarily registrable

---

52 *Fuller v. Bemis*, 50 F. 926 (C.C. S.D.N.Y. 1892).

53 Ibid.

54 *Kalem Co. v Harper Bros.*, 222 U.S. 55 (1911).

55 Ibid. at 61.

56 U.S. Copyright Office, Compendium of U.S. Copyright Office Practices § 804.1 (3d ed. 2014).

in and of itself'.[57] Plot is described as 'the storyline, plan, or sequence of events in a dramatic work'.[58] A 'character' is 'a person, animal, or even an inanimate object through which the content of a dramatic work is portrayed'.[59] Characters in a dramatic work are typically not separately registrable for copyright, but 'the Office may issue a registration based on the authorship describing, depicting, or embodying the characters. The more delineated a character is, either textually or visually, the more likely copyright protection will be available for those specific attributes of a character'.[60] Stage direction might be a copyrightable component of a dramatic work, but simple 'stage business' – 'activity performed by the actors to illustrate character or create dramatic effect'[61] – and 'blocking' – 'the positioning and movement of actors onstage or in frame'[62] – are generally not registrable since they comprise common bodily and other movements.[63] Other elements of a dramatic work, e.g., illustrations of costumes, scenery, sets, props, and lighting, may also be covered by its registration.[64] Although limited in scope, a 'treatment' for a dramatic work – 'a written description of a dramatic work or television show, which outlines and describes the scenes and/or characters and often includes sample dialog'[65] – may also be registered.

4.2 Choreography

What if there were no drama, but only movement connecting a series of magic pieces? Although cases under copyright law prior to the current act only protected choreography if it qualified as a dramatic work,[66] that is, it told a story, choreography is now included in the statute as a distinct type of work and is copyrightable without a story,[67] and abstract choreographic works, which are, of course, original and fixed, are copy-

---

57 Ibid. § 804.3.
58 Ibid. § 804.3(A).
59 Ibid. § 804.3(B).
60 Ibid.
61 Ibid. § 804.3(D)(1)
62 Ibid. § 804.3(D)(2).
63 Ibid. § 804.3(D)(1)–(2).
64 Ibid. § 804.3(F).
65 Ibid. § 804.4(F)(2).
66 See, Melville B. Nimmer and David Nimmer, *Nimmer on Copyright* § 2.07[B] (2007); Borge Varmer, Copyright in Choreographic Works, Copr. Off. Rev'n Study No. 28, reprinted in 1 Studies on Copyright (1963).
67 U.S. Copyright Office, Compendium of U.S. Copyright Office Practices § 805.2(D).

rightable.[68] Although it need not tell a story, 'the dance movements are expected to convey some thematic or emotional concept to an audience'.[69] Choreography is defined as 'the composition and arrangement of dance movements and patterns, and is usually intended to be accompanied by music. Dance is static and kinetic successions of bodily movement in certain rhythmic and spatial relationships'.[70] The important features are 'a related series of dance movements and patterns organized into a coherent whole';[71] 'a work of authorship cannot be registered as a choreographic work unless it is comprised of dance steps, dance movements, and/or dance patterns'.[72] Social dance steps and simple routines, as with other basic building blocks of expression, are not protectable,[73] nor are 'mere exercises'.[74] Choreography is generally performed by 'skilled dancers', one characteristic that distinguishes it from mere social dance steps.[75]

Modern magic shows sometimes include dancers and obvious choreography. Without dancers, it is unclear that a magician's movements could be viewed as choreography under the Copyright Office's interpretation as reflected in the Compendium (Third). On the one hand, such a performance could consist of a related series of static and kinetic bodily movements organized into a coherent whole. If the performance is enacted rhythmically, ideally set to music, it would seem to be

---

[68] Ibid.

[69] Ibid.

[70] U.S. Copyright Office, Compendium of U.S. Copyright Office Practices § 805.1 (3d ed. 2014) (quoting U.S. Copyright Office, Compendium of U.S. Copyright Office Practices § 450.01 (2d ed. 1984)); Compendium (3d ed.) § 805.4(A).

[71] Ibid. (quoting U.S. Copyright Office, Compendium of U.S. Copyright Office Practices § 450.03(a) (2d ed. 1984)).

[72] U.S. Copyright Office, Compendium of U.S. Copyright Office Practices § 805.2(E) (3d ed. 2014).

[73] Ibid.; H.R. Rep. No. 94-1476, at 54 (1976).

[74] U.S. Copyright Office, Compendium of U.S. Copyright Office Practices § 805.5(B)(3). For a time, the Copyright Office would accept registration of 'compilations' of simple exercises for copyright, and to publicly perform such a work presumably could infringe copyright. However, it changed its policy, determining that 'A work may be precluded from registration as a functional system or process if the particular movements and the order in which they are performed purportedly improve one's health or physical or mental condition' (citation omitted). 'Examples of functional physical movements that cannot be registered with the Office include exercise routines, aerobic dances, yoga positions, and the like', ibid.

[75] U.S. Copyright Office, Compendium of U.S. Copyright Office Practices § 805.2(F) (3d ed. 2014).

potentially copyrightable as a work of choreography.[76] On the other hand, the Compendium (Third)'s emphasis on the inclusion of 'dance steps'[77] might, if endorsed by courts, be found to limit choreographic protection for magic routines unless they involve rhythmic movements of the magician's (or her assistant's) body. Indeed, the Compendium (Third) specifically states that the Copyright Office 'cannot register feats of physical skill or dexterity or other choreographed productions that do not involve the rhythmic movement of a dancer's body',[78] including 'magic acts' among those types of works. That categorical exclusion is overbroad, and does not reflect the fact that some 'magic acts' could qualify as choreographic works; a magic act can be far more creative than many of the other specifically identified examples of uncopyrightable material, such as American football plays and juggling. However, if the performance is not sufficiently choreographic, it still could qualify for copyright as a pantomime.

4.3 Pantomime

Although early cases acknowledged that drama could be portrayed through pantomime alone, pantomime is now recognized as a distinct category of copyrightable work.[79] Unlike dramatic works, under current copyright law pantomime need not tell a story.[80] The Compendium (Third) defines it as 'the art of imitating or acting out situations, characters, or some other events with gestures and body movement'.[81] Consistent with the general concepts of copyrightability, a pantomime copyright does not extend to 'stock gestures' or 'common techniques'[82] and must contain more than 'a few stock gestures, movements, or facial expressions with

---

[76] Of course, any work must be fixed in a tangible medium of expression to qualify for federal copyright. This could be accomplished most easily by videotaping or filming the performance, although a written description sufficiently descriptive to permit its enactment may suffice. If unfixed, one should consider potential state (often called 'common law') copyright. See Cal. Civ. Code § 980.

[77] U.S. Copyright Office, Compendium of U.S. Copyright Office Practices § 805.5(B)(3), which states 'Because choreography is a subset of dance, a work of authorship cannot be registered as a choreographic work unless it is comprised of dance steps, dance movements, and/or dance patterns'.

[78] Ibid.

[79] 17 U.S.C. § 102(a) (2007).

[80] U.S. Copyright Office, Compendium of U.S. Copyright Office Practices § 806.2(D) (3d ed. 2014).

[81] Ibid. at § 806.1 (3d ed. 2014).

[82] Ibid. at § 806.4(D).

minor linear or spatial variations'.[83] That said, a compilation of such separately uncopyrightable elements could receive copyright protection. The Compendium (Third) acknowledges that 'Pantomimes that incorporate stock gestures, ordinary motor activities, or even athletic exercises may be protected by copyright, provided that the work as a whole contains a sufficient amount of original authorship', citing *Teller*.[84] Creating uncertainty about the standard of protection for magic performances as pantomimes, the Compendium (Third) states that because of functionality, 'the Office cannot register feats of physical skill or dexterity that do not involve the physical movement of a performer's body in an integrated, coherent, and expressive compositional whole, such as magic acts, circus acts, juggling, fight sequences, or the like'.[85] The Compendium creates confusion by lumping magic acts together with sports plays and functional yoga sequences, while also referring to a magic performance as a type of potentially copyrightable set of uncopyrightable elements. Reflecting some of the early 'dramatic work' cases discussed above, it is likely that the Office would also reject applications for copyright in mere 'tableaux employing less than a minimum amount of action'.[86]

Until *Teller*, there were no reported decisions involving infringement purely of pantomime, but at least those that satisfy the criteria discussed above should clearly be copyrightable. One could certainly consider at least some magic performances to imitate or act out situations, characters or other events through movement of the body in an expressive compositional whole. If sufficiently original (not copied from other sources), such a performance should qualify as a copyrightable pantomime once it is fixed. *Teller* accorded copyright protection to a single illusion as a 'dramatic pantomime', as will be discussed in more detail below. Teller was also successful in registering his illusion for copyright as such. According to Compendium (Third), 'The forms of fixation for choreography and pantomime typically include dance notation, motion pictures, textual descriptions, as well as drawings, illustrations, and/or photographs'.[87]

---

83 Ibid. at § 806.5(A).

84 See *Teller v. Dogge*, 110 U.S.P.Q.2d 1302, 1306 (D. Nev. 2013).

85 U.S. Copyright Office, Compendium of U.S. Copyright Office Practices § 806.5(B) (3d ed. 2014). A revised draft being considered by the Copyright Office as this work goes to print would add 'wrestling matches' to the list.

86 See U.S. Copyright Office, Compendium of U.S. Copyright Office Practices § 461 (2d ed. 1984).

87 U.S. Copyright Office, Compendium of U.S. Copyright Office Practices § 1509.2(D) (3d ed. 2014).

In Teller's case, a textual description including drawings was used as the deposit copy.

In sum, a magic performance should be treated as any other type of performance work. If it contains a plot – a 'thread of consecutively related events' – and at least one 'character', with or without dialogue, it could be a 'dramatic work'. If it has no plot or characters, but includes successions of rhythmic bodily movement organized into a 'coherent whole,' especially if accompanied by music, it may qualify as a 'choreographic work'. If it contains no such movement, but acts out 'situations, characters or some other events with gestures and body movements', it should be protectable as 'pantomime'. Of course, like any other work, to be copyrightable, it must be original, that is, not copied from prior work, and fixed in some tangible medium, which could include either a written form sufficient to permit performance or a film or other audiovisual recording of a performance. Surely, among the many varieties of contemporary magic acts, at least some of them will qualify as one or more of those types of works, or incorporate portions that qualify as such works.

It is clear that a magic play such as those produced by Maskelyne could be protected by copyright, whether or not the plot or characters were complex and sophisticated, and whether or not there was dialogue. If it tells a story, or if it evokes emotion through action rather than just narration, it is a dramatic work. The copying of even a special effect scene from such a play, such as the early 'railroad scene', could be an infringement of such a work, so long as what was copied was original with the author and fixed in a tangible form.[88] Similarly, a show like DelGaudio's 'In and Of Itself' may well contain copyrightable dramatic, choreographic or pantomime material.

5. COMPILATIONS

Copyright can also extend to 'compilations' of other material, that is, an original selection, coordination or arrangement of preexisting material.[89] If the selection, ordering or arrangement are original (and, of course, fixed and otherwise sufficient for copyright protection), then copyright is avail-

[88] See also *Chappell & Co. v. Fields*, 210 F. 864 (2d Cir. 1914) (affirming injunction against a play that used a single scene from a prior play); *Casino v. Vitaphone Corp.*, 163 Misc. 403 (Sup. Ct. N.Y. 1937) (common-law copyright in revue infringed by use of material similar to a single sketch 'The Postman Always Rings Twice' in another play).

[89] 17 U.S.C. § 103 (2007).

able, even if the components compiled are in the public domain. If a magic performance failed to qualify as a copyrightable dramatic, choreographic or pantomime work and involved more than a few components that were selected, ordered or arranged in an original way, that selection, ordering or arrangement would be protectable as a compilation.[90]

Some magic performances are simply that – a magician in front of an audience performs a series of pieces. If such a performance comprised a particular compilation – an original selection, ordering or arrangement of pieces – it could be protectable as such.[91] For a time, the Copyright Office would register compilations of exercise routines.[92] One court was persuaded that a compilation of unprotectable yoga postures could be copyrightable and potentially infringed by public performance, i.e., by a teacher and students engaging in the series of poses in a place open to the public.[93] As discussed above, the Office has changed its policy as to functional compilations such as yoga sequences. In a more recent case involving Bikram yoga, the Ninth Circuit panel correctly analyzed this type of compilation, affirming the District Court's finding that the sequence was not protectable.[94] Recognizing that copyright in a work doesn't extend to methods or processes depicted in the work, the court found that the selection, ordering and arrangement of Bikram's sequence

---

[90] See U.S. Copyright Office, Compendium of U.S. Copyright Office Practices § 806.4(D) (3d ed. 2014).

[91] 17 U.S.C. § 102(a) (2007).

[92] For example, Tae Bo, a form of kick boxing movements compiled into exercise programs, has been registered for copyright. See also *Open Source Yoga Unity v. Choudhury*, 74 U.S.P.Q.2d 1434, 2005 WL 756558 (N.D.Cal.) (not reported in F.Supp.2d) (noting that in 2002, Bikram modified his copyright registration to apply to the sequence itself as a compilation).

[93] *Open Source Yoga Unity v. Choudhury*, 74 U.S.P.Q.2d 1434, 2005 WL 756558 (N.D.Cal.) (not reported in F.Supp.2d). This author believes that the decision was erroneous. This is because Bikram had himself described the sequence itself as functional – achieving human health. Moreover, Bikram had said that there was only one sequence of poses by which to achieve that utilitarian result – his sequence. This would lead to no protection for the sequence, under the complex 'merger' doctrine. See also J. Sussman, Your Karma Ran Over My Dogma: Bikram Yoga and the (Im)possibilities of Copyrighting Yoga, 25 Loy. L.A. Ent. L. Rev 245 (2005); K. Machan, Bending Over Backwards for Copyright Protection: Bikram Yoga and the Quest for Federal Copyright Protection of an Asana Sequence, 12 UCLA Ent. L. Rev. 29 (2004). Neither functionality of the sequence nor merger should typically apply to a magic compilation.

[94] *Bikram's Yoga Coll. of India, L.P. v. Evolation Yoga, LLC*, 803 F.3d 1032 (9th Cir. 2015).

was itself a 'process' for achieving good health. That functionality precluded copyright protection.

Arguably, a magic performance compilation is not functional in the way that an exercise routine or series of yoga poses would be. Unlike those types of compilations, the goal of magic is to entertain; i.e., its purpose is aesthetic and not utilitarian. Therefore, it would seem that magic routines – a selection of magic pieces, perhaps combined with costumes, masks, choreography and/or pantomime elements – would similarly be protectable, even if each element were not protectable as any form of movement art or other material discussed above. Although there is not much precedent, in one case, a court found that a country music television program comprised of uncopyrightable elements could be copyrightable as a compilation.[95]

A related concept is that some decisions have indicated that a work that is not formally a 'compilation' could be infringed by a work that copies numerous substantial multiple components of that work. In *Metcalf v. Bochco*,[96] the plaintiff's work was a television script. The defendant's show apparently included several elements that in themselves would not be protectable, but in combination, the court felt there could be sufficient similarity proved that summary judgment should be denied. In *Satava v. Lowry*,[97] the court seemed to follow a similar approach. In both cases, liability was ultimately not found, but the basic conceptual approach to infringement analysis remains viable.

5.1 'Now You Don't . . .' – The Failure of the Mystery Magician

In the only modern reported decision asserting infringement of copyright by use of a similar magic performance prior to *Teller*, the plaintiff was unsuccessful.[98] A review of that case will illustrate the concepts discussed above and provide ideas about how elements of magic performances might receive protection. Magicians have long attempted to protect their

---

95 *Apple Barrel Productions, Inc. v. Beard*, 730 F.2d 384 (5th Cir. 1984); See also U.S. Copyright Office, Compendium of U.S. Copyright Office Practices § 806.4(D) (3d ed. 2014) (using Teller's pantomime as an example of a work consisting of a set of standard gestures and activities that may, as a whole, constitute copyrightable authorship).

96 294 F.3d 1069 (9th Cir. 2002).

97 323 F.3d 805 (9th Cir. 2003).

98 *Rice v. Fox Broad. Co.*, 148 F.Supp.2d 1029 (C.D. Cal. 2001), aff'd in part, rev'd in part, 330 F.3d 1170 (9th Cir. 2003). There were also claims for false advertising and unfair competition in the case, but this article will focus on the copyright claims.

devices and methods by maintaining secrecy. The ethical code of professional magicians has discouraged revealing how an illusion is performed, but there are many instances of competitors seeking to determine how an illusion was created, and numerous works have been published purporting to reveal the secrets of successful magic pieces. The *Rice* case is a modern example, ironically involving a claim by a magician named Robert Rice, who had created a video called 'The Mystery Magician',[99] revealing the secrets behind some well-known magic tricks ('Rice Video'), against Fox Broadcasting's similarly themed series of television programs, entitled 'Breaking the Magician's Code: Magic's Biggest Secrets Finally Revealed' ('Fox Programs').

In the Rice Video, Rice is a masked man, the 'Mystery Magician', purporting to be a professional magician, dressed in the classic black tuxedo tails, a white tuxedo shirt, black bow tie and red-lined black cape, who performs in a black stocking mask and hood to disguise his identity because he fears retribution from the magician community for his disclosures. He is assisted by two men and two women, also masked. Rice selected 12 illusions that were at least fifty years old, but still popular. He revealed the 'secrets' behind six of those selected illusions. Notwithstanding those revelations, the character asks his viewers to continue to be inspired to believe in magic and the skill of the magician.

About ten years later, Fox began to develop the Fox Programs. The first of the four Programs contained 11 tricks, five of which were also in the Rice Video (only four of which were revealed in both programs). Unlike the Rice Video, in the Fox Programs, all of the tricks are revealed. The tricks in the Fox Programs were generally in a different order than that in the Rice Video.[100]

Rice asserted that there were similarities between the characters, plot, setting, sequence of events, dialogue, mood, tone and pace, and the themes (if not expression) of the opening and closing monologues. There were issues over whether the Fox Programs could have been actually copied from the Rice Video, but the court found that there was sufficient evidence that some of the individuals involved in the Fox Programs may have had an opportunity to see the Rice Video that the case couldn't be rejected on lack of access.[101] Because the court is in the Ninth Circuit, it

---

[99] The full title was 'The Mystery Magician – He Dares to Expose the Secrets Behind Magic's Most Mystifying Illusions', *Rice*, 148 F.Supp.2d at 1036.

[100] *Rice*, 148 F.Supp.2d at 1039.

[101] Ibid. at 1049–1050. It is possible that the weak evidence of access hurt Rice's position by leading the court to look for a greater degree of similarity between the works to find infringement, although the court claimed not to do so (ibid.).

applied that Circuit's 'extrinsic', analytical test for actionable 'substantial similarity'.[102] The court first applied many of the limiting doctrines in copyright discussed above to 'filter out' unprotectable aspects of the Rice Video – non-protection of ideas, facts and other non-original or public domain material, the merger doctrine and scènes à faire. Based on its application of those concepts to this case, the court determined that Rice's copyright should be 'thin', protected only against almost a substantially identical work.[103] The court then compared the elements listed above and found no substantial similarity.

The Masked Magician character who performs and reveals the tricks in the Fox Programs was also dressed in black, with a mask/hood that covered his head and concealed his identity, but more casually attired than the Mystery Magician, wearing a black turtleneck, jacket and gloves. The court describes the Masked Magician's mask as looking very different from the simple mask worn by the Mystery Magician.[104] The settings of the two works were apparently somewhat different.

---

[102] The Ninth Circuit and the Second Circuit, two of the most important Circuits for copyright cases, have different ways of describing how they analyze copyright infringement cases. Of course, both require ownership of a valid copyright. The Ninth Circuit then examines whether there was 'copying', shown by 'access' (proof of defendant's opportunity to view the plaintiff's work) and 'substantial similarity'. For a plaintiff to succeed, it must prove substantial similarity under both of two tests: an analytical 'extrinsic' test, and a more subjective 'intrinsic' test. At the summary judgment stage, Ninth Circuit courts look only at the extrinsic test, because if a plaintiff cannot prove substantial similarity under the extrinsic test, there is no actionable substantial similarity. Although those courts are arguably not always consistent in applying the extrinsic analysis, the court here very thoroughly 'filtered out' what it viewed as unprotectable elements and compared the remainder, using a demanding 'virtual identity' standard, i.e., only nearly identical copying is actionable. Broadly speaking, the Second Circuit first looks for evidence of 'actual copying', often shown by a combination of 'access' and 'probative similarities', i.e., similarities that tend to show the defendant's work was actually copied from the plaintiff's rather than independently created. The probative similarity analysis is objective and analytical, but even unprotectable elements in common can tend to be probative of copying and will be considered. If a plaintiff cannot succeed in showing copying, summary judgment for the defendant is appropriate. If copying is found, then the factfinder compares the works more subjectively to see if there is 'substantial similarity' of only protectable elements, i.e., wrongful appropriation. Under that approach, it seems more likely that Rice would have prevailed at the summary judgment level, requiring a finder of fact to then assess the substantial similarity of the two works.

[103] *Rice*, 148 F.Supp.2d at 1061.

[104] Ibid. at 1038.

Rice claimed on appeal that the lower court erred in its approach and conclusions.[105] He argued that the lower court erred in treating the Rice Video like a factual or functional work, where protection is 'thin', rather than as a dramatic, creative work, where less than virtual identity can be actionable. Generally, where there are only a few ways to express an idea or method, giving copyright protection to any of the expressions could indirectly permit ownership of the idea or method. Therefore, courts will find that any way of expressing that idea or method is merged with the idea or method and is not protectable by copyright, or, at most, protected only against near verbatim copying. This complex merger concept is often applied as to very simple or functional works, and is less likely to be applied in connection with creative, artistic works.[106] Indeed, the district court found that the Rice Video was 'essentially a compilation of facts equally available to any subsequent author'.[107] The appellate court found that 'there are only a finite number of ways to reveal the secrets behind magic tricks, and the perform and reveal sequence is the most logical expression of this idea'.[108] Thus, they rejected the claim that the plot and sequence of events here were substantially similar.

While this approach was deadly for Rice, it does suggest that a more dramatic form of presentation of magic tricks, one involving a more creative plot than performing, then revealing, tricks, could be viewed less as a mere compilation of facts and could be protected from similar dramatic presentations of similar tricks. For example, Maskelyne's famous 19th-century magic play, *Will, the Witch and the Watchman*, while being a vehicle for the presentation of several popular illusions of the day, was far more elaborate in its sequence of events than simply performing, then revealing, tricks.[109] Moreover, it can be argued that the Rice Video was a compilation, but not only of facts. Rather, it was a compilation of magic

---

[105] See Appellant's Opening Brief, *Rice v. Fox Broad. Co.*, 330 F.3d 1170 (9th Cir. 2003).

[106] But see *Rosenthal Jewelry v. Kapakian*, 446 F.2d 738 (9th Cir. 1971) (seemingly applying merger doctrine to jeweled bee pins).

[107] *Rice*, 148 F.Supp.2d at 1054.

[108] *Rice*, 330 F.3d at 1177.

[109] See the script of *Will, the Witch and the Watchman*, reproduced in S. Sharpe, *The Magic Play* (Magic, 1976). See J. Steinmeyer, *Hiding the Elephant: How Magicians Developed the Impossible and Learned to Disappear* (Carroll & Graf, 2003) 102–103. Although not as successful with audiences, Maskelyne developed a play entitled *The Coming Race*, based on a satirical novel, in which the illusions seem more like today's special effects than a simple magic show. See M. and M. Christopher, *The Illustrated History of Magic* (updated edn, Avalon, 2006) 170.

illusions, movement, character, and other material that merited protection against a substantially similar compilation.

Much of the argument and discussion on appeal in Rice related to the central character of each of the programs. Similarity between characters in two works is one element that is considered in an infringement analysis. But it is a doctrinally messy area of copyright law, particularly in the Ninth Circuit. The basic problem again relates to a court's quite justifiable reluctance to permit control of an idea by means of copyright.[110] A literary character often represents an idea – a character type. If a particular character is fleshed out, that is, more specifically delineated by details of character in terms of appearance, characteristics and behavior, then many courts will treat that character itself as protected by the copyright in the work in which it is so described. This may well be the rule in the Ninth Circuit, but a 1950s' case involving the character Sam Spade suggested a more difficult hurdle for literary characters in the Ninth Circuit.[111] Although that case was really a contract interpretation, rather than copyright infringement, case, the court suggested that, in order for a literary character to be protected by copyright, it must 'constitute the story being told'. It is not entirely clear when or how that standard is satisfied. But many character cases involve characters depicted in visual or audiovisual works, and the Ninth Circuit courts appear to find the standard more easily satisfied in that context.[112] Because of that doctrinal uncertainty, however, courts in the Ninth Circuit will sometimes apply both tests, which is what the courts did here. The latest approach in the Ninth Circuit, at least as to characters depicted in visual form, is set forth in a decision involving a claim of copyright in Batman's world-famous 'Batmobile' as a 'character'.[113] In that decision, the court endorsed a three-part test for character copyrightability derived in part from *Rice*, as discussed further below. First, it must have 'physical' as well as 'conceptual' qualities.[114] Second, it must be sufficiently delineated to be recognizable as 'the same character', regardless of where it appears, i.e., 'it must display consistent,

---

[110] *Nichols v. Universal Pictures Corp.*, 45 F.2d 119 (2d Cir. 1930), cert. denied, 282 U.S. 902 (1931).

[111] *Warner Bros. Pictures, Inc. v. Columbia Broadcasting*, 216 F.2d 945 (9th Cir. 1954).

[112] See, e.g., *Walt Disney Prods. v. Air Pirates*, 581 F.2d 751 (9th Cir. 1978) (Disney cartoon characters); *MGM v. Honda*, 900 F.Supp. 1287 (Cal. 1995) (film version of James Bond).

[113] *DC Comics v. Towle*, 802 F.3d 1012 (9th Cir. 2015), cert. denied, 136 S.Ct. 1390 (2016).

[114] *Towle*, 802 F.3d at 1021.

identifiable character traits and attributes'.[115] Third, it must be 'especially distinctive', containing 'unique elements of expression'.[116] It is submitted that the proposed approach was unnecessary and failed to reflect basic copyright principals. First, it is difficult to see why the Batmobile should be considered a 'character'. Unlike some other works (e.g., 'Cars'), it doesn't really have a personality or other characteristics of a more typical human or anthropomorphized 'character'. The same result could have been obtained by simply finding it to be a protected sculptural work, with identifiable aesthetic qualities. Second, copyright is not limited to 'unique', 'especially distinctive' material. Although such novelty is a requirement for patents, it sets too high a bar for copyright. Third, to the extent the first requirement merely requires a visual ('physical') depiction, and the second element requires expressive delineation, it is merely a confusing restatement of the previously developed requirements. As discussed below, if the second element requires something more – 'consistency' or being 'the same character' in multiple settings, it fails to recognize that a character may in fact change across works without losing its identity, or may appear in only a single work and still merit copyright protection.

Of course, since *Rice* predated *Towle*, the court applied the law as it preceded *Towle*.[117] Because the Mystery Magician was depicted in an audiovisual work, one might expect the Ninth Circuit to be more willing to find it a copyrightable element of the Rice Video. On the other hand, it is true that Rice's 'Mystery Magician' was dressed in fairly 'standard magician garb',[118] and that his role was simple, 'performing and revealing the magic tricks'.[119] He may in fact have not been sufficiently well delineated to achieve much copyright protection. Indeed, other than the mask, he does appear to be rather a stock character magician, without much character detail. But the court also suggested a more demanding test for delineation, endorsed again in *Towle*, that is troublesome – that a character must have 'displayed consistent, widely identifiable traits'.[120] Citing cases dealing with famous characters who have appeared in at least several works – Godzilla, James Bond and Rocky Balboa – the court implied that

[115] Ibid. at 1022.
[116] Ibid. at 1022.
[117] In explaining the new 'uniqueness' requirement, the court, citing *Rice v. Fox Broad. Co.*, 330 F.3d 1170, 1175 (9th Cir. 2003) decision, said that to achieve protection, a character 'cannot be a stock character such as a magician in standard magician garb', *Towle*, 802 F.3d at 1021.
[118] Rice, 330 F.3d at 1175.
[119] Ibid.
[120] Ibid.

a character who only appears in a single video selling 17,000 copies could not satisfy that test. Although it is true that often characters that are the subject of copyright suits have in fact appeared in multiple works, that is irrelevant to the delineation test. Indeed, copyright, if it exists at all in a work, subsists in works from first fixation, without any distribution. Tying copyright protection for a work to the number of copies sold has no basis in copyright law.

It is also hard to see how the fact that a character appears in multiple stories satisfies the 'constitutes the story being told' test. The Ninth Circuit's only citation of authority for the 'consistent, widely identifiable' test was to the three district court decisions regarding the aforementioned well-known characters, and it is unpersuasive. The concern underlying character copyrightability is to prevent monopoly of an abstract character idea or stock character type. How 'consistent' or 'widely identifiable' the character is does not address that concern. Indeed, a stock or non-delineated character could be very 'consistent' and 'widely identifiable'. However, Rice's character may well have been insufficiently delineated and a stock magician character, so the outcome may be correct. But future audio-visual magician characters should be protected by copyright if they are original and well-delineated, or, perhaps in the Ninth Circuit if they are also an important part of the 'story'. This should be true, regardless of how many shows they appear in and how many copies of their works are sold.

Rice did not fare well on the appeal as to similarity in other elements of his video either. As to the dialogue, the court found that, although there were some similarities in the opening and closing monologues in both works, they were only non-infringing 'generic parallels in overall tone'.[121] Although both programs were filmed in an empty building, the Rice Video took place in a theatre and the Fox Programs in a warehouse, leading the court to conclude that the similarity of setting was also generic.[122] As to the plot and sequence of events, the court found the introductions to the two works to be different. Both shows involved performing a trick and then revealing how it was done, but the court correctly noted that copyright doesn't protect the idea of revealing magic tricks, and that the sequence was 'merged' with the idea and was a 'scène à faire', hence unprotectable.[123] The overall mood of the two shows was similar – a mood of secrecy and mystery. But both the district and appellate courts found that such a mood

[121] *Rice*, 330 F.3d at 1176.
[122] Ibid. at 1177.
[123] Ibid.

was too generic and considered the mood and pace to be scènes à faire in this context. And because the mood was so closely related to the idea of revealing the mysteries of magic, merger also precluded protection.[124]

Rice also asserted that his work should be protected as a compilation. Of course a compilation would only be infringed by the exercise of a right of copyright in that compilation, or a substantial part of it.[125] Selecting a few pieces and ordering them in a different way probably would not infringe the prior compilation. Hence, in *Rice* the actual tricks and their order of presentation was different in the two works. There were five illusions of the 11 in the first episode of the Fox Programs that were also in the Rice Video, but they were performed in a different order.[126] Therefore, the court found no infringement of a compilation in this case.

5.2 'Or Do You? . . .': *Teller v. Dogge*

The magician plaintiff was successful in the other of the two modern reported decisions asserting infringement of copyright by use of a similar magic performance.[127] A review of that case will illustrate the concepts discussed above and perhaps suggest how future cases would be resolved.

The ethical code of professional magicians has discouraged revealing how an illusion is performed, but there are many instances of competitors seeking to determine how an illusion was created, and numerous works have been published purporting to reveal the secrets of successful magic pieces. In one such exposé, a Belgian performer named Gerard Dogge posted two YouTube videos of himself performing and offering to sell an illusion 'strikingly similar' to 'Shadows,' the signature illusion of Teller from the world-renowned magic duo Penn and Teller.[128]

Teller started performing 'Shadows' in 1976, and registered it as a 'dramatic pantomime' with the United States Copyright Office on

---

[124] Ibid.

[125] *Feist Publications, Inc. v. Rural Telephone Service Co., Inc.*, 499 U.S. 340 (1991).

[126] *Rice*, 330 F.3d at 1177.

[127] *Teller v. Dogge*, 8 F.Supp.3d 1228 (D. Nev. 2014). Teller also brought an unfair competition claim, but this article will focus on the copyright claims. See also *Rice v. Fox Broad. Co.*, 148 F.Supp.2d 1029 (C.D. Cal. 2001), aff'd in part, rev'd in part, 330 F.3d 1170 (9th Cir. 2003) (rejecting copyright protection for magic performance).

[128] *Teller*, 8 F.Supp.3d at 1231.

January 6, 1983.[129] Since then, 'Shadows' has been 'an integral part of Teller's act'.[130]

> [Teller's illusion] consists of a spotlight trained on a vase containing a single rose. The light falls in such a manner that the shadow of the rose is projected onto a white screen positioned some distance behind it. Teller then enters the otherwise still scene, picks up a large knife, and proceeds to use the knife to dramatically sever the leaves and petals of the rose's shadow on the screen slowly, one-by-one, whereupon the corresponding leaves of the real rose sitting in the vase fall to the ground, breaking from the stem at the point where Teller cut the shadow. The scene closes with Teller pricking his thumb with the knife, and holding his hand in front of the canvas. A silhouette of a trail of blood appears, trickling down the canvas just below the shadow of Teller's hand. Teller then wipes his hand across the 'blood' shadow, leaving a crimson streak upon the canvas.[131]

On March 15, 2012, Dogge posted two YouTube videos in which he performed an illusion entitled 'The Rose and Her Shadow',[132] which bore many similarities to Teller's illusion. 'After all of the petals are severed from the rose, Dogge removes the stem from the bottle and pours the water from the bottle into a drinking glass.'[133]

On March 21, 2012, Teller informed Dogge that he owned the copyright to 'Shadows' as a registered dramatic work.[134] When Dogge failed to take down the YouTube videos, Teller sued for copyright infringement and unfair competition.[135] Although the District Court of Nevada ruled on Teller's motion for summary judgment as to both claims, only the copyright claim will be discussed here (the court found genuine issues of material fact as to the unfair competition claim).

The court began its copyright analysis by noting that in order to prevail, Teller must show that he owned a valid copyright in 'Shadows' and that Dogge copied the 'constituent elements . . . that are original'.[136] Although Teller's registration for 'Shadows' constituted prima facie evidence of a valid copyright, Dogge attacked the validity of that registration on four grounds. First, Dogge argued, correctly, that magic tricks, in and of themselves, are not copyrightable. As noted above, however, copyright

---

129 Ibid. at 1231, 1233.
130 Ibid. at 1231.
131 Ibid.
132 Ibid.
133 Ibid.
134 Ibid. at 1236.
135 Ibid. at 1231.
136 Ibid. at 1233.

protection does extend to dramatic works and pantomimes.[137] 'The mere fact that a dramatic work or pantomime includes a magic trick, or even that a particular illusion is its central feature does not render it devoid of copyright protection'.[138] Because 'Shadows' appears 'as a series of stage directions acted out by a single performer', the court determined that it was copyrightable as a dramatic work.

Next, Dogge argued that Teller had abandoned his copyright to 'Shadows' by failing to prevent other infringements of the work.[139] Because Dogge provided no evidence of an affirmative act by Teller indicating Teller's intent to abandon his copyright to 'Shadows', the court quickly dismissed this claim.[140] Undeterred, Dogge then argued that Teller had waived copyright protection of 'Shadows' by challenging others to copy the work. The 'challenge' Dogge pointed to was a statement made by Teller's partner, Penn Jillette, that 'no one knows how "Shadows" is done and no one will ever figure it out'.[141] The court also rejected this argument, pointing out that copyright protection is not waived by 'challenging others' and, even if it did, that Penn's statement in no way encouraged others to publicly perform the work.[142] In his last attempt to undermine the validity of Teller's copyright, Dogge argued that Teller is invalid because notice of the copyright was not given before Teller's performance of 'Shadows'.[143] The Copyright Act, however, does not require notice, so the court dismissed this claim.

The court then turned to whether Dogge, in fact, copied Teller's work. In the Ninth Circuit, plaintiffs must show that the infringer had access to the work and that the works are 'substantially similar in their protected elements'.[144] In this case, access was clear: Dogge's caption for the videos

---

[137] 17 U.S.C. § 102(a)(3)–(4)

[138] *Teller*, 8 F.Supp.3d at 1233. Although 'Shadows' was validly registered as a dramatic work, this statement seems to imply that it would also be copyrightable, and presumably registrable, as a pantomime. It should be noted, however, that the court's subheading for this section reads 'Registration as a Pantomime Drama'. Together, these references indicate that the court may have conflated pantomimes with dramatic works. The Copyright Act, however, is clear that pantomimes and dramatic works are distinct and independent works of authorship. Compare 17 U.S.C. § 102(a)(3) ('dramatic works, including accompanying music'), with 17 U.S.C. § 102(a)(4) ('pantomimes and choreographic works').

[139] *Teller*, 8 F.Supp.3d at 1233.

[140] Ibid.

[141] Ibid. at 1233–1234.

[142] Ibid. at 1234.

[143] Ibid.

[144] *Cavalier v. Random House, Inc.*, 297 F.3d 815, 822 (9th Cir. 2002).

stated 'I've seen the great Penn & Teller performing a similar trick' and Dogge, who represented himself without the benefit of a lawyer, testified that he 'saw Shadows from Mr. Teller . . . on the internet'.[145]

To determine whether the works were 'substantially similar', the court first applied an 'extrinsic test' to objectively identify the 'articulable similarities between the plot, theme, dialogue, mood, setting, pace, characters, and sequence of events' of the works.[146] The court identified numerous such similarities.

The only real differences between the works were the props employed (a vase versus a bottle) and the last few seconds of the performances (smearing blood versus pouring water).[147] Such minor differences were 'inconsequential' compared to the 'overwhelming similarities' between the two works. Accordingly, the court determined the two works were 'substantially similar' under the extrinsic test.[148]

Next, the court applied the more subjective 'intrinsic test' to determine 'whether the ordinary, reasonable audience would find the works substantially similar in the total concept and feel of the works'.[149] In this regard, the court stated that the underlying concept of a performer cutting a rose's shadow while an actual rose is severed was 'incredibly unique'.[150] As a result, the performances would be 'indistinguishable from each other in the mind of an ordinary observer'.[151] Accordingly, the court held that the works were substantially similar under the intrinsic test as well.[152]

Because Teller provided compelling evidence that he owned an original work of authorship fixed in a tangible medium of expression and that Dogge's 'The Rose and Her Shadow' copied original elements of Teller's 'Shadows', the court entered summary judgment on his copyright infringement claim.

---

145 *Teller*, 8 F.Supp.3d at 1235.

146 Ibid. at 1235 (quoting *Kouf v. Walt Disney Pictures & Television*, 16 F.3d 1042, 1045 (9th Cir. 1994)).

147 Ibid. at 1235.

148 Ibid. at 1235–1236.

149 Ibid. at 1235 (quoting *Cavalier*, 297 F.3d at 822).

150 Ibid. at 1236.

151 Ibid. at 1236.

152 Ibid. at 1236.

6. U.S. PERFORMERS' RIGHTS? COMMON LAW COPYRIGHT, UNFAIR COMPETITION AND RIGHT OF PUBLICITY

In addition to federal copyright protection, there may be some potential state claims available to the magician. State protection for works of expression is often referred to as 'common law' copyright, which is somewhat of a misnomer, because it is sometimes derived from state statutes rather than judicial decisions.[153] Prior to 1978, works could be protected by common law copyright both when not fixed in a tangible medium and even after fixation, so long as they were not 'published' or registered with the federal Copyright Office.

Courts have sometimes struggled with the question of what constitutes 'publication'. Certainly once physical copies of a work have been distributed to the general public, a publication has taken place. For the most part, courts have long found that a public performance, no matter how large the audience, does not publish a work.[154]

Unfortunately for the magician, in another of the few cases specifically dealing with a magic performance,[155] the Supreme Court of Florida seems to have misunderstood this concept. In the 1930s, Charles Hoffman, professionally known as 'Think-a-Drink' Hoffman, performed an illusion in which he appeared to produce numerous real alcoholic and other beverages from apparently empty cocktail shakers and beakers seemingly filled with water. He preceded his performance with an address to the audience upon which he secured federal copyright. Glazer, a competing magician calling himself 'Think-a-Drink Count Maurice' or 'Have-a-Drink Count Maurice', performed a similar act, and Hoffman sued. Hoffman asserted copyright infringement and unfair competition claims, as well as common law copyright infringement in his sleight of hand performance. Hoffman succeeded in the lower court. The court found that Hoffman was the exclusive owner of the performance and the trade name, and enjoined Glazer, who appealed. The Florida Supreme Court affirmed the injunction as to the copyrighted address and the trade name, but not the performance. Based on *Serrana*, the court found the mechanical equipment not copyrightable, and based on *Fuller*, the court decided that Hoffman's act was not sufficiently dramatic to be copyrightable. Although the court

---

[153] See, Cal. Civ. Code § 980.

[154] See, *Ferris v. Froham*, 223 U.S. 424 (1912); *Estate of Martin Luther King, Jr., Inc. v. CBS, Inc.*, 194 F.3d 1211, 1223 (11th Cir. 1999) (Cook, J., concurring in result, dissenting in part).

[155] *Glazer v. Hoffman*, 16 So.2d 53 (Fla. 1943).

seemingly recognized that a common law right could cover broader subject matter, namely, 'intellectual productions', such protection is lost once publication takes place or if there is a 'dedication to the public'. The court found that Hoffman's many public performances constituted both a publication and a 'dedication'. That decision is inconsistent both with a prior U.S. Supreme Court ruling[156] and with modern decisions.[157]

The scope of subject matter for common law copyright is uncertain. But it is possible that a magician's performance, or at least elements of that performance, could be protected by state law, if not 'fixed'. For example, in one New York case, a sketch from a musical revue was found to be protected by common law copyright and infringed by a similar performance.[158]

Apart from common law copyright in an expressive performed work, there is some precedent that an unauthorized use of a performance could be a form of unfair competition. Some cases have found the unauthorized making of phonorecords from live orchestra performances to constitute actionable unfair competition.[159] In one case, well-known magician Horace Goldin, succeeded on an unfair competition claim and obtained an injunction against a motion picture entitled *Sawing a Lady in Half* that utilized his famous illusion of that title.[160]

Finally, many states recognize that a person has a right to control certain commercial uses of her name, likeness and other aspects of identity.[161] This right is sometimes called a right of publicity. A full discussion of the potential application of that right to magic performers and performances is beyond the scope of this chapter. However, it is worth noting that the U.S. Supreme Court has only heard one case involving the right of publicity.

---

156 *Ferris v. Frohman*, 223 U.S. 424 (1912).

157 See, e.g., *Estate of Martin Luther King, Jr., Inc. v. CBS, Inc.*, 194 F.3d 1211, 1223 (11th Cir. 1999) (Cook, J., concurring in result, dissenting in part).

158 *Casino Prod'ns, Inc. v. Vitaphone Corp.*, 163 Misc. 403 (Sup.Ct. N.Y. 1937) ('The Postman Always Rings Twice' sketch).

159 *Waring v. WDAS Broad. Station, Inc.*, 194 A. 631 (Pa. 1937); *Metro. Opera Ass'n., Inc. v. Wagner-Nichols Recording Corp.*, 101 N.Y.S.2d 483 (Sup. Ct. N.Y. County 1950), aff'd, 107 N.Y.S.2d 795 (N.Y. App. Div. 1951). But see, *National Basketball Ass'n v. Motorola*, 105 F.3d 841 (2d Cir. 1997) (limiting misappropriation claim to 'hot news' situation between competitors, suggesting preemption of other types of misappropriation claims). See generally F. Jay Dougherty, Not a Spike Lee Joint? Issues in the Authorship of Motion Pictures Under U.S. Copyright Law, 49 UCLA L. Rev. 225, 300–306 (2001) (arguing that motion picture performers may be 'authors' under U.S. copyright law in some circumstances).

160 *Goldin v. Clarion Photoplays, Inc.*, 202 A.D. 1 (Sup. Ct. N.Y. 1922).

161 See generally T. McCarthy, *The Rights of Publicity and Privacy* (Clark Boardman Callaghan, 2018).

That was a case in which a 'human cannonball's' entire performance was filmed and broadcast on a news program.[162] The Court found that such a taking of a performer's 'entire act' could violate the performer's state law right of publicity, and would not be protected by Constitutional freedom of speech, even when used as part of a news report. Although most cases involving the right of publicity deal with the unauthorized use of persona in advertising or in packaging of products, and application of the right of publicity against media portrayals has been criticized,[163] it would appear that a magician would be no worse off than a human cannonball if his 'entire act' were to be appropriated and exploited without his permission.

Many countries in the world other than the United States recognize a form of legal protection for performers, known unsurprisingly as 'performers' rights'.[164] This is a distinct legal regime from copyright, and generally is quite a bit more limited than copyright in certain ways. For example, this right would ordinarily only protect a performer against unauthorized recording of her performance and certain forms of exploitation of an unauthorized recording. Hence, it would not normally protect a magician against another's imitation of his act,[165] but might be violated by an unauthorized 'bootleg' recording of a magician's performance or exploitation of the recording, e.g., by posting such a bootlegged recording on YouTube. At least some countries would cover magicians as a type of 'variety artiste' protected by this schema.[166] Moreover, not only has United States law occasionally found an unauthorized exploitation of a live performance to be actionable unfair competition, the U.S. has also recently become a signatory to at least two international treaties that require legal protection of musical performers,[167] as a minimum, perhaps signaling eventual support for more expansive recognition of other performers' rights. Indeed, as a result of those treaties' obligations, the United States

---

[162] *Zachini v. Scripps-Howard Broad. Co.*, 433 U.S. 562 (1977).

[163] See, e.g., F. Jay Dougherty, All the World's Not a Stooge: The 'Transformativeness' Test for Analyzing a First Amendment Defense to a Right of Publicity Claim Against Distribution of a Work of Art, 27 Col. J. Law & Arts 1 (2003).

[164] See generally Richard Arnold, *Performers' Rights* (3rd edn, Sweet & Maxwell, 2004); Samuel Ricketson and Jane Ginsburg, *International Copyright and Neighboring Rights* (2nd edn, OUP, 2006).

[165] It has been argued that performers' rights should be expanded to be more equivalent to copyright in this regard. See R. Arnold, *Performers' Rights*, ibid. at § 1.113.

[166] See, e.g., the United Kingdom. See R. Arnold, *Performers' Rights*, supra note 164, § 2.16.

[167] TRIPs Agreement; WIPO Phonograms & Performers Treaty.

now provides performers with statutory protection against making and trafficking in unauthorized recordings of their musical performances.[168] Finally, although not yet in effect, the U.S. is among parties who have signed the recent WIPO Performers' Rights treaty. This area may provide fertile ground for magicians to seek federal protection in the future for some aspects of their performances.

7. CONCLUSION

Although magic illusions may not receive copyright protection as such, the cases discussed in this chapter suggest that some protection is available in appropriate cases. Elements such as sufficiently delineated and unusual characters and embedding magic illusions within larger works telling stories, or even merely portraying characters and events in a pantomime, could be protected. Some 'useful articles' included in a magic act may include pictorial, graphic or sculptural material that could be copyrightable, and some such items may not be 'useful articles' at all, if their 'use' is merely to portray the appearance of a useful article. Even a magic act using only uncopyrightable elements could receive protection as a 'compilation' against a literal or near-literal copy, if there are a sufficient number of those elements and they are selected, ordered or arranged in an original manner. Many other elements of a performance may reflect copyright authorship, as discussed above. Magicians and their representatives can take these ideas into account when designing magic performances if they desire to have exclusive rights.

---

[168] 17 U.S.C. § 1101 (2007). There are contentious arguments about Congress's power to grant legal protection as to performances not fixed in a tangible medium of expression. Although Congress's power to legislate copyright protection is limited to 'writings', implying some tangible embodiment, it appears that Congress may have power to legislate as to live performances pursuant to the interstate commerce power. See, *U.S. v. Moghadam*, 175 F.3d 1269 (11th Cir., 1999).

12. Copyright protection of sport moves

Péter Mezei*

1. INTRODUCTION

The International Gymnastics Federation (FIG) honored Krisztián Berki, the champion of the 2012 London Olympic Games in pommel horse, by adding a new element named after him to the FIG's code of points for Men's Artistic Gymnastics in 2016.[1] The FIG's decision acknowledges both the fact that he started his Olympic Games winning routine with this move and, more generally, Berki's creativity and professionalism within the sport.[2] A question might arise: does Berki (and/or his trainer) deserve copyright protection for developing this specific move? Similarly, does copyright protection extend to other

* The author is grateful for Christina Angelopoulos and Ana Ramalho for their support in the collection of sources. Further, members of the InfoCopy research group, who participated in the research introduced in section 3 deserve special appreciation for their enthusiasm toward sports and copyright law. These graduates and undergraduates were the following (names in alphabetical order): Ottília Ágoston, Villő Bujdosó, Szilveszter Csernus-Lukács, Fanni Gulyás, István Harkai, Melinda Kaskovits, László Németh, Zoltán Oláh, Attila Sovány, Jenő Spitzer and Mónika Tóth.

[1] The official description of the element is as follows: '1/1 spindle w. legs straddle or together on the pommels inside max. 2 circles. (Berki) (E)'. See: FIG CoP Changes 20172020, 3, available at http://www.figgymnastics.com/publicdir/rules/files/mag/CoP_MAG_20172020_ICIe.pdf. Berki is the second Hungarian Olympic champion of the pommel horse after whom an element has been named. The International Gymnastics Hall of Famer Zoltán Magyar – the champion of the 1976 and 1980 Olympic Games – performed the Magyar travel (crosswise circling travel down the horse). Another move was named after Magyar, the Magyar spindle (turning the body in the opposite direction from the circling legs). The Sivadó travel – which is a reverse Magyar travel – was named after János Sivadó, who did not win any Olympic medals, however. Both the Magyar and the Sivadó travel were developed under the supervision of trainer László Vígh.

[2] The Berki element is a combination of the so-called Thomas flair, developed by another innovative gymnast, Kurt Thomas. One of the novelties of the Berki element is that the Thomas flair is executed on the pommels. The visualized Berki element is available at https://vine.co/v/51mhulzTqI9.

iconic sporting moves such as Bob Cousy's behind-the-back pass, Mitsuo Tsukahara's vault, Antonín Panenka's penalty kick, Werner Rittberger's loop jump, Dick Fosbury's flop or even to the Yawgoons snowboard features?[3] If one looks more generally at the interrelationship between sports and intellectual property law it is clear that a certain degree of crossover already exists. Organizing, broadcasting, performing, displaying and retransmitting sport events; managing sport organizations; selling/merchandising brand names, mascots, memorabilia; marketing athletes' (sometimes celebrities') personality rights; endorsement contracts; registering trademarks of teams or individuals; protecting databases of sport fixtures; sports betting; producing sport-related software, etc. are only a few of the fascinating examples of the intertwining of sports and IP.[4]

Such IP protection is understandable when one considers the importance

---

[3] 'In a sport where bigger, higher, longer, and more spinning and flipping define the boundaries, the Yawgoons get creative. They use the natural terrain (rocks, trees, grass) as well as the unnatural (buildings, snowcats, pipes) to construct landscapes with one I've-never-seen-that-before feature after another. Then they shoot video of their runs and let the snowboarding world watch in awe.' See: Matt Ruby, Coolest Scene in Snowboarding? The Coast of Rhode Island, *The New York Times* (December 30, 2016) http://www.nytimes.com/interactive/2016/12/30/sports/yawgoons-snowboarding.html (last accessed November 25, 2017).

[4] See on these questions for example: Michael Suppappola, Is Tiger Woods's Swing Really a Work of Art? Defining the Line Between the Right of Publicity and the First Amendment, 28 Western New England Law Review 57–88 (2005); Julia Johnson, Database Protection a Reality? How the Professional and Fantasy Sporting World Could Benefit from a Sui Generis Intellectual Property Right, 27 Intellectual Property Journal 237–263 (2015); Thomas Margoni and Ben Van Rompuy, UK Horserace Betting Right: At Odds with EU Law?, 37 European Intellectual Property Review 479–483 (2015); Zach Schonbrun, Keep Your Eye on the Balls to Become a Better Athlete, *The New York Times*, January 4, 2017, available at http://www.nytimes.com/2017/01/04/sports/neurotracker-athletic-performance.html. From the case law see especially: *England and Wales Cricket Board Ltd & Anor v. Tixdaq Ltd & Anor* [2016] EWHC 575 (Ch) (18 March 2016) (whether 8-seconds-long video clips represent a substantial amount of cricket games that last for more than two hours); Case C-46/02 *Fixtures Marketing Ltd v. Oy Veikkaus Ab*, 9 November 2004, ECLI:EU:C:2004:694; Case C-203/02 *British Horseracing Board Ltd. v. William Hill Organisation Ltd.*, 9 November 2004, ECLI:EU:C:2004:695; Case C-338/02 *Fixtures Marketing Ltd v. Svenska Spel AB*, 9 November 2004, ECLI:EU:C:2004:696; Case C-444/02 *Fixture Marketing v. Organismos prognostikon agonon podosfairou AE (OPAP)*, 9 November 2004, ECLI:EU:C:2004:697; Case C-279/13 *C More Entertainment AB v. Linus Sandberg*, 26 March 2015, ECLI:EU:C:2015:199.

of sports from a business perspective. For example, the 2016 Rio de Janeiro Olympic Games were followed by approximately 3.5 billion viewers via licensed broadcasts, and consequently 'the IOC says TV revenue is up a solid 7.1 percent over the previous four years to $4.1 billion'.[5] Manchester United football club sold the most jerseys among European soccer teams in 2016: 2.85 million.[6] The aggregated salary cap of the 32 American football teams of the National Football League (NFL) could hypothetically amount to US$4.968 billion in 2016.[7]

With regard to the specific focus area for the present chapter, it is noted that much has been written on the protectability of sport moves across the last two decades.[8] Weber argued back in 2000 that

> while copyright (and other intellectual property) protection for athletes' original 'moves' would have been unthinkable – and unnecessary – twenty or more years ago, conditions and circumstances in today's sports world now make such

[5] Scott Roxborough, Rio Olympics Worldwide Audience to Top 3.5 Billion, IOC Estimates, *The Hollywood Reporter* (August 18, 2016) http://www.hollywoodreporter.com/news/rio-olympics-worldwide-audience-top-920526 (last accessed November 25, 2017).

[6] Totalsportek2, Top Selling Football Jerseys of Players and Clubs (Revealed), Totalsportek (October 23, 2016), http://www.totalsportek.com/football/clubs-with-most-shirt-sales/ (last accessed November 25, 2017).

[7] Sportrac, NFL Team Salary Cap Tracker, http://www.spotrac.com/nfl/cap/2017/ (last accessed November 25, 2017). For further, similar – but older – statistics see: W. Tucker Griffith and Ekaterina Gordeeva, Beyond the Perfect Score: Protecting Routine-Oriented Athletic Performance with Copyright Law, 30 Connecticut L. Rev. 675–676 (1998).

[8] Carl A. Kukkonen III, Be a Good Sport and Refrain from Using my Patented Putt: Intellectual Property Protection for Sports Related Movements, 80 J. Pat. & Trademark Off. Soc'y 808, 829 (1998); Griffith and Gordeeva, supra note 7, at 675, 730; Loren J. Weber, Something in the Way She Moves: The Case for Applying Copyright Protection to Sports Moves, 23 Colum.-VLA J.L. & Arts 317, 361 (2000); Proloy K. Das, Note. Offensive Protection: The Potential Application of Intellectual Property Law to Scripted Sports Plays, 75 Ind. L.J. 1073 (2000) at 1073, 1101; F. Scott Kieff, Robert G. Kramer and Robert M. Kunstadt, It's Your Turn, But It's My Move: Intellectual Property Protection for Sports 'Moves', 25 Santa Clara Computer & High Tech. L.J. 765, 785 (2009); Alexander Bussey, Stretching Copyright to its Limits: on the Copyrightability of Yoga and Other Sports Movements in Light of the U.S. Copyright Office's New Characterization of Compilations, 20 Jeffrey S. Moorad Sports L. J. 1, 33 (2013); Viola Elam, Sporting Events as Dramatic Works in the UK Copyright System, 13 The Ent. and Sports L. J 1, 90 (2015); Thomas Margoni, The Protection of Sports Events in the EU: Property, Intellectual Property, Unfair Competition and Special Forms of Protection, 47 Int. Rev. Intellect. Prop. Compet. Law 386, 417 (2016).

> protection a desirable addition to the bundle of valuable rights athletes have secured in their own performances over the years.[9]

However, the prevalent view amongst scholars has remained conservative on the issue, with most disagreeing with Weber's assertion and continuing to exclude sport moves from the protectable subject matters under copyright law. Yet, interestingly, recent scholarly work has gradually become more accepting of the idea of choreographies in aesthetic or routine-oriented sports, like figure skating, synchronized swimming or wrestling, potentially satisfying the threshold of originality. Therefore, in light of the above departure from the traditional/conservative view, this chapter will explore whether copyright law is ready to embrace sport moves more generally.

It is important to note from the outset that this chapter will not discuss the fundamentals of copyright protection (especially the requirement of originality) as this has been excellently covered by Enrico Bonadio and Nicola Lucchi in the Introduction to this book. Moreover, this chapter will not touch upon business-related aspects of the sports industry nor the protection of sport moves under intellectual property laws other than copyright law.[10] Similarly, performances or other achievements (including broadcasters' acts) also fall outside of the scope of the present chapter, because they are protected differently in different countries and such an examination would be better suited to a separate piece. However, in short, some countries provide such performances with copyright protection as works of authorship, whilst others treat them as related or neighboring rights. The scope of these two forms of protection show, however, significant differences.[11]

Finally, the copyrightability of sport events as a whole is excluded from the scope of this chapter. This is due to the fact that this chapter aims to examine only atomic sport moves and the choreographies of them in isolation, rather than a broader evaluation of entire sport events. Further,

---

[9] Weber, supra note 8, at 318. Later on, Weber stressed that 'the sports industry lacked the ability to capitalize to the extent possible today on the potential for advertising revenues derived from the association of products and services with athletes' performances, and as such far fewer fan dollars circulated in the world of sports at that time (as evidenced for instance by pro athletes' salaries, which were meager compared to today's near ubiquitous multi-million dollar contracts)'. See ibid. at 320–321 (footnotes omitted).

[10] See, e.g., Kukkonen, supra note 8, at 814, 829; Das, supra note 8, at 1076, 1083 and 1088,1089; Kieff et al., supra note 8, at 768, 776 and 781, 784.

[11] On the neighboring rights aspects of sports moves see: Margoni, supra note 8, at 394 and 396.

multiple court decisions have convincingly stressed the uncopyrightable nature of sport events. A U.S. federal district court ruling confirmed that organizing an international high school all-star game did not represent more than an unprotectable idea.[12] Another quite old ruling from the U.S. confirmed that organizing a roller-skating competition did not infringe the copyright of the book that contained the rules of the competition.[13] The Federal Court of Appeals of Canada noted that 'unlike dance, a sporting event is for the most part a random series of events. The unpredictability of the action is inconsistent with the concept of choreography'.[14] This view is reinforced by the Court of Justice of the European Union (CJEU), who commented on football matches in a much less sophisticated and more direct fashion: 'sporting events cannot be regarded as intellectual creations classifiable as works within the meaning of the Copyright Directive. That applies in particular to football matches, which are subject to rules of the game, leaving no room for creative freedom for the purposes of copyright'.[15]

Yet surprisingly, the Paris Court of Appeals concluded in September 2011 that the sailing race called 'la route du rhum' (that is, 'the route of rum') is protected by copyright law. The court of first instance noted (and the court of appeals confirmed) that copyright protection of a sport event requires that the latter follows a screenplay and it is original in nature. The given sailing race takes place as a live performance, where

---

[12] '[T]he idea for the FLG is not protectable through copyright.' See: *Hoopla Sports & Entertainment, Inc. v. Nike, Inc.*, 947 F. Supp. 347, 354 (1996). Here, however, the plaintiff also failed to meet another prerequisite of copyright protection. As the court noted: 'Hoopla nowhere alleges that the FLG was recorded or otherwise "fixed in any tangible medium of expression." Thus, even leaving aside the question of whether sports events fall within the protection of the Copyright Act, the Court finds as a matter of law that Hoopla does not possess a copyright under the Copyright Act in the FLG itself'. See ibid. at 354–355 (footnotes omitted). On the decision see Bussey, supra note 8, at 7.

[13] 'The official program of the defendants is in no way an infringement of plaintiff's copyrighted books.' See: *Seltzer v. Sunbrock*, 22 F.Supp. 621, 626 (1938). On the decision see Bussey, supra note 8, at 7.

[14] *FWS Joint Sports Claimants v. The Copyright Board*, 36 CRR (3d) 483 (Fed. CA of Canada), 1991. On the decision see Gregory C. Ludlow & Mark Le Blanc, Survey of Intellectual Property Law Part V – Copyright & Industrial Designs (1987–1999), 31 Ottawa Law Review 117 (1999); Leslie Chong, Copyright and Innovation from Yoga to Football (October 25, 2010) http://www.iposgoode.ca/2010/10/copyright-and-innovation-in-football-playbooks/ (last accessed November 25, 2017).

[15] Joined cases C-403/08 and C-429/08, *FAPL v. QC Leisure and Others*, ECLI:EU:C:2011:631, para. 98. On the decision see Margoni, supra note 8, at 389.

each sailor competes individually, the competition takes place every four years (during the fall), it is open to all types of boats, and the route is fixed (departure from Saint-Malo, arriving at Pointe-à-Pitre).[16] French scholars have, however, correctly criticized the above decision, as none of the features of the sailing race seem to bear the level of originality that is required for copyright protection.[17]

In light of the aforementioned jurisprudence on the copyrightability of sporting events, it is important to once more emphasize the delineation between the above exclusions relating to entire sporting events and this chapter's analysis on certain sporting moves in isolation. The purpose of this chapter is to specifically analyze whether athletes or their trainers deserve copyright protection for the atomic elements of sport events or the combination (including choreographies) of sport moves. Thus, the focus for this chapter centers on a fundamental question of copyright law: originality.

The structure of this chapter is as follows. Section 2 will elaborate on the fundamentals of sports. Section 3 will examine the outcome of a unique research – conducted under the supervision of the author of the present chapter and carried out by the members of the InfoCopy research group of the Szeged Law School (hereinafter InfoCopy research). This research allowed for an empirical categorization of sports and their atomic elements. Based on the results of the research and considering the relevant international case law and academic writings, section 4 draws some conclusions on the protectability of sport moves and choreographies. This part suggests that while sport moves do not (and will not) enjoy a blanket protection under copyright law, artistic choreographies of sport exercises are, however, copyrightable. Finally, section 5 collects those arguments that speak against such protection and will expand on why the practical exercise/enforcement of that protection would have a chilling effect on the prosperity of sports in general.

2. WHAT IS SPORT?

As the present chapter aims to focus on sport moves and choreographies from a copyright perspective, it seems necessary to begin with the following question: 'what is sport'? Views differ on the precise meaning of sports.

---

[16] Cour d'appel, Paris, Pôle 5, chambre 1, 21 Septembre 2011 – no 09/06928.

[17] Michel Vivant and Jean-Michel Bruguiere, *Droit d'auteur et droits voisins* 1059, footnote 1 (2nd edn, Dalloz, 2013).

SportAccord, the International Sports Federation, takes the view that 'the sport . . . should have an element of competition; the sport . . . should in no way be harmful to any living creatures; the sport should not rely on equipment that is provided by a single supplier; the sport should not rely on any 'luck' element specifically designed into the sport'.[18] According to the above federation, sports might be classified in the following categories: 'primarily physical; primarily mind; primarily motorized; primarily coordination; primarily animal-supported'.[19] Others have offered a different classification. Weber, Griffith and Gordeeva, as well as Elam have separated head-to-head or adversarial sports, aesthetic or routine-oriented sports and routine-oriented non-competitive activities.[20] They all have suggested that aesthetic/routine-oriented sports are capable of earning copyright protection.[21]

However, neither of the aforementioned definitions will be followed in the present chapter. First, all sports that involve the movement of the participants' body shall be covered by the research, irrespective of whether the activity is of a professional nature or leisure-oriented. Thus, generalized categories like those introduced above limit the analysis of the copyrightability of sport moves. Second, aesthetic merits are not prerequisites to the protection of works. Thus, even adversarial sports might allow for some creativity on the side of athletes, which consequently might lead to the protection of their creations. If one takes the sensational rise of Connor McGregor within the Ultimate Fighting Championships as an example: a creative punch from an unexpected angle in an unexpected moment secured him victory in his championship bout in under 13 seconds.

Jesien and Bussey offer a different and ultimately more detailed categorization of what constitutes sport.[22] Jesien has distinguished between the following four different groups, into which sports (and sports moves) might fall, and has arranged these categories in order from the least to

---

18 Retrieved from SportAccord's archived website via the Internet Archive Wayback Machine (Noc. 14, 2016) http://www.sportaccord.com/en/members/index.php?idIndex=32&idContent=14881 (last accessed November 25, 2017).

19 Ibid.

20 Griffith and Gordeeva, supra note 7, at 677–678; Weber, supra note 8, at 321–322; Elam, supra note 8, at para. 7.

21 Griffith and Gordeeva, supra note 7, at 708–723; Elam, supra note 8, at paras 40–51. Weber has, however, stressed that a case-by-case analysis is required to decide the protectability of each and every sports move. See: Weber, supra note 8, at 356–358.

22 Karolina Jesien, Don't Sweat It: Copyright Protection for Yoga . . . Are Exercise Routines Next?, 5 Cardozo Pub. L. Pol'y & Ethics J. 635 (2007); Bussey, supra note 8, at 32.

the most possibly copyrightable ones: '[1] sports events; [2] scripted sports plays; [3] routine-oriented competitive sports; and [4] routine-oriented non-competitive sports'.[23] She has afforded a high possibility of copyrightability to groups [3] and [4], where the latter included routine-oriented sports, like figure-skating and gymnastics, if performed outside of a competition, as well as yoga sequences. Bussey supports the above definition and develops it further by distinguishing between five groups: '(1) Sports events; (2) Scripted sports plays; (3) Function-based athletic routines; (4) Routine-oriented competitive sports; and (5) Routine-oriented non-competitive performances'.[24] Bussey has justified the expansion of Jesien's list with a fifth (actually, the third) category due to the development of case law and policy considerations regarding yoga.[25] Thus he has removed yoga postures and similar exercises from the most probably copyrightable group of sports moves into a lower ranked group. Based on these groupings Bussey has concluded that the first three categories are outside of the scope of copyright protection. Group 4 surpasses the originality threshold and therefore – at least theoretically – deserves protection under copyright law. Finally, group 5 would include figure skating and gymnastics pro-tours that are simply designed to entertain a crowd and should certainly be copyrightable as a choreographic authorship 'organized into an integrated, coherent, and expressive whole'.[26]

The present chapter adopts the logic developed by Bussey regarding the categorization of sports, yet uses a different method in reaching its end results. It follows a special pragmatic approach. Under this approach several cornerstones of sports are highlighted that have practical significance from a copyright perspective. Relying on these fundamentals, a general balancing test on the protectability of moves of the different sports will be provided.

These cornerstones are as follows: first, all sports aim at achieving success. More precisely, athletes aim to reach given purposes in accordance with given rules. Any activities that are done outside of the scope of regulations shall not be treated as sports for the purpose of the present assessment.

Secondly, the present chapter focuses solely on those sports that might be executed with the help of the bodies of athletes. Consequently, only

---

[23] Jesien, supra note 22, at 635.

[24] Bussey, supra note 8, at 32.

[25] Library of Congress, Copyright Office, Registration of Claims to Copyright, 37 CFR Part 201, Docket No. 2012-6, Federal Register, Vol. 77, No. 121, Friday, June 22, 2012, 37607 [hereinafter Library of Congress]; *Bikram's Yoga College of India, L.P., v. Evolation Yoga*, LLC, et al., 803 F.3d 1032 (2015).

[26] Bussey, supra note 8, at 32 (footnotes omitted).

those sports where moves are limited by the boundaries of the physical world are covered. Such boundaries are for example the size/length of an average sport field or track; or the timeframe of an exercise or a whole sport event. There are further limits that humans cannot exceed.[27] So for example, no matter how great an athlete Usain Bolt is, he will never be able to run 100 meters in 8 seconds. It similarly seems to be impossible to swim the 100 meters freestyle without breathing. In light of the above, e-sports (like League of Legends competitions) or fictional sports (like Quidditch, the magical sport of the Harry Potter universe) are excluded from the scope of this chapter (although they are fascinating and have a significant economic value). Indeed, the 'moves' of the avatars of the players (that is, the e-athletes) are executed in accordance with the code of the computer programs and the feats of fictional Quidditch players are created by J.K. Rowling. In both cases these moves do not rely on players' activities.

Although this chapter does not categorically state that all sports are fully functional, in light of the first two cornerstones indicated above, the present author takes the view that all sports have more or less functionality.[28] Those views, which categorically reject the functionality of sports on the ground that their main purpose is to entertain spectators, shall not be followed.[29] In addition, this chapter aims to discuss whether and how much creativity is allowed under the functionality of the given sport.

Finally, the consumption of sports differs heavily from the conventional consumption of other creative works. Pablo Picasso's paintings, Coldplay's music, Steven Spielberg's movies or Kurt Vonnegut's books might be enjoyed at any time. The importance and curiosity of sports lies, however, in their newsworthiness. The consumption of sports is therefore more limited in time. Similarly, the ability of athletes to create something unique is confined to the limited time span of the athlete's career.

3. ORIGINALITY/CREATIVITY VERSUS FUNCTIONALITY: AN EMPIRICAL RESEARCH OF SELECTED SPORTS

This part will present the findings from some recent research into the question of functionality (of sports) and creativity/originality (of sport moves).

---

27 Compare to Weber, supra note 8, at 341.

28 '[A] routine may be both functional and expressive.' See: Bussey, supra note 8, at 17.

29 See, e.g., Kieff et al., supra note 8, at 779.

Based on the work of the members of the InfoCopy Student Research Group, a statistically non-significant, still properly descriptive research was conducted during the spring semester of 2016. Participants were asked to select different sports according to their tastes and personal interests, and to collect the following information on their selections: first, the written (and maybe unwritten) rules/codes of points (hereinafter: rules) of the given sport published by the governing international federation of the sport; second, the conventional or regular (and most probably functional) moves, which are either precisely listed in or generally/abstractly regulated ('allowed') by the rules; and third, the non-conventional but legitimate moves, which are not expressly regulated by the rules, but equally are not excluded by the same norms either. The latter element is crucial, as it is this area which might leave space for athletic creativity – and maybe copyrightability as a consequence. The collection of audio-visual examples for the conventional and non-conventional moves was also requested from the participants.

In sum, the participants comparatively analyzed the above criteria against almost twenty different sports. These included adversarial team sports, such as American football, soccer, baseball and water polo; adversarial individual sports, like swimming, diving and archery; combat sports, like boxing; and choreography-based or routine-oriented sports, like figure skating, synchronized swimming and gymnastics. Wrestling was also analyzed, but it is hard to categorize properly, as it is based on a loose or 'flexible choreography', but it is presented as an 'entertaining combat sport'. Consequently, wrestling is a great example for sports that allow space for both functionality and creativity. All these examples were analyzed by the participants as average lay observers.

Two general questions were discussed by the participants of the research. The first being how functional the sport is. More precisely, how detailed or restrictive the rules of the games are and what are the exact aims of the sports, i.e. how do you succeed. The second overarching question was in stark contrast to the former, namely how much space is allowed for creativity, within which athletes might rely on legitimate, but unique or unexpected moves? In general, the research aimed to discover how predetermined the possibilities or choices of athletes are. So, for example, adversarial team sports are generally based on scoring points/goals/touchdowns, etc. In these cases, the aesthetics or creative nature of the execution of the moves is generally irrelevant, as they do not lead to more points and are not honored by the referees. It is conceded that unexpected behavior on the field might allow for some advantage against your opponents, however, such advantages are relevant solely from the perspective of the end results: scoring and generally winning the

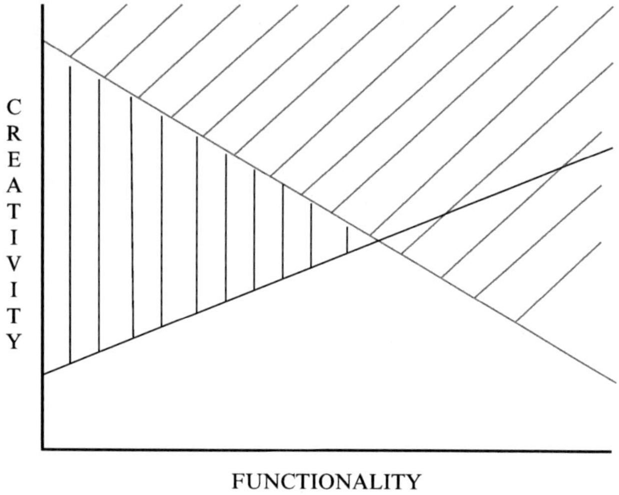

Figure 12.1 The empty frame of reference of sport moves

game/competition. Contrary to this, in gymnastics, figure skating and synchronized swimming, success depends upon both the technical and aesthetic execution of routines.

These two aspects are summarized on a frame of reference (see Figure 12.1). The horizontal axis represent functionality, whilst the vertical axis represents creativity. Two further propositions were applied. An inverse ratio rule (indicated with white lines) highlights that the more functional a sport is the less space it allows for creativity. Another inverse ratio rule (indicated with black lines) focuses on originality, as the prerequisite of copyright protection. This inverse ratio rule indicates that more creativity is needed in functional sports to become truly creative. On the contrary, less creativity suffices in routine-oriented sports. Finally, the frame of reference includes a triangle on the left, indicated with black vertical lines. Theoretically, this works as the space whereby less functional and more creative sports might allow for moves that are creative enough to become copyrightable. The research aimed to figure out which, if any, sports would fit in to this territory. The results are depicted in Figure 12.2.

The results of the research confirm that the majority of sports do not allow enough space for athletic creativity. This trend is clearly visible with respect to individual and team adversarial sports. In most of the cases, the end result (success) might be reached through the execution or performance of the most effective functional moves. This is especially true for rowing, archery or swimming, where the execution of

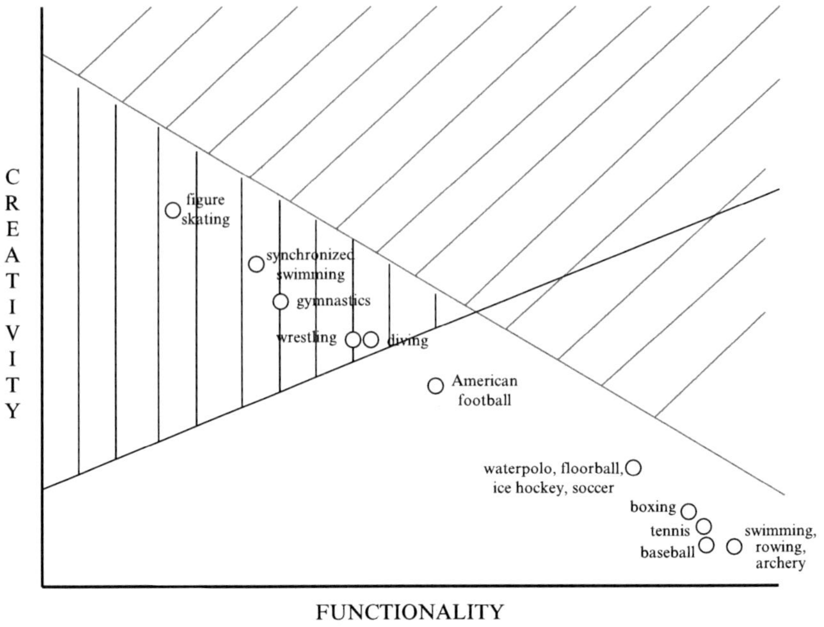

Figure 12.2 The results of the research

movements clearly resembles the nature of the given sport. In archery, the posture of the archer, the eye dominance, as well as the releasing of the string all directly affect the end result. In rowing, the harmony and power output of rowers – that might derive from the perfect use of oars – is the prerequisite for a successful competition. In swimming, breaststroke, backstroke and butterfly are all events which abide by a strict specified swimming movement. In freestyle, the swimmers use whatever technique they prefer. They, however, almost inevitably use the front crawl as that is generally the fastest (and thus most effective) stroke. Some slight creativity is still allowed in freestyle swimming, as evidenced by the 'Ryan Lochte Turn'.[30] Here, 'Ryan Lochte used a new turn in the 200 IM world championship final [in 2015] where he pushed off on his back and kicked underwater for 10 meters. He then surfaced

30 '[Ryan Lochte's] slightly altered method of flip-turning has shaved off almost a second from his lap time, a huge difference in competitive swimming'. See Drew Pulig, See How Ryan Lochte's Innovative Swimming Technique Helped Earn His Olympic Medals, Hypebeast (August 10, 2016) https://hypebeast.com/2016/8/ryan-lochte-swimming-innovation-olympics (last accessed November 25, 2017).

and swam what is commonly known as freestyle'.[31] Although the above turn was declared illegal for medleys,[32] it is still acceptable in freestyle competitions.

In other sports, the rules and nature of the games allow more freedom in selecting different moves. In boxing, stances, punches and defense include a variety of options, the following of which leads to the reliance on different styles ('boxer/out-fighter', 'boxer/puncher', 'counter puncher', 'brawler/slugger' or 'swarmer/in-fighter'). In baseball, pitchers might use different types of pitching, depending generally on the opponent's strengths and weaknesses or the strategy of the pitcher's team. In tennis, classic shots (serve, forehand, backhand, volley, half-volley, overhead smash, drop shot, lob) are applied by players. Some might, however, develop effective attacking techniques, like Roger Federer has invented the 'Sneak Attack by Roger' (SABR).[33]

In adversarial team sports, especially in soccer, water polo, American football or basketball, the rules of the games set the general limits of body movements. Still, there is considerable space for creativity (or skill) on the side of trainers or athletes. This is evidenced by the many innovative moves developed and applied during games. Such notable examples are the Zidane trick, the Panenka penalty kick or René Higuita's scorpion kick in soccer; Tamás Faragó's back hand flip-and-roll in water polo; playbooks and formations in American football; pick-and-roll developed to perfectionism by Karl Malone and John Stockton, Bob Cousy's behind-the-back-pass in basketball. These moves, however, aim to reach a given purpose: to outmaneuver an opponent more effectively and to

---

[31] FINA Officially Makes 'Ryan Lochte Turn' Illegal in IM Races (September 8, 2015) https://www.swimmingworldmagazine.com/news/fina-officially-makes-ryan-lochte-turn-illegal-in-im-races/ (last accessed November 25, 2017).

[32] 'According to SW 9.1 each of the strokes must cover one quarter (1/4) of the distance. Being on the back when leaving the wall for the Freestyle portion of the Ind. Medley is covering more than one quarter of the distance in the style of Backstroke and is, therefore, a disqualification. Backstroke swimming is only defined as being on the back.' See: Memorandum – Interpretation FINA Rule SW 9.1 – Medley Swimming (September 7, 2015) http://www.swimmingworldmagazine.com/news/wp-content/uploads/2015/09/FINA-IM-rules.pdf (last accessed November 25, 2017).

[33] 'SABR . . . features Federer moving way up on an opponent's second serve to hit a half-volley in an attempt to dictate pace and frustrate opponents. . . . The move helps him position himself after he hits a deep, chipped return and puts his opponents, like Richard Gasquet on Wednesday night, off balance.' See: Chris Chase, What is Roger Federer's New SABR Move?, For The Win – USA Today Sports, September 9, 2015, available at http://ftw.usatoday.com/2015/09/what-is-roger-federers-new-sabr-move.

score more easily. Therefore, they do not offer any alternate value during the game.

The research has also evidenced, however, that in some sports enough freedom is provided for athletes to express their creativity and personality and thus to 'create' sport moves or a combination thereof that might meet the fundamental requirements of copyrightability; although, these sports are solely routine-oriented or aesthetic sports. Diving, synchronized swimming or gymnastics might serve as good examples here, yet it is figure skating that allows for the greatest possibility to evidence substantial creativity amongst athletes and their trainers.

With that in mind, let's now take a closer look as to why figure skating affords a greater degree of creativity compared to other sports. Rules 610–612 and 619 of the International Skating Union's (ISU) Special Regulations and Technical Rules[34] outline the general framework for skating exercises. These rules expressly list the required elements for short programs,[35] the maximum amount of different elements allowed to be incorporated within free skating,[36] the time limit of the programs,[37] as well as movements which are illegal.[38] Similar rules apply for pair skating.[39] Multiple other rules – not discussed at length here – are set to regulate the general contours of each performance. Nonetheless, the ISU allows for a very broad and flexible interpretation in the creation of the exact program. To name a few: '[t]he sequence of the elements is optional';[40] '[s]tep sequences may include any unlisted jumps'; [41] '[f]ree Skating consists of a

---

[34] International Skating Union, Special Regulations & Technical Rules – Single & Pair Skating and Ice Dance 2016 as accepted by the 56th Ordinary Congress (June 2016) http://static.isu.org/media/1003/2016-special-regulation-sandp-and-ice-dance-and-technical-rules-sandp-and-id_final.pdf (last accessed November 25, 2017).

[35] 'The Short Program for Single Skating (Senior and Junior) consists of seven (7) required elements.' See ibid. at Rule 611.1.a). Those elements are listed in Rules 611.2 and 611.3.

[36] 'A well balanced Free Skating program for Men must contain: maximum of eight jump elements (one of which must be an Axel type jump); maximum of three spins, one of which must be a spin combination, one a flying spin or a spin with a flying entrance and one a spin with only one position; maximum of one step sequence; maximum of one choreographic sequence.' See ibid. at Rule 612.1. Similar rules apply for other (ladies, junior) competitions.

[37] Ibid. at Rule 502.

[38] Ibid. at Rule 610.

[39] Ibid. at Rules 619–621.

[40] Ibid. at Rule 611.1.a).

[41] Ibid. at Rule 611.4.f).

well balanced program of Free Skating elements';[42] and '[t]he Competitor has complete freedom to select the Free Skating elements, the sum of which will comprise the program'.[43] Likewise, trainers and skaters are generally free to select the music and the clothes for their programs.[44] All these rules guarantee that the trainers and the skaters have a very broad remit when deciding upon the choreography/composition of their program. Further, ISU's Regulations expressly stress that it treats each type of skating as a form of artistic expression. So, for example, the score of the program is partially based on the skaters' performance,[45] the composition of the program[46] and the interpretation of the music.[47] Further, ice dancing is categorically treated as a form of dance,[48] where pattern dances might be selected from a pre-set list of classic dances.[49]

4. SPORT MOVES VERSUS CHOREOGRAPHIES

In light of the results of the above research, as well as the relevant case law and academic writings on the topic, this chapter will now turn to offering some conclusions on the protectability of sport moves and

---

42 Ibid. at Rule 612.1.

43 Ibid. at Rule 612.2.

44 'At ISU Championships, the Olympic Winter Games and International Competitions, the clothing of the Competitors must be modest, dignified and appropriate for athletic competition – not garish or theatrical in design. Clothing may, however, reflect the character of the music chosen. The clothing must not give the effect of excessive nudity inappropriate for the discipline. Men must wear full length trousers and must not wear tights. In addition, in Ice Dance, Ladies must wear a skirt. Accessories and props are not permitted.' See ibid. at Rule 501.1. Although the above quote includes multiple limitations, all of them reflect the artistic nature of figure skating, and thus inherently allow the competitors to select creative clothes for the choreography.

45 'Involvement of the Skater/Pair/Couple physically, emotionally and intellectually as they deliver the intent of the music and composition.' See ibid. at Rule 504.3.

46 'An intentionally developed and/or original arrangement of all types of movements according to the principles of musical phrase, space, pattern, and structure.' See ibid.

47 'The personal, creative, and genuine translation of the rhythm, character and content of music to movement on ice.' See ibid.

48 The expression 'dance' is included 497 times in the Regulations. Further, the Regulations stress expressly: 'Short Dance is a dance created by an Ice Dance Couple to dance music with designated rhythm(s) and/or theme(s) . . .'. See ibid. at Rule 709.1.a).

49 Ibid. at Rule 707.2.

choreographies. All sports seem to provide some space for athletes to invent new – either useful or entertaining – moves.[50] The Seventh Circuit in the U.S. held in dictum in the famous *Baltimore Orioles* case that 'the Players' performances possess the modest creativity required for copyrightability. . . . That the [baseball] Players' performances possess great commercial value indicates that the works embody the modicum of creativity required for copyrightability'.[51] In adversarial sports, such creativity might be referred to more precisely as 'athletic skill and judgment'. Here, athletes might freely pick and choose from a multitude of available moves. This selection reflects improvisation and interaction on the part of the player, whose judgment of the situation and reaction to their opponents' moves (players' positions, line-ups etc.) stimulates a particular response. Similarly, the reliance on team strategy also plays a significant part. Therefore, it is fair to say that all of these decisions by athletes form a predominantly functional purpose. Weber has correctly noted that:

> The proper test . . . of functionality would be whether the move can conceivably have been developed for a purely aesthetic purpose (in which case the move is not functional for copyright purposes), or whether the very rationale of the move's creation was necessarily to increase directly the chances of an athlete prevailing in an athletic contest (rendering the move functional and uncopyrightable).[52]

Further, all the atomic moves are trivial, that is, de minimis. Consequently, individual sport moves represent nothing more than ideas and surely do not amount to expressions in terms of copyright law. This chapter favors Kukkonen's analysis that '[l]ike other improvisational performers, athletes will almost always rely on a core library of moves and strategies. An athlete's individual moves can be analogized to ideas . . . which by themselves are uncopyrightable, but, when combined, may constitute a sufficient copyrightable expression'.[53]

A further argument against the protectability of sport moves is submitted under U.S. copyright law, namely the 'merger' doctrine. The Ninth Circuit clarified the importance of the merger doctrine as follows:

---

[50] Elam, supra note 8, at para. 45.

[51] *Baltimore Orioles, Inc. v. Major League Baseball Players Association*, 805 F.2d 663, 669 (1986) note 7. On the decision see Kukkonen, supra note 8, at 812–813; Griffith and Gordeeva, supra note 7, at 701–702; Weber, supra note 8, at 343–344; Bussey, supra note 8, at 7–8.

[52] Weber, supra note 8, at 360.

[53] Kukkonen, supra note 8, at 811.

'[w]hen the "idea" and its expression are . . . inseparable, copying the "expression" will not be barred, since protecting the "expression" in such circumstances would confer a monopoly of the "idea" upon the copyright owner free of the conditions and limitations imposed by the patent law'.[54] With respect to sport moves, the above doctrine means in practical terms that there is a strictly limited number of ways to express an idea during a competition. This argument is generally true for adversarial sports. However, some commentators have tried to reject the merger doctrine. Das notes that the

> [a]pplication of the merger doctrine to such cases is inappropriate. The theory behind the merger doctrine is to prevent monopolization. To claim that a coach would be able to copyright every single possible play that would result in the monopolization of a particular route or end position is farfetched. Given the infinite combinations available from just two patterns (run by two players), the inclusion of eleven separate patterns (run by eleven separate players) within each play provides a field of options that borders on unlimited. As such, the merger doctrine should not limit the copyrightability of scripted sports plays.[55]

Despite Das' reservations, the present author still favors those academic views that have accepted the applicability of the merger doctrine to atomic sport moves.[56]

Moreover, the present author would further submit that this analysis on the unprotectable nature of atomic sports moves seems to be valid for individual skating, gymnastics, synchronized swimming and wrestling, as well as for diving movements. This is due to the fact that they do not represent anything else on their own and remain simply an idea that is executed with aesthetic value, in accordance with the rules of the games and physics. Further, they are trivial/de minimis, no matter how many turns one athlete performs in an Axel in figure skating.

Nevertheless, the choreographies of routine-oriented sports seem eligible for copyright protection. They seem so despite the U.S. Copyright Office stressing in its 2012 Statement of Policy on the Registration of Compilations that 'a selection, coordination, or arrangement of functional physical movements such as sports movements, exercises, and other ordinary motor activities alone do not represent the type of authorship intended to be protected under the copyright law as a choreographic

---

54 See *Herbert Rosenthal Jewelry Corp. v. Kalpakian*, 446 F.2d 738, 742 (1971).

55 See Das, supra note 8, at 1097.

56 See Weber, supra note 8, at 339–340; Maja Murza, *Urheberrecht von Choreografen: Eine rechtsvergleichende Studie* (De Gruyter, 2012) 133–134; Bussey, supra note 8, at 16.

work'.[57] This opinion remains unconvincing in the present author's view with respect to routines of aesthetic sports, as they are far more than ordinary motor activities. Thus the statement of policy leaves enough space for interpretation.[58]

The copyrightability of artistic choreographies of routine-oriented sports is best evidenced through figure skating, because this sport operates in a much less functional fashion than adversarial sports. This is exhibited through the fact that most rules tend to focus on the creation and performance of artistic compositions.[59] Although the 'story being told' by these choreographies might be difficult to understand for lay observers, this will not bar the protectability of the exercises under copyright law. In support of this, Griffith and Gordeeva – focusing solely on U.S. copyright law – have correctly noted that '[a]thletic routines exhibit composite creativity',[60] where individual elements of the choreography – like the different (and compulsory) elements (jumps, spins, creative skate-work), the selection of music, costume, make-up – may not warrant copyright protection on their own, yet the creative combination of these elements surpass the level of originality and thus should be protected as a choreographed piece or dramatic production.[61] The carefully thought-out composition of these elements will allow a broader story to emanate from the skaters' and their coaches' choreography. Second, even if this story remained hard to follow by spectators, the *Norowzian* decision from the UK has convincingly confirmed that even 'storyless' choreographies might be protected as dramatic works.[62]

In addition, the scope of creativity – and thus originality – with respect to choreographies of routine-oriented sports might be broad enough to surpass the threshold of copyrightability. Choreography – which is designed to protect dances – is generally accepted as a protected subject matter under the Berne Convention and accordingly in most countries.[63]

---

57 Library of Congress, supra note 25, at 37607.

58 Compare to Bussey, supra note 8, at 22.

59 See Margoni, supra note 8, at 390.

60 Ibid. at 713.

61 Ibid. at 713–714.

62 *Norowzian v. Ark Ltd (No. 2.)* [2000] FSR 363 (CA). On the decision see Francis Yeoh, Choreographers and Copyright Ownership: Investigating an Apparent Dysfunction, 10 Journal of Intellectual Property Law & Practice 913–915 (2015).

63 '[D]ramatic or dramatico-musical works; choreographic works and entertainments in dumb show'. See art. 2(1) of Berne Convention for the Protection of Literary and Artistic Works, art. 6bis(1), September 9, 1886, as revised at Paris on

As indicated above, ice dancing is per se treated as a form of dancing.[64] This treatment shall be analogically applied to single and pair skating, too. Finally, fixation, where required, might be easily met by writing down notations of the composition, or alternatively through visual or audiovisual reproduction of the performance.

The above argumentation seems applicable to other routine-oriented sports as well, such as competitive dance, gymnastics or synchronized swimming.[65] In this sense, the present author adheres to Elam's view, who notes '[i]n figure skating, acrobatic gymnastics and synchronised swimming each individual routine is choreographed and practiced repeatedly in a rehearsal space'.[66] The same could be said of diving, which also requires a broad modicum of creativity from divers. Another example, although not covered by the InfoCopy research, but discussed at length by Weber, is the selection of feats in freestyle extreme sports. Weber cites skateboarding, stunt cycling and big-air snowboarding as sports representing more than simple athletic skill and judgment, which would arguably surpass the requisite level of creativity.[67] Weber's opinion seems to be less convincing, however, with respect to extreme sports like white-water kayaking, barefoot water-skiing, wakeboarding, sky surfing and power kiting. This chapter would submit that in these cases, the selection of feats depends much more on the physical limits of the given sport and the individual sportsmen.

By way of contrast, wrestling works somewhat differently. Wrestling performances have an agreed schedule. These 'choreographies' are, however, loose or flexible and are certainly far from being similar to dance, which often forms the backbone for most choreographies. Wrestling programs allow wrestlers, as well as the referee and the organizers of the event to amend/expand the performance slightly in light of the reaction of the audience in order to enhance their entertainment.[68] All in all, although less

---

July 24, 1971, and amended in 1979, S. Treaty Doc. No. 99-27 (1986) [hereinafter 'Berne Convention'].

[64] '[R]outine-oriented performances are heavily influenced by the fundamentals of dance'. See Griffith and Gordeeva, supra note 7, at 715.

[65] Kukkonen, supra note 8, at 810; Murza, supra note 56, at 129–130; Bussey, supra note 8, at 22.

[66] Elam, supra note 8, at para. 40.

[67] Weber, supra note 8, at 321–323.

[68] 'The wrestlers know what they're supposed to do up to a point, but they're not the writers, and they're not in charge – there has to be someone directly in communication with the guys organizing the match to deliver instructions into the ring (such as in-match changes to the script), and that someone is the referee. He gets directions from the agents who set up the match (or in the case of the WWE,

convincing, wrestling seems to also fit into the category of choreographies that are capable of attracting copyright protection.

On the other hand, no copyright protection seems to exist for 'choreographies' of adversarial sports, which has been confirmed for example in the Netherlands.[69] Such conclusions are, however, challenged by Das, who – whilst focusing mainly on American football – has opined that:

> A scripted sports play is nothing more than a group routine with individuals performing a predetermined set act in cohesion with one another. It follows that a scripted sports play would carry the same intrinsically choreographed steps as theatrical play or ballet. Logic dictates that, as such, it should be copyrightable.[70]

The present author harbors reservations over Das' analysis. Choreographies of routine-oriented sports that are similar to dance (like figure skating or synchronized swimming) or that represent significant aesthetic value (like gymnastics and occasionally even extreme sports) differ heavily from adversarial sports, where the actions of athletes are random, unpredictable and depend heavily upon the reaction of their opponents. For example, a scripted offensive play by an American football team might be annulled by the defensive line's blitz, where line-backers and defensive backs cross the line of scrummage to sack the quarterback of the offensive team. A scripted play is thus nothing more than a command and consequently an idea of how to move on the field. Elsewhere Das has wisely noted that scripted sports plays 'generally consist[] of rather simple movements',[71] and that 'a play rarely originates entirely with a coach'.[72] Both of these comments stand in opposition to his earlier opinion. Finally, some might argue that an American football formation was already successfully registered at the U.S. Copyright Office in 1984.[73] But the then-examiner of the U.S. Copyright Office, Julia Huff, clearly emphasized the Office's point of

---

Vince McMahon himself) through the earpiece.' See Evan V. Symon, 5 Behind-the-Scenes Facts I Learned as a Pro Wrestler, Cracked, March 16, 2014 available at http://www.cracked.com/personal-experiences-1284-5-behind-the-scenes-facts-i-learned-as-pro-wrestler.html (last accessed November 25, 2017).

[69] On football games see *Hoge Raad* (October 23, 1987), NJ 1988, 310 (KNVB v. NOS), available at http://jure.nl/ECLI:NL:HR:1987:AD0055. On the decision see H. Cohen Jehoram, Prestatieontlening II – Vrijheid van nieuwsgaring (NOS – KNVB), 37 Ars Aequi 461–468 (1988).

[70] Das, supra note 8, at 1086.

[71] Ibid. at 1092.

[72] Ibid.

[73] 'With Lozano's encouragement Smith paid $10 to the U.S. Copyright Office in 1984 to register the I-Bone formation, which the two men claim Smith invented'.

view, when she noted that '[g]ame plays are themselves not copyrightable. . . . They're considered ideas'.[74]

Further, the fact that a 'work' is registered by the U.S. Copyright Office (here: the printed copies of the I-Bone formation) does not mean that the underlying concept would survive an examination of copyrightability by judges. This was held in the recent case concerning the 'hot yoga sequence' developed by Bikram Choudhury. In short, the book that contained and described the 26 asanas performed in a fixed order in a room heated up to 105 degrees Fahrenheit was registered at the U.S. Copyright Office in 1979. In 2002, Choudhury also registered the 'compilation of exercises' contained in the book by a supplementary registration.[75] In 2012 the U.S. Copyright Office published a statement of policy on the registration of compilations, where the Office excluded the exercises per se from the scope of protected subject matter, although it left the door open for registration of other subject matters, such as photographs that depict/describe the exercises.[76] Following this statement of policy, the Ninth Circuit declared that Choudhury's hot yoga sequence is solely an idea and therefore not a compilation or a choreographic piece of work.[77] The exclusion of the Bikram Yoga sequence from the scope of copyright protection clearly evidences that the registration of a sports-related expression does in no way guarantee that the underlying idea is also protected.

5. CONCLUSION: MUCH ADO ABOUT NOTHING?

The bespoke research alluded to earlier, in conjunction with the leading international cases and academic opinion, seems to indicate that none of the atomic sport moves, neither in adversarial nor in routine-oriented sports, deserve copyright protection. This is generally due to their functional and de minimis nature, as well as the fact that they do not represent expressions and that they are bound by the rules of the given game and of physics.

The unprotectable nature of atomic sports moves is also evidenced by an a contrario argument. Any protection granted to a de minimis body

---

See: Craig Neff (ed.), Whose Bone Is It, Anyway?, Sports Illustrated (January 23, 1989) at 7.

[74] Ibid.

[75] *Bikram's Yoga*, 803 F.3d at 1035–1036.

[76] Library of Congress, supra note 25, at 37607.

[77] *Bikram's Yoga*, 803 F.3d at 1036–1044. See Bussey, supra note 8, at 10–17; Elam, supra note 8, at paras 75–79.

movement would run against the most fundamental purpose of copyright law: to exclude the existence of monopolies in connection with trivial matters.[78] Copyrights are based on the exclusive protection of the interests of the creators; yet, that protection is rationally limited in multiple forms. If, however, the use of a given move was constantly subject to the requirement to request permission from the rightholder then that would certainly run against the prosperity of the given sport. As Weber noted:

> This protest comes from professional baseball player Matt Williams, who has stated with respect to IP protection for sports moves: 'You've got to catch a ball any way you can possibly catch it. If I dive for a ball at third base, am I going to have to pay [former player] Graig Nettles money? That's ridiculous.'[79]

The NBA decision also highlighted that:

> If the inventor of the T-formation in football had been able to copyright it, the sport might have come to an end instead of prospering. Even where athletic preparation most resembles authorship – figure skating, gymnastics, and, some would uncharitably say, professional wrestling – a performer who conceives and executes a particularly graceful and difficult – or, in the case of wrestling, seemingly painful – acrobatic feat cannot copyright it without impairing the underlying competition in the future. A claim of being the only athlete to perform a feat doesn't mean much if no one else is allowed to try.[80]

Some might argue that such a form of protection would force coaches to invent newer and newer routines or formations.[81] However, this chapter disagrees with that logic. No doubt, there is a huge variety of formations in American football (which was the example used by Das above). These formations are, however, tightened by the strict rules on the possible behavior of the players in different positions, the physical boundaries of the field, as well as the formation used by the other team.

On the other hand, choreographies of routine-oriented/aesthetic sports do surpass the threshold of creativity and originality and are therefore copyrightable subject matter. Notwithstanding the above, a secondary

---

[78] Weber, supra note 8, at 326; Murza, supra note 56, at 133.

[79] Weber, supra note 8, at 341.

[80] *National Basketball Association v. Motorola, Inc.*, 105 F.3d 841, 846 (1997). See Griffith and Gordeeva, supra note 7, at 702–708; Weber, supra note 8, at 344; Das, supra note 8, at 1090–1091; Bussey, supra note 8, at 8–9.

[81] 'Given the infinite combinations available from just two patterns (run by two players), the inclusion of eleven separate patterns (run by eleven separate players) within each play provides a field of options that borders on unlimited'. See Das, supra note 8, at 1097.

question arises: even if these choreographies are copyrightable, should that protection be granted to athletes and/or their trainers at all? To put it differently: what would be the drawbacks of providing protection to unique sports choreographies?

Firstly, as athletes perform choreographies, they are the ones who might commit a copyright infringement. Yet, they act under the guidance of a coach or trainer and they reproduce/perform the exercise within the frames of a sport event organized by professionals (mainly leagues and federations). Consequently, there are several actors who might face claims of liability, not just the athlete in question. Therefore, complicated questions arise such as: Who shall pay the license fee to the author of the choreography? Kieff et al. have argued that:

> [D]irect payment for performance, other than prize money, is not the norm. If skiers were to envision themselves more like actors in an improvised drama, contributing their copyrightable performances to a collective work, this might open the way for negotiations with broadcasters and sponsors to include compensation (such as a share in television proceeds, or residuals on rebroadcasting).[82]

Such an inclusion of the license fee in the compensation paid by broadcasters seems to be superficial. As broadcasters would transmit the exercises to the TV spectators, they might need to obtain licenses for this purpose, but this would be an extremely complicated task to complete.[83] Further and above all, athletes – as direct performers or copiers of choreographies – could not escape requesting permission from the right holders either.

Secondly, it could be the case that the license to use has been properly requested, but the choreography is not used during the competition. This can easily happen if an athlete gets ill and is unable to perform properly, or if the trainer changes the exercise to another that has a higher value. These hypothetical examples further highlight the difficulties that licensing in advance would have and compounds the fact that it would be hard to imagine such practices actually operating in sports. An immediate counter-point to this could be for example the combined method proposed by Das. On the one hand, he has claimed that:

> League rules should allow for a more limited copyright or patent in terms of duration. Limiting the reach of a scripted play for a season provides coaches with the necessary incentive to develop new creative plays, as envisioned behind the purpose of the laws, while ensuring the competitiveness of the game. Such

---

82 Kieff et al., supra note 8, at 780.
83 Bussey, supra note 8, at 24–25.

an act would increase the competition and the ingenuity behind the game itself.[84]

On the other hand, he has supplemented the above system with the idea of an on-site enforcement ('on-site legal counsel').[85]

However, what has been proposed by Das, in relation to American football, poses more drawbacks than it offers advantages when applied to routine-oriented sports, where copyright protection might exist. The proposed 'extra-statutory IP regime' sounds unique, which means that it is not impossible, but does not seem to be practical either. If intellectual property protection (be it copyright or patent) might exist, it exists under the statutory framework of the given field. For example, limiting the term of protection would certainly work against such statutes. Further, any 'league rule' could only be enforceable by the given league against its own players/teams. Consequently, nothing actually guarantees that another league from the same country or another country, or another federation would respect such provisions. Similarly, no guarantee exists that the 'on-site legal counsels' would recognize all protected choreographies, especially those developed by foreign athletes. (It might be admittedly much less imaginable in routine-oriented sports than in adversarial sports.) Indeed, as different countries apply different rules on copyrightability, it might be imaginable that a given choreography is protected in country A, but not in country B. Such a discrepancy would clearly raise difficulties in the international enforcement of copyrights. All these concerns collectively highlight what a chilling effect this form of protection would have on the normal flow of competitions.

Thirdly, it might be imaginable that choreographies are protected by copyright law, but their use is limited under the concept of limitations and exceptions, or under the fair use/fair dealing provisions. The fair use doctrine of the U.S. copyright law[86] seems to be useless in these situations.

---

[84] Das, supra note 8, at 1098.

[85] On the idea of the 'on-site legal counsel' see ibid. at 1099–1100.

[86] 'Notwithstanding the provisions of sections 106 and 106A, the fair use of a copyrighted work, including such use by reproduction in copies or phonorecords or by any other means specified by that section, for purposes such as criticism, comment, news reporting, teaching (including multiple copies for classroom use), scholarship, or research, is not an infringement of copyright. In determining whether the use made of a work in any particular case is a fair use the factors to be considered shall include –

(1) the purpose and character of the use, including whether such use is of a commercial nature or is for nonprofit educational purposes;
(2) the nature of the copyrighted work;

The use of choreographies by fellow competitors would run against all factors of the doctrine.[87] These uses are neither transformative (as athletes purposely aim to present the same exercise), nor educational nor research oriented. Indeed, they might be indirectly commercial in nature. The choreography used by the competitor is expressive and artistic in nature. The second performance utilizes the whole of the source material. Finally, the performance of the fellow athlete might potentially substitute for the original author and his/her work on the market. Limitations and exceptions work differently in Europe. The InfoSoc Directive enumerates the possible examples in a semi-closed list.[88] According to this list, all limitations and exceptions are set by the Directive. Member States are, however, allowed to keep those provisions that already exist under their national law; they are cases of minor importance, and only concern analogue uses and they do not affect the free circulation of goods and services within the EU.[89] The present author is not aware of any limitation or exception under the national laws of the Member States that would allow for the free use of choreographies on sports competitions. In addition, limitations and exceptions are always viewed through the lenses of the 'three-step test'. The test was originally introduced by the Berne Convention with respect to the limitations on the rights of reproduction.[90] Later, the TRIPS Agreement and the World Copyright Treaty extended its application to other economic rights as well. In the EU the test is now codified by the InfoSoc Directive.[91] A limitation or exception on the economic rights of the choreographer would be allowed under this test only if the use is a certain special case, which does not conflict with a normal exploitation of the choreography and does not unreasonably prejudice the legitimate interests of the choreographer. The arguments introduced above with respect to the

---

(3) the amount and substantiality of the portion used in relation to the copyrighted work as a whole; and
(4) the effect of the use upon the potential market for or value of the copyrighted work.

The fact that a work is unpublished shall not itself bar a finding of fair use if such finding is made upon consideration of all the above factors.' See: 17 U.S.C. § 107 (hereinafter USCA).

87 Bussey, supra note 8, at 25–26.

88 Directive 2001/29/EC, of the European Parliament and of the Council of 22 May 2001 on the harmonisation of certain aspects of copyright and related rights in the information society, 2001 O.J. (L 167), art. 5(1)–(4). (Hereinafter: InfoSoc Directive.)

89 InfoSoc Directive, art. 5(3)(o).

90 Berne Convention, art. 9(2).

91 InfoSoc Directive, art. 5(5).

fair use doctrine seem to indicate that any use of the choreography by a fellow athlete would be unable to satisfy the second and third step of the test.

Fourthly, some have argued that moral rights sufficiently protect the interests of athletes.[92] This idea seems to be dysfunctional, too. On the one hand, the current U.S. moral rights regime protects solely works of the visual arts[93] and therefore these rules are inapplicable for sports choreographies. The fact that the U.S. Copyright Office has recently showed some interest in developing the moral rights regime[94] does not guarantee that there will be significant changes in this field in the near future. Continental European countries, as well as some common law countries, such as the U.K., provide for broader moral rights. Nevertheless, moral rights do not seem to provide substantive support for creators of sports choreographies. Similarly, in the case of atomic sports moves, if choreographies were named after athletes or trainers, other professionals – fellow competitors, federations and commentators – would certainly refer to the names of the exercises properly, as requested by the traditions of sports. Likewise, the integrity of the choreographies would be kept by the follow competitors as this would be the only chance to perform a – one can assume – perfect routine. Finally, Bussey properly noted that 'it may be reasonable to provide moral rights to the creator of an exercise routine, but these rights would not give him the option to enjoin others from performing his routine'.[95]

All the above points were used to reject the idea of the practical protection of sports moves, yet the most compelling argument remains the last one. It comes directly from the rationale of sports. In several amateur and professional sports developing and presenting new elements (if possible on a significant tournament or event) guarantees a higher chance of success. This is especially true for aesthetic sports. On the contrary, copying someone else's choreography might lead to a 'professional hara-kiri'. To put it differently: in sports, being the first is always a merit; but copying others might be embarrassing. There is a greater chance that athletes would not rely on fellow competitors' choreographies, as slavish appropriation

---

92 Griffith and Gordeeva, supra note 7, at 723–729.

93 USCA § 106A.

94 The United States Copyright Office and The George Mason University School of Law and its Center for the Protection of Intellectual Property co-sponsored a symposium on moral rights – titled Authors, Attribution, and Integrity: Examining Moral Rights in the United States – on April 18, 2016. The transcript of the event is available at http://www.georgemasonjicl.org/wp-content/uploads/2016/08/Summer-Issue-2016.pdf.

95 Bussey, supra note 8, at 29.

of others' exercises could easily be detected (by the jury or reporters). Consequently, there is a strong incentive not to copy others, but in fact to develop something new, something unexpected. In other cases – especially in adversarial sports – including traditional sport moves in an exercise is expected. Indeed, it is the most effective method to achieve success. In these situations, functionality (and the federation's code of points) trumps creativity.

The overprotection of sport moves would appear to be detrimental to the normal functioning of amateur and professional sports. As an alternative to copyright law, the most effective 'protection' of choreographies and – even more – individual sport moves would be an extra-legal 'professional honoring method'. This method might include the naming of a movement or routine after the particular athlete or trainer who invented or developed it. Such 'protection' would represent both professional appreciation and a unique, extra-legal form of the right of attribution. Further, athletes or trainers might be awarded professional titles in order to honor their athletic skills and creativity.[96] Finally, and in the case of professional athletes, signing endorsements or any other form of contract would reflect their achievements to a far greater extent than copyright law is able to.

[96] Multiple honors exist already (mainly) in adversarial sports, like 'most valuable players', 'rookie of the year', 'trainer of the year', etc. All these awards represent a given sportsman's skills and abilities presented during the competition.

PART III

Industry and science

13. Copyright and typefaces

Arul George Scaria and Mathews P. George*

1. INTRODUCTION

Typefaces, which are treated by many as an art form, refers to the manner in which a set of letters and/or numbers is presented, whether on paper or on a digital screen.[1] Some of the prominent examples of typefaces include Arial, Calibri and Times New Roman. Very often people tend to use the terms 'typeface' and 'font' interchangeably. However, from a technical perspective, a 'font' is an article in which a typeface resides and plays a critical role in the way typefaces appear on paper or digital screen.[2]

Typefaces and fonts have been in existence since the invention of the printing press and the first typefaces, 'Donatus-Kalendar', were a set of new, reusable letters engraved on wood or metal blocks which were then imprinted on paper.[3] In the digital context, 'font' technically refers to the code or program that instructs a processor or printer how to render or print a certain 'typeface' on a digital screen or a piece of paper.[4] Though the terms 'typefaces' and 'fonts' are different technical expressions, we will be treating them as synonyms for the purpose of this chapter, in view of the extensive usage of these terms as synonyms in the digital context. We will be using the term 'typeface' hereafter to signify both typefaces and fonts.

The proliferation of diverse software which can be used for creating

* The authors thankfully acknowledge the research inputs provided by Karthik Tayur and Pragya Jha. We would also like to thank Aayush Kashyap, Enrico Bonadio, Kavya Susan Mammen, Nicola Lucchi, Shreyashi Ray and Zakir Thomas for their insightful comments on the draft version of this chapter.

[1] Emily N. Evans, Fonts, Typefaces and IP Protection: Getting to Just Right, 21 J. Intell. Prop. L. 307, 309–310.

[2] Ibid.

[3] Ibid. at 311.

[4] Jacqueline D. Lipton, To © or Not to ©? Copyright and Innovation in the Digital Typeface Industry, 43 U.C. Davis L. Rev. 143, 148 (2009) (citing H.R. 1990, 102d Cong. § 1001(b) (4) (1991)).

typefaces has enabled more people to engage in the designing of typefaces than ever before. However, the fact that in the digital context a typeface can also be easily copied at negligible costs has prompted some scholars to propose justifications for intellectual property protection for typefaces.[5] While different forms of intellectual property rights can be considered in this context, including copyright and designs, this chapter analyses the issue from the copyright perspective only, by asking two fundamental questions: (a) whether typefaces are copyrightable; and (b) whether typefaces ought to be copyrightable. The chapter has used the doctrinal method for answering these questions.

2. COPYRIGHTABILITY OF TYPEFACES

Though set in a different context concerning the separation of law and morality, H.L.A Hart, in his critically acclaimed article in the Harvard Law Review, differentiates 'core' and 'penumbra' areas of law.[6] While the former deals with settled positions of law, the latter deals with unsettled positions of law.[7] Common law jurisdictions, unlike civil law jurisdictions,

---

[5] Kevin Garnett, Gillian Davies and Gwilym Harbottle, *Copinger and Skone James on Copyright* (16th edn, Sweet & Maxwell, 2011) 631.

[6] H.L.A. Hart, Positivism and the Separation of Law and Morals, 71 Harvard Law Review 593 (1957).

[7] Ibid. (He discusses an interesting hypothetical fact situation which is also relevant for our chapter: 'A legal rule forbids you to take a vehicle into the public park. Plainly this forbids an automobile, but what about bicycles, roller skates, toy automobiles? What about airplanes? Are these, as we say, to be called "vehicles" for the purpose of the rule or not? If we are to communicate with each other at all, and if, as in the most elementary form of law, we are to express our intentions that a certain type of behaviour be regulated by rules, then the general words we use – like "vehicle" in the case I consider – must have some standard instance in which no doubts are felt about its application. There must be a core of settled meaning, but there will be, as well, a penumbra of debatable cases in which words are neither obviously applicable nor obviously ruled out. These cases will each have some features in common with the standard case; they will lack others or be accompanied by features not present in the standard case. Human invention and natural processes continually throw up such variants on the familiar, and if we are to say that these ranges of facts do or do not fall under existing rules, then the classifier must make a decision which is not dictated to him, for the facts and phenomena to which we fit our words and apply our rules are as it were dumb. The toy automobile cannot speak up and say, "I am a vehicle for the purpose of this legal rule," nor can the roller skates chorus, "We are not a vehicle." Fact situations do not await us neatly labeled, creased, and folded, nor is their legal classification written on them to be simply read off by the judge. Instead, in

place considerable reliance on interpretations and precedents. Arguably, the copyrightability of typefaces is not a legal issue which falls into the 'core' area of copyright law of (at least) some of the jurisdictions. Considering this legal position, this chapter will examine the approach adopted in three common law jurisdictions – viz., the United Kingdom, the United States of America and India – to understand and appreciate the nuances of the copyrightability of typefaces.[8] Further, we shall also examine the extant international legal regime. Our endeavour, through this comparative analysis, is to provide insights which can enrich debates on the desirability of copyright protection for typefaces.

2.1 United Kingdom

The Statute of Anne, also known as the Copyright Act, 1710, was the first statute to provide for copyright protection.[9] It provided fixed terms of copyright protection for the purpose of balancing public and private interests. The Statute of Anne is frequently cited while discussing the utilitarian underpinnings of copyright law.[10] There have been other enactments at later junctures and these include the Copyright Act of 1911, the Copyright Act of 1956, and the Copyright, Designs and Patents Act of 1988 ('CDPA').

In the UK, typefaces are generally protected as artistic works.[11] But the CDPA also provides certain specific exceptions to copyright infringement

applying legal rules, someone must take the responsibility of deciding that words do or do not cover some case in hand with all the practical consequences involved in this decision').

[8] Compilation of the replies to the questionnaire on graphical user interface, icon and typeface/type font designs ('Compilation of the replies'), Standing Committee on the Law of Trademarks, Industrial Designs and Geographical Indications, Thirty-Sixth Session Geneva, 17 October–19 October, 2016, SCT/36/2, 27 January 2017, http://www.wipo.int/edocs/mdocs/sct/en/sct_36/sct_36_2_rev.pdf (last accessed October 13, 2017). In their submissions to WIPO, both the United States of America and the United Kingdom maintained that they offer some kind of protection to typefaces. India did not make any submission.

[9] Yale Law School, The Avalon Project, The Statute of Anne, http://avalon.law.yale.edu/18th_century/anne_1710.asp (last accessed October 13, 2017).

[10] Rahul Mathan and Nikhil Narendran, Fair Dealing of Computer Programs in India, 7 Ind. J. L. & Tech. 91 (2011). See also Lior Zemer, 'Three Arguments on Locke, Authorship, Communication and Solitude, 7(1) W.I.P.O.J. 40–50 (2015).

[11] Lionel Bently and Brad Sherman, *Intellectual Property Law* (4th edn, Oxford University Press, 2014) 266; Garnett et al., supra note 5 at 631; and Compilation of the replies, supra note 8 at p. 13 (United Kingdom offers multifold protection for typefaces through industrial design law, copyright law and trademark law).

with respect to typefaces. These include: (a) using the typeface in the ordinary course of typing, composing text, typesetting or printing; (b) possessing an article for the purpose of such use; or (c) doing anything in relation to material produced by such use.[12] Interestingly, some scholars are of the opinion that while the policy objective behind these exemption provisions is to save users from infringement liability in such contexts, these provisions cannot exempt people who deal in typefaces.[13] In other words, the exemption provisions may not save people who are selling, letting for hire, distributing, etc., from infringement liability.[14] Interestingly, unlike the rest of the copyrightable subject matter that enjoys an extended period of copyright protection, the term of protection for articles incorporating the typefaces has been limited to 25 years from the end of the calendar year in which the article was marketed.[15]

Copyright is a legal right that derives its legal status from the respective statute. In other words, its jural genesis lies in the relevant statute and not common law.[16] Accordingly, it can be logically argued that the explicit provisions relating to typefaces under CDPA, unlike the earlier regime, indicate an implied exclusion of copyright protection for typefaces under the antecedent copyright law regime.[17] Its logical corollary also points at the necessity for express statutory provisions in the case of extension

---

12 § 54 of the CDPA.

13 Garnett et al., supra note 5, 632–633.

14 Ibid. and Compilation of the replies, supra note 8 at p. 9 ('As such, it is not an infringement of copyright to use a font when writing text, printing that text or having a computer upon which that text is composed, as long as you have legally obtained the font in the first place (e.g. through a licence via word processing software). It is, however, infringing to import or deal with equipment that is specifically adapted or designed for producing material in a particular typeface, without permission or a licence. For example, it would be infringing to sell a printing press comprising a protected font, font packages, or word processing software that includes said font, without a licence').

15 § 55 of the CDPA.

16 *Donaldson v. Becket* (1774) 2 Brown's Parl. Cases (2d ed.) 129, 1 Eng. Rep. 837; 4 Burr. 2408, 98 Eng. Rep. 257; 17 Cobbett's Parl. Hist. 953 (1813).

17 P.R. Maxwell, *The Interpretation of Statutes* (12th edition by P. St. J. Langan, N.M. Tripathi Publishers, 1976) at p. 40 ('In Heydon's case in 1584, it was resolved by the Barons of the Exchequer "that for the sure and true interpretation of all statutes in general (be they penal or beneficial, restrictive or enlarging of the common law), four things are to be discerned and considered: (1st) What was the common law before making of the Act? (2nd) What was the mischief and defect for which the common law did not provide? (3rd) What remedy the Parliament hath resolved and appointed to cure the disease of the Commonwealth. And (4th) the true reason of the remedy"; Although judges are unlikely to propound formally in their judgments the four questions in Heydon's Case, consideration of the mischief

of copyright protection for typefaces.[18] Moreover, as highlighted in *re Ananda Expanded Italics*, the specific inclusion of exceptions to infringement are also relevant for maintaining a fine balance between the interests of the rights holders and the users, when the legislature extends copyright protection to typefaces.[19]

One may note that CDPA extends protection to typographical arrangements of published editions also.[20] However, it is important to note that the protection for typographical arrangements of published editions is different from protection for typefaces. In the case of typographical arrangements, the protection is limited to the overall appearance of a page or set of pages, and not to a specific typeface.[21] While the reproduction of specific layouts of a work may be prevented under the rights given for typographical arrangements of published editions, it is important to note that it does not prevent the reproduction of that work itself.[22] In this context, it is also important to note that the underlying work may not even have to be under copyright protection to obtain the rights provided for typographical arrangements.[23] Arguably, this may plausibly lead to an ironical legal scenario wherein one may be able to extend copyright protection for typographical arrangements by simply changing the typeface provided it meets the requisite criteria under the CDPA.

2.2 United States

The Copyright Act of 1976 currently provides the statutory framework for copyright law in the US. It extends copyright protection to 'pictorial, graphic and sculptural works', which include the following:

> [T]wo-dimensional and three dimensional works of fine, graphic, and applied art, photographs, prints and art reproductions, maps, globes, charts, diagrams, models, and technical drawings, including architectural plans. Such works

---

or object of the enactment is common, and will often provide the solution to a problem of interpretation').

[18] Ibid.

[19] 2002 (24) PTC 42, Paragraph 46.

[20] See § 1 of CDPA, which reads as follows: 'Copyright is a property right which subsists in accordance with this part in the following description of work:

(a) original, literary, dramatic, musical or artistic works,
(b) sound recordings, films broadcasts or cable programmes,
(c) the typographical arrangement of published editions'.

[21] Bently and Sherman, supra note 11 at 90.

[22] Ibid.

[23] Ibid. and Garnett et al., supra note 5 at 131.

> shall include works of artistic craftsmanship insofar as their form but not their mechanical or utilitarian aspects are concerned; the design of a useful article, as defined in this section, shall be considered a pictorial, graphic, or sculptural work only if, and only to the extent that, such design incorporates pictorial, graphic, or sculptural features that can be identified separately from, and are capable of existing independently of, the utilitarian aspects of the article.[24]

This definition entails inherent limitations if one attempts to extend copyright protection to the area of typefaces. For instance, a design that is also a useful article is considered a pictorial, graphic or sculptural work only to the extent that such design incorporates pictorial, graphic or sculptural features that can be identified separately from, and are capable of existing independently of, the utilitarian aspects of the article. Copyright protection in the case of useful articles is determined by the 'separability test' as set out in the statutory provision.[25] An article is considered useful if it has an intrinsic utilitarian function beyond its appearance, or if the design elements show a merger of aesthetic and functional considerations.[26] Under such circumstances, copyright protection is extended only when 'pictorial, graphic or sculptural work' has aesthetic elements which are not dictated by their functionality, and further, copyright protection is extended only to those aesthetic elements. Legibility is clearly a utilitarian function associated with the design choices for a typeface and this poses considerable challenges for the acceptance of typefaces as a copyrightable subject matter.

The US Supreme Court, in its recent decision in *Star Athletica, LLC v. Varsity Brands*, has set out certain guidelines on copyrightability with respect to useful articles.[27] The case originated as a copyright infringement suit filed by Varsity Brands against Star Athletica, the petitioner accusing the respondent of marketing cheerleading uniforms which were substantially similar to designs created by the petitioner.[28] According to

---

[24] § 101 of the Copyright Act of 1976.

[25] See Blake Fry, Why Typefaces Proliferate Without Copyright Protection, 8 J. Telecomm. & High Tech. L. 425, 433 (2010).

[26] Ibid.

[27] *Star Athletica L.L.C. v. Varsity Brands Inco., et al.*, 580 U.S., (2017), https://www.supremecourt.gov/opinions/16pdf/15-866_0971.pdf (last accessed October 13, 2017).

[28] The petitioner argued that the respondent's copyrights were invalid since the designs were for non-copyrightable useful articles; and further, the pictorial, graphic or sculptural elements of respondent's designs were not physically or conceptually separable from the uniforms. The District Court entered a summary judgment in favour of petitioner. But this decision was vacated and remanded by the U.S. Court of Appeals for the Sixth Circuit in August 2015. The opinion

the majority decision in this case, such artistic features would be eligible for copyright protection, only if they meet a two-step test: firstly, if they can be perceived as a two- or three-dimensional work of art separate from the useful article; and secondly, if they would qualify as a protectable pictorial, graphic, or sculptural work either on their own or in some other medium if imagined separately from the useful article.[29] The same test may be applied in the context of the copyrightability of typeface designs. As one can imagine, only a small fraction of typefaces existing today may satisfy the two-step test laid down by the United States Supreme Court in this instant judgment.

Even before the *Star Athletica* decision, different courts and scholars have attempted to use different tests in this area, indicating the constantly evolving character of this complex area of law. The resulting interpretive disarray in the field has been highlighted by different scholars.[30] However, it is highly probable that most typefaces may fail to secure copyright protection even under most of those approaches. For example, in *Kieselstein-Cord v. Accessories by Pearl*, where the articles in question were two ornamental belt buckles, the majority had observed that the buckles had a separate ornamental use and that the utilitarian function was 'subsidiary' to the primary ornamental aspect.[31] Although the court did not formulate their analysis as a test, this observation prompted the majority to reach the conclusion that the buckles in question can be a subject matter of copyright.[32] As to applying this test to a typeface, it is necessary to consider both the expressive/ornamental and utilitarian aspects of the typeface in question. If the expressive/ornamental component of a typeface is taken as the secondary aspect, just complementing the utilitarian aspect of

---

handed down by the Sixth Circuit entered a summary judgment victory in favour of the respondent, finding that the graphic features of respondent's designs are more like fabric designs than dress designs and are thus protectable subject matter under the Copyright Act. In January 2016, the petitioner petitioned the Supreme Court for a writ of certiorari, reasoning that the appeal must be granted to resolve the conflict amongst Circuit Courts on the position of law with respect to the copyrightability of typefaces.

[29] Star *Athletica L.L.C. v. Varsity Brands Inco., et al.*, 580 U.S.___(2017) at 3, https://www.supremecourt.gov/opinions/16pdf/15-866_0971.pdf (last accessed October 13, 2017).

[30] See, for example, William F. Patry, *Patry on Copyright (Vol. 2)* (Thomson/West, 2012) 3:136–3:147 and Shubha Ghosh, Remapping Copyright Functionality: The Quixotic Search for a Unified Test of Separability for PGS Works (8 September 2016) https://papers.ssrn.com/sol3/papers.cfm?abstract_id=2836445.

[31] *Kieselstein-Cord v. Accessories by Pearl, Inc.*, 632 F.2d 989 (2d Cir. 1980).

[32] Ibid. at 993–994.

conveying the words in print, typeface designs may not be considered as a copyrightable subject matter.[33]

Similarly, in *Brandir Int'l v. Cascade Pacific Lumber*, the Second Circuit had adopted a test proposed by Professor Robert Denicola.[34] While this test intends to minimize the protection afforded to the functional components of the utilitarian article, it also tries to protect the aspects which are truly expressive.[35] According to the majority in this decision, if the '[d]esign elements reflect a merger of aesthetic and functional considerations, the artistic aspects of a work cannot be said to be conceptually separable from the utilitarian elements. Conversely, where the design elements can be identified as reflecting the designer's artistic judgment exercised independently of functional influences, conceptual separability exists'.[36] Typeface designing can be a highly artistic endeavour and a typeface designer may make numerous artistic choices in the design process. But legibility is a utilitarian function which is constantly associated with the design choices for a typeface, thereby challenging the possibilities of typefaces becoming a copyrightable subject matter, even under the aforesaid *Brandir* test. Some of the other prominent decisions which have dealt with the issue of copyrightability in this area include *Carol Barnhardt v. Economy Cover Corporation*,[37] *Hart v. Dan Chase Taxidermy Supply Co.*,[38] and *Chosun International, Inc. v. Chrisha Creations Ltd.*[39]

Scholars like Nimmer have also contributed to the jurisprudential debates in this area. According to the so-called Nimmer/Poe test, the core question to be asked is whether the article in question will be marketable to some significant segment owing to its aesthetic qualities, even if the article has no utilitarian use.[40] As discussed earlier, in the case of typefaces,

---

[33] Clive Lewis and Peter Walker, Typographical Influences on Reading, 80 Brit. J. Psychol. 241, 243 (1989).

[34] *Brandir Int'l, Inc. v. Cascade Pacific Lumber Co.*, 834 F.2d 1142, 1145 (2d Cir. 1987). Robert C. Denicola, Applied Art and Industrial Design: A Suggested Approach to Copyright in Useful Articles, 67 Minn. L. Rev. 707 (1983).

[35] Ibid. at 741.

[36] *Brandir Int'l, Inc. v. Cascade Pacific Lumber Co*, supra note 34. Applying this test, the Court denied copyright protection to a bicycle rack constructed of metal tubing and modelled after a wire sculpture of continuous undulating wire.

[37] *Carol Barnhardt v. Economy Cover Corporation*, 773 F. 2d 411 (2nd Cir. 1985).

[38] *Hart v. Dan Chase Taxidermy Supply Co.*, 86 F. 3d 320 (2nd Cir. 1996).

[39] *Chosun International, Inc. v. Chrisha Creations Ltd.*, 413 F. 3d 324 (2nd Cir. 2005).

[40] Melville B. Nimmer and David Nimmer, *Nimmer on Copyright (Vol. 1)* (LexisNexis, 2010) § 2.08[B][3].

the legibility aspect often plays an important role in their selection and only under certain exceptional circumstances, will a typeface also be able to meet the requirements of this test.

In the specific context of typefaces, the Fourth Circuit Court of Appeals in *Eltra v. Ringer* has held that typefaces are not copyrightable.[41] In this case, the Eltra Corporation, a manufacturer of typesetting equipment, had sought copyright registration for typeface design under the provisions of Copyright Act of 1909.[42] Citing Copyright Regulations, the Register refused to grant copyright registration.[43] When the matter finally reached the Court of Appeals for the Fourth Circuit, the Court examined the regulations and held that '[i]t is patent that typeface is an industrial design in which the design cannot exist independently and separately as a work of art'.[44]

The extant legal position that we have narrated so far may be read in conjunction with the submission of the United States government before the World Intellectual Property Organization wherein it submitted that the typefaces do not enjoy copyright protection under the extant municipal law regime.[45] But it needs to be noted that irrespective of the copyrightability of the typefaces, the associated software code may be copyrightable in the US.[46] The United States Copyright Office suggested in a 1992 statement that the '[c]omputer programs designed for generating typeface . . . may involve original computer instructions entitled to protection under the Copyright Act'.[47] This position was also confirmed by the District Court in *Adobe v. Southern Software*.[48] Computer programs

---

41 *Eltra Corp. v. Ringer*, 579 F.2d 294, 298 (4th Cir. 1978).

42 § 5(g) of Copyright Act of 1909 allowed registration on 'works of art; models or designs for works of art'.

43 The relevant regulation reads as follows: 'If the sole intrinsic function of an article is its utility, the fact that the article is unique and attractively shaped will not qualify it as a work of art. However, if the shape of a utilitarian article incorporates features, such as artistic sculpture, carving, or pictorial representation, which can be identified separately and are capable of existing independently as a work of art, such features will be eligible for registration'.

44 Supra note 41 at 298.

45 Compilation of the replies, supra note 8 at p. 3. It was submitted that the typefaces do not enjoy protection under the Copyright Law and the Trademark Law. On the other hand, they enjoy protection under the Design Patent Law.

46 Compilation of the replies, supra note 8 at p. 20.

47 Registrability of Computer Programs that Generate Typefaces, 57 Fed. Reg. 6201, 6201-02 (Feb. 21, 1992).

48 *Adobe Sys. v. Southern Software, Inc.*, 1998 U.S. Dist. LEXIS 1941, at 16–17 (holding font programs eligible for copyright protection because their design requires creativity, and selection of code is not solely dictated by function).

are protectable as literary works.[49] The copyrightability of a computer program is not contingent on the copyrightability of its outputs, including the resultant typefaces.[50]

2.3 India

The Indian copyright law framework is governed by the Copyright Act, 1957. The Copyright Act, 1957 is silent on providing copyright protection for typeface designs. There have been no reported judgments so far from either the Supreme Court or the High Courts on the copyrightability of typefaces. An incisive analysis of the copyright legislation in India suggests non-copyrightability of typefaces under the Indian law. There are different factors which one may take into consideration in this regard. Firstly, the Copyright Act 1957 explicitly mentions that there shall be no copyright or any similar right, except in accordance with the provisions of the Act.[51] In other words, copyright protection in India is limited to subject matter and rights specifically enumerated under the Copyright Act, and, further, any subject matter or right not specifically enumerated under the Act is outside the purview of copyright protection in India. This aspect of copyright law is no longer *res integra*. The recent decision of the Delhi High Court in *The Chancellor, Masters & Scholars of University of Oxford & others v. Rameshwari Photocopy Services & others*, which arose in the context of course packs, is an example in this regard.[52] In this case, the High Court clarified that the copyright is not a natural or common law right, but a statutory right, and the scope of rights under the Indian copyright law is subject to the contours and framework of the statute.[53]

Secondly, the copyrightability of typefaces was explicitly rejected by the Registrar of Copyright in India in *re Ananda Expanded Italics*.[54] In

---

49 *Apple Computer, Inc. v. Franklin Computer Corp.*, 714 F.2d. 1240, 1247–1248 (3d Cir. 1983).

50 *AdobeSys. v. Southern Software, Inc.*, 1998 U.S. Dist. LEXIS 1941, at 10.

51 § 16 of the Copyright Act of 1957.

52 2016 SCC Online Del 5128, also available at http://lobis.nic.in/ddir/dhc/RSE/judgement/16-09-2016/RSE16092016S24392012.pdf (last accessed October 1 3, 2017).

53 Ibid. at Paragraph 28 ('The position becomes unequivocally clear from Section 16 which provides that no person shall be entitled to copyright or any similar right in any work otherwise than under and in accordance with the provisions of the Copyright Act. The same unequivocally prescribes that there is no copyright except as prescribed in the Act, converting copyright from a natural or common law right to a statutory right').

54 *Re Ananda Expanded Italics*, 2002 (24) PTC 42.

this case, M/s. ABP Ltd. had sought copyright registration for typeface designs. The Copyright Office initially rejected the application for registration stating that the works do not qualify as an 'artistic work' as per the definition of 'artistic work' set out under Section 2(c) of the Copyright Act, 1957. Since the applicant was not given an opportunity to present their case before the decision was taken, the Calcutta High Court set aside the impugned order. The Court directed the Registrar to pass orders after affording the petitioner an opportunity to be heard. Citing various grounds, including: (i) non-inclusion of typeface design under copyrightable subject matter as per the Copyright Act, 1957; (ii) absence of automatic common law copyright protection in favour of typeface design; and (iii) non-recognition of typeface design as copyrightable subject matter under international conventions which India has signed and ratified, the Registrar held against copyright protection for typefaces. Further, the Registrar also observed that even if typefaces are assumed to be copyrightable, typeface design can be registered as a 'design' under the Designs Act; and by virtue of the fact that the petitioners had already reproduced it more than 50 times, copyright protection, if any, had already been exhausted.[55]

2.4 International Treaties

As highlighted in the order in *re Ananda Expanded Italics*, non-copyrightability of typefaces consonance with various international conventions and agreements.[56] International conventions such as the Berne Convention for the Protection of Literary and Artistic Works ('Berne Convention') and the Agreement on Trade-Related Aspects of Intellectual Property Rights (TRIPS), are silent on copyright protection for typefaces. It also needs to be highlighted that there exists a separate international agreement on the protection of typefaces, viz. the 'Vienna Agreement

---

[55] The potential overlaps between copyright protection and design protection have been addressed in § 15 of the Copyright Act, 1957, which reads as follows: '15. Special provision regarding Copyright in designs registered or Capable of being registered under the Indian Patents and Designs Act, 1911: (1) Copyright shall not subsist under this Act in any design which is registered under the Indian Patents and Designs Act, 1911(5 of 1911). (2) Copyright in any design, which is capable of being registered under the Indian Patents and Designs Act, 1911(2 of 1911), but which has not been so registered, shall cease as soon as any article to which the design has been applied has been produced more than fifty times by an industrial process by the owner of the copyright or, with his license, by any other person'.

[56] *Re Ananda Expanded Italics*, supra note 54, Paragraphs 37–42.

for the Protection of Type Faces'.[57] But not many countries have become parties to that agreement and among the three jurisdictions we discussed in this chapter, the UK is the only jurisdiction that has signed this treaty.[58] Even the UK has not ratified the treaty and more importantly, the treaty itself has not come into force.[59]

Finally, the Berne Convention, one of the most important copyright conventions, gives member countries the discretion to determine the extent of application of their laws to works of applied art or industrial designs or models.[60] This implies the necessity for express legislative intention if typefaces are to be provided with copyright protection. Further, the Convention bestows the parties with the discretion to determine the term of protection for applied art.[61] If any country desires to protect works of applied art, it is required to expressly provide for it, including the term of protection. For example, as we saw earlier in this chapter, the CDPA in the UK expressly provides for 25 years of copyright protection for typefaces.[62]

Before concluding the discussions on relevant legal positions in various

---

57 Vienna Agreement for the Protection of Type Faces and their International Deposit and Protocol Concerning the Term of Protection, June 12, 1973, 1974 Gr. Br. Misc. No. 27 (Cmnd. 5754).

58 The treaty has not yet come into force since it requires ratification by five or more countries to do so. The Preamble of the Agreement sets out the intention to encourage the creation of typefaces and provide protection for the same. The Agreement is conscious of the role played by typefaces in the dissemination of culture.

59 As for the Contracting Parties and declarations, visit http://www.wipo.int/export/sites/www/treaties/en/documents/other_treaties/vtf-parties.pdf (last accessed October 13, 2017).

60 Article 2, Paragraph 7 of the Berne Convention provides as follows: 'Subject to the provisions of Article 7(4) of this Convention, it shall be a matter for legislation in the countries of the Union to determine the extent of the application of their laws to works of applied art and industrial designs and models, as well as the conditions under which such work, designs and models shall be protected. Works protected in the countries of origin solely as designs and models shall be entitled in another country of the Union only to such special protection as is granted in that country to design and models; however, if no such special protection is granted in that country, such works shall be protected as artistic works'. See also, supra note 54, Paragraph 40.

61 Article 7, Paragraph 4 of the Berne Convention deals with 'Terms of protection for Photographic works and Works of Applied Art': 'It shall be a matter for legislation in the country of the Union to determine the term of protection of photographic works and that of works of applied art in so far as they are protected as artistic works; however, this term shall be at least until the end of a period of twenty-five years from the making of such a work'.

62 § 15 of CDPA.

jurisdictions, irrespective of their contentious nature, we may also examine some plausible arguments which may be raised in favour of the copyrightability of typefaces. Since typeface designs are not per se included in the definition of 'artistic works', it may be argued that certain typefaces which have aesthetic/ornamental or expressive elements will fall under 'any other work of artistic craftsmanship'. As per the doctrine of *noscitur a sociis*, the meaning of a questionable word or phrase in a statute may be ascertained by referring to the meaning of words or phrases associated with it.[63] Accordingly, 'any other work of artistic craftsmanship' in the definition of 'artistic works' may be interpreted to include those typefaces which have separable aesthetic components which go beyond their utilitarian functions, and they may be protected as artistic work.[64] But an in-depth analysis of these arguments will reveal their shallow nature for the following reasons, particularly: (i) typefaces are not included among the specific subject matter in most copyright laws, including Indian/US copyright law; and (ii) attempts to establish a separate agreement for the protection of typefaces, i.e., the Vienna Agreement for the Protection of Type Faces, illustrates that the intention was to keep them separate from the Berne convention and the TRIPS Agreement, and one may contrast this with the approach taken with regard to the protection of computer programs under these treaties.

Plausibly, copyright protection for computer programs may *ipso facto* lead to copyright claims over typefaces when unique codes are a prerequisite for generating a particular kind of typeface. Applying the canons of interpretation, this amounts to colourable exercise of right, as one cannot circumvent an implied prohibition through indirect means. If copyright protection for computer programs leads to copyright protection of typefaces (especially those which have only a functional and not an aesthetic/expressive component), it may be making the implied prohibition otiose.[65]

---

[63] Maxwell, supra note 17, at 289–293.

[64] The Indian Supreme Court pegged the originality standards of Indian Copyright Law in *Eastern Book Company & Ors v. D.B. Modak & Anr* (2008) 1 SCC 1. (It endorsed the standard of the Canadian Supreme Court (that the appropriate standard is neither one of 'sweat of the brow' nor of 'creativity') and held that: 'Creative works by definition are original and are protected by copyright, but creativity is not required in order to render a work original. The original work should be the product of an exercise of skill and judgment and it is a workable yet fair standard . . . novelty or invention or innovative idea is not the requirement for protection of copyright but it does require a minimal degree of creativity').

[65] Maxwell, supra note 17, at 40, 105–106 (quoting Lord Reid in *Gartside v. I.R.C.* [1968] A.C. 553, at 612, cf. *Coutts & Co. v. I.R.C.* [1964] A.C. 1393: ('It is always proper to construe an ambiguous word or phrase in light of the mischief

Moreover, it also needs to be emphasized that if a programmer writes a different (original) code which produces a similar typeface, it will not result in violation of any copyright in the software, and this is one of the primary reasons why typeface makers are lobbying extensively for copyright protection of typefaces per se.

The scope of copyright protection for typeface-related software may be tested before a court of law in one of the pending litigations before the United States District Court for the Eastern District of New York – *Font Brothers Inc. v. Hasbro Inc.*[66] In this case, the plaintiff has alleged copyright infringement under 17 U.S.C. § 501 pertaining to the plaintiff's typeface software, viz. the 'Generation B Type Font software'. Allegedly, the defendant has created unauthorized and infringing copies of the software and distributed the same to third parties. The plaintiff has *inter alia* prayed for preliminary and permanent injunction so as to restrain the defendant from infringing the copyrighted work. While the case is yet to be decided, the outcome of this decision may lead to greater clarity on the contours of copyright protection for typefaces in the digital era.

3. DESIRABILITY OF COPYRIGHT PROTECTION FOR TYPEFACES

One of the most influential theories in the area of copyright law, particularly in common law jurisdictions, is welfare theory (also known as 'incentive theory'). Welfare theory, which is based on the utilitarian approach, is prospective in nature and seeks to frame law in a manner that will increase overall happiness or welfare. As far as copyright is concerned, the supporters of this theory note that most of the products of intellectual creativity have two important characteristics of public goods – a non-rivalrous

---

which the proviso is obviously designed to prevent, and in light of the reasonableness of the consequences which follow from giving it a particular construction.' Quoting Lord Reid in *Gill v. Donald Humberstone & Co. Ltd.* [1963] 1 W.L.R. 929 at p. 34: 'If the language is capable of more than one interpretation, we ought to discard the more natural meaning if it leads to an unreasonable result, and adopt that interpretation which leads to a reasonably practicable result').

[66] See *Font Brothers Inc. v. Hasbro Inc.* (Case 1:16-cv-00320-CBA-PK). The complaint filed by the plaintiff before the United States District Court for the Eastern District of New York available at https://torrentfreak.com/images/pony complaint.pdf (last accessed October 21, 2017).

character in consumption and relative non-excludability.[67] According to the proponents of the theory, owing to these characteristics, people may hesitate to invest in the creation of such goods.[68] According to welfare theory, temporary exclusivity in the form of copyright can address this issue by incentivizing creativity.[69]

But one of the most important insights, which is often overlooked, is that welfare theory does not advocate the extension of IP protection to cases where creativity is thriving in spite of the public goods characteristic of the good in question.[70] In other words, welfare theory suggests IP protection only in cases where incentivization (in the form of temporary exclusivity) is required for promoting the creation of public goods. As many of the recent studies have shown, creativity may thrive in certain industries even in the absence of strong IP rights or even with the non-existence of IP rights.[71] Many of these industries are now referred to as IP's 'negative spaces'.[72] The term 'negative space' refers to certain situations or areas of operation wherein IP protection may not be required for promoting innovation/creativity, or situations where IP protection may even disincentivize creativity.[73] Some of the prominent examples where creativity and innovation appear to be thriving despite minimal IP protection (or even with the complete absence of IP protection) are discussed in this book and they include street and graffiti art,[74] culinary arts,[75] jokes[76] and magic tricks,[77] although the authors of these chapters also stress with

---

[67] William Fisher, Theories of Intellectual Property, pp. 4–5, https://cyber.harvard.edu/people/tfisher/iptheory.pdf (last accessed October 13, 2017).

[68] Ibid.

[69] Ibid.

[70] See, generally, Mark A. Lemley, Property, Intellectual Property, and Free Riding, 83 Texas L. Rev. 1031 (2005).

[71] See, generally, Kal Raustiala and Christopher Sprigman, *The Knockoff Economy: How Imitation Sparks Innovation* (Oxford University Press, 2012).

[72] Kal Raustiala and Christopher Sprigman, The Piracy Paradox Revisited, 61 Stan. L. Rev. 1201, 1201–1202 (2009).

[73] Kal Raustiala and Christopher Sprigman, *The Knockoff Economy*, supra note 71, 153, 6–7. See, also, Kate Darling and Aaron Perzanowski, *Creativity without Law: Challenging the Assumptions of Intellectual Property* (NYU Press, 2017).

[74] See Chapter 4, 'Street art, graffiti and copyright', authored by Enrico Bonadio.

[75] See Chapter 6, 'Copyright in culinary presentations', authored by Cathay Smith.

[76] See Chapter 10, 'Copyright protection for modern comedic material', authored by Trevor Gates.

[77] See Chapter 11, 'Now you own it, now you don't – or do you? Copyright and related rights in magic productions and performances', authored by Jay Dougherty; as far as typefaces are concerned see Fry, supra note 25.

varying degrees of conviction that such subject matters may or should deserve copyright protection.

It is a fact that the number and diversity of typefaces have increased greatly over the years, even in countries which do not offer explicit copyright protection for typefaces.[78] There can be several inherent factors which incentivize the creation of typefaces without IP protection and we would like to highlight some of them.[79]

Firstly, the digital world has substantially reduced the entry barriers into the typeface industry.[80] Earlier, designing a particular typeface was a function of highly skilled manpower and the entry barriers into this field were really high. However, in the digital world it is relatively easy to create new typefaces. This has resulted in a substantial increase in the number of typefaces in existence, and the sheer number of typefaces available itself may bring down the appeal for piracy.[81] Secondly, the growing digital market wherein different technological products require different features for letters, continuously pushes for more creative typefaces. For example, when a newspaper disseminates their information through both print and digital media, they may have to use different typefaces for their print, desktop and smartphone users.[82] While the print versions of newspapers and books may use conventional typefaces, smartphones may require typefaces that have different legibility features. Since typefaces are not always designed for entirely aesthetic purposes and are also designed for utilitarian purposes, there will always be an impetus for more typefaces to be created to respond to the market demands.[83] Thirdly, typefaces are also subject to trends and thus often susceptible to having a shorter life-span. The industry trends *ipso facto* encourage the creation of new designs. This becomes all the more important in the advertising industry where the main goal is to stand out in a sea of words.[84] Finally, many creators of typefaces are also reported to be more interested in investing resources in creation, rather than protection or enforcement, as protection is often expensive and ineffective.[85]

---

[78] Fry, supra note 25, at 443–446.

[79] Kal Raustiala and Christopher Sprigman, supra note 71 at 145–155.

[80] Ibid. at 151.

[81] Ibid. ('Availability of cheaper . . . fonts does not eliminate piracy, but it helps to blunt its appeal. . .').

[82] Ibid. at 152.

[83] Ibid. at 152–153.

[84] Fry, supra note 25, at 479.

[85] Evans, supra note 1.

4. CONCLUSION

As is evident from the discussions in this chapter, jurisprudence pertaining to copyright protection for typefaces is still evolving. Both United States and Indian copyright law frameworks do not explicitly deal with copyright protection for typefaces. In the absence of conclusive judicial pronouncements and statutory provisions, the issue continues to be in the 'penumbra' area of law.[86] But the issue of copyright protection in this area is still quite pertinent, particularly in view of litigations like *Font Brothers, Inc. v. Hasbro, Inc.*, which we have discussed earlier. More evidence-based discussions and policy making are highly desirable in this area so as to guide courts and legislatures. As we saw in the preceding section, typefaces meet most of the characteristics associated with the negative spaces of IP, and creativity appears to be thriving in this area even in the absence of copyright protection. Under such circumstances, there is a strong argument in favour of not extending copyright protection for typefaces. As discussed earlier, it is also pertinent to note that copyright is not a natural or inalienable right.[87] It is a right granted by the state towards the enhancement of social welfare. If the act of incentivizing creativity per se can be de-hyphenated from the realm of copyright protection, then it is a policy imperative to not grant copyright protection for typefaces.

---

86 Hart, supra notes 6 and 7.

87 See *Donaldson v. Becket*, supra note 16; Copyright Act of 1957, supra note 51.

14. The press publishers' right in the European Union: an overreaching proposal and the future of news online

Stavroula Karapapa

1. INTRODUCTION

Since 2000, revenues in the press publishing sector and newspaper advertising sales have dropped[1] and they are expected to fall even further. According to the figures offered in the Commission's Impact Assessment on the Modernisation of EU Copyright Rules,[2] print circulation of daily newspapers has been constantly declining for years,[3] whereas at the same time the digital audiences for newspapers and magazines have been growing exponentially. What is more, web traffic has doubled over the past five years,[4] and it is estimated that the websites and apps of newspapers and magazines are the main services used to access news for 42% of users in the EU. In its global entertainment and media outlook for 2016–2020, Price Waterhouse Coopers (PwC) estimates that global

[1] Gareth Price, Opportunities and Challenges for Journalism in the Digital Age: Asian and European Perspectives, Chatham House – The Royal Institute of International Affairs 3 (August 25, 2015), http://www.asef.org/images/docs/Final%20Chatham%20House%20Monograph-ERT7-20150825.pdf (last accessed June 18, 2018).

[2] See Commission Staff Working Document: Impact Assessment on the Modernisation of EU Copyright Rules, accompanying the document Proposal for a Directive of the European Parliament and of the Council on Copyright in the Digital Single Market and Proposal for a Regulation of the European Parliament and of the Council laying down Rules on the Exercise of Copyright and Related Rights applicable to Certain Online Transmissions of Broadcasting Organisations and Retransmissions of Television and Radio Programmes, SWD(2016) 301 final, Vol. 1, 155 (September 14, 2016).

[3] By 17% in the period 2010–2014 in eight EU Member States.

[4] From 248.4 to 503.4 million unique users between 2011 and 2015.

newspaper revenues are expected to decline at 1.5% compounded annually, and global newspaper advertising is expected to drop at 2.9% compounded annually.[5] Similarly, Magna's global advertising forecast indicates that print media advertising sales – excluding digital revenues from publishing companies – decreased by 10% in 2016.[6] In the UK alone, as reported in *The Financial Times*, revenues from print advertising were expected to fall by between 15 and 20% in 2016, following a 15% decline in 2015.[7] To a large extent, this drop in revenues and advertising sales is due to the competition that press publishers face from a multiplicity of alternative sources, most of which are available online for free. Indeed, as the PwC outlook stresses, 'the general trend is that the more digitized a country's newspaper industry, the faster its overall revenue[s] are projected to decline'.[8]

In order to empower press publishers vis-à-vis news aggregators and online media monitoring services, a number of initiatives were introduced across Europe. In some Member States, such as Belgium, France and Italy, Google and national press publishers concluded special agreements on the use of newspaper articles in Google News;[9] whereas other Member States, notably Germany and Spain, adopted legislative initiatives with regards to the online exploitation of news content (the so-called 'Google tax').

Following a public consultation,[10] on 14 September 2016, the Commission published a Proposal for a Directive on Copyright in the

---

[5] PwC, Global Entertainment & Media Outlook 2016–2020, Newspapers & Magazines: Transitioning from a Print Past to a Digital Future – Triggering New Strategies and Wide Divergences Between Markets, PwC 1 (2016), https://www.pwc.com/gx/en/entertainment-media/pdf/newspapers-and-magazines-outlook-article.pdf (last accessed June 18, 2018).

[6] Daily newspapers –11% to $12.4 billion, magazines –9% to $8.5 billion. See Vincent Letang and Luke Stillman, Global Advertising Forecast: Winter Update, Magna 7 (December 5, 2016), https://www.magnaglobal.com/wp-content/uploads/2016/12/MAGNA-December-Global-Forecast-Update-Press-Release.pdf (last accessed June 18, 2018).

[7] David Bond, UK Newspapers Team Up to Combat Falling Revenues: Owners Explore Ways of Putting Old Rivalries Aside to Work Together, *The Financial Times* (October 23, 2016), https://www.ft.com/content/8c3e8c70-979a-11e6-a1dc-bdf38d484582?mhq5j=e1 (last accessed June 18, 2018).

[8] PwC, Global Entertainment & Media Outlook 2016–2020, supra note 5.

[9] See e.g. Rick Mitchell, France: Copyrights – Businesses Welcome Google Deal with Government, Publishers in France, 27(3) W.I.P.R. 13 (2013).

[10] European Commission, Public Consultation on the Role of Publishers in the Copyright Value Chain and on the 'Panorama Exception', Digital Single Market (March 23, 2016), https://ec.europa.eu/digital-single-market/en/news/public-con

Digital Single Market, which included – inter alia – the provision affording press publishers the exclusive rights of reproduction and making their press publications available for digital use. The provision is included in Article 11 of the proposal, under the heading 'Protection of press publications concerning digital uses':

1. Member States shall provide publishers of press publications with the rights provided for in Article 2 and Article 3(2) of Directive 2001/29/EC for the digital use of their press publications.
2. The rights referred to in paragraph 1 shall leave intact and shall in no way affect any rights provided for in Union law to authors and other rightholders, in respect of the works and other subject-matter incorporated in a press publication. Such rights may not be invoked against those authors and other rightholders and, in particular, may not deprive them of their right to exploit their works and other subject-matter independently from the press publication in which they are incorporated.
3. Articles 5 to 8 of Directive 2001/29/EC and Directive 2012/28/EU shall apply mutatis mutandis in respect of the rights referred to in paragraph 1.
4. The rights referred to in paragraph 1 shall expire 20 years after the publication of the press publication. This term shall be calculated from the first day of January of the year following the date of publication.

On 11 July 2017, the proposed press publishers' right received the supportive votes of Members of the European Parliament voting in the European Parliament's Committees on Culture and Education (CULT) and Industry, Research and Energy (ITRE); and it also received the votes of the Committee on Legal Affairs (JURI) in a narrow 13:12 vote on 20 June 2018.[11]

These votes came after the intense opposition from various parliamentary Committees and national delegations and they are not yet final, particularly in light of the Parliament's rejection of the proposed Copyright Directive

sultation-role-publishers-copyright-value-chain-and-panorama-exception (last ac cessed June 18, 2018).

[11] See European Parliament (Rapporteur: Zdzisław Krasnodębski), Opinion of the Committee on Industry, Research and Energy for the Committee on Legal Affairs on the Proposal for a Directive of the European Parliament and of the Council on Copyright in the Digital Single Market (COM(2016)0593 – C8-0383/2016 – 2016/0280(COD)), PE592.363v03-00, ITRE_AD(2017)592363 (August 1, 2017). It should be noted that although the draft Opinion of the ITRE Committee (PE 592.363v01-00, ITRE_PA(2017)592363) did not include any amendments on the press publishers' right, the right and its scope were heavily debated during the stage of suggesting amendments for inclusion in the final Opinion of the Committee (PE 592.364v01-00, ITRE_AM(2017)592364), with the position of numerous members of the Committee being to delete Article 11 and accompanying recitals altogether. The prevailing view, however, was to include a right to press publications without limiting it to digital publications only.

on 5 July 2018, by 318 votes to 278, and 31 abstentions. Just over a year after the proposal was published, the proposal faced opposition from Parliamentary Committees. For instance, the Committee on the Internal Market and Consumer Protection of the European Parliament (IMCO) suggested in its draft Opinion the deletion of Article 11, on the basis that it lacked sufficient justification and was not appropriate in achieving its objectives[12] – a position that was not reiterated in the Committee's final Opinion.[13] The CULT Committee had opined that the proposed right should be subject to fundamental revisions if it was to be accepted by the Parliament,[14] and the Committee on Legal Affairs (JURI) had also put forward substantive amendments to the proposed right.[15] Various Member States, such as Estonia and the United Kingdom, voiced reservations on the proposed right, while only France, Italy and Spain offered their unambiguous support.[16]

The proposal, which is sent back to the drawing board after the

---

[12] European Parliament (Rapporteur: Catherine Stihler), Draft Opinion of the Committee on the Internal Market and Consumer Protection for the Committee on Legal Affairs on the Proposal for a Directive of the European Parliament and of the Council on Copyright in the Digital Single Market, COM(2016)0593 – C8-0383/2016 – 2016/0280(COD), PE 599.682v01-00, IMCO_PA(2017)599682, amendment 61 (February 20, 2017).

[13] European Parliament (Rapporteur: Catherine Stihler), Opinion of the Committee on the Internal Market and Consumer Protection for the Committee on Legal Affairs on the Proposal for a Directive of the European Parliament and of the Council on Copyright in the Digital Single Market, COM(2016)0593 – C8-0383/2016 – 2016/0280(COD), PE 599.682v02-00, IMCO_AD(2017)599682 (June 14, 2017).

[14] European Parliament (Rapporteur: Marc Joulaud), Opinion of the Committee on Culture and Education for the Committee on Legal Affairs on the Proposal for a Directive of the European Parliament and of the Council on Copyright in the Digital Single Market, COM(2016)0593 – C8-0383/2016 – 2016/0280(COD), PE595.591v03-00, CULT_AD(2017)595591 (September 4, 2017); European Parliament (Rapporteur: Marc Joulaud), Draft Opinion of the Committee on Culture and Education for the Committee on Legal Affairs on the Proposal for a Directive of the European Parliament and of the Council on Copyright in the Digital Single Market, COM(2016)0593 – C8-0383/2016 – 2016/0280(COD), PE595.591v01-00, CULT_PA(2017)595591 (February 6, 2017).

[15] See European Parliament (Rapporteur: Therese Comodini Cachia/ Rapporteur for the opinion: Catherine Stihler), Draft Report of the Committee on Legal Affairs on the Proposal for a Directive of the European Parliament and of the Council on Copyright in the Digital Single Market, COM(2016)0593 – C8-0383/2016 – 2016/0280(COD), PE 601.094v01-00, JURI_PR(2017)601094 (March 10, 2017).

[16] Matthew Karnitschnig and Chris Spillane, Plan to Make Google Pay for News Hits Rocks: The Adoption of a So-called Publisher's Right would Reverberate around the World, Politico (February 15, 2017, 7:36 PM CET; Updated Feb. 17, 2017, 3:34 PM CET), http://www.politico.eu/article/plan-to-

Parliament's rejection, has also received intense criticism from scholars across Europe,[17] even before the proposed Directive was published,[18] in that it is unnecessary, undesirable, fundamentally misconceived and unlikely to achieve anything apart from adding to the complexity and cost of operating in the copyright environment.

---

make-google-pay-for-news-hits-rocks-copyright-reform-european-commission/ (last accessed June 18, 2018).

[17] See indicatively Lionel Bently et al., Response to Article 11 of the Proposal for a Directive on Copyright in the Digital Single Market, entitled 'Protection of Press Publications concerning Digital Uses' on behalf of 37 Professors and Leading Scholars of Intellectual Property, Information Law and Digital Economy, 1 (December 5, 2016), https://www.cipil.law.cam.ac.uk/sites/www.law.cam.ac.uk/files/images/www.cipil.law.cam.ac.uk/documents/ipomodernisingipprofresponse presspublishers.pdf (last accessed June 18, 2018); also see P. Bernt Hugenholtz, Say Nay to the Neighbouring Right, Kluwer Copyright Blog (April 14, 2016), http://kluwercopyrightblog.com/2016/04/14/say-nay-to-the-neighbouring-right/ (last acce ssed June 18, 2018); Copyright Reform: Open Letter from European Research Centres (February 24, 2017), http://www.create.ac.uk/wp-content/uploads/2017/02/OpenLetter_EU_Copyright_Reform_24_02_2017.pdf (last accessed June 18, 2018); Raquel Xalabarder, Press Publisher Rights in the Proposed Directive on Copyright in the Digital Single Market, CREATe Working Paper 2015/16, December 2016, 17 et seq., https://zenodo.org/record/183788/files/CREATe-Working-Paper-2016-15.pdf (last accessed June 18, 2018); Matthew Karnitschnig and Chris Spillane, Plan to Make Google Pay for News Hits Rocks, supra note 16; Christophe Geiger, Oleksandr Bulayenko and Giancarlo F. Frosio, The Introduction of a Neighbouring Right for Press Publishers at EU Level: The Unneeded (and Unwanted) Reform, 39(4) European Intellectual Property Review 202 (2017); Martin Senftleben, Maximilian Kerk, Miriam Buiten and Klaus Heine, New Rights or New Business Models? An Inquiry into the Future of Publishing in the Digital Era, 48(5) International Review of Intellectual Property and Competition Law 538 (2017); a more recent initiative includes the Statement from EU Academics on Proposed Press Publishers' Right, https://www.ivir.nl/academics-against-press-publishers-right/ (last accessed June 18, 2018).

[18] See e.g. Martin Kretschmer, Severine Dusollier, Christophe Geiger and P. Bernt Hugenholtz, The European Commission's Public Consultation on the Role of Publishers in the Copyright Value Chain: A Response by the European Copyright Society, 38(10) European Intellectual Property Review 591 (2016); Martin Senftleben, Copyright Reform, GS Media and Innovation Climate in the EU – Euphonious Chord or Dissonant Cacophony?, 5 Tijdschrift voor Auteurs-, Media- en Informatierecht 130–133 (2016); Ana Ramalho, Beyond the Cover Story – an Enquiry into the EU Competence to Introduce a Right for Publishers, 48(1) International Review of Intellectual Property and Competition Law 71 (2017); Reto M. Hilty, Kaya Köklü and Valentina Moscon, Position Statement of the Max Planck Institute for Innovation and Competition on the 'Public Consultation on the Role of Publishers in the Copyright Value Chain' (June 15, 2016), http://www.ip.mpg.de/fileadmin/ipmpg/content/aktuelles/MPI_Position_st atement_15_6_2016_def.pdf (last accessed June 18, 2018).

Indeed, the proposed press publishers' right, as incorporated in Article 11 of the proposed Directive, is not appropriate in addressing the concerns of press publishers in the light of digital technologies and, at the same time, it is not adequately limited to addressing such concerns. As I explain below, the proposed right covers too broad a subject matter; its scope and duration are excessive; it introduces an additional layer of protection without clear and sufficient evidence justifying the need for it; and it affects copyright law without addressing issues of licensing and enforcement, which are the source of concerns on the future of press publishing. It is for these reasons that it is herein argued that the proposed press publishers' right should not be included in EU law.

2. THE RATIONALE FOR A PRESS PUBLISHERS' RIGHT

The proposal for a related press publishers' right followed a public consultation on the role of publishers in the copyright chain launched by the Commission in early 2016. Its purpose was – among other things – to collect views on the desirability of granting an EU neighbouring right to publishers and on whether the need for EU intervention should be different in the press sector compared to other sectors. Indeed, the Impact Assessment on the Modernisation of EU Copyright Rules states that '[t]he shift from print to digital has enlarged the audience of press publications but made the exploitation and enforcement of the rights in publications increasingly difficult. In addition, publishers face difficulties as regards compensation for uses under exceptions'.[19]

In order to address this issue, the Proposal for a Directive on Copyright in the Digital Single Market included two relevant provisions: one provision for press publishers (Article 11) and another for publishers in general (Article 12). The former provision lays out a related right and the latter a share in the authors' compensation, so that when authors are compensated for an exception or limitation, the publishers may also be entitled to that revenue.

As Recital 31 indicates, it is the right to have access to information that justifies the publishers' need for a related right under Article 11.[20] Indeed, according to this Recital:

[19] European Commission, SWD(2016) 301 final, supra note 2, Vol. 1, 5.3.1, 155.

[20] For an analysis of the justification of the proposed right see: Mireille M.M. Van Eechoud, A Publisher's Intellectual Property Right: Implications for Freedom of Expression, Authors and Open Content Policies, Open Forum Europe

> A free and pluralist press is essential to ensure quality journalism and citizens' access to information. It provides a fundamental contribution to public debate and the proper functioning of a democratic society. In the transition from print to digital, publishers of press publications are facing problems in licensing the online use of their publications and recouping their investments. In the absence of recognition of publishers of press publications as rightholders, licensing and enforcement in the digital environment is often complex and inefficient.

The fundamental right to have access to information has been applied in other legislative instruments on copyright too, such as the Berne Convention, but with the view to justifying the exception for press summaries,[21] instead of an exclusive right. The Commission's Impact Assessment offers further explanations for the need to establish an exclusive right for publishers: that it would address the need to strengthen their bargaining position vis-à-vis online platforms, facilitate licensing and help the development of new business models.

However, even though the Commission offers statistical evidence on the extent of the so-called 'newspaper crisis',[22] the claims on the causal relationship between the introduction of a press publishers' intellectual property right and the increase in the revenues of the press leading to media diversity are neither supported not substantiated with data. For instance, the evidence presented by the Commission does not justify how the restriction of news communication will address the problem of declining advertising revenues of traditional newspapers. Neither the Impact Assessment[23] nor

---

(January 2017), http://www.openforumeurope.org/wp-content/uploads/2017/01/OFE-Academic-Paper-Implications-of-publishers-right_FINAL.pdf (last accessed June 18, 2018); Alexander Peukert, An EU Related Right for Press Publishers Concerning Digital Uses. A Legal Analysis, Goethe University Frankfurt am Main, Research Paper of the Faculty of Law No. 22/2016 (December 16, 2016), https://www.eco.de/wp-content/blogs.dir/copyright_-legal-analysis.pdf (last accessed June 18, 2018); Richard Danbury, Is an EU Publishers' Right a Good Idea? Final Report on the AHRC Project: Evaluating Potential Legal Responses to Threats to the Production of News in a Digital Era, Cambridge: Centre for Intellectual Property and Information Law (June 15, 2016), https://www.cipil.law.cam.ac.uk/sites/www.law.cam.ac.uk/files/images/www.cipil.law.cam.ac.uk/documents/copyright_and_news/danbury_publishers_right_report.pdf (last accessed June 18, 2018).

21 See e.g. Article 10(1) of the Berne Convention for the Protection of Literary and Artistic Works of September 9, 1886 as last amended on September 28, 1979.

22 See European Commission, SWD(2016) 301 final, supra note 2, Vol. 1, 155.

23 Ibid. Vol. 3, at 175–176, Annex 13.

the Commission Communication[24] explain how the introduction of an additional layer of rights would facilitate rights clearance for online uses and reduce transaction costs for all stakeholders concerned.

If the main concerns are about making licensing easier and enforcement more effective, the creation of an additional layer of rights is certainly not the most appropriate solution, despite the initial supporting votes of some Parliamentary Committees. It has been convincingly argued that an amendment of Article 5 of the Enforcement Directive,[25] indicating that press publishers are entitled to bring proceedings to enforce the copyright in materials of which they are the identified publishers, would be a more effective way of addressing difficulties in enforcement.[26] A proposal along these lines has made its way into the draft Report of the JURI Committee.[27] The amendment states:

> Member States shall provide publishers of press publications with a presumption of representation of authors of literary works contained in those publications and the legal capacity to sue in their own name when defending the rights of such authors for the digital use of their press publications.[28]

The controversial nature of the Commission's proposal is reflected in the tabled amendments, most of which suggest deleting Article 11 altogether,[29] whereas others propose allowing a press publisher right as an option for Member States.[30]

At the same time, although it is often said that news aggregation services are primarily responsible for the 'newspaper crisis', studies indicate that

---

[24] Communication from the Commission to the European Parliament, the Council, the European Economic and Social Committee and the Committee of the Regions: Promoting a Fair, Efficient and Competitive European Copyright-based Economy in the Digital Single Market, COM(2016)592 (September 14, 2016).

[25] Directive 2004/48/EC of the European Parliament and of the Council of April 29, 2004 on the Enforcement of Intellectual Property Rights, OJ L 157 (April 30, 2004).

[26] Copyright Reform, supra note 17, 3; Bently et al., supra note 17, 2.

[27] European Parliament, Draft Report, PE 601.094v01-00, JURI_PR(2017)60 1094, supra note 15.

[28] Ibid. amendment 52. According to amendment 53, the proposal specifies by way of a new Article 11(1a) that the presumption does not apply in criminal proceedings.

[29] PE604.543 and PE604.544, tabled amendments 631, 732, 733, 734, 735, 736, 743, 744, 745, 765, 766, 767, 771, 772, 773, 774, 777, 778, 779, 780 and corresponding amendments deleting recitals (April 28, 2017).

[30] PE603.010, tabled amendments 299, 334, 348 (April 28, 2017).

news aggregators are more likely to have a complementary[31] rather than a substitution effect on the number of homepage visits received by newspapers.[32] There is indeed evidence that news aggregators increase web traffic to news publishers' websites.[33] Independent studies also indicate that the proposed right will have a negative impact on publishers, authors and journalists, and consumers and citizens alike. For instance, according to a study by the European Digital Media Association (EDiMA),[34] the introduction of a publishers' right in Spain has caused publishers – particularly smaller ones – to lose as much as 14% of their web traffic, estimated to cost the Spanish news publishing industry €10 million a year. According to the same study, the proposed right will also increase the search costs for citizens as access to news from aggregators and applications will become more difficult. In Spain, for instance, the introduction of the publishers' right resulted in a loss of €1.85 billion a year for consumers, whereas in Germany, 57% of consumers find text 'snippets' helpful.

---

31 See also in this regard Eleonora Rosati, Neighbouring Rights for Publishers: Are National and (Possible) EU Initiatives Lawful?, 47(5) International Review of Intellectual Property and Competition Law 571, 572 (2017).

32 Chrysanthos Dellarocas, Juliana Sutanto, Mihai Calin & Elia Palme, Attention Allocation in Information-Rich Environments: The Case of News Aggregators, 62(9) Management Science 2543–2562 (December 10, 2015).

33 See Jason M.T. Roos, Carl F. Mela and Ron Sacher, The Effect of Links and Excerpts on Internet News Consumption, S.S.R.N. (Sept. 24, 2015), https://papers.ssrn.com/sol3/papers.cfm?abstract_id=2678938 (last accessed June 18, 2018); other sources of empirical data include Susan Athey, Mark Mobius and Jeno Pal, The Impact of News Aggregators on Internet News Consumption: The Case of Localization, Stanford Business, Working Paper No. 3353 (January 11, 2017), https://www.gsb.stanford.edu/faculty-research/working-papers/impact-news-aggregators-internet-news-consumption-case-localization (last accessed June 18, 2018); Joan Calzada and Ricard Gil, What Do News Aggregators Do? Evidence from Google News in Spain and Germany, S.S.R.N. (September 11, 2016), https://papers.ssrn.com/sol3/papers.cfm?abstract_id=2837553; The Impact of Web Traffic on Revenues of Traditional Newspaper Publishers: A Study for France, Germany, Spain, and the UK, Deloitte (March 2016), https://www2.deloitte.com/content/dam/Deloitte/uk/Documents/technology-media-telecommunications/deloitte-uk-impact-of-web-traffic-on-newspaper-revenues-2016.pdf (last accessed June 18, 2018). A contrario, a German press publisher who objected to the use of regular snippets in Google's general search service experienced a 40% reduction of traffic on its websites. See Landgericht Berlin 92 O 5/14 (February 19, 2016).

34 See Directive Copyright in the Digital Single Market: The Impact of Article 11 – Publisher Rights, EDiMA & Digital Europe, http://edima-eu.org/pdfs/latest_news/EDiMA%20DE%20Policy%20Brief%20on%20Publisher%20Rights.pdf (last accessed June 18, 2018).

3. THE (NOT SO SUCCESSFUL) NATIONAL EXAMPLES

The European Commission's proposal for a press publishers' right follows a number of ineffective national initiatives introduced with a view to recouping revenues that press publishers allegedly lose due to digitisation and the development of neighbouring distribution channels.[35]

Germany introduced a press publishers' right with effect from 1 August 2013. Sections 87f, 87g and 87h of the Urheberrechtsgesetz (German Copyright Act) offer press publishers the right to exploit their publications at a commercial level for one year, thereby preventing third parties from making excerpts from newspaper articles available without obtaining a licence.[36] Section 87f(2) of the Urheberrechtsgesetz defines press products as the technical, editorial determination of journalistic contributions that periodically appear under a title on any medium and include, in particular, articles and illustrations, which serve to convey information, opinion, or entertainment. It is only when individual words or very small text excerpts are copied that press publishers do not have a claim.[37] This is an internal limit of the ancillary right and the Arbitration Board of the German Patent and Trade Marks Office has specified that the ancillary right covers only publications that are longer than seven words, excluding the search terms.[38] The ancillary right lasts for one year only,[39] it is transferrable,[40] and it creates an entitlement to remuneration.[41] This is a unique aspect of the right, in that authors too are provided with the right to participate in remuneration, according to section 87h: German law does not only compensate press publishers' investment, but it also ensures that authors

---

35 Igor Barabash, Ancillary Copyright for Publishers: The End of Search Engines and News Aggregators in Germany?, 35(5) European Intellectual Property Review 243 (2013). See in general Rosati, supra note 31; for the proposed legislative solutions see Silvia Scalzini, Is there Free-Riding? A Comparative Analysis of the Problem of Protecting Publishing Materials in Europe, 10(6) Journal of Intellectual Property Law and Practice 454, 461–463 (2015).

36 Art. 87f(1), Urheberrechtsgesetz (German Copyright Act).

37 Ibid.

38 Deutsches Patent- und Markenamt (September 24, 2015), https://www.dpma.de/service/dasdpmainformiert/hinweise/tarifpresseverleger/index.html (last accessed June 18, 2018); also see Jakob Kucharczyk, A 'Legal' Snippet in Germany could Mean. . . Seven Words, Maximum, DISCO (October 27, 2015), http://www.project-disco.org/intellectual-property/102715-a-legal-snippet-in-germany-could-mean-seven-words-maximum/#.WY2g0xiZOCQ (last accessed June 18, 2018).

39 Article 87g(2), Urheberrechtsgesetz, supra note 36.

40 Ibid at Article 87g(1).

41 Ibid at Article 87h.

have a reasonable participation in remuneration for the use of their press products on the internet.

The German ancillary right was to a large extent rendered meaningless as – on the day that it entered into force – Google decided to make its Google News service 'opt-in' instead of 'opt-out'.[42] This means in practical terms that Google News would only feature results from those press publishers that have expressly opted in, and hence consented to, Google's indexing and showing to the public their data in its news aggregator. Therefore, clearing a licence or paying remuneration according to the new German law would not be necessary. Various German press publishers have since authorised Google to index their publications free of charge and to feature them in Google's News and Search services.[43] Smaller news aggregation services operating in Germany delisted press publishers or stopped using snippets.[44]

Following Google's refusal to negotiate with VG Media and pay a reasonable compensation for the display of snippets, VG Media filed a complaint before the Copyright Arbitration Board of the German Patent and Trade Mark Office. The Board initially decided that Google ought to pay as the ancillary right applied.[45] Google, however, categorically refused to recognise the applicability of the right and to pay the rightholders accordingly, making enforcement by the Berlin Regional Court necessary. By virtue of Directive 98/34/EC (as amended by Directive 98/48/EC), Member States ought to notify the Commission of 'technical regulations' they intend to adopt[46] so as to allow the Commission to assess the impact of such a right on the internal market.[47] Because of impending elections in Germany, however, the German Government did not notify the Commission of their intention to introduce the ancillary right. In May

---

[42] Google News Bleibt Offene Plattform für Alle Deutschen Verlage, Der offizielle Google Produkt-Blog (June 21, 2013), https://germany.googleblog.com/2013/06/google-news-bleibt-offene-plattform-fuer-verlage.html (last accessed June 18, 2018).

[43] See Andreas Becker, *German Publishers vs. Google*, Deutsche Welle (October 30, 2014), http://dw.com/p/1DeXc (last accessed June 18, 2018).

[44] See Kretschmer et al., supra note 18, 594.

[45] Deutsches Patent- und Markenamt, supra note 38.

[46] See in this regard Bo Vesterdorf, The Effect of Failure to Notify the Spanish and German Ancillary Copyright Laws, 37(5) European Intellectual Property Review 263 (2015).

[47] Antitrust questions regarding the abuse of market power by Google and the legality of its forced securing of consent are being considered in separate proceedings before the Berlin Court of Appeal and the European Commission. These issues play a subordinate role in the course of the copyright-related lawsuit.

2017, the Berlin Court decided to stay proceedings[48] and make a reference to the Court of Justice of the European Union (CJEU) for guidance on the lawfulness of the German ancillary right.[49]

In 2014, the Spanish Parliament adopted a law, which – inter alia – reformed the quotation exception available under Article 32 of the Ley de Propiedad Intelectual (Intellectual Property Law).[50] Under the new law, it is permissible to quote 'non-significant fragments of content available to the public' in cases where the content is available from periodicals or regularly updated websites, and where the content at issue has the purpose of providing information, creating public opinion, or entertainment. Even though it is not necessary to clear a licence for such a use, the latter is subject to the payment of compensation to editors and other rightholders. Unlike the German ancillary right, the Spanish law introduces a copyright limitation in the form of an entitlement to equitable remuneration that cannot be waived.

Just before the Spanish law came into force, Google announced that it would discontinue its News service in Spain.[51] Interestingly, recent empirical evidence demonstrates the positive impact of news aggregators

---

48 Landgericht Berlin, 16 O 546/15 (May 8, 2017).

49 *VG Media Gesellschaft zur Verwertung der Urheber- und Leistungsschutzrechte von Medienunternehmen mbH v. Google Inc*, Case C-299/17, OJ C 309, 21–22 (September 18, 2017). The questions are the following:

1. Does a national rule which prohibits only commercial operators of search engines and commercial service providers which edit content, but not other users, including commercial users, from making press products or parts thereof (excluding individual words and very short text excerpts) available to the public constitute, under Article 1(2) and (5) of Directive 98/34/EC (as amended by Directive 98/48/EC), a rule which is not specifically aimed at the services defined in that point; and, if that is not the case:
2. Does a national rule which prohibits only commercial operators of search engines and commercial service providers which edit content, but not other users, including commercial users, from making press products or parts thereof (excluding individual words and very short text excerpts) available to the public constitute a technical regulation within the meaning of Article 1(11) of Directive 98/34/EC (as amended by Directive 98/48/EC), namely a compulsory rule on the provision of a service?

50 For a critique of this provision see Xalabarder, supra note 17, 17 et seq.; Raquel Xalabarder, The Remunerated Statutory Limitation for News Aggregation and Search Engines Proposed by the Spanish Government – Its Compliance with International and EU Law, Infojustice (October 3, 2014), http://infojustice.org/archives/33346 (last accessed June 18, 2018).

51 An Update on Google News in Spain, Google Europe Blog (December 11, 2014), https://europe.googleblog.com/2014/12/an-update-on-google-news-in-spain.html (last accessed June 18, 2018).

on online news sites in Spain. In particular, in the first three months of 2015, the closing of news aggregation services, including Google News, has resulted in a decline of internet traffic to Spanish newspapers of over 6%, with the decline having a stronger impact on small publications.[52]

In its Impact Assessment,[53] the Commission accepts that the initiatives in Germany and Spain have proved to be ineffective, seemingly attributing their failure to their domestic scope. The Impact Assessment suggests that the proposed press publishers' right is, on the other hand, likely to be more effective and lead to greater legal certainty because of its pan-European effect.[54] However, a group of press publishers have opposed the introduction of an ancillary right at EU level claiming that, as in the German and Spanish examples, such legislation 'is a step away from a forward-looking, modern and diverse European press. It will only make it harder for [them] to grow and develop innovative models'.[55]

Moreover, unlike the German ancillary right that lasts for one year only and the Spanish provision that does not lay down a fully-fledged exclusive right, the Commission's proposal is even more radical and far-reaching than the national approaches in that it stipulates an exclusive right of 20 years' duration.

---

52 See Pedro Posada de la Concha, Alberto Gutiérrez García and Hugo Hernández Cobos, Impacto del Nuevo Artículo 32.2 de la Ley de Propiedad Intelectual, Informe para la Asociación Española de Editoriales de Publicaciones Periódicas, NERA (July 9, 2015), http://www.aeepp.com/pdf/InformeNera.pdf (last accessed June 18, 2018).

53 European Commission SWD(2016) 301 final, supra note 2, Vol. 1 at 159–160.

54 Ibid at 166–167; contra: Maria Lillà Montagnani, The EU Consultation on Ancillary Rights for Publishers and the Panorama Exception: Modernising Copyright Through a 'One Step Forward and Two Steps Back' Approach, Kluwer Copyright Blog (September 20, 2016), http://kluwercopyrightblog.com/2016/09/20/the-eu-consultation-on-ancillary-rights-for-publishers-and-the-panorama-exception-modernising-copyright-through-a-one-step-forward-and-two-steps-back-approach/ (last accessed June 18, 2018).

55 Arsenio Escolar et al., Ancillary Copyright: Group of Press Publishers Write Letter to the European Commission, International Federation of Reproduction Rights Organisation (December 4, 2015), http://ifrro.org/content/ancillary-copyright-group-press-publishers-write-letter-european-commission (last accessed June 18, 2018).

4. THE EXCESSIVE SCOPE OF THE PROPOSED RIGHT

Article 11 outlines a related right of a 20-year term in the form of a fully-fledged exclusive right to reproduce and make available to the public 'for digital use' any publisher's 'press publications', i.e. a right that is derived from, or subordinate to, copyright. The proposed right is meant to be compulsory for Member States to implement and it consists of more than mere compensation to the publishers.[56]

Even though the proposed right purports to increase the returns of the investments made by press publishers, and to simplify enforcement and licensing, the relevant provision raises both doctrinal and normative concerns. From a doctrinal point of view, the definition of press publication is ill-suited to its purpose and creates uncertainty due to its extremely broad scope. Under the definition included in Article 2(4), facts or information are also covered. This is against established copyright principles, according to which protection does not extend to mere data.[57] From a normative perspective too, the very creation of an exclusive right to press publications could lead to an unacceptable monopolisation of information through the expansion of copyright protection. As explained below, the proposed right could cover snippets and headlines embedding links that may not be covered by copyright protection.

The related right has hence not been endorsed as such by the various committees of the European Parliament, and it has been rejected on 5 July 2018 by the Parliament's plenary, leaving the Parliament's position up for debate, amendment, and a vote during the plenary session of September 2018. In

[56] Other publishers, according to Article 12, will have a claim to share any compensation that the author receives for any exceptions and limitations that are available under national laws. Such exceptions and limitations cover reprography, private copying, and possibly the rental remuneration right. Article 12 of the Proposal reads: 'Member States may provide that where an author has transferred or licensed a right to a publisher, such a transfer or a licence constitutes a sufficient legal basis for the publisher to claim a share of the compensation for the uses of the work made under an exception or limitation to the transferred or licensed right'. Unlike Article 11, Article 12 is not compulsory for Member States to implement ('Member States may provide . . .'). See, however, the relevant revision included in the Opinion of the European Parliament, where it is suggested that the provision of Article 12 is made mandatory for Member States to implement. See European Parliament, Opinion of the Committee on Culture and Education, PE595.591v03-00, CULT_AD(2017)595591, supra note 14, amendment 70.

[57] See e.g. Article 9(2) of the Agreement on Trade-Related Aspects of Intellectual Property Rights 1994 (TRIPS Agreement).

the initial discussions on the wording of the proposed right, it was only the ITRE Committee that took a protectionist approach in its Opinion, expressing support for the view that print editions are worth as much protection as digital editions, and therefore proposing the expansion of the right to all press publications.[58] Yet an interesting aspect of the ITRE proposal is that it recognises that there should be limits to the right by expressly excluding its application to acts of hyperlinking 'as they do not constitute communication to the public'.[59] The IMCO Committee, on the other hand, took a more radical approach and suggested the deletion of Article 11.[60] This was recommended due to the lack of sufficient justification in support of the right, with the Committee indicating that a revision of the Enforcement Directive would be more appropriate. It is worth noting that there was intense debate in both Committees over these and other amendments, including on whether the right should be deleted.[61] These discussions may have strengthened the position of press publishers in recognising an argument for enlarging the scope of protected uses to cover both digital and print uses.[62]

The CULT Committee set forth two main revisions that have to do with the scope of the proposed right – that should cover only press

---

58 European Parliament, Opinion of the Committee on Industry, Research and Energy, PE592.363v03-00, ITRE_AD(2017)592363, supra note 11, notably amendments 43 and 44.

59 Ibid. amendment 45.

60 European Parliament, Draft Opinion of the Committee on the Internal Market and Consumer Protection, PE 599.682v01-00, IMCO_PA(2017)599682, supra note 12, amendment 61.

61 See European Parliament, Amendments 12–259 of the Committee on Industry, Research and Energy for the Committee on Legal Affairs on the Proposal for a Directive of the European Parliament and of the Council on Copyright in the Digital Single Market (COM(2016)0593 – C8-0383/2016 – 2016/0280(COD)), PE 592.364v01-00, ITRE_AM(2017)592364 (April 5, 2017); see European Parliament, Amendments 63–303 of the Committee on the Internal Market and Consumer Protection for the Committee on Legal Affairs on the Proposal for a Directive of the European Parliament and of the Council on Copyright in the Digital Single Market, COM(2016)0593 – C8-0383/2016 – 2016/0280(COD), PE 602.819v01-00, IMCO_AM(2017)602819 (April 5, 2017); European Parliament, Amendments 304–576 of the Committee on the Internal Market and Consumer Protection for the Committee on Legal Affairs on the Proposal for a Directive of the European Parliament and of the Council on Copyright in the Digital Single Market, COM(2016)0593 – C8-0383/2016 – 2016/0280(COD), PE 602.820v01-00, IMCO_AM(2017)602820 (April 5, 2017).

62 European Parliament, Opinion of the Committee on the Internal Market and Consumer Protection, PE 599.682v02-00, IMCO_AD(2017)599682, supra note 13, amendment 38.

publications of a professional nature – and its duration.[63] In particular, the Parliamentary Committee clarified that the relevant provision should include only business-to-business relationships and not private and non-commercial uses.[64] What is more, it also suggested reducing the duration of the right to eight years,[65] even though the initial proposal in its draft Opinion was for three years.[66] The reason is that the average lifespan of press publications is considerably shorter than other types of copyright protected works, and a shorter duration would reflect the need to strike a balance with citizens' right to access information. An analogy can here be drawn to unregistered Community designs. Unlike registered Community designs, which can be protected for a maximum of 25 years, unregistered Community designs are protected for a period of three years from the date on which the design was first made available to the public within the territory of the EU. As acknowledged in the preamble of Regulation 6/2002: '[s]ome . . . sectors produce large numbers of designs for products frequently having a short market life where protection without the burden of registration formalities is an advantage and the duration of protection is of lesser significance'.[67] These include fast-moving industries, such as the fashion industry. The same rationale could apply to press publications and news items.

An interesting approach was the one proposed by the Committee of Legal Affairs (JURI) in its draft report. Stressing the need for proportionality, the draft report envisages not a new right, but instead a presumption of ownership.[68] This alternative option is less intrusive and more proportionate in dealing with the problems that press publishers face in the transition from print to digital. It aspires to simplify licensing mechanisms through the presumption that a press publisher has the relevant rights in the copyright protected content without having to demonstrate contractual transfers for each and every newspaper article.

The aforementioned proposals aim to offer more balanced approaches to what seems to be a broadly conceived right in terms of its scope and

---

[63] European Parliament, Opinion of the Committee on Culture and Education, PE595.591v03-00, CULT_AD(2017)595591, supra note 14.

[64] Ibid. amendments 26 and 73.

[65] Ibid. amendment 78.

[66] European Parliament, Draft Opinion of the Committee on Culture and Education, PE595.591v01-00, CULT_PA(2017)595591, supra note 14, amendment 69.

[67] Council Regulation (EC) No 6/2002 of December 12, 2001 on Community designs, Recital 16. Also see Recital 25.

[68] European Parliament Draft Report, PE 601.094v01-00, JURI_PR(2017)601094, supra note 15, amendment 52.

protectable subject matter. In the light of the Parliament's rejection of the proposed Copyright Directive on 5 July 2018, it is likely that the broad definition offered to press publications in Article 11 will be subject to debate and amendment.

4.1 The Definition of Press Publication

The broad scope of the Commission's proposal is stressed by the very definition of protectable subject matter. According to Article 2(4) of the proposal, protectable subject matter includes 'a fixation of a collection of literary works of a journalistic nature'. This phrase seems to have a rather broad meaning, as demonstrated by the decision of the CJEU in Case C-73/07 *Satamedia*.[69] In this case, the Court of Justice found that:

> activities . . . may be classified as 'journalistic activities' if their object is the disclosure to the public of information, opinions or ideas, irrespective of the medium which is used to transmit them. They are not limited to media undertakings and may be carried out for profit-making purposes.[70]

The reference to 'journalistic nature' appears to be an inclusive phrase that covers literary works, except those that are meant for literary enjoyment only, such as literature or poetry.[71] Some commentators indicate that, according to this phrase, it should be the works as such that have a journalistic nature, and not – for instance – the publisher or the medium of publication. This means that the publication of collections of certain kinds of subject matter, such as fixture lists, should be excluded from the concept of press publication.[72]

The definition of press publications also includes a second category of protected subject matter, covering 'other works or subject matter and constitut[ing] an individual item within a periodical or regularly-updated publication under a single title'. This appears to expand the subject matter beyond 'literary works of a journalistic nature', for instance, to works such as photographs or drawings. Recital 33 states that 'periodical publications which are published for scientific or academic purposes, such as scientific journals, should not be covered by the protection granted to press publications',[73] yet this exclusion does not seem to align with the

69 *Tietosuojavaltuutettu v. Satakunnan Markkinapörssi Oy, Satamedia Oy*, Case C-73/07, [2008] ECLI:EU:C:2008:727 (*Satamedia*).

70 Ibid. [64].

71 See Bently et al., supra note 17, 9–10.

72 Ibid. at 9.

73 European Commission SWD(2016) 301 final, supra note 2, Vol. 1 at 158.

broad definition offered in the Article that affords protection to press publications 'having the purpose of providing information related to news *or other topics*'.[74] Such other topics could include scientific news or relevant information. This is broader than what the Commission seems to envisage.

The CULT Committee offered a revised definition of press publication under Article 2(4):

> 'press publication' means a professional fixation under a single title of a collection of literary works of a journalistic nature produced by one or several authors, which may also comprise other works or subject-matter and constitutes an individual item where:
> (a) it occurs within a periodical or regularly-updated publication under a single title, such as a newspaper or a general or special interest magazine;
> (b) its purpose is to provide information related to news or other topics; and
> (c) it is published in any media under the editorial responsibility and control of a service provider.[75]

The most interesting addition of the CULT Committee's proposed revision – which does not feature in the JURI report – is the introduction of the element of the professional nature of a protected press publication. Indeed, the CULT Committee's main concern is that the press publishers' right should be limited to business-to-business relationships and that private and non-commercial uses should not be covered by the new right. Another aspect that is worth mentioning is that the three satellite conditions incorporated in the definition of press publication are listed in a cumulative way, thereby further limiting its scope.

4.2 Snippets, Linking, and Other Digital Uses

According to Article 11 of the proposed Directive, press publishers are given the exclusive right of reproduction and the right of making available to the public. Both rights are granted for the digital use of press publications. As Recital 34 to the proposed Directive indicates, the rights granted to the publishers of press publications should have the same scope as the rights of reproduction and making available to the public provided for in

---

[74] Emphasis added. See in this regard Bently et al., supra note 17, 12.

[75] European Parliament, Opinion of the Committee on Culture and Education, PE595.591v03-00, CULT_AD(2017)595591, supra note 14, amendment 43; the Committee retained the very similar definition it had suggested in its draft Opinion (PE595.591v01-00, CULT_PA(2017)595591, supra note 14, amendment 40.

the Information Society Directive (2001/29/EC), in so far as digital uses are concerned.

It is not clear, however, what kinds of uses the proposed right will cover. Recital 33 indicates that protection does not extend to acts of hyperlinking that do not constitute communication to the public. In *Svensson*[76] and other cases of the CJEU,[77] hyperlinking was found to not amount to an act of exploitation when the communication of the work was not addressed to a new public.[78] This was notably the case of links to materials that were freely available to the public at large and not subject to access restrictions. This position was revisited in 2016, in *GS Media*,[79] where the CJEU held that, where the material has not been made available with the consent of the rightholders, links to such content would amount to infringement if the person posting the links had knowledge of the status of the material. In paragraph 51 of the judgment, the CJEU held that:

> . . . when the posting of hyperlinks is carried out for profit, it can be expected that the person who posted such a link carries out the necessary checks to ensure that the work concerned is not illegally published on the website to which those hyperlinks lead, so that it must be presumed that that posting has occurred with the full knowledge of the protected nature of that work and the possible lack of consent to publication on the internet by the copyright holder.

Knowledge is presumed for the persons posting the links, as commercial operators are expected to investigate the status of the materials to which they link. This could be read to mean that, when for-profit posting of materials takes place, the persons placing the links should be diligent in ensuring lawful and consensual publication, possibly by consulting press publishers, otherwise their activities will amount to infringement. Those who are not placing links for profit are not liable however, absent knowledge on the status of the material.[80]

If linking that does not constitute communication to the public is not to be included within the scope of the press publishers' right, there are two

---

[76] *Nils Svensson and Others v. Retriever Sverige AB*, Case C-466/12, [2014] ECLI:EU:C:2014:76 (*Svensson*).

[77] *BestWater International GmbH v. Michael Mebes, Stefan Potsch*, Case C-348/13, [2014] ECLI:EU:C:2014:2315 (order of the Court) (*BestWater*).

[78] *Svensson*; *BestWater*; also see Stavroula Karapapa, The Requirement for a 'New Public' in EU Copyright Law, 1 European Law Review 63 (2017).

[79] *GS Media BV v. Sanoma Media Netherlands BV, Playboy Enterprises International Inc., Britt Geertruida Dekker*, Case C-160/15, [2016] ECLI:EU:C:2016:644 (*GS Media*).

[80] Ibid. [47].

other instances that could be covered by the right. These are snippets and digital uses other than mere linking, such as showing headlines enabling links. According to EU data,[81] most online users read only the headlines and a featured snippet instead of clicking on the link that would take them to publishers' websites. Those readers would not increase the press publishers' web traffic. Under the new right, snippets and active-link headlines would only be exempt from infringement when covered by the exemption available for temporary copies under Article 5(1) of the Information Society Directive (2001/29/EC). The exemption covers acts of temporary reproduction that are an integral and essential part of a technological process aimed at availing of a lawful use of the work, providing that the act has no independent economic significance.

In *Infopaq I*,[82] Article 5(1) was found to apply to some, but not all, phases of the process of extracting snippets from newspaper articles, namely those phases where the deletion of the temporary copy is automatic and not dependent upon discretionary human intervention. In *Infopaq II*,[83] however, the CJEU decided that acts of temporary copying may be exempt even if they involve human intervention, provided that they are an integral and essential part of a technological process. To the extent that the temporary copying exemption applies, search engines and news aggregators would not be affected by the proposed related right for press publishers. It would only be a matter of how the meaning of 'no independent economic significance' will be construed in this context. Snippets are likely to benefit from Article 5(1) when they are merely part of a technological process of searching.[84] When, however, search engines and news aggregators make editorial choices by reference to the news items, the exception will not apply.

---

[81] See the European Commission, Flash Eurobarometer 437 (fieldwork March 2016, published September 2016), http://ec.europa.eu/COMMFrontOffice/publicopinion/index.cfm/Survey/getSurveyDetail/instruments/FLASH/surveyKy/2123 (last accessed June 18, 2018).

[82] *Infopaq International A/S v. Danske Dagblades Forening*, Case C-5/08, [2010] F.S.R. 495 (Infopaq I).

[83] *Infopaq International A/S v. Danske Dagblades Forening*, Case C-302/10, [2012] ECLI:EU:C:2012:16 (*Infopaq II*).

[84] Snippets could be covered by Article 10(1) of the Berne Convention that sets an exception for quotations made for press summaries. According to this provision: 'It shall be permissible to make quotations from a work which has already been lawfully made available to the public, provided that their making is compatible with fair practice, and their extent does not exceed that justified by the purpose, including quotations from newspaper articles and periodicals in the form of press summaries'.

Article 11(3) and Recital 34 also state that both rights should be subject to the same provisions on exceptions and limitations as those applicable to the rights provided for in the Information Society Directive (2001/29/EC), including the exception on quotation for purposes such as criticism or review laid down in Article 5(3)(d) of that Directive.[85] Digital uses, such as scanning and indexing, which are at the core of the activity of many news monitoring services or libraries, and are not fully covered by any exception or limitation, will be included within the scope of the proposed press publishers' right.

4.3 Towards the Monopolisation of Information?

As we have seen above, press publications mainly involve news items. News, however, has remained for a long time outside the scope of copyright laws. For instance, Article 2(8) of the Berne Convention clearly states that '[t]he protection of this Convention shall not apply to news of the day or to miscellaneous facts having the character of mere items of press information'. If snippets, headlines incorporating links, and other digital uses, such as non-permitted text mining, are to be covered by the proposed press publishers' right, it is very likely that the protection thereby afforded would be more akin to protecting information than offering protection to original subject matter as such.

This raises the question as to how far copyright law should go. Should each and every digital use of the work be restricted by copyright? The Proposal for a Directive seems to imply that this is to be answered in the affirmative. It is not a surprise that the European Parliament's suggested revisions indicate that the proposed press publishers' right ought to be more clearly limited in terms of scope and duration or, alternatively, removed altogether from the proposed Directive. Copyright has never been about the protection of information and the proposed Directive tends to confuse copyright protected subject matter and mere information by seemingly extending protection to snippets, headlines and some forms of text mining.[86]

---

[85] See however the decision of the Court of Justice in *Hewlett-Packard Belgium SPRL v. Reprobel SCRL*, Case C-572/13, [2015] ECLI:EU:C:2015:750 (*Reprobel*) at [47], where the Court found that not only are publishers not among the reproduction rightholders listed in Article 2 of the Information Society Directive (2001/29/EC), but they are also not entitled to any fair compensation for the harm that copyright exceptions may cause.

[86] What is more, in light of the generally established principle that copyright arises as a matter of originality rather than length, the decision of the German

5. WHY THE PROPOSED PRESS PUBLISHERS' RIGHT SHOULD BE DELETED OR REPLACED WITH A MEANINGFUL ALTERNATIVE

Since the press publishers' right was announced, it has been subject to intense criticism.[87] In its draft Opinion, the IMCO Committee suggested the deletion of Article 11 on the basis that it arguably lacked sufficient justification and was not appropriate in achieving its objectives.[88] The proposed right has also been the focus of legal scholarship. For instance, a group of 37 intellectual property professors has stated that 'the proposed right is unnecessary, undesirable, would introduce an unacceptable level of uncertainty and be unlikely to achieve anything apart from adding to the complexity and cost of operating in the copyright environment'.[89]

A main criticism of the proposed press publishers' right has to do with the lack of clear evidence justifying the creation of such a right. There is no evidence to support the claim that search engines and news aggregators are behind the loss of press publishers' revenue. As explained earlier, the data that has been included in the Impact Assessment do not convincingly demonstrate how the proposed press publishers' right would have an impact on the declining revenue of press publishers.[90] Neither the Impact Assessment[91] nor the Commission Communication[92] explain how the introduction of an additional layer of rights would facilitate rights clearance for online uses and reduce transaction costs for all stakeholders concerned.

Part of the argument is that press publishers have been derivative owners of the authors' exclusive rights for many years. If the problems facing press publishers are related to licensing and enforcement as indicated in the Impact Assessment, a more direct solution to the problem would be more efficient than the introduction of a new right. The JURI draft report and certain scholars acutely note that the aim of simplifying enforcement could have been achieved by amending Article 5 of the Enforcement Directive

---

Patent and Trade Marks Office that the German ancillary right covers only publications that are longer than seven words seems to be equally confusing. Deutsches Patent- und Markenamt, supra note 38.

[87] See indicatively Bently et al., supra note 17, 1; Xalabarder, supra note 17.

[88] European Parliament, Draft Opinion of the Committee on the Internal Market and Consumer Protection, PE 599.682v01-00, IMCO_PA(2017)599682, supra note 12, amendment 61.

[89] See Bently et al., supra note 17, 1.

[90] Van Eechoud, supra note 20, at 11.

[91] SWD(2016) 301 final, supra note 2, Vol. 3 at 175–176, Annex 13.

[92] COM(2016)592, supra note 24.

with a view to creating a presumption that press publishers are entitled to enforce copyright in any item that they publish. A defendant would have to rebut such a presumption by demonstrating that the material was in the public domain or licensed by its author.[93] What is more, publishers can rely on the database right[94] to prevent extraction and re-utilisation of their protected content, to the extent that their online news site falls within the definition of database.

Another aspect of the right that is worth further attention is that it is not likely to meet its primary objective, namely to empower press publishers, especially vis-à-vis its overlap with authorial entitlements. Article 11(2) of the proposal states that author's rights incorporated in the press publication 'may not be invoked against those authors and other rightholders and, in particular, may not deprive them of their right to exploit their works and other subject matter independently from the press publication in which they are incorporated'. Something similar is indicated in Recital 35.[95] However, press publishers already acquire the authors' copyright through employment contracts or contracts with freelance journalists and, in this light, the proposed right establishes two layers of rights for the same creation.[96] In this regard, the proposed right will simply duplicate existing entitlements, without making a meaningful addition to the publisher's portfolio of intellectual property protection.

6. CONCLUSION

The proposed related right of press publishers aims to offer an additional layer of protection to press publications with a view to ensuring sustainability of the press and simplifying the process of concluding licences and

---

93 See European Parliament, Draft Report, PE 601.094v01-00, JURI_PR(2017)601094, supra note 15, amendment 52; Lionel Bently et al., Strengthening the Position of Press Publishers and Authors and Performers in the Copyright Directive, Study for the JURI Committee, Policy Department for Citizens' Rights and Constitutional Affairs Directorate General for Internal Policies of the Union, PE 596.810 (September 2017); Bently et al., supra note 17, at 22.

94 See Directive 96/9/EC of the European Parliament and of the Council of 11 March 1996 on the legal protection of databases, OJ L 77 (March 27, 1996).

95 This Recital indicates that 'publishers of press publications should not be able to invoke the protection granted to them against authors and other rightholders. This is without prejudice to contractual arrangements concluded between the publishers of press publications, on the one side, and authors and other rightholders, on the other side'.

96 Copyright Reform, supra note 17.

enforcing rights. Although digitisation has been marked by a decline of revenues in the press publishing sector and by the drop of newspaper advertising sales in Europe, the proposed right is neither an adequate nor a proportionate measure in dealing with the so-called 'newspaper crisis'. There is indeed no hard evidence that the right can achieve its stated objectives, and its design is questionable both in terms of scope and duration, with the likely effect that the protection thereby afforded would be more akin to the protection of information instead of original subject matter.

It is true that press publishers depend on the enforcement of their derivative rights in order to protect the investment in their publications and to assert the rights they hold by law or by means of assignment, licence or any other contractual arrangement. In this light, measures to strengthen their enforcement position are necessary. Such measures, however, should not disrupt other industries and they should be proportionate with respect to the achievement of the stated objectives, including the need for an open internet and a free, pluralist press. At a time when some news aggregators have entered into voluntary licensing agreements with major publishers, an adequate policy solution would require the simplification of licensing mechanisms at the pan-European level.

Even though there may not be sufficient evidence to support the introduction of a press publisher right, there is enough to support a solution that would empower press publishers by protecting their legal entitlement, whilst safeguarding the plurality of news and opinions on the internet. Licensing could be simplified through a presumption of representation of authors, according to which press publishers are entitled to enforce copyright in the items they publish. European policies should adapt the regulatory framework to the emergent transition from print to digital affecting the news sector. It is perhaps in this way that the parameters for quality journalism and citizens' access to information may not only be set but also reinforced as essential values in the proper functioning of a democratic society.

15. Law and odor: elusive copyright and other IP protection for fragrances

Charles Cronin*

1. INTRODUCTION

Like the media and entertainment industries whose principal marketable 'product' is information, the fragrance industry has become unusually vulnerable to new technologies that enable the appropriation of once-proprietary knowledge. This information, particularly fragrance formulas and manufacturing processes, is among the perfume industry's most valuable assets, and the most costly to generate. Accordingly, in recent years, like the media and entertainment sector, the fragrance industry has vigorously pursued various forms of legal protection for its information assets.

Our discussion briefly reviews why, for centuries, the fragrance industry was not concerned about the appropriation of its intellectual property. Then we consider how changes in the fragrance industry, and associated technologies, have prompted its recent drive to seek protection through trade secret, trademark, patent, and copyright.

We focus particularly on the tenability of recent litigation in France and the Netherlands that has turned on the question of whether fragrances may qualify as copyrightable works. Courts entertaining these claims have arrived at strikingly contradictory opinions on this question. These contrary opinions, however, lead one to the ultimate question of whether the relatively limited capacity of human olfaction, and specifically humans' exiguous capacity to perceive the originality of scents, should determine whether copyright should be afforded to works of fragrance created through human ingenuity and creativity.

The discussion concludes by considering whether existing intellectual property law provides adequate protection for fragrances. And, if not,

* My chapter title is a shameless, albeit authorized, version of that of my colleague Claire Guillemin's recently published doctoral dissertation, *Law & Odeur: Fragrance Protection in the Fields of Perfumery and Cosmetics* (Nomos, 2016).

whether the fragrance industry's ongoing viability may depend upon enactment of sui generis intellectual property protections along the lines of those provided to other industries that have claimed danger of atrophy without such exceptional treatment.

2. INTELLECTUAL PROPERTY PROTECTION FOR FRAGRANCES

2.1 Historical Perspective

Only over the last fifty years or so has the fragrance industry become significantly concerned about intellectual property protection. For centuries, this industry was comprised of hundreds of small family firms. These businesses were not overly concerned about competitors appropriating their formulas or manufacturing techniques because these were kept under lock and key, and typically shared among a small cadre of related individuals. This economically valuable information was recognized and kept secret among a limited number of people who shared the same economic objective.

Moreover, the value of the raw materials used in their fragrances tended to be higher than the associated research, development and manufacturing costs. Through the early 20th century, obtaining essences of jasmine and tuberose, for instance, required extravagant investment of painstaking manual labor. Accordingly, even if a competitor acquired the formula of another firm's successful fragrance, the advantage of that acquisition would be tempered by the fact it would not significantly decrease the cost of producing a competing product.

The production of certain essences used by the fragrance industry is still labor intensive and costly – for example the jasmine essence produced in France – but much of the financial investment now is directed towards the research and development that is involved in creating and marketing new fragrances. This is true in part because today successful new fragrances must not only appeal to consumers, but also comply with increasingly rigorous safety and disclosure requirements promulgated by the International Fragrance Association (IFRA) and government agencies.

Transparency and safety regulations challenge fragrance houses' ability to maintain the confidentiality of the formulas and processes used in their proprietary blends. More challenging still has been the development of gas-chromatography-mass spectrometry (GC-MS) technology. GC-MS, developed in the 1960s, enables one to analyze with increasing accuracy, the chemical composition of a substance by segregating by weight the molecules that comprise it. This technology not only identifies the chemicals, but also their relative quantity, within a substance. In other words, it is now possible to obtain highly accurate information about the most valuable component of a fragrance formula – a list of the various chemicals comprising it. Even worse, from the perspective of those who do not want such information revealed, using GC-MS technology to reverse engineer a fragrance is not prohibited by intellectual property law.[1]

Like the media and entertainment sectors that have been bedeviled by digital technologies in recent decades the fragrance industry, facing a similar technologically produced quandary, has cast about seeking new means of legal protection for its IP assets. These include patent, trademark/trade dress, and copyright.

2.2 Patent Protection

A handful of multinational fragrance and flavor companies develop and manufacture most of the fragrances now sold under the brands and names of luxury products and couture houses, fashion designers, and celebrities. These fragrances are deployed in various purely discretionary products like candles, potpourris and, of course, liquid perfumes typically packaged in distinctive flasks.

While the industry emphasizes to the public its fundamental role in the creation of these glamorous products, the perfumes used in them constitute on average only twenty percent of the fragrances produced by these

---

[1] Between 2012 and 2016 a number of large European and American corporations in various industrial sectors such as heavy industry and computing hardware lobbied the Government of the European Union for enactment of Directive (EU) 2016/943 on the protection of undisclosed know-how and business information (trade secrets) against their unlawful acquisition, use and disclosure ('Trade Secret Directive'). They pooled these efforts through the Trade Secrets and Innovation Coalition (TSIC), established specifically for this purpose. The TSIC membership comprised a dozen or so industrial giants like Michelin, Du Pont, and Intel, but also, curiously, the International Fragrance Association (IFRA), a trade organization in Brussels that promotes the industry worldwide, and establishes safety standards with which members must comply. The principal aim of the Trade Secret Directive is to standardize by 2018, the national laws of EU member states by harmonizing their definitions of trade secrets, and what actions constitute misappropriation of them. IFRA's participation in the promotion of the Directive did not result in legislation that would curb the industry's loss of intellectual property through legitimate reverse engineering, but it did provide an opportunity for the fragrance industry to publicize its particular vulnerability to the loss of such information.

companies.[2] Most of the fragrances that this industry produces are used in lack-luster goods like laundry detergents and cleaning agents. Many products are used to camouflage offensive smells that would otherwise exist in certain goods, or even to suppress the perceptibility of odors to render 'scent-free' products.

Despite the arguably useful attributes of many of its new products, the fragrance industry has typically not relied upon patents to protect them. Some of the most valuable fragrances it produces are also the most persistent, with market lives much longer than the 20-year term patents provide. In fact, profitable fine fragrances like Chanel's 'No. 5' and Hermès's 'Calèche' continue to be marketable many decades after their introduction. Their appeal depends in part on their use exclusively in costly and entirely discretionary products. Classification as, or public association with, useful, patentable products would compromise this appeal.

Some manufacturers have successfully obtained utility patents for fragrances based upon claims of their therapeutic efficacy.[3] However, the fragrance industry relies upon patents mainly to protect newly created scent molecules known as 'captive odorants'. These new molecules enable the creation of fragrances whose uniqueness is based on the presence of the new molecules. Accordingly, the owner of the patent for the molecule may hold the odorant 'captive' and use it exclusively in fragrances it manufactures. The patent holder of a captive odorant might initially limit use of it to its own products. Once its potential has been demonstrated, however, the owner typically licenses its use to others before it is 'released' into the public domain at the end of its 20-year term of patent protection.

Patents effectively protect captives and innovative and useful products and processes, but neither patents, nor any longer trade secrets, protect the industry's main output: thousands of original fragrances utilized in a vast number of end products. Might copyright provide legal protection, hitherto untapped, for these creative works?

2.3 Copyright

2.3.1 Improvising at the orgue à parfums

Perfumers creating new fragrances work seated before a 'perfume organ' (Figure 15.1). This 'instrument' has a striking resemblance to the consoles

---

[2] See, Huggard Consulting Group, *The Socio-Economic Contributions of Fragrances* (2011).

[3] See, e.g., U.S. Patent 7,169,746, granted January 13, 2007 to Shiseido Co., Ltd. covers a fragrance used in aromatherapy; U.S. Patent 6,089,953, granted July 18, 2000 to Chin-Tang Chen of Taipei, Taiwan, protects a fragrance-imbued brassiere.

Figure 15.1 Orgue à parfums

of sprawling 'romantic' or 'symphonic' pipe organs that comprise several keyboards and dozens of stops.[4] Like an improvising musician who pulls stops to create various combinations of sound, the perfumer draws dabs of scented essences from the semicircular array of glass vials before him, which he combines to compose an original fragrance.

Like a musician, the perfumer works with a gamut of 'notes' identified as base, middle or top.[5] By combining these notes the perfumer builds harmonious 'chords'. As with a musical work, the longer sustained base notes – often woody or musky scents – are the foundation for the more fleetingly perceived middle and top herbaceous and floral notes. The intensity of a particular component of a fragrance is referred to as its 'volume', and the perfumer's improvisations are tested using paper strips known as 'touches' or piano keys. Not surprisingly, some of the final results of these combinations of scents have been named with musical terms like *Arpège* and *Sonata*.

---

4 'Stops' are the small tablets or knobs on an organ console by which the player engages particular families or 'ranks' of pipes (flutes, reeds, etc.).

5 Of course musicians work with 'bass' – not 'base' – notes.

2.3.2 Fragrance as copyrightable expression

Musicians typically noodle at a piano or other instrument, experimenting with combinations of pitches, rhythms and harmonies to create original works. To the extent these compositions, once recorded as sound or as visual symbols, contain expression originating from the musician himself, they are protected by copyright.[6]

It seems reasonable, therefore, given the enormous expansion over several centuries of the scope of expression protected by copyright, that man-made fragrances should be eligible for such protection. Highly trained perfumers create new fragrances, seeking to evoke an olfactory evocation of a mood or environment described in a verbal brief. This is true of all new fragrances, whether used in high-end perfumes or laundry detergents. In this respect the work of perfumers is akin to that of musicians who write scores to accompany movies or advertisements. The commercial nature of the end products in which the original fragrances or scores are deployed has no bearing on their copyrightability.

Unlike chemical engineers who combine substances to devise, for example, purely useful pharmaceuticals, perfumers do so to create primarily aesthetic works. Moreover, perfumers believe that their compounds contain olfactory indications of their authorship, like signatures on paintings.

The technique of creating original fragrances through experimentation, to manifest an intellectual conception of a scent, has not changed much since the establishment of the modern perfume industry in France in the 18th and 19th centuries. Only late in the 20th century, however, have perfumers raised the question of whether original fragrances should qualify as copyrightable works of authorship.[7]

Copyright now extends to original works disparate as stuffed animals and action characters, which have little in common with the literary works to which its protection was originally limited. Copyright is a particularly attractive form of protection for industries characterized by rapid turnover of products embodying original expression. It is immediately available

---

[6] U.S. copyright law limits protection to works fixed in a tangible format, whereas civil law countries like France also offer copyright protection to ephemeral works like improvised music and dance that has not been recorded.

[7] In fact, the existing classifications under which works may be registered in the United States would not accommodate an application for registration of a work of fragrance. The U.S. Copyright Office accepts registrations for works of: literature, visual arts, performing arts, sound recordings, and single serials. See, eCO Frequently Asked Questions, U.S. Copyright Office, http://copyright.gov/eco/faq.html (last visited October 6, 2017).

once the expression has been fixed, has a low threshold of eligibility, costs nothing to obtain, and lasts about a century or longer.

There seems little question that perfumers qualify as 'authors' given the ever-increasing catholic understanding of this term in connection with copyrightable expression. Moreover, with the surge in the development of bespoke and celebrity fragrances, perfumers increasingly consider their *métier* akin to that of writers and artists. Perfume purveyor Frédéric Malle, for instance, maintains a stable of perfumers whose fragrances are sold under the names of their creators as 'editions', and Kilian Hennessy, scion of the cognac family, prominently labels all of his fragrances with 'By Kilian'.[8]

2.3.3 Odor in the courts! Copyright disputes about fragrances

France is no longer the center of the fragrance industry, yet it continues to play a significant role, particularly in the creation and marketing of luxury perfumes. It is not surprising, therefore, that it has been primarily French courts that have explored the question of whether fragrances may be copyrightable.

In the late 1990s, for example, fashion designer Thierry Mugler sued the august French fragrance house of Molinard for copyright infringement. Mugler claimed that Molinard had manufactured and sold a 'smell-alike' product of his popular fragrance 'Angel'.[9] The success of this perfume has been attributed to its unusual 'gourmand fragrance' with strong culinary notes to conjure cotton candy and caramel scents that Mugler recalled from having visited fairgrounds as a child.

The Tribunal de Commerce in Paris agreed with Mugler's claim that 'Angel' was a copyrightable work. The court likened the creation of a fragrance to the composition of a musical score, and suggested that varying perceptions of, and reactions to, 'Angel' were analogous to a similar diversity among those engendered by a musical work.

Molinard ultimately avoided liability for infringement, however, because Mugler did not actually own any copyrightable interest in the disputed fragrance. Mugler did not create 'Angel'; master perfumers Olivier Cresp and Yves de Chiris created it while working for Quest International, a large fragrance house. Mugler had commissioned Quest to develop and produce the perfume, based on a verbal script referencing his impressionist recollections of sugary treats.

---

[8] See, Frédéric Malle, Editions de Parfumes, http://www.fredericmalle.com/perfumers (last visited October 6, 2017); also see, By Kilan, Kilan, https://www.bykilian.com/us/biography.php (last visited October 6, 2017).

[9] *Mugler v. Molinard*, Tribunal de commerce [T. Com.] [Commerce Ct.] Paris, 15e ch., September 24, 1999, Gaz. Pal. 2001, 17-18.01, no. 17-18.

Several years after the resolution of Mugler's dispute, the French cosmetics giant l'Oréal sued the Belgian company Bellure, claiming it had infringed the copyright in l'Oréal's highly popular perfume 'Trésor'.[10] In 2006 the Regional Court of Appeals in Paris upheld the lower court's determination that 'Trésor' was copyrightable expression, and that Bellure had infringed upon it with its 'smell-alike' product branded 'Female Treasure'.

While l'Oréal was litigating this case in France, its subsidiary, Lancôme was pursuing a similar copyright infringement claim over Trésor in the Netherlands. Lancôme claimed that Kecofa, a small Dutch distributor, had infringed on Lancôme's copyright by selling 'Female Treasure' for a fraction of the price of Trésor. This dispute was ultimately reviewed by the Netherlands' Supreme Court, which agreed with Lancôme's argument that Trésor is copyrightable expression.[11]

Rationalizing its determination that fragrances could be copyrightable, the court emphasized the Dutch copyright statute's deliberately open-ended designation of what constitutes a copyrightable work. Under Dutch law the only prerequisites for copyright are that a work is perceptible to humans, and that it is original insofar as it bears the personal stamp of its author. Unlike, for instance, creators of pharmaceuticals who develop anonymous new formulas for purely useful ends, perfumers aver that they devise formulas that convey their identity and individual expression. Curiously, while the Netherlands' Supreme Court decision fleetingly alludes to the challenge of humans' limited olfaction to the copyrightability of fragrances, it brushes away this concern, suggesting that the nature of the creator's work, and not the outcome of that work, is key in determining copyright eligibility.

Returning to France, the Court of Appeals' 2006 decision in *l'Oréal* underscored how the work and output of a perfumer are like those of creators of other original works that copyright protects. It emphasized that under French law all works of human intellection are eligible for copyright protection – even those that might also be patentable, and even those that are not fixed – as long as they are perceptible, and original insofar as they reveal the imprint of the creator's personality.

In 2007 the same appellate court in Paris affirmed a lower court's determination that the fragrance 'Le Male' [sic], a fragrance branded by

---

[10] *Bellure v. l'Oréal, et al.*, Cour d'appel (C.A.) [Regional Ct. App.] Paris, 4e ch. A, January 25, 2006, D. 2006, at 580, J. Daleau, aff'g *Bellure v. L'Oréal et al.*, Tribunal de Grande Instance [T.G.I.] [Ordinary Ct. of Original Jurisdiction] Paris, May 26, 2004, D. 2004, note Galloux.

[11] See, *Lancôme/Kecofa, Hoge Raad der Nederlanden* [H.R.] [Sup. Ct. of the Netherlands], 16 June 2006, AMI 2006/5, at 161 (ann. Quaedvlieg) (Neth.).

designer Jean-Paul Gaultier, was copyrightable expression that had been infringed by a similarly scented fragrance sold by a competitor under the name 'Inmate for Men'.[12] The defendant was Senteur Mazal, a small fragrance company on the outskirts of Paris.

Mazal argued that fragrances did not meet copyright's threshold of originality. The wide variation in human perception of a single fragrance, Mazal claimed, makes it impossible for those perceiving it to identify it consistently as the expression of a particular creator. The court discounted this argument, observing that there are similar divergences in the perceptions of readers and audiences of literary and musical works. The *mélange* of reactions this diversity produces, however, does not mean a book or song cannot be consistently associated with a particular author.

The shortcomings of the *Mugler*, *l'Oréal*, *Gaultier* and *Lancôme* judgments do not lie in their determinations that the work of perfumers can be an intellectual and aesthetic endeavor akin to writing and painting. Rather, these decisions are ungrounded because they do not squarely address the incapacity of human olfaction to perceive the complexity of the work rendered from a perfumer's intellectual effort. The question whether a fragrance can be considered a copyrightable work devolves to the question of *how perceptible* original expression must be to qualify for copyright protection. This issue was ultimately considered by France's Cour de cassation in 2013 in yet another dispute involving l'Oréal's 'Trésor'.[13]

L'Oréal's subsidiary Lancôme prosecuted this case attempting to set a precedent that would help counter the proliferation of knockoff fragrances sold in flea markets and on urban sidewalks alongside counterfeit Hermès scarves and Ray-Ban sunglasses. The defendant, Patrice Farque

---

[12] See, *Beauté Prestige Int'l. v. Senteur Mazal, Cour d'appel* [C.A.] [Regional Ct. App.] Paris, 4e ch., February 14, 2007, D. 2007, at 735, J. Daleau.

[13] *Société Lancôme v. Patrice Farque*, Cass. com., December 10, 2013 [pourvoi no 11-19872] available at http://www.cecoa.eu/images/cecoa/artdroit201312001.pdf. Seven years earlier the Cour de cassation reviewed an infringement claim that also turned on whether copyright could protect fragrances. See *Bsiri-Barbir v. Haarman & Reimer*, Cour de cassation [Cass. 1e civ.] [Ultimate Ct. App.] Paris, June 13, 2006, D. 2006, Somm. 1741, J. Daleau. This dispute involved a claim by a perfumer who had been dismissed by Haarman & Reimer that she was owed royalties from the sale of 'Dune' based on the copyrightable expression contained in the fragrance she helped to create. As the same court would later hold in the Lancôme dispute, the Cour de cassation found that fragrances were not copyrightable expression. In this earlier case of *Bsiri-Barbir*, however, the court based its decision entirely on a determination that the work of perfumers does not rise to the level of authorship, but is merely the application of technical knowledge or 'know-how'.

had been arrested in the scrappy outskirts of Paris while selling bottles of counterfeit 'Trésor'. When criminal charges against him were ultimately dropped, Lancôme then claimed that Farque was civilly liable for infringing its copyright in 'Trésor'.

The Cour de Cassation disagreed. Notwithstanding the question of the ethicality of Farque's conduct, he could not be liable for copyright infringement because the work he was said to have infringed was not protectable in the first place. The court reached this conclusion by separately considering the creation and perception of fragrances. The development of a perfume may involve creative and original thought beyond the technical know-how we might associate with, for instance, draftsmanship and pharmacology. Because humans cannot adequately discern the complexity of this original intellection, however, the particular work in which it is contained is not copyrightable expression because it cannot be broadly communicated.

An analogy to literary works may help convey the court's reasoning. Proust's writings were once copyrightable expression despite the facts that many French readers did not fully comprehend or appreciate their complexity, and that many others could not appreciate them at all because they could not read French. They were protectable in the first half of the 20th century as Proust's original expression, however, because millions of readers were able to comprehend them substantially, and associate them with a particular author.

If, however, Proust had recorded his work using a symbolic code comprehensible only to him, and perhaps a few close associates, the visible recording of this work might be broadly communicated, but not Proust's expression contained within it. From the visible document, we might be able to discern some vague meaning from the repetition or patterns of the symbols, in much the same way we can detect salient aspects of a complex fragrance containing hundreds of scent components. But we would be unable to perceive enough information that would enable us to identify the work as Proust's.

3. SUI GENERIS INTELLECTUAL PROPERTY PROTECTION FOR FRAGRANCES?

The fact that intellectual property law now provides scant protection to works of fragrance raises the question whether and how this paucity may affect the industry's sustainability. Will reverse engineering, and the production of 'copycat' fragrances that it enables, undermine innovation in this sector, bankrupt its core trendsetting members, and ultimately lead to the demise of this industry? If this dire scenario is likely to ensue, should governments avert it by enacting bespoke legislation providing protection for fragrances that is not available under existing intellectual property law?

In 1984, and again in 1998, the U.S. Congress enacted such legislation to protect the interests of semiconductor chip manufacturers, and the boat-building industry, respectively.[14] The semiconductor chips and the boat hulls that this legislation protects are useful, non-copyrightable goods. Manufacturers of these goods claimed that patents and trade secrets did not effectively protect the information generated by the costly research and development that is embodied in these products. Accordingly, they lobbied Congress to legislate protection specific to these products to avert the potentially ruinous effects on these industries in the United States from unfettered reverse engineering.[15] In both instances Congress accommodated these concerns by appending to the Copyright Act limited terms of protection for these non-copyrightable works. In 1996 The European Commission similarly accommodated commercial data aggregators' concerns when it promulgated sui generis legislation to protect non-copyrightable data collections.[16]

In the United States the clothing industry, and high-end fashion designers in particular, have also sought legislation providing copy protection for non-copyrightable designs for arguably useful apparel. They have observed the legal protection offered to these designs in European nations, and a purported disadvantage suffered by the American fashion industry without similar protection.[17] The most recent unsuccessful attempt on this

---

14 Semiconductor Chip Protection Act of 1984, 17 U.S.C. §§ 901–14 (2012); Vessel Hull Design Protection Act, 17 U.S.C. ch. 13 (2012).

15 See, Leon Radmonsky, Sixteen Years after the Passage of the U.S. Semiconductor Chip Protection Act: Is International Protection Working? 15 Berkeley Tech. L. J. 1049, 1053 (2000) (discussing why patent, copyright, and trade secret laws provided insufficient protection for semiconductor chips).

16 Commission Directive on the Legal Protection of Databases, 1996 O.J. (L 77) 39.

17 Council Regulation (EC) No. 6/2002 of 12 December 2001 O.J. (L 3, 1.2002, p.), amended by Council Regulation (EC) No. 1891/2006 of 18 December 2006 O.J. (L 386, 29.12.2006, p. 4), available at http://oami.europa.eu/en/design/pdf/6-02-CV-en.pdf. See Silvia Beltrametti, Evaluation of the Design Piracy Protection Act: Is the Cure Worse than the Disease? An Analogy with Counterfeiting and a Comparison with the Protection Available in the European Community, 8 Nw. J. Tech. & Intell. Prop. 147 (2010) (suggesting that the European Union, and members France and Italy particularly, have maintained robust fashion markets by legislating international and national protection for fashion designs).

front was the Innovative Design Protection Act of 2012.[18] This legislation would have piggybacked onto the U.S. Copyright Act's boat hull provision, to provide to a wide range of apparel designs a three-year term of protection against substantially identical unauthorized copies.

In many respects the fragrance industry's concerns about intellectual property protection align with those of the fashion industry. Both industries produce 'classics' like Patou's 'Joy' and Chanel's suits, but much of the output of both industries are new products with brief market lives. Most of the economic potential, for instance, of the design of the outfit Melania Trump wore to her husband's inauguration, will be realized in the short term. Once her powder blue Ralph Lauren dress and gloves have been copied and reproduced in cheaply rendered knockoffs they will have lost their cachet, and fashion industry leaders will have long moved on to promote newer and more profitable designs.

The same is true of the fragrance industry, particularly the division of fine fragrances, which releases hundreds of new perfumes annually.[19] Like new apparel designs, most new perfumes have a brief period of marketability, especially 'celebrity' perfumes (e.g., Paris Hilton, Britney Spears) whose appeal is yoked to the star's volatile popularity. By the time the celebrity's renown, or notoriety, wanes, and the market for his or her perfume has become saturated, fragrance creators will already have formulated and distributed new fragrances linked to more fashionable and marketable celebrities.

Knockoff, low-cost, versions of well-known perfumes that have remained marketable for decades may injure the financial interests of the sellers of the original fragrances. This potential injury is tempered, however, by the fact that much of the value of luxury goods like fine fragrances is tied to consumers associating them with specific sources. A replica of an Hermès suit will never be as economically valuable as an original with Hermès's branding. The premium that consumers pay for Hermès clothing is, in part, the price of assurance that the merchandise comports with long-established standards of quality associated with a particular brand.

The same is equally, if not more, true of fragrances. Because perfumes are volatile, and evolve while being worn, consumers cannot readily confirm whether knockoffs of even well known perfumes like 'Shalimar' or 'Opium' comport with the originals in ingredient quality and

---

[18] S. 3523, 112th Cong. (2012) (providing designers three years of protection against knockoffs of new fashion designs).

[19] The industry creates over 60,000 new proprietary fragrance blends each year, and over 600 of these are developed in connection with briefs for new fine fragrances. See, http://www.ifraorg.org.

composition. In other words, as Porsche once memorably claimed about their automobiles, '. . . there is no substitute'.[20]

4. CONCLUSION

Given the French high court's recent unequivocal rejection of copyright for fragrances, it is unlikely the fragrance industry will pursue elsewhere this form of legal protection. Patents provide a 20-year term of protection for new scent molecules and innovative manufacturing processes. They cannot, however, be used to protect producers' non-useful proprietary blends, many of which are ultimately sold by retailers as uniquely named perfumes.

Likewise, trade secrets protect a fragrance manufacturers' commercially valuable information, but only as long as its secrecy is maintained. Because the composition of new fragrances can be handily and legally obtained through reverse engineering, the secrecy of this valuable information is compromised the moment a sample of the fragrance leaves the manufacturer's physical plant. The European Union's recently enacted trade secret legislation explicitly accommodates such reverse engineering and does not provide any new security for fragrance formulas.

Despite the ostensibly bleak outlook for intellectual property protection, the fragrance industry is surprisingly vibrant, with market analysts projecting continual growth in the years ahead.[21] In this respect the fragrance sector is unlike, for instance, the popular music and print publishing industries that have been crippled economically by digital reproduction and distribution technologies. Despite the lack of effective protection for its most valuable intellectual property under existing law, and widespread use of technologies that enable the legal acquisition of its trade secrets, the fragrance industry has not retrenched.

The industry's vitality in the face of weakened intellectual property protection can be attributed not only to a burgeoning market in developing nations like China, but also innovative uses of fragrances in the marketplace. To an increasing extent, new fragrances are created as part of larger branding campaigns by purveyors of goods and services.

---

[20] Readers of a certain age will recall the character played by Tom Cruise using this aphorism following the car chase scene in the movie *Risky Business* (Geffen Company 1983).

[21] See, e.g., Global Industry Analysts, Inc., Fragrances and Perfumes – A Global Strategic Business Report (2018) available at: http://www.strategyr.com/MarketResearch/Fragrances_and_Perfumes_Market_Trends.asp.

Clothiers like Abercrombie & Fitch and Loro Piana, for instance, diffuse throughout their retail shops olfactory 'logos' created especially for their brands. Hotel chains Omni, Hyatt, and an increasing number of others, do likewise to provide customers a brief hedonic sensation, and to cultivate a subliminal association between a particular fragrance and the hotel(s) in which one encounters it.[22] The fragrance itself might not be protectable as intellectual property, but trademark may protect a particular use of it once it is integrated into trade dress that consumers associate with a particular provider of a good or service.[23]

In some respects, the fine fragrance industry today is becoming more like the industry of centuries ago when hundreds of small firms manufactured and sold their luxury products in nearby urban markets. In recent years, there has been a proliferation of 'niche' or 'artisanal' fragrance creators and manufacturers.[24] Their success reflects a more generalized predilection, among well-heeled consumers in particular, for products that are not widely available, and whose production is associated with a particular individual and location. The expansion of the niche fragrance market has made possible the development of a market for bespoke personal fragrances.

Fragrances branded with the names of celebrities like Sarah Jessica Parker are developed by large corporations seeking maximum profit through sales of a star-branded fragrance that market tests have indicated will appeal to millions. The objective of bespoke personal fragrances, on the other hand, is to appeal to, and flatter, mainly the people for whom they are formulated. Bespoke personal fragrances are marketed exclusively to individuals; they are never introduced into the mass retail market and therefore do not risk competition from knockoffs or counterfeits.

---

22 In an interesting twist on using fragrance as trade dress, Hyatt Hotels scents the air of the public spaces of each of its sumptuous 'Park Hyatt' properties with a different and unique bespoke fragrance, which is also used to scent the lotions, soaps, and candles liberally sprinkled throughout the particular hotel's public and private spaces. Hyatt hopes that because customers will encounter a particular bespoke fragrance *only* at the hotel in which it is deployed; they will associate it not with 'Hyatt' – which has properties ranging from relatively austere to luxuriant – but with a particular top-tier Hyatt hotel that is purposefully distinct from all other Hyatt properties.

23 See, Charles Cronin, Lost and Found: Intellectual Property of the Fragrance Industry; From Trade Secret to Trade Dress, 5 N.Y.U. J. Intell. Prop. & Ent. L. 256 (2015).

24 French perfumers find 'niche' a regrettable designation given that the term denotes 'dog house' in their language.

In fact, the production and distribution of counterfeit fragrance products, and the misappropriation of trade secrets – e.g., supplier lists, formula inventories, proprietary production methods – may pose greater threats to the fragrance industry than do legitimate reverse engineering of fragrances, and the production of 'smell-alike' products. And both the European Union and the United States have legislated strong anti-counterfeiting and trade secret protection measures that mitigate these threats.

Works of fragrance are on the margins of copyright not because of a lack of original expression in their creation but rather our inability to comprehend that expression. This does not imply that fragrances do not have a role in human understanding and enjoyment of the material world, but rather that this role occurs mainly in our unconscious. As Mozart once wrote about the capacity of the 'less learned' to appreciate several of his piano concertos: 'There are passages here and there from which the connoisseurs alone can derive satisfaction; but these passages are written in such a way that the less learned cannot fail to be pleased, though without knowing why'.[25] Essentially all humans are 'pleased, though without knowing why' by fragrances, just as we are by an attractive shade of paint color, or the taste of a particular wine, the production of both which likely involved considerable human creative intellection to produce.

Intellectual property law in general, and copyright in particular, does not protect the creative investment of vintners blending new wines, colorists blending new tints, florists devising new arrangements, etc. This lack of protection has not inhibited innovation in these and many other fields, in part because the economic value of the intellectual investment in such innovations is closely tied to the material objects in which they are embodied. The same is true for new fragrances because the value of their formulas depends upon the availability, affordability, and use of particular raw materials. This is not true of more purely informational, and copyrightable, works of literature, poetry, music, choreography, etc., whose value transcends instantiation in a particular medium.

---

[25] Emily Anderson (ed. & trans.), *The Letters of Mozart and His Family* (2nd ed., St. Martin's Press, 1966) 833.

16. Subsistence of copyright over CAD files in 3D printing: the Canadian, the U.S. and European outlook

Teshager Dagne

1. INTRODUCTION

This chapter discusses the subsistence of copyright over computer-aided design (CAD) in three-dimensional (3D) printing files under Canadian, United States and European Union copyright laws. 3D printing is a new and continuously developing area of technology which is becoming more popular and accessible. 3D printing, also known as additive manufacturing, is a process of laying materials on top of each other to build a 3D solid object. 3D printing starts with the virtual model of the product, held in a CAD file. This file holds the instructions that, when sent to the printer, guide the device in building up the output through layering successive deposits of material until the model is complete.

3D printing has the potential to touch on many areas of intellectual property, leading to diverse issues.[1] This chapter focuses on issues that arise in copyright law in relation to the CAD files used by the printers. It is unclear whether copyright protection may exist over these design files, and if so, under what circumstances. Thus, this chapter examines the subsistence of copyright over CAD files to be used in 3D printing: whether copyright exists and applies to the work.

The discussion starts with a brief description of the phenomenon of 3D printing. The section that follows explores the intersections between copyrights law and 3D printing. Section 3 analyses the issue of originality of CAD files under the legal systems of Canada, the U.S. and the European Union.

[1] See Teshager W. Dagne, Overview of Implications of 3D Printing upon Canadian Intellectual Property Law, 31 Canadian Intellectual Property Review 101–133 (2015).

2. 3D PRINTING

3D printing is a manufacturing method in which materials, such as plastic or metal, are deposited on to one another in thin layers to produce a three-dimensional object. From the manufacturing of guns to food preparation, it has the potential to revolutionize (and improve) many aspects of our lives, in much the same way the Internet has revolutionized communication. The process of 3D printing involves the preparation of a computer-assisted digital (CAD) file, which may be derived from pictures or drawings, scanned from goods using a 3D scanner, or downloaded from websites. Such a file can easily be distributed, copied, modified and then 'printed' by a printer device. This makes it possible to turn digital bits into physical objects at the press of a button. The technology's potential as a game changer, in this respect, presents challenging legal questions, which need to be addressed before the technology becomes more commonplace.

Developments in three aspects of 3D printing have brought technology to the forefront of discussions at the intersection of law and technology. These developments consist of: the accessibility of 3D printing machines; the exchangeability of computer assisted design (CAD) files; and the availability of materials used to print goods. First, 3D printing machines have become accessible at consumer prices. It is envisaged that 'soon, probably in the next few years, the market will be ready for a mainstream 3D printer sold by the millions at Walmart'.[2] Even when individuals do not own a 3D printing machine, they can access the technology from online 3D printing platforms, merely by uploading CAD files over the Internet.[3] The US Postal Service Office predicts a significant increase in its revenue, as CAD files are ordered online for printing and the product is then shipped for same-day delivery.[4] A CAD file contains the instructions for a printer to create a tangible object. A CAD file can be designed from scratch using a 3D modelling program. Moreover, a 3D scanner can be used to translate an existing product into a digital file, while a photograph

---

[2] Chris Anderson, The New MakerBot Replicator Might Just Change Your World (September 19, 2012). Retrieved from Wired Magazine website: http://www.wired.com/2012/09/how-makerbots-replicator2-will-launch-era-of-desktop-manufacturing/ (last accessed October 16, 2017).

[3] For example, Shapeways, had printed more than one million products for over 300,000 users by the middle of 2013.

[4] United States Postal Service Office of the Inspector General, If It Prints, It Ships: 3D Printing and the Postal Service (July 7, 2014), https://www.uspsoig.gov/sites/default/files/document-library-files/2015/rarc-wp-14-011_if_it_prints_it_ships_3d_printing_and_the_postal_service_0.pdf (last accessed October 16, 2017).

or drawing can be converted into a CAD file, to be modified and distributed with ease via an increasing number of file-sharing websites. Another aspect of 3D printing is the material used to make the output, namely the tangible 3D-printed object. Just as an inkjet printer prints documents with ink, a 3D printer uses molten or powdered plastics, glass, ceramics, or metal sintering powders, amongst other materials, depending on the area of application.[5] There are also liquid baths and post-production chemicals used in 3D printing that have become available at consumer-friendly prices.

The simplified process of producing almost any kind of object as a result of the above factors means there are no limits to what 3D printing can produce. The rise of the technology has led to a process of decentralization whereby manufacturing has shifted from companies to consumers.[6] The rapid growth of 3D printing is comparable to that of computers and the Internet. As the entry of home computers and the Internet into the mainstream caused legislators to quickly realize that laws needed to be redefined and created to regulate them, 3D printing has the potential to provoke significant legal concerns. Given that 3D printing technology is already readily available to consumers worldwide and CAD files are easily altered and shared online, 3D printing poses great challenges to the current IP legal system, particularly copyright law.

3. COPYRIGHT AND 3D PRINTING

Copyright gives authors the ability to control the copying and reproduction of their works. Copyright laws are designed to 'protect the owners of creative works' while balancing this with the public interest (e.g., accessibility). Specifically, copyright laws regulate copying. As the exact purpose of a 3D printer is to copy and distribute, 3D printing creates significant challenges for copyright.

In copyright law, the primary question regarding 3D printing involves the CAD file that serves as a blueprint to produce a physical product. This chapter focuses on the subsistence of copyrights over CAD files in consumers' use of 3D printing. CAD files are required components of the 3D printing process, as the file holds the instructions on which the printer

---

[5] J. F. Hornick, 3D Printing and the Future (or Demise) of Intellectual Property, 1 Mary Ann Libert, 34–43 (2014).

[6] R. Maier, Patents in a 3D World: The Challenge of the Second File-Sharing Revolution, Intellectual Property Today, 24–27 (2014).

relies. Consequently, the exploration and analysis of subsistence of the CAD file is related to and affected by the method in which the CAD file is created. In consumers' use of 3D printing, a CAD file can be created based on existing objects in two ways: an object can be scanned, or have a picture taken of it. In the first, a 3D scanner can capture the physical dimensions of an object, by placing data point coordinates on the surface. Each data point will hold location coordinate information about the shape, similar to a pixel in a digital image, which is then rendered into the digital design.

Among a number of copyright questions that arise regarding the CAD files, the first is whether a design file that is inspired by a copyright-protected underlying object will extend its copyright protection over the CAD file. For example, a sculpture created by a living artist is capable of copyright for the life of the author, plus a period of 50 years, under Canadian law. If a CAD file is created of this sculpture during this term of copyright protection, what is the law that applies to the design? Would copyright protection automatically attach, as the originating object was protected? The second issue that emerges is related to CAD files derived from articles in which it is unclear whether copyright protection has attached. In such cases, how does the law apply to CAD files that are based on objects that have both functional features and artistic expressions? The underlying question in these cases is: When do CAD files hold copyright on their own, regardless of any copyright protection of the originating article that inspired it? This introduces the question of the threshold of copyrightability under the law, and an evaluation of whether a CAD file meets the threshold for copyrightability.

Copyrightability of a work is determined through standards of originality, which vary across jurisdictions. In determining whether a work attracts copyright protection, standards of originality balance the value of labour, the public interest, and the concept of authorship; and each method of determining originality favours a different factor.[7] For example, jurisdictions such as the United Kingdom embrace a 'sweat of the brow' approach, which tips the scale in favour of the author, at the expense of the public interest. Other jurisdictions opt for a public interest approach, as demonstrated in cases like Canada's *CCH*[8] case, in which originality was determined on the basis of the public interest in protecting works, rather than solely on the author's effort.[9] Further, jurisdictions

---

[7] In *Walter v Lane*, All ER Rep. Ext. 1666 (1903).

[8] *CCH Canadian Ltd v Law Society of Upper Canada*, 2004 SCC 13, 1 SRC 339 (2004).

[9] See, Daniel J. Gervais, Feist Goes Global: A Comparative Analysis of the Notion of Originality in Copyright Law, 49 J. Copyright Society of the USA 949–981 (2002).

such as France centre their determinations of originality on the concept of 'genuine authorship', maintaining protection of the public interest by requiring more originality in a protected work than the 'sweat of the brow' labour-based threshold.[10]

The copyrightability of CAD files in the above scenarios may vary according to the standard of originality in the relevant jurisdiction. The following discussion will examine the copyrightability of CAD files by examining the thresholds of originality for subsistence of copyright in different jurisdictions.

4. THRESHOLD OF ORIGINALITY AND CAD FILES

4.1 Canada

In Canada, copyright will attach to any original creative work, provided the work meets the appropriate legislative requirements and common law guidelines. The Copyright Act contains the requirement that copyright will subsist in every original work.[11] Originality is further defined in section 2 as including:

> . . . every original production in the literary, scientific or artistic domain, whatever may be the mode or form of its expression, such as compilations, books, pamphlets and other writings, lectures, dramatic or dramatico-musical works, musical works, translations, illustrations, sketches and plastic works relative to geography, topography, architecture or science.[12]

Whether or not copyright may subsist over the CAD file itself will depend on whether the file meets this requirement. Generally speaking, copyright will attach to a work that originates from an author, is not a mere copy, and was made using minimal intellectual effort.[13] The Supreme Court confirmed these standards of originality in *CCH*.[14] The Court decided that the standard under Canadian copyright law is that original work needs to originate from the author and be 'more than a mere copy', yet there is no threshold requirement of creativity, novelty, or uniqueness.[15] However,

---

10 Ibid.

11 Copyright Act, RSC c. C-42, Pub L No 105-304 112 Stat 2860, § 5(1) (1998).

12 Ibid., § 2.

13 *CCH*, supra note 8; David Vaver, *Intellectual Property Law* (2nd edn, Irwin Law, 2011) 58.

14 *CCH*, supra note 8.

15 *CCH*, supra note 8, paras 16, 25.

there does need to be some exercise of skill and judgment exerted during this process. This means that the work must originate from the author, and there must be some level of personal contribution:

> The exercise of skill and judgment required to produce the work must not be so trivial that it could be characterized as a purely mechanical exercise. For example, any skill and judgment that might be involved in simply changing the font of a work to produce 'another' work would be too trivial to merit copyright protection as an 'original' work.[16]

Originality does not necessarily mean something new, nor does it require a certain amount of creativity.[17]

When we assess the originality of CAD files in the light of this criterion, their originality will depend on the amount of effort exuded, or the level of modifications applied. Can a CAD file, inspired by an existing object, attract copyright protection – or is it simply a replica or copy of the underlying object? The answer to this question may correspond to the method by which the object is adapted into the CAD file. How a design file comes into existence may affect its ability to acquire copyright protection.

For example, simply taking a scan or picture of a work may not be a sufficiently original process to attract copyright in itself.[18] Questions can emerge as to whether the creator actually did come up with the content of the work, despite having influenced the format.[19] Artistic contributions can become less clear when an object is scanned or captured through an image.[20] Capturing a physical project is a more functional process. When an object is scanned or photographed, the CAD software is a tool used by the author to assist in producing the final result in the digital design. The question in this case is not whether there has been human intervention, but rather, to what extent has a person exercised skill and judgment in the

[16] *CCH*, supra note 8, para 16.

[17] Vaver, supra note 13, 58.

[18] Matt Simon, When Copyright Can Kill: How 3D Printers Are Breaking the Barriers Between 'Intellectual' Property and the Physical World, 3 Pace IP Sports & Ent Law Forum 72, 74 (2013); Michael Weinberg, What's the Deal with Copyright and 3D Printing, 15, 18, https://www.publicknowledge.org/files/What's%20the%20Deal%20with%20Copyright_%20Final%20version2.pdf (last accessed October 16, 2017).

[19] Andrew J. Wu, From Video Games to Artificial Intelligence: Assigning Copyright Ownership to Works Generated by Increasingly Sophisticated Computer Programs, AIPLA Q J (1997).

[20] Simon, supra note 18, 74.

process. The user may become an author if the machine has provided a way for the author to manipulate the expression.[21]

Simply using a scanner or camera to capture an object may be similar to a computer-generated work. The design file is not being created with the assistance of the scanner or camera, but rather they are the mediums that capture the object, and the software is translating them into digital designs.[22] The user took the initiative to capture the object; however, there were no creative decisions made during the production of the work.[23]

Furthermore, if an output is easily repeated, this typically signifies the limited nature of human intervention.[24] Scanners specifically 'try to reproduce the scanned object as faithfully as possible'.[25] The user plays a limited role in fixation, without any further manipulation of a design. Users are 'not creatively interpreting the object in any way'.[26] This implies that using a scanner or camera is insufficient to attract copyright protection in Canada.

Canadian courts have dealt incidentally with a discussion of the minimum skill required to possess ownership over CAD files in *Oakcraft*.[27] In *Oakcraft*, the ownership of floor plans was found to belong with the designer, who had consulted with the homeowners and conceived the plans. Despite the fact that a secondary group of technicians had converted the work into the CAD file, the technicians were found to have simply operated the machines while the designer had 'employed intellect and experience along with skill and judgment in creating his plan'.[28] The technicians' role in *Oakcraft* could be compared to the use of a scanner or camera. Simple operation of the machine is not a sufficient exercise of skill and judgment to create an original work sufficient to attract copyright.

Due to these above discussed parameters, some have claimed that even if a CAD file were made entirely from scratch, if it is based on an existing object, it would not be entitled to copyright protection.[29] This is due to the fact that the file, despite being independently created, is still a direct

---

[21] Wu, supra note 19, 152.

[22] Wu, supra note 19, 142.

[23] James Grimmelmann, Indistinguishable from Magic: A Wizard's Guide to Copyright and 3D Printing, 71 Wash & Lee L Review 8 (2014); Simon, supra note 18, 72–74; Weinberg, 'Copyright,' supra note 18, 15.

[24] Wu, supra note 19, 152.

[25] Weinberg, 'Copyright', supra note 18, 15.

[26] Ibid., 15–16.

[27] *Oakcraft Homes Inc v Ecklund*, CanLII 41981 (2013).

[28] Ibid., para 50.

[29] Grimmelmann, supra note 23, 10.

replication of a real-life object, with no changes. An exact replica is not sufficient to attract copyright protection under Canadian copyright law.

4.2 United States

In the US, the Supreme Court decided in *Feist v Rural Telephone* that copyright protection applies only to the original selection and arrangement of the facts – the expression aspect of databases. This case moved the discussion away from mere industriousness and labour as a rationality for protection towards a creativity standard for originality in American law, requiring a modicum of creativity, or creative spark, however low that threshold may be interpreted substantively for the originality assessment. In the US, therefore, a higher threshold test for originality has led courts to find no copyright in digital designs, which indicates that there is also likely no copyright over the majority of CAD files in this jurisdiction: if a 3D design file merely represents an accurate 3D depiction of a pre-existing physical object, then the file may not be protectable subject matter.[30]

In *Meshworks, Inc v Toyota Motor Sales, USA, Inc,* the US State Court found that 3D design files of a manufacturer's vehicles were not sufficiently original and therefore not copyrightable.[31] In *Meshworks*, the plaintiffs started with the actual object (a car), collected data points on the object, and then rendered it into a digital file – similarly to how a 3D scanner operates. The Court found that the result was just a mere copy. The digital image was meant to depict the car, and that was the result. There was no originality in using the software to produce the replication, the files were 'merely copies of Toyota's products, not sufficiently original to warrant copyright protection' and the 'intent was to replicate, as exactly as possible, the image'.[32]

The Court noted that if the Plaintiffs had made decisions about lighting, shading, or background, those decisions might have been sufficient to warrant protection. However, in *Meshworks*, the Plaintiffs had assigned that task to another party, and it was this third party who made those decisions as to 'digitally applied color, texture, lighting, and animation'.[33] Therefore, had the Plaintiffs made any changes, or added any creative elements to the original digitalizations, this process of selection and

---

30 Simon, supra note 18, 74.
31 *Meshworks*, 528 F (3d) 1258 (10th Cir 2008).
32 Ibid., 1260.
33 Ibid., 1261.

arrangement would have created elements that could 'be said to owe their origins' to the designer.[34]

It is important to note that in comparison to Canada, the USA requires a greater degree of creativity in the process and result in order to be considered sufficiently original to warrant copyright protection. Therefore, with the lower threshold of originality in Canada, where an exercise of skill and judgment with no creativity is required, the suggestions of design changes in *Meshworks* would more than likely be sufficient to meet the standard of an original object under Canadian copyright.

In Canada, the requirement to make some decisive changes to the digital image may also not even be an issue, as again, creativity is not required. Designers in Canada will just need to demonstrate that they have used some sort of minimal intellectual effort. This may become more difficult when the designer is just scanning or taking a picture of an object, and using this to render the object into a digital file, as the standard for originality is that of being more than a mere copy.[35]

However, designers may be able to rely on the statement of the court in *Théberge*, in which transforming an object into a digital medium and vice versa was an infringement, despite no new copies being created.[36] Therefore, it is possible that digitalization of an object into a 3D file would be considered to be producing a secondary item. As such, therefore, the CAD file is a new work covered under copyright protection as part of the primary object, and not capable of copyright protection on its own.

4.3 The European Union

There has been limited harmonization of the copyright laws of European Union Member States and many debates about the practicality and desirability of achieving full copyright harmonization, which could be obtained through several methods, such as the creation of an EU copyright code, or codification of the present body of copyright directives.[37] Although these steps have not been taken, the Court of Justice of the European Union (CJEU)[38] has been moving towards harmonization of copyright law. For example, the CJEU's decision in *Infopaq* in 2009 provided an

---

34 Ibid., 1264.

35 *CCH*, supra note 8, paras 16, 25.

36 *Théberge*, 2 SCR 336, para 47 (2002).

37 Eleonora Rosati, *Originality in EU Copyright: Full Harmonization through Case Law* (Edward Elgar, 2013).

38 Commonly 'ECJ', referred to throughout this chapter as CJEU.

EU-wide understanding of the threshold of originality for subsistence of copyright.[39] The court further clarified and elaborated on this threshold in the subsequent cases, *BSA*,[40] *FAPL*,[41] *Painer*,[42] and *Football Dataco*.[43] Before *Infopaq*, there was no widespread understanding of the threshold of originality, and different EU Member States had different approaches to determining this criterion for copyright protection. The UK embraced the 'sweat of the brow' approach, while continental European countries chose a higher standard of originality.[44]

The current threshold of originality under copyright law in the European Union is that a work must be an 'author's own intellectual creation'.[45] This definition is applied widely to incorporate all categories of works under the Berne Convention into European copyright law.[46] In the five landmark decisions from the CJEU, *Infopaq*, *BSA*, *FAPL*, *Painer*, and *Football Dataco*, the court defines and elaborates on the EU threshold of originality. These cases demonstrate that:

1. A work must be the author's own intellectual creation to be considered original (this applies to all subject matter covered by EU copyright directives);[47]
2. When the author can make creative and free choices so that the work reflects his personality, the work is considered original;[48] and

---

[39] Case C-5/08, *Infopaq International A/S v Danske Dagblades Forening*, E.C.R. I-06569, (2009); henceforth referred to as '*Infopaq*'.

[40] Case C-393/009, *Bezpečnostní softwarová asociace v Ministerstvo kultury*, E.C.R. I-13971, (2010); henceforth referred to as '*BSA*'.

[41] Cases C-403/08 and C-429/08, *Football Association Premier League Ltd and Karen Murphy*, E.C.R. I-09083, FAPL (2011); henceforth referred to as '*FAPL*'.

[42] Case C-45/10, *Eva-Maria Painer v Standard Verlags GmbH and others*, E.C.R. I-12533, (2011); henceforth referred to as '*Painer*'.

[43] Case C-604/10, *Football Dataco v Yahoo! UK Ltd and others*, 2 C.M.L.R. 24 (2012); henceforth referred to as '*Football Dataco*'.

[44] See, Estelle Derclaye, Assessing the Impact and Reception of the Court of Justice of the European Union Case Law on UK Copyright Law: What Does the Future Hold?, Revue Internationale du Droit d'auteur 5–117 (2014), http://eprints.nottingham.ac.uk/3613/2/RIDA_article_derclaye_April_2014_eprints.pdf (last accessed October 16, 2017).

[45] *Infopaq*, supra note 39.

[46] Ibid.

[47] Ibid., 36; Thomas Margoni, The Digitization of Cultural Heritage: Originality, Derivative Works and (Non) Original Photographs, IVIR, 14, http://www.ivir.nl/publicaties/download/1507 (last accessed October 16, 2017).

[48] Margoni, ibid., note 12; *Football Dataco*, supra note 43, 38 and 39; *Infopaq* supra note 39, 45; *BSA* supra note 40, 50; Painer supra note 42, 89 and 92.

3. When there is only one way to express an idea, or an expression is already determined by a goal or constricted by narrow rules which leave no room to make creative choices, the work is not original.[49]

Applying the originality of CAD files under the EU law, the aspect of the CAD file to be examined will be the CAD file made up of just scanning or taking a picture of an object that is copyright-protected. This type of CAD file can be considered an author's own intellectual creation simply because it originated from the person scanning the object or taking a picture and rendering a CAD file. However, the author's creative and free choices must be demonstrated so that the work reflects his personality. It is difficult to say that a CAD file depicting an object that is simply scanned or whose picture is taken demonstrates creative and free choices. The extent to which the CAD file accurately depicts the scanned object depends on the quality and sophistication of the 3D scanner and not on the intervention of the author.

The implication is that for a CAD file that is inspired by a copyright-protected underlying object, the object extends its copyright protection over the CAD file. However, if another person makes such CAD file, the file does not attract copyright protection in its own right because there is no sufficient degree of creative and free choices that distinguish the CAD file from its underlying object. If there are modifications to the CAD file that demonstrate the author's creative and free choices, such a modified CAD file attracts copyright protection in its own right. Thus, CAD files that are inspired by copyright-protected objects but have a sufficient degree of an author's creative and free choices to distinguish them from their source can be sufficiently original to warrant copyright protection.

A CAD file can be considered original if it has been created from scratch using design software. This would be a design based on a concept from the user's imagination, or one that is not meant to replicate an existing object – an independent creation. Under EU law, CAD files created from scratch using CAD software may be sufficiently original to warrant copyright protection.

5. CONCLUSION

3D printing is a continuously developing area of technology. There are many questions and legal challenges that arise as the technology becomes

---

49 *FAPL*, supra note 41, 98; *BSA*, supra note 40, 49; *Football Dataco*, supra note 43, 39; supra Margoni, supra note 47, p. 4.

more common and accessible. Copyright protection of the design files is an especially significant issue due to the digital nature of CAD files, as they are transmitted across the Internet with ease, and can be effortlessly modified with the click of a button. Users of the technology wanting to assert protection over their creative endeavours will have to demonstrate that there is sufficient originality in the work to bring it out of the public domain. This can be done by demonstrating the degree of creativity used in creating the work. If the CAD file has been created through a simple scan or photograph, the user will need to intervene in the process and exercise creative choices to distinguish it from the originating object. Modifications of existing files may be unlikely to attract copyright until they resemble new works, similar to an independent creation.

17. Copyrightability of engineered DNA sequences

Nicola Lucchi

1. INTRODUCTION

The extremely fast technological progress in the field of biotechnology has brought about new legislative and judicial attempts to restructure the global regulatory regime especially with regard to intellectual property rights, trying to balance the interests of both rights-holders and consumers. The success of the biotechnology industry has not come about spontaneously. On the contrary, it has been the result of several *influential* and favorable government policies toward the sector in combination with a liberalized regime for the governance of biomedical research.[1]

Debates and 'concerns about the need to regulate the disruptive potential of biological manipulation were apparent almost from the moment when genetic engineering became feasible in the 1970s'.[2] The emergence of all these concerns has also shed light on the problem of how to regulate the use, access, distribution and appropriation of individual genetic information as well as the information content of DNA.

The purpose of this chapter is to consider whether engineered nucleic acid sequences – artificially created by scientists – can be protected through copyright instead of patents. This idea is not really new, but originated in the early 1980s at a time when the patentability of biotechnology inventions was still an open question. This approach has seen little development, at least until two recent US Supreme Court

---

[1] See Steven Vallas, Daniel Lee Kleinman and Diana Biscotti, Political Structures and the Making of U.S. Biotechnology, in Fred Block and Matthew Keller, eds, *State of Innovation: The U.S. Government's Role in Technology Development* (Routledge, 2011) 57.

[2] Ibid.

decisions (*Mayo*[3] and *Myriad*[4]) completely changed US patent eligibility criteria for biotechnology, reversing more than thirty years of case law and questioning the enforceability and validity of patents on naturally occurring genetic material (even if isolated). In particular, in *Myriad*, the United States Supreme Court held that naturally occurring DNA is not eligible for patenting just because it has been isolated from its natural state. Naturally occurring DNA remains a 'product of nature' even after isolation and, therefore, falls under the 'laws of nature' exception to patent eligibility. In other words, the isolation of naturally occurring DNA is not enough for the naturally occurring DNA to be considered man-made. At the same time, in *Mayo* the United States Supreme Court held that certain diagnostic methods are not patentable because they involve standard conventional steps. In particular, the Court held that the process by which the drug is metabolized in the body is a natural process, and so the relationship between the amounts of metabolites and the effectiveness of drug dose is a law of nature. In other words, simply observing such a naturally occurring relationship – without adding anything else – was not enough to make the claims patentable.

These two decisions narrowed patent-eligible protection over living organisms and created significant ambiguity among both patent applicants and patent examiners. In order to mitigate this confusion, the United States Patent and Trademark Office (USPTO) arranged a series of guides on subject matter eligibility. These new challenges faced by biotechnology innovators after *Mayo* and *Myriad*, have also led to reconsidering the idea of using the copyright system as a form of intellectual property protection for engineered nucleic acid sequences, in particular when these sequences are the result of an innovative branch of biological sciences called synthetic biology. Considering this more complex scenario, the chapter explores the possible legal arguments in support of, or against, the copyright protection for engineered DNA sequences.

2. WHAT IS SYNTHETIC BIOLOGY?

Synthetic biology is a relatively young research field that seeks to create or reprogram existing cellular functions with other useful properties, applying

---

3 *Mayo Collaborative Services v. Prometheus Laboratories, Inc.*, 132 S. Ct. 1289 (2012).

4 *Association for Molecular Pathology v. Myriad Genetics, Inc.*, 133 S. Ct. 2107, 186 L. Ed. 2d. 124 (2013).

principles of mechanical engineering to biological design.[5] Generally speaking, it can be explained as the application of engineering knowledge and principles to the essential components of biology.[6] All living organisms are – in fact – based on a set of instructions indicating what they look like and what they can do. These instructions are contained in the organism's DNA within every cell. The DNA is the molecule that contains the genetic material and which then functions as an information carrier.[7] The DNA is composed of genes, which are – essentially – traits or fragments of DNA containing the genetic information necessary to build a specific protein. The 'totality of genetic information belonging to a cell or an organism and in particular the DNA that carries this information' is called a 'genome'.[8] For centuries, humans have been altering the genetic code of animals and plants in order to generate organisms with desired features. As scientists have gained more knowledge on deciphering and artificially changing the genetic code, they have also developed the ability to transfer genetic information from one organism to an unrelated one. Recent developments in the emerging field of synthetic biology – particularly in the manipulation of organisms' genomes – have allowed biologists to build from scratch or re-engineer the genomes of micro-organisms.[9] By combining these advances with engineering principles, synthetic biologists are now able to design cells and even organisms with completely new features, involving the creation of novel DNA sequences 'that may have never existed before in living organisms'.[10] Synthetic biology therefore stretches far beyond mere genetic engineering, which is about the simple

---

5 See Vijai Singh, Recent Advancements in Synthetic Biology: Current Status and Challenges, 535 Gene 1 (2014); Patrick M. Boyle and Pamela A. Silver, Parts Plus Pipes: Synthetic Biology Approaches to Metabolic Engineering, 14 Metabolic Eng. 223 (2012); Eric A. Davidson et al., Building Synthetic Systems to Learn Nature's Design Principles, 751 Evolutionary Syst. Biology 411 (2012); Jay D. Keasling, Synthetic Biology and the Development of Tools for Metabolic Engineering, 14 Metabolic Eng. 189 (2012); Andrew W. Torrance, Synthesizing Law for Synthetic Biology, 11 Minn. J.L. Sci. & Tech. 629, 633 (2010).

6 See European Commission, Synthetic Biology: Applying Engineering to Biology, Report of a NEST High-Level Expert Group, 5 (2005); Royal Academy of Engineering, *Synthetic Biology: Scope, Applications and Implications* (2009) 13.

7 See W. Klug, M. Cummings et al., *Essentials of Genetics* (12th edn, Pearson, 2012) 4.

8 See Bruce Alberts et al., *Molecular Biology of the Cell* (6th edn, Garland Science, 2014).

9 See Jeffrey M. Perkel, Cell Engineering: How to Hack the Genome, 547 Nature 477 (2017).

10 See Gregory N. Mandel and Gary E. Marchant, The Living Regulatory Challenges of Synthetic Biology, 100. Iowa L. Rev. 155, 160 (2014).

manipulation of natural organisms, and tends to focus more heavily on the creation of entirely new synthetic life.[11] In particular, this innovative research technology is built on the understanding that 'DNA sequences can be assembled together like building blocks, producing a living entity with any desired combination of traits, much as one can assemble a car by putting together many individual pieces with different functions'.[12] This huge paradigm shift in the biological sciences gives rise to many new legal issues and rights, including the fact that – among other things – synthetic DNA sequences, because of their very nature, might be considered as falling within the bounds of copyright protection.

3. COPYRIGHT PROTECTION FOR ENGINEERED DNA: REQUIREMENTS AND SUBJECT MATTER

As mentioned, the idea of using copyright protection for engineered DNA sequences is not new. A number of law scholars have already discussed the applicability of copyright law as an alternative to patent protection for nucleic acid sequences and biotechnology works in general.[13] These pro-

---

[11] See Stefanie B. Seitz, Let's Talk About. . . Synthetic Biology – Emerging Technologies and the Public, in K. Hagen et al., eds, *Ambivalences of Creating Life: Societal and Philosophical Dimensions of Synthetic Biology* 157, 158 (Springer, 2016).

[12] See Gregory N. Mandel and Gary E. Marchant, The Living Regulatory Challenges of Synthetic Biology, 100. Iowa L. Rev. 155, 159 (2014).

[13] See e.g., Irving Kayton, Copyright in Living Genetically Engineered Works, 50 Geo. Wash. L. Rev. 191 (1982); Duncan M. Davidson, Common Law, Uncommon Software, 47 U. Pitt. L. Rev. 1037, 1104–1105 (1986); Donna Smith, Copyright Protection for the Intellectual Property Rights to Recombinant Deoxyribonucleic Acid: A Proposal, 19 St. Mary's L.J. 1083 (1988); Dan L. Burk, Copyrightability of Recombinant DNA Sequences, 29 Jurimetrics J. 469, 531–532 (1988–89); Jorge A. Goldstein, Copyrightability of Genetic Works, 2 Bio/Technology 138 (1984); Donna Smith, Comment, Copyright Protection for the Intellectual Property Rights to Recombinant Deoxyribonucleic Acid: A Proposal, 19 St. Mary's L.J. 1083, 1096–1108 (1988); M. Scott McBride, Bioinformatics and Intellectual Property Protection, 17 Berkeley Tech. L.J. 1331 (2002); Sue Coke, Copyright and Gene Technology, 10 J. of Law and Med. 97 (2002); Gunnar W. G. Karnell, Protection of Results of Genetic Research by Copyright or Design Rights? 17 Eur. Intellect. Prop. Rev. 355 (2005); Tani Chen, Can a Biological Sequence Be Copyrighted?, Intell. Prop. & Tech. L.J., March 1 (2007); Doreen M. Hogle, Copyright for Innovative Biotechnological Research: An Attractive Alternative to Patent or Trade Secret Protection, 5 High Tech. L.J. 75 (1990); James G. Silva, Copyright Protection of Biotechnology Works: Into the Dustbin of History?, B.C. Intell. Prop. & Tech. F. (January 28, 2000), http://bciptf.org/

posals were mainly developed in the 1980s, but now – after the US Supreme Court addressed the issue of patent subject matter and the emerging field of synthetic biology in *Mayo* and *Myriad*[14] – the concept has returned to the spotlight, once more attracting the attention of legal academics and practitioners.[15] In particular, the idea of copyrighting DNA was first put forward in 1982 in a law review article written by Professor Irving Kayton and published in the George Washington Law Review.[16] The central thesis of this article was that a biotechnology work is automatically copyrighted when it is original and when the genetic information is fixed in a tangible medium because it satisfies exactly all the statutory and constitutional requirements for copyright protection. In reality, one may note that DNA sequences should be considered non-original and non-protectable facts.[17] However, redesigning and restructuring new biological entities (such as enzymes, genetic circuits and cells) or the genomes of existing organisms

---

wp-content/uploads/2011/07/43-Silva_IPTF11.pdf; Matthew Rimmer, Beyond Blue Gene: Intellectual Property and Bio-informatics, 34 Int'l Rev. Intell. Prop. & Competition L. 31 (2003); Stephen R. Wilson, Copyright Protection for DNA Sequences: Can the Biotech Industry Harmonize Science with Song? 44 Jurimetrics J. 409 (2004).

[14] See *Mayo Collaborative Services v. Prometheus Laboratories*, supra note 3 and *Association for Molecular Pathology v. Myriad Genetics*, supra note 4.

[15] See Dan L. Burk, DNA Copyright in the Administrative State, 51 UC Davis Law Review (2018); Christopher M. Holman, Charting the Contours of a Copyright Regime Optimized for Engineered Genetic Code, 69 Okla. L. Rev. 399 (2017); Christopher M. Holman, Copyright for Engineered DNA: An Idea Whose Time Has Come, 113 W. Va. L. Rev. 699 (2011); Christopher M. Holman, Developments in Synthetic Biology are Altering the IP Imperatives of Biotechnology, 17 Vand. J. Ent. Tech. L. 385 (2015); Christopher M. Holman, Copyright for Engineered DNA, GQ Life Sciences (February 23, 2016), https://www.gqlifesciences.com/copyright-for-engineered-dna/; Andrew W. Torrance, Synthesizing Law for Synthetic Biology, 11 Minn. J.L. Sci. & Tech. 629 (2010); Andrew W. Torrance, DNA Copyright, 46 Val. U. L. Rev. 1 (2011); Michael D. Murray, Post-Myriad Genetics Copyright of Synthetic Biology and Living Media, 10 Okla. J. L. & Tech. 71 (2014); David Walker, Patent Protection or Copyright for Nucleic Acid Sequences? 36 The Licensing Journal 1 (2016); Jorge Roig, Can DNA Be Speech? 34 Cardozo Arts & Ent. L.J., 163 (2016); Jimmy J. Zhuang, Copyright: Better Fitting Genes, 97 J. Pat. & Trademark Off. Soc'y 442 (2015). See also Joseph N. Michelotti, Genes as Intellectual Property, 11 Mich. St. U. J. Med. & L. 71 (2007); Tani Chen, Can a Biological Sequence Be Copyrighted, 19 Intell. Prop. & Tech. L.J. 1 (2007).

[16] See Irving Kayton, Copyright in Living Genetically Engineered Works, 50 Geo. Wash. L. Rev. 191 (1982).

[17] See Stephen R. Wilson, Copyright Protection for DNA Sequences: Can the Biotech Industry Harmonize Science with Song?, 44 Jurimentrics J. 409, 463 (2004).

to make them more efficient or programming them to carry out new functions can really satisfy the originality requirement. This is even more so because, presumably, the synthetic biologist would have to make a creative decision regarding which core components (parts of enzymes, genetic circuits, metabolic pathways, etc.) must be modelled, understood, redesigned and modified in order to 'meet specific performance criteria, and the assembly of these smaller parts and devices into larger integrated systems to solve specific problems'.[18]

Then, it has been argued that the written representation of an artificially engineered sequence of DNA or protein – fixed in a tangible form – may be protected as an original literary work[19] because the statutory definition of 'literary work' includes every production in the literary, scientific and artistic domain, irrespective of the mode or form of its expression (so also any representation of words, figures or symbols).[20] The written description of a genetic sequence – being a string of letters or an alphabet representing the four nucleotides, adenine, thymine, guanine and cytosine (A, T, G and C) – may be considered a 'literary work' within this meaning.[21] In addition, since DNA is made up of these four chemical nucleotides, engineered 'DNA sequences should easily meet' also the fixation requirement.[22] In particular – as studies of DNA have shown – a DNA sequence may completely meet the fixation requirement because it 'possesses definite sequences of nucleotides that can easily be determined; copies of DNA may be synthesized routinely

---

[18] Jay Keasling, The Promise of Synthetic Biology, 35 The Bridge 18, 20 (2005) (illustrating the elements that distinguish synthetic biology from traditional molecular and cellular biology).

[19] See Sue Coke, Copyright and Gene Technology, 10 J. of Law and Med. 97 (2002).

[20] See Article 2 (1) of the Berne Convention for the Protection of Literary and Artistic Works, Sept. 9, 1886, as revised at Paris on July 24, 1971, and amended in 1979, S. Treaty Doc. No. 99-27 (1986) [hereinafter 'Berne Convention'].

[21] At the same time, other scholars have questioned this approach, arguing that copyright may not actually be applied to a written description of a genetic sequence because there is only one established way to express or describe a certain sequence of nucleotides or amino acids. In this case, there is no distinction between idea and expression. See e.g. Gunnar Karnell, Protection of Results of Genetic Research by Copyright or Design Rights? 17 Eur. Intell. Prop. Rev. 355, 357 (1995); David B. Resnik, *Owning the Genome: A Moral Analysis of DNA Patenting* (SUNY Press, 2004) 51 (arguing that 'since DNA sequences are chemical compounds that exist in nature, attempting to copyright a sequence would be like trying to copyright a fact, such as any other chemical formula').

[22] See Andrew W. Torrance, Synthesizing Law for Synthetic Biology, supra note 5, at 642.

and in effectively unlimited quantities; and molecular DNA has been known to last for at least many thousands of years with its nucleotide sequence intact'.[23] Concerning the authorship requirement, some elements of human creativity must have occurred in order for the DNA sequences to be copyrightable. This requirement can be met by products of synthetic biology because they involve 'the design and construction of new, human-designed DNA sequences'.[24] Specifically, synthetic biologists design and construct artificial biological systems that do not currently exist in nature, writing DNA sequences that instruct a cell or engineering organism to behave according to design specifications.[25] Since here there is an author, such DNA sequences can be effectively qualified as 'original works of authorship'.[26] Therefore – from a theoretical perspective – it would be the choice of the scientist to decide whether to enforce her copyright or not.

As far as the identification of the 'copyrightable subject matter box' is concerned, engineered genetic works could fall into the category of literary works. Indeed, following the wording of the Berne Convention, the expression 'literary and artistic works' includes every production in the literary, scientific and artistic domain, whatever may be the mode or form of its expression. In other words – according to the US Copyright Act – 'literary works are works expressed in words, numbers, or other verbal or numerical symbols or indicia, regardless of the nature of the material objects in which they are embodied'.[27] A digital computer program is a literary work as it can be expressed in 'indicia', such as magnetic impulses or other electronic signals and it is also fixed in a tangible medium.[28] Analogously, 'genetically engineered works are expressed in "indicia," the nucleotides that make up DNA'.[29] All the genetic information in the DNA contains a set of instructions for producing something (typically

---

[23] See ibid., at 643 (citing Fred Sanger et al., DNA Sequencing with Chain-Terminating Inhibitors, 74 Proc. Nat'l Acad. Sci. U.S.A. 5463 (1977)); Joseph Sambrook and David W. Russell, *Molecular Cloning: A Laboratory Manual*, Vol 2 (3rd edn, Cold Spring Harbor Press, 2001) 8.4; Eske Willerslev and Alan Cooper, Ancient DNA, 272 Proc. R. Soc'y B 3, 3–5 (2005).

[24] See ibid, at 643.

[25] See Natalie Kuldell et al., *BioBuilder: Synthetic Biology in the Lab* (Shroff Publishers, 2015) 1, 2.

[26] See Andrew W. Torrance, Synthesizing Law for Synthetic Biology, supra note 5, at 644.

[27] 17 U.S.C § 101.

[28] See Irving Kayton, Copyright in Living Genetically Engineered Works, supra note 13, at 199.

[29] Ibid.

a protein), just as a piece of computer software contains a set of coded instruction designed to cause a computer to perform a specific task. There are therefore many similarities between genetic and computer code and some companies have already tried – even if unsuccessfully – to register an engineered 'biological artifact' with the US Copyright Office.[30] In a very specific case,[31] some researchers in the field of synthetic biology have utilized novel techniques for creating completely new segments of DNA in order to obtain living organisms that behave differently from the way they would in the natural state.[32] DNA, in fact, could be recombined into arrangements not found in existing genomes and, in addition, synthetic biology – designing and manipulating sophisticated synthetic cellular circuits – can create or reprogram new living organisms completely from scratch.[33] These elements of scientific argumentation seem to support the analogy between synthetic biology artifacts and computer programs, looking into alternative protection options such as copyrights, which are more straightforward than patents – as they entail no substantive examination – and with the concrete possibility of fostering a feasible open source regime for accessing and using essential public knowledge assets in the life sciences.[34]

4. THREE THEORIES FOR THE COPYRIGHTABILITY OF ENGINEERED DNA SEQUENCES

Legal scholars have developed three main theories under which a synthetic DNA sequence may be considered capable of copyright protection. The first theory, and the most straightforward, focuses on the analogies between computer programs and DNA sequences (mentioned earlier). Here the main argument is that because synthetic genetic sequences – and the information content encoded in them[35] – are

---

[30] See Heidi Ledford, Bioengineers Look Beyond Patents, 499 Nature 16 (2013).

[31] This example has been also cited in Pamela Samuelson, Evolving Conceptions of Copyright Subject Matter, 78 U. Pitt. L. Rev. 17, 19 (2016).

[32] See Andrew Torrance, DNA Copyright, 46 Val. U. L. Rev. 1, 25 (2011).

[33] See Arjun Bhutkar, Synthetic Biology: Navigating the Challenges Ahead, 8 J. Biolaw & Bus. 19, 20 (2005).

[34] See Heidi Ledford, Bioengineers Look Beyond Patents, 499 Nature 16 (2013).

[35] See Christopher M. Holman, Charting the Contours of a Copyright Regime Optimized for Engineered Genetic Code, 69 Okla. L. Rev. 399, 420 (2017).

a sort of biological computer program, and computer programs are considered copyrightable subject matter, then DNA sequences should also be – for analogical interpretation – copyrightable.[36] According to the second theory, the protection of synthetic biological creations may be provided under the concept of a 'compilation copyright': a mix of properly manipulated, selected and arranged gene sequences may indeed generate an original work of authorship. In particular, engineered systems of genes and other molecular components created through synthetic biology could be considered as the result of DNA sequences assembled together like building blocks. Finally, the third theory treats DNA sequences as an ordinary literary work. Under this scenario, an engineered DNA sequence can be considered similar to a set of written instructions given to a biological machine, and as a consequence this sequence can be copyright protected because 'someone worked it out and wrote it down – so the order of the letters is copyrightable, like a string of letters in a book'.[37]

Each of these theories can be corroborated with some good arguments, but only one of them may really be able to meet all the legal requirements and overcome all the possible obstacles that an intellectual creation has to meet in order to qualify for copyright protection.

Copyright law generally protects original works of authorship consisting in a literary, dramatic, musical or artistic work.[38] These apparently narrow categories are however interpreted quite broadly and may cover many works created by human ingenuity. For example, the category of literary works may be deemed wide enough to embrace also human-designed DNA sequences. Copyright laws generally provide a wide and abstract definition of 'literary work':[39] any intellectual creation expressed in print or writing format – regardless of its specific features or nature – can potentially be placed into this category. In addition, we have already seen that the requirements of originality, fixation and authorship may be satisfied by synthetic DNA sequences developed for synthetic biology. Regardless of these supportive arguments, there remain a few obstacles to copyright protection. As argued by some scholars, for example, 'the three-dimensional reproduction of the biological product would not resemble the two-dimensional literary work in form of the DNA sequence,

---

36 See James G. Silva, Copyright Protection of Biotechnology Works: Into the Dustbin of History?, supra note 13, at 1.

37 See Leslie Roberts, Who Owns the Human Genome, 237 Science 358, 359 (1987).

38 See e.g. Berne Convention Art. 2; Copyright Act, 17 U.S.C § 101.

39 See Art. 2(1), Berne Convention.

which would prevent finding copyright infringement in copying of the three-dimensional work'.[40]

The obstacle for recognizing an engineered DNA sequence as a literary work can possibly be overcome under the second theory, which views DNA sequences as compilations of literary works. Compilations of facts (such as catalogs, financial reports, dictionaries and even telephone directories) can indeed receive copyright protection.[41] Under US copyright law, a compilation is defined as 'a work formed by the collection and assembling of preexisting materials or data that are selected, coordinated or arranged in such a way that the resulting work as a whole constitutes an original work of authorship'.[42] Obviously, in these cases, only the expression of those facts can be copyrighted. Thus, it may be possible for biotech companies to secure copyright protection of genetic information compiled in a database.[43] This theory could also be used to justify copyright protection for a combination of different DNA fragments – not found together naturally – and their insertion into a host cell in order to compose a collection and assembly of pre-existing material, facts or data.[44] Yet, this theory has also some weaknesses. One of the central principles of copyright law is that protection cannot be offered to 'utilitarian works', namely works that do not have creative or artistic expression as their primary purpose. Since the purpose of creating a specific DNA sequence is clearly utilitarian or functional – indeed, it is normally not made for purely aesthetic reasons, but to produce a certain

---

40 Robert Tomkowicz, *Intellectual Property Overlaps: Theory, Strategies, and Solutions* (Routledge, 2013) 138.

41 In the copyright law in the United States, see 17 U.S.C. § 106(2); see also Copyright in Derivative Works and Compilations, U.S. Copyright Off. (October 2013), http://copyright.gov/circs/circ14.pdf. In the European Union, copyright in compilation is provided by the Directive 96/9/EC of the European Parliament and of the Council of March 11, 1996 on the legal protection of databases, 1996 O.J. (L 77) 20.

42 17 U.S.C. § 101.

43 David B. Resnik, *Owning the Genome: A Moral Analysis of DNA Patenting* (SUNY Press, 2004) 51 (referring to the case of Celera genomics that developed an electronic genomics database with the purpose of commercializing genomic information). However, the problem, in similar situations, is that a strict proprietary regime for genomic databases may conflict with the principle of public access to genetic resources.

44 Roberto Dias et al., Metagenomics: Library Construction and Screening Methods, in Camilla Benedetti, ed., *Metagenomics: Perspectives, Methods, and Applications* (Elsevier, 2014) 45; Doreen M. Hogle, Copyright for Innovative Biotechnological Research: An Attractive Alternative to Patent or Trade Secret Protection, 5 Berkeley Tech. L.J. 75, 88 (1990).

result – protecting a DNA sequence under copyright law would mean abandoning this fundamental tenet.[45] Under these circumstances, only 'useless' engineered DNA sequences, developed for mere artistic pleasure, could receive protection.[46]

The most coherent and convincing theory for copyright protection of engineered DNA sequences seems to be the one that equates gene function to a sort of biological software. In effect, current technology developed by synthetic biology has allowed genes and cells to be programmed as computers.[47] The codified definition of 'computer program' which the US Congress added by amendment to 17 U.S.C. § 101 in 1980 describes it as 'a set of statements or instructions to be used directly or indirectly in a computer in order to bring about a certain result'.[48] This definition is then materially identical to the definition reported in the preparatory documents of the European Directive on the legal protection of computer programs.[49] Also, it has been internationally recognized that computer programs qualify as literary works even though they are not written using words. These definitions, along with the emerging technologies encompassed by synthetic biology, seem to be consistent with the assumption that 'living organisms are capable of being programmed in a manner analogous to programming computers'.[50] In other words, synthetic DNA sequences can 'act like application software designed to reproduce certain proteins or other biological material'.[51]

5. IS ENGINEERED DNA REALLY COPYRIGHTABLE SUBJECT MATTER?

Since the patentability of biotechnological inventions has initially encountered several resistances, it seems plausible that the possible

---

[45] See James G. Silva, Copyright Protection of Biotechnology Works: Into the Dustbin of History? 2000 B.C. Intell. Prop. & Tech. F. 12801 (2000).

[46] Ibid.

[47] See Andrew W. Torrance, Synthesizing Law for Synthetic Biology, supra note 5, at 636.

[48] 17 U.S.C. § 101.

[49] COM (88) 816 final – SYN 183 (1989) O.J. C91/9.

[50] See Andrew W. Torrance, Synthesizing Law for Synthetic Biology, supra note 5, at 636.

[51] See Robert Tomkowicz, *Intellectual Property Overlap*, supra note 40, at 139 (proposing that this argument can also overcome limitations of copyright law on the protection of functional works, as copyright statutes expressly include software in the definition of copyrightable subject matter despite its utilitarian nature).

extension of copyright to biotechnology works may not be easy to 'digest'. The patentability of genetic information[52] was indeed gradually recognized only after a number of decisions of courts including the US Supreme Court in 1980[53] and by the board of appeal of the European Patent Office[54] in the 1990s.[55] The question of whether copyright protection can be extended to engineered DNA sequences has not been judicially considered. Only the US Copyright Office has taken an express position. It has indeed clarified that it will not grant copyright registration to gene sequences or DNA molecules because they cannot be considered a copyrightable subject matter.[56] But this position – as it has been stressed by other legal scholars – does not preclude US courts 'from declaring engineered DNA copyrightable'.[57] The only other official position on the subject has been made by the Australian Reform Commission in the report examining the laws and practices governing

---

[52] Here the term includes genetic materials and gene fragments, such as expressed sequence tags (ESTs) and single nucleotide polymorphisms (SNPs).

[53] *Diamond v. Chakrabarty*, 447 US 303 (1980).

[54] Decision T 19/90-3.3.2, 1990 O.J. Eur. Pat. Off. 476.

[55] See Dominique Guellec and Bruno van Pottelsberghe de la Potterie, *The* Economics *of the* European Patent System: IP Policy *for* Innovation *and* Competition (OUP, 2007) 123.

[56] See Office of Technology Assessment, U.S. Congress, Publ'n No. OTA-BA-370, New Developments in Biotechnology: Patenting Life – Special Report 43 (1989) (reporting the Copyright Office's unofficial position that nucleic acid sequences are not copyrightable). More recently, the USPTO has explicitly rejected a copyright claim in a genetically engineered fish that an applicant had genetically altered so that the fish 'fluoresces' when it is exposed to artificial light; see US Copyright Office, Re: GloFish Red Zebra Danio Glowing in Artificial Sunlight (5 September 2013) available at http://ipmall.law.unh.edu/sites/default/files/hosted_resources/CopyrightAppeals/2013/GloFishRedZebraDanioGlowing.pdf (cited in Pamela Samuelson, Evolving Conceptions of Copyright Subject Matter, 78 U. Pitt. L. Rev. 17, 19 (2016)).

[57] See Christopher M. Holman, Copyright for Engineered DNA (Part 3), GQ Life Sciences (February 23, 2016), https://www.gqlifesciences.com/copyright-for-engineered-dna/ (illustrating how the US Copyright Office originally refused to register videogames display, but this was appealed and the courts overruled the Office, finding that a videogame display can be the subject of copyright protection). See also Christopher M. Holman, Claes Gustafsson and Andrew W. Torrance, Are Engineered Genetic Sequences Copyrightable? The U.S. Copyright Office Addresses a Matter of First Impression, 35 Biotechnology L. Rep. 103 (2016) (reporting the outcome of a recent submission of an engineered DNA sequence to the US Copyright Office for registration). For a discussion on the consequences of an appeal to federal court, see Dan L. Burk, DNA Copyright in the Administrative State, 51 UC Davis Law Review 1298 (2018).

intellectual property rights over genetic materials and related technologies, with a particular focus on human health issues.[58] This Australian body seemed more open vis-à-vis this issue, noting that 'copyright could potentially subsist in the representation of a genetic sequence provided sufficient skill, labour and effort is involved in creating that expression'.[59]

Also, supporters of copyrightability for engineered biotechnology works agreed on the fact that copyright law may be broad enough to consider the engineered genetic code as copyrightable subject matter. Recent advances in genetic engineering indeed allow the creation of synthetic organisms that incorporate non-canonical or non-natural amino acids into the complete set of proteins encoded by a genome (namely the so-called proteome).[60] It means that the current progress in genetic code engineering is characterized by entirely synthetic genes. Modern synthetic biology aspires to re-write genomes, encoding natural biological systems and 'producing engineered surrogates that might be usefully supplant some natural biological system'.[61] In particular, the recent advances in the development of engineered DNA are capable of reprogramming 'biological machines', applying exactly 'the same principles of engineering currently used in the

---

[58] Austl. L. Reform Comm'n (ALRC), Genes and Ingenuity: Gene Patenting, and Human Health, ALRC Report 99 (August 2004), § 28, available at https://www.alrc.gov.au/publications/report-99.

[59] Ibid. at 28.21. In addition, the Commission stressed how 'Whether or not copyright subsists in the written representation of a gene or amino acid sequence under the Copyright Act, it appears unlikely that this would hinder the use of that sequence in genetic research. As copyright protects only the expression of ideas – rather than the ideas or information expressed – researchers could continue to use the molecule itself in their research, despite the protection of its written representation. If copyright does subsist in the written representation of a sequence, this would be subject to the fair dealing exception to copyright infringement for research or study. In addition, if another researcher independently established the genetic sequence and articulated it in any identical way to that in which it had been first expressed, that independently researched expression would not infringe the copyright in the original expression'. See ibid. at 28.23.

[60] See Xiao Lin, Allen Chi Shing Yu and Ting Fung Chan, Efforts and Challenges in Engineering the Genetic Code 7 Life 12 (2017); Christopher M. Holman, Developments in Synthetic Biology are Altering the IP Imperatives of Biotechnology, 17 Vand. J. Ent. & Tech. L. 385 (2015) (noting how engineered DNA sequences are no longer mere recombinations of DNA sequences derived from nature); Andrew W. Torrance, DNA Copyright, 46 Val. U. L. Rev. 1, 6 (2011) (illustrating the evolving concepts of DNA).

[61] See Drew Endy, Foundations for Engineering Biology, 438 Nature 449 (2005).

development of the software used to program non-biological machines, i.e., computers'.[62]

Should copyright be recognized as subsistent in engineered biotechnology works, which rights would the owners be able to enforce? They would probably be able to prevent others from reproducing genetically engineered information fixed in an organism. In other words, they could preclude other scientists from using the DNA sequence they have created and incorporating and fixing it into a 'microorganism to make an identical or substantially similar copy'.[63]

6. CONCLUSION

This chapter has expanded on whether copyright may be flexible enough to protect genetic material created by biotechnological processes. This might not be an impossible task. As has been argued, for copyright regimes to be ready to take such a bold step, 'contemporary judges, practitioners, and scholars must reframe and, in some instances, reimagine the proper contours of copyrightability in order to bring living works under copyright protection'.[64] This is not just a matter of whether protection 'may' be available. It is also a matter of whether such protection 'should' be offered. Indeed, availability of copyright in this field may produce the socially desirable outcome of guaranteeing areas of freedom for the users of biotechnological products and processes, such as scientists, researchers and the like. It may achieve this balance by replacing the strict liability regime of patent law with the more flexible fair use defense of copyright regimes and the fostering of a feasible open source regime.

Copyright protection of engineered DNA sequences could therefore be seen as a new way of framing discussions between genetic information and all attempts to abridge the full access to it. In particular, it would allow us to apply to genetic information a similar regulatory regime as the one used for 'free code': an intangible resource of public utility, which must be given paternity but not full and absolute ownership, whose exploitation

---

[62] See Christopher M. Holman, Developments in Synthetic Biology are Altering the IP Imperatives of Biotechnology, 17 Vand. J. Ent. & Tech. L. 385, 420 (2015).

[63] Irving Kayton, Copyright in Living Genetically Engineered Works, supra note 13, at 194.

[64] See Murray, Post-Myriad Genetics Copyright of Synthetic Biology and Living Media, supra note 15, at 8.

is subject to transparency, not opposed to but distinct from the market.[65] Scientific knowledge is indeed supposed to be open and free. It should also benefit the general public, with a twofold purpose: on the one hand, it should be available to the scientific community; on the other hand, it should be used for further research.

It is a matter of fact that 'the recognition of copyright protection for software only occurred after many years of serious debate and discussion', thus we cannot rule out that the same thing may happen with engineered DNA sequences.[66]

---

65 See Michael J. Madison et al., Constructing Commons in the Cultural Environment, 95 Cornell L. Rev. 657 (2010); Jeorge L. Contreras, Data Sharing, Latency Variables and the Science Commons, 25 Berkeley Tech. L. J. 1601 (2010).

66 See Holman, Gustafsson and Torrance, Are Engineered Genetic Sequences Copyrightable? The U.S. Copyright Office Addresses a Matter of First Impression, supra note 57, at 104.

18. Artificial Intelligence, computer generated works and copyright

Massimo Maggiore

1. INTRODUCTION

It is common knowledge that the fundamentals of intellectual property law in general and copyright in particular are constantly challenged by new technologies. This assumption is less dull than it seems if one considers it from a specific angle that engages with the structural design of the law. There are indeed situations where the traditional concepts of 'originality' or 'creativity' of works as requirements for copyright protection appear hard to espouse with the generative process behind certain types of works.

I refer here to the use of computers and Artificial Intelligence ('AI') in the making of creative works. There is no doubt that technology has, for a long time, been used to create authorial works. The advent of photography in the 19th century, for example, very soon attracted courts' scrutiny about its compatibility with copyright. It was thus plainly established that the deployment of machines in the creative process would be no obstacle to the recognition of copyright where they only mediate, without negating or entirely absorbing, the process of artistic production. This is the position taken in 1884 by the US Supreme Court in *Burrow-Giles Lithographic Co. v. Sarony*,[1] when the Court responded in the affirmative to the question of the copyrightability of photographs. On that occasion, the Court defined authorship in terms of causation – the author being 'the cause of the picture' and 'the man who gives effect to the idea'.[2] In this view, the camera was clearly not something interrupting that causal link between the picture and the person making creative choices such as determining the framing, lighting, and the subjects being portrayed.

The same reasoning would hold true today when applied to computers

---

[1] 111 U.S. 53, 58–59 (1884).

[2] Cited in Annemarie Bridy, Coding Creativity: Copyright and the Artificially Intelligent Author, 2012 Stan. Tech. L. Rev. 5 (2012).

used by authors as instruments in the making of works of authorship. The resulting works are generally referred to as Computer Assisted Works ('CAWs'), as the technology is merely a tool for the author, like a brush in the hands of a painter. Examples of computer assisted works would be endless nowadays[3] but certainly none of them would be problematic from a copyright perspective.

Technology can, however, take the lead and supplant the human creator in the generative process of works. This is the case for Computer Generated Works ('CGWs') by which an AI-enabled machine learning software is the real originator of the final works. In CGWs the programmer establishes the rules and instructions according to which the program functions. The input given by the programmer, or by the software's user, may also entail feeding the machine general parameters; however, unlike CAWs, the creative outcome shall bear no (or at best a very feeble) causal link with the programmer or the software's users. In CGWs the creative choices are akin to random – or at least unpredictable from a human perspective – as they depend on the machine and not on the humans behind it.

In the realm of CGWs, notable recent examples are provided by attempts to have the computer systems mirror the highest expressions of human arts in history. The Next Rembrandt project (http://www.nextrembrandt.com) is a remarkable manifestation of the wonders of AI in this respect. In a nutshell, the project's goal was to have a machine produce a brand new Rembrandt painting as if the Dutch genius himself had painted it. This is epitomized by the enticing question on the project's website: 'Can the great master be brought back to create one more painting?' In order to realize the new Rembrandt painting, computer engineers and art experts collaborated and fed the system massive volumes of data concerning the characteristics of Rembrandt's works, including certain biometric parameters around the individuals portrayed by Rembrandt. As a result, the computer, through a process somewhere between data mining and machine learning, created recurrent artistic (both aesthetic and technical) patterns in Rembrandt's style, which an algorithm elaborated and transformed into a creative output of the system itself (i.e. a new Rembrandt painted with a 3D printer portraying a Caucasian male with facial hair, 30–40 years of age, in dark clothing wearing a white collar and

---

[3] Examples may include the use of word processing programs to produce texts; computer animation to make animation movies or digitize manually created drawings; music sampling where computer programs and vinyl emulation software have replaced samplers but where the musician would still choose the music samples and their combination; and computer aided design (also known as CAD), widespread among architects and other design professionals.

a hat and facing to the right). Who is the author of this work and would this work merit copyright protection? Would it be sufficient to attribute the authorship to the researchers who extracted data from Rembrandt's paintings and programmed them into the machine?

The New Rembrandt experiment is just one example of the rise of computers and AI producing works that belong in the domain of copyright. Other examples include Ray Kurzweil's Cybernetic Poet, which, again by using a type of machine-learning logic, distilled the poetic style of a number of authors and produced its own original poem. Yet more examples include the translation of texts by computer programs and the composition of music by computers to create full albums like 'Song of the Neurons' (which is available and can be played on iTunes) that are entirely composed and played by computers reunited in a system of artificial neural networks known as the Creativity Machine, designed by Imagination Engines Inc. It is noteworthy that the project's webpage enthusiastically flaunts the fact that one of the project's main achievements is that no 'humans were in the loop to decide what the most appropriate note was to follow another. It all happened rather spontaneously!'[4]

The inherent expressive quality of CGWs (as described above) is such that they may comfortably fit into descriptions of literary or artistic works that would normally be eligible for copyright protection. However, this does not necessarily mean that CGWs meet all the legal requirements for copyright. An authorship and right to ownership conundrum is created when works reflect creative choices attributable to machines. The dilemma stems from the human-centric conception of copyright law that is structured around the idea that only human beings are the source of creativity and may produce original works in a copyright sense. As it has been aptly put, 'when contemplating the creative, images of Beethoven, Joyce and Monet come to mind, not images of machinery',[5] and this is in its essence what copyright is about. The settled position is that the author is the natural person making the creative choices and, as such, infusing her own personality into a given work. This is, with varying nuances, the meaning of 'author' or 'artist' which traditionally sits at the root of copyright law. Therefore, it was no surprise that at the EU level *Infopaq*[6]

---

4 See http://imagination-engines.com/iei_musical_composition.php (visited January 16, 2018).

5 See Ralph D. Clifford, Intellectual Property in the Era of the Creative Computer Program: Will the True Creator Please Stand Up, 71 Tul. L. Rev. 1675, 1976 (1997).

6 *Infopaq International A/S v. Danske Dagblades Forening* (C-5708) EU:C:2009:465; [2012] Bus. L.R. 102.

interpreted 'originality' as the 'author's own intellectual creation'. This definition establishes a bi-univocal relationship between the act of creating a copyrightable work and a human acting as its creator so that there is no originality and hence no copyright where there is no natural person behind a work.

AI leading to CGWs is destabilizing the copyright construct built around the traditional concept of identifiable, human 'authors' to the extent that there are now works that appear to deserve consideration for copyright law purposes, while being 'author-less' from the legacy copyright perspective. The dilemma becomes even more intricate when focusing further on the originality threshold requirement. Originality is often interpreted so as to require the demonstration of a creative spark, as opposed to a purely industriousness (or sweat of the brow) inspired test, which affords protection to the demonstration of some effort or skill, labour, and judgment. Thus, originality is viewed as a component of creativity, with the two concepts conflating with each other. The dominance of the human factor appears even clearer here the extent that creativity is not intended as a function of the artistic quality of works, but rather as the footprint of the author's own personality and intellect. Neither of these elements are machine attributes. This is certainly the case in civil law jurisdictions where copyright law is concerned primarily with the protection of the authors' personality, rather than of their works. However, human creativity has been recognized as the primary justification of copyright in the US – at least after the US Supreme Court's judgment in *Feist Publications, Inc. v. Rural Telephone Service Co.* ('*Feist*')[7] – as well as in the EU after the above mentioned *Infopaq* decision.

The above analysis provides background to the present legal situation in which copyright is generally not available for CGWs. There are, however, a handful of jurisdictions where exceptions apply, such as the UK, where CGWs are in fact copyright protected. Section 2 of this chapter will outline the traditional human-centric nature of copyright law, which constitutes an obstacle for the protection of CGWs, and expand on the originality and creativity requirements as regulated in common and civil law jurisdictions as well as at EU level. Section 3 seeks to deconstruct the human author ideology and expands on whether copyright laws may shift their focus from a need to protect authors to the need to protect the works themselves. Section 4 will then set out certain policy considerations that militate in favour of offering CGWs copyright protection, and suggests

---

[7] *Feist Publications, Inc. v. Rural Telephone Service Co.*, 499 U.S. 340 (1991).

that such works are offered a sui generis right, building upon the legislative solution adopted in the UK.

2. THE HISTORICAL BEACON OF COPYRIGHT LAW: THE AUTHOR AS THE CREATIVE HUMAN BEING

Copyright laws around the world have historically been based on the centrality of the 'author' as the flesh and blood individual behind the work, by displaying a 'human' infrastructure that pays tribute to the 'ideology of authorship'[8] and is rooted in the Romantic idea that associates human genius with authorship. What is relevant here is the historic conception of authorship which informed the original structure of many copyright laws and at least formally has not changed: the author is idealized as a creative, lone artist, inspired directly from nature.

The human author as the *raison d'être* of copyright is also reflected in one of its fundamental principles – the idea-expression dichotomy. Works always originate from ideas which, however, will not be protected by copyright until they are expressed in a form. The issue is particularly relevant for expressions created by AI software where the computer generates the expression even though it might not have been the source of the underlying idea. As has been correctly observed, the difficulty here is the task of identifying who (the computer or the individual) has generated what (the idea or the expression). As a result, if the idea and the expression cannot be readily distinguished, then copyright protection should not be available.[9]

Other obstacles to copyright protection for CGWs stem from the originality/creativity requirements, which I will focus on in the next sub-section.

---

[8] Peter Jaszi, Toward a Theory of Copyright: The Metamorphoses of 'Authorship', 42 Duke L. J. 455, 502 (1991). See also Michael J. Madison, Beyond Creativity: Copyright as Knowledge Law, 12 Vand. J. Ent. & Tech. L. 817, 829 (2010) (identifying a progressive slide of copyright law from a law originally intended to protect knowledge to a law protecting creativity, as a result of a rising ideology of the creative authorship that developed in the 19th century).

[9] Timothy L. Butler, Can a Computer be an Author – Copyright Aspects of Artificial Intelligence, 4 Comm/Ent L.S. 707, 728 (1981) (noting that it would be 'a herculean task to determine if the expression is one of an underlying idea (and thus of human origin and therefore possibly copyrightable) or is only "apparently" an expression of an idea (the existence of which is anthropomorphically read into the expression by the viewer) and thus not within the contemplation of copyright law'.

2.1 Originality and Creativity as Requirements for Protection

In the US a work is copyrightable if it 'owes its origin to the author'.[10] This reflects a minimalist interpretation of the originality requirement (in other words that a work should not be a copy of a pre-existing one), which would no doubt help in considering CGWs as protectable. Yet, the concept of originality as a threshold requirement has progressively overlapped with, and then has been replaced by, that of creativity as a purely human quality. A work shall therefore be considered creative if it carries the (even low) imprint of the personality of an individual.

This is certainly the characterization of the requirement in other countries, such as Italy, where copyright law does not use the somewhat neutral word 'original', but requires the work to have 'creative character', and defines the creation of a copyrightable material as the 'peculiar expression of intellectual work'.[11] Italian courts have constantly found that a work meets the creative character requirement when it is 'the result of the expressive endeavour of the author, mirroring the author's personal way of representing facts, ideas, situations and feelings'.[12] An almost identical approach can be found in German copyright law, which reminds us that 'only the author's own intellectual creations constitute [copyrightable] works'.[13]

Unlike Italy and Germany, France and Spain do not define 'originality' or 'creativity,' although both statutes set them out as thresholds requirements. The French Intellectual Property Code speaks about an 'exclusive incorporeal right enforceable against all persons' as a result 'of the mere fact of its creation'.[14] Spanish Intellectual Property law, on the other hand, simply identifies the threshold requirement in the 'originality' of the work.[15] Both in Spain and France there should be no doubt, however, that the source of creativity or originality as requirements for protection can only

---

[10] See also Pamela Samuelson, Allocating Ownership Rights in Computer Generated Works, 47 U. Pitt. L. Rev. 1185, 1198 (1985).

[11] Article 6 of the Italian copyright law, I.21 April 1941, n.633, G.U. 16 July 1941, n.166 (It.), as amended.

[12] Tribunal of Milan, Sezione specializzata in materia di impresa, ordinanza February 4, 2015, AIDA 1743, 1745 (2016).

[13] See Article 2(2) of the Gesetz ber Urheberrecht und verwandte Schutzrechte [Copyright Act] September 9, 1965, BGBI, p. 273 (Ger.).

[14] See Article L111-1 of the French Code de la Propriété Intellectuelle [Intellectual Property Code] enacted by Loi 92-597 du 1er juillet 1992 relative au code de la propriété intellectuelle (partie legislative) [Law 92-597 of July 1, 1992 on the Intellectual Property Code, Journal Officiel de la République française] [J.O.] [Official Gazette of France], February 8, 1994, last consolidated August 27, 2016.

[15] See Article 5, Ley de Propiedad Intelectual (B.O.E. 1996, 8930) (Spain).

be human. This is confirmed by the French Code defining protectable subject matter as 'oeuvres de l'esprit',[16] or, in the English translation provided by the French government, as 'works of the mind'.[17] Both 'mind' or 'esprit' connote the human and not the artificial author, as that 'part of a person that makes it possible for him or her to think, feel emotions, and understand things'.[18] As far as Spain is concerned, the foundational elements of copyright protection law are explained as '. . . the rights that correspond to the author, that is the person who realized the purely human and personal effort of creating the work and that, for that reason, constitute the essential nucleus of the subject matter'.[19] Spanish law also states that the author of a work is the natural person that creates it (Article 5).

Such nuanced declinations of the 'originality' and 'creativity' concepts are nowadays harmonized under the test of originality devised by the Court of Justice of the European Union (CJEU) in *Infopaq*. After this seminal case a work in the EU is considered original only if it is the 'author's own intellectual creation', a qualification that the InfoSoc Directive[20] has generalized as applicable to all copyright works (the requirement had been initially set out in the Software and Database Directives). The impact of *Infopaq* on the issue of the copyrightability of CGWs appears clear. As has been correctly noted, this case 'clarifies that a human author is necessary for a copyright work'.[21]

The *Infopaq* test is in line with the traditional conception of copyright as a law protecting authors and their creativity, which is common among civil law countries, such as Italy, Germany, France, and Spain. Elsewhere the threshold originality requirement has been historically shaped around the somewhat more impersonal test of 'labour and skill', which, per se, demands less of a human nexus for works to be copyrightable. This is the

---

[16] See Article L112-1 of the French Code de la Propriété Intellectuelle, supra note 14.

[17] See the English text at: https://www.legifrance.gouv.fr/content/download/1959/13723/. . ./3/. . ./Code_35.pdf.

[18] See the definition of 'Mind' at https://dictionary.cambridge.org/dictionary/english/mind.

[19] See Preámbulo, Ley de Propiedad Intelectual, supra note 15 (stating '. . . los derechos que corresponden al autor, que es quien realiza la tarea puramente humana y personal de creación de la obra y que, por lo mismo, constituyen el nucleo esencial del objeto de la presente Ley').

[20] Directive 2001/29/EC, of the European Parliament and of the Council of 22 May 2001 on the harmonisation of certain aspects of copyright and related rights in the information society, 2001 O.J. (L 167).

[21] Christian Handig, The Copyright Term "Work" – European Harmonisation at an Unknown Level, 40 Int'l Rev. Intell. Prop. & Competition L. 665, 668 (2009).

case in the UK, where originality is required to subsist in copyright works under Section 1(1)(a) CDPA. Across the Channel, however, precedents such as *Temple Islands Collections*[22] and *Football Dataco v. Yahoo! UK*[23] have led scholars to hold that the UK's 'skill and labour' standard has merged with the EU *Infopaq*'s 'intellectual creation' test, meaning that the UK has also internalized a more personal and less labour and effort inspired test of originality.[24]

The importance of creativity as a component of originality has also been affirmed in the US Supreme Court's ruling in *Feist*. Pursuant to § 102 of 17 U.S.C. 17, copyright protection shall subsist for works of authorship that are 'original'. *Feist* put an end to the variable approaches to originality that had arisen in the US[25] and set a standard of originality according to which 'copyright protects only those constituent elements of a work that possess more than a de minimis quantum of creativity'.[26] By so ruling, Feist killed off the 'sweat of the brow' and 'industriousness' requirements that had previously underpinned the copyrightability of compilations.

Australia is another jurisdiction where the copyright act requires that the work originates from a human author, to the extent that the 'two expressions "author" and "original work" have always been correlative: the one connotes the other'.[27] Particularly, the case law in Australia relates originality to a human source and creativity to the human intellect. Such findings have, for example, been confirmed in *Acohs v. Ucorp*,[28] a case concerning copyright protection for the output of a computer consisting of a Material Safety Data Sheet ('MSDS') (manufacturers, importers and suppliers of certain hazardous substances are required by law to prepare and set out information about the relevant substance). The MSDS is automatically filled in by the

---

22 *Temple Island Collections* [2012] EWPCC 1; [2012] E.C.D.R. 11.

23 *Football Dataco Ltd. v. Yahoo! UK Ltd* (C-604/10) EU:C:2012:115; [2012] Bus. L.R. 1753.

24 Andres Guadamuz, Do Androids Dream of Electric Copyright? Comparative Analysis of Originality in Artificial Intelligence Generated Works, 2 Intell. Prop. Q. 169, 180 (2017).

25 Elizabeth F. Judge and Daniel Gervais, Of Silos and Constellations: Comparing Notions of Originality in Copyright Law, 27 Cardozo Arts & Ent. L.J. 375 (2009).

26 *Feist Publications, Inc. v. Rural Telephone Service Co.*, 499 U.S. 340, 363 (1991).

27 Jani McCutcheon, The Vanishing Author in Computer-Generated Works: A Critical Analysis of Recent Australian Case Law, 36 Melb. U. L. Rev. 915, 935 (2013) (with reference to case-law).

28 *Acohs Pty Ltd. v. Ucorp Pty Ltd* [2010] FCA 577, confirmed on appeal.

Acohs system upon request on the basis of the extraction of data stored in a database.[29] Acohs asserted copyright in the HTML source code for each particular MSDS and claimed that the respondent had copied the MSDS by extracting the HTML code. The Court referred to *IceTV v. Nine Network*,[30] which reaffirmed that 'The "author" of a literary work and the concept of "authorship" are central to the statutory protection given by copyright legislation', and on this basis found that no copyright would subsist in the computer generated HTML source code of the MSDS, on the grounds that '. . . the source code as a work – i.e. as a complete entity – was not written by any single human author. It was generated by a computer program'.[31]

Finally, the 2004 Supreme Court of Canada's (SCC) judgment in *CCH v. Law Society of Upper Canada* ('*CCH*')[32] shed light on the originality requirement in Canada. In *CCH* the SCC took an approach which, although explicitly denying that creativity in a civil law sense is required, applied a 'skill and judgment' test defined in relation to activities which only humans can perform: 'the use of one's knowledge, developed aptitude or practised ability in producing the work and . . . the use of one's capacity for discernment and ability to form an opinion or evaluation by comparing possible options in producing the work'.[33] As noted, although the court did not ever use the word human, it probably did not feel the need to do so because it is implied in the definition of 'skill and judgment' that it gave.[34]

The above overview of the threshold requirements in a number of jurisdictions confirms that copyright protects authorial works created by humans. The human tie is reinforced when creativity or originality in a copyright sense are codified and interpreted by the case law as synonyms

---

29 See para 24 of the judgment, where the way the Acohs system, also named Infosafe System, works is described: 'The technical centre of the Infosafe System is a computer program – written in Visual Basic code – which has the functions of receiving a request for the generation of a particular MSDS, of calling up the data and other elements from the Central Database, of compiling the relevant HTML source code, and of sending that code to the user's computer. To do this, the program uses "routines" called "GETHTML", "Generate_MSDSReport" and "GenerateHTML". These routines are in turn run by a fourth routine, "WebClass_Start". The source code will then cause the MSDS to appear as intended on the user's screen, in the conventional way of web pages generally'.

30 *IceTV Pty Ltd v. Nine Network Australia Pty Ltd* (2009) 239 CLR 458.

31 *Acohs Pty Ltd. v. Ucorp Pty Ltd*, supra note 28, at para 50.

32 Cited in Mark Perry and Thomas Margoni, From Music Tracks to Google Maps: Who Owns Computer-generated Works?, 26 Comp. Law and Security Rev. 621, 623 (2010).

33 Ibid. at 625.

34 Ibid.

with imagination, inspiration or artistry – none of which spring to mind as attributes of computers or of CGWs. The longstanding theoretical underpinnings of copyright laws bode ill for an extension of protection to CGWs, as such works are characterized by the absence of human intervention in the creative process.

3. FROM THE LAW OF AUTHORS TO THE LAW OF COPYRIGHTABLE WORKS?

It remains to be seen, however, if the ideology of the Romantic human author as the historical justification for copyright protection is, in reality, seamlessly reflected in every aspect of copyright law. Against this background, this section outlines some contradictory features of copyright laws, being mindful of a few examples of author-less or dispersed authorship works (other than CGWs) that may be helpful in deconstructing the human author pillar of copyright.

Upon closer examination, human authorship may not be an a priori of the law. It is indeed nothing more than it should obviously be expected to be – a legal construct designed around policy considerations. This assumption may find support by considering, for example, how copyright law deals with the level of creativity required for works to be protected. Indeed, the ideology of the Romantic author intuitively seems to imply a hierarchy of creative productions, where those that are artistically meritorious would displace, as the only ones deserving protection, those that are devoid of any artistic value. However, a departure from the original abstraction of the concept of authorship and originality becomes apparent here, to the extent that it is a settled concept that copyright law is not concerned with the artistic merit of the work. Anything falling within the categories of copyrightable works may be protected, regardless of how modest the artistic quality. An author, taking stock of this principle, as explained in the US Supreme Court's landmark case *Bleinstein v. Donaldson Lithographic Co.*,[35] has held that the *Bleinstein* opinion, with its rejection of the court's role as adjudicator of the aesthetic value of a work, left the category of authorship 'with little or no meaningful content and none of its traditional associations'[36] in its original Romantic ideal.

It cannot be taken for granted that in *Bleinstein* the Court did actually play down the role of authorship in copyright, as it seemed more concerned

---

[35] See *Bleinstein v. Donaldson Lithographing Co.*, 188 U.S. 239 (1903).

[36] See Jaszi, supra note 8, at 483.

with avoiding the risk that requiring a threshold of aesthetic merit could result in making protection dependent on the taste of the judge.[37] On the other hand, the Court insisted on the uniqueness that any work may display, which is strictly linked to the personality and singularity of each individual creator and independent from the artistic gradient. Thus, *Bleinstein* cannot be taken to offer an opening to author-less or creativity-less works, which would instead be needed for CGWs to be copyrightable.

Yet it is undeniable that the very low threshold of creativity makes the necessary human nexus requirement correlatively faint. A paradox becomes apparent when one considers that CGWs may display a very high level of creativity, if viewed in terms of their intrinsic quality, and yet are deprived of copyright protection for want of a human creator.

Nevertheless, the current legal copyright framework seems to gravitate around the 'creative human author'. In spite of this, it is uncertain who the author is, or, to put it differently, what qualities a person should have to be recognized as author. This is because neither national, nor international copyright laws provide a definitive definition of authorship. If one looks at the primary source of international copyright law, the Berne Convention, s(he) realizes that it 'provides only limited guidance: while it lists a series of works in article 2 that each Union country is to protect, it does not contain any definition of the term author'.[38]

National laws may provide indirect hints around how authors should be defined. This is the case, for example, in the UK Copyright, Designs and Patents Act 1988 ('CDPA'). Section 9(1) in particular identifies the author, in relation to the work, as the person who creates it. Similarly, some European copyright statutes provide that the original title giving rise to copyright ownership shall be the 'creation' of the work.[39] However,

---

[37] See *Bleinstein v. Donaldson Lithographing Co.* supra note 35, at 251–252 (Justice Holmes noted that 'It would be a dangerous undertaking for a person . . . to constitute themselves final judges of the worth . . . At one extreme some works of genius would be sure to miss appreciation. Their very novelty would make them repulsive until the public has learned the new language in which their author spoke').

[38] Sam Rickeston, *The Berne Convention 1886–1986* (Sweet & Maxwell, 1984) chapter 6.4. Yet, as has been observed, Professor Rickeston had regardless no doubt that there was a basic agreement between contracting states as to the meaning of the term 'author', which should impliedly be taken to refer to the person who created the work – see Jane C. Ginsburg, People Not Machines: Authorship and What it Means in the Berne Convention, IIC (2018) 49: 131–135, at p. 31.

[39] This is the case of the Italian Copyright law (Article 6); the French Intellectual Property Code (Article L111-1); and the Spanish Copyright Law (Article 5).

none of these laws define 'creation' and therefore continue to trouble what is meant by creative authorship in a copyright context.

Thus, the concept of authorship, although central to copyright law, lends itself to ambiguity and instability. Possibly, the most bewildering example of how the concept of authorship may be twisted to accommodate specific policy goals is provided by the doctrine of 'work made for hire', under Section 201(b) of the US Copyright Act 1976. According to this doctrine, the entity (individual or undertaking) who paid to have the work created, whether by an employee or an independent contractor, is considered the author unless otherwise agreed. Now, it is clear that such a doctrine entirely separates the act of creation from the ensuing legal interest in the work and gives rise to a complete legal artefact whereby authorship and endowment of the exclusive rights of economic exploitation deriving from copyright ownership are conflated. This doctrine exemplifies the vanishing of the author as a real central concept of copyright law. Here, an 'author' is created by virtue of a simple contractual covenant. No input into the creative process is required of the 'author', not even that of directing the work of the person hired for the purpose. Hence the concept of authorship as a function of a creative act is traded for the concept of 'deemed authorship' justified by the payment of a consideration.

The above is one part of a much wider debate on whether it is not somewhat of an optical illusion to keep looking at the 'author' and at 'creativity' as the sole and real source of justification of the global copyright law construct. Indeed, copyright law as such contains clear indications of its loose, or at least unstable, dependence on the concept of authorship. Contrary to conventional wisdom, a closer look may reveal that copyright is not the law of authors, but the law of copyrightable works. The list of protectable subject matter, with its idea-expression dichotomy, objectifies the (obscure) concept of author, by epitomizing the intellectual abstraction into a tangible manifestation, that is to say the work. The example of 'works made for hire' shows that copyright law can do without the author (meaning that the law may tolerate a framework whereby the statutory author is divorced from the actual creator), but cannot do without the 'work' or 'subject matter' which may appear as the ultimate target and reason for protection.

Other examples in the copyright arena can be given to show how a refocusing of its goals may be achieved, shifting from a law protecting creativity to a law protecting works and the knowledge encapsulated in them. Products such as Wikipedia or Open Source ('OS') software are two prominent examples. The first issue is one of scale and of a clash between copyright law and the norms and expectations of the communities that contribute to Wikipedia or to OS software lines of code. These are cases of mass collaboration of countless individuals who are inspired solely by their will to share their creative inputs.

The above phenomena are at odds with the straitjacket of traditional copyright laws, which provide default rules requiring the identification of the author(s) and automatically vesting in them certain exclusive rights of ownership. For example, under UK copyright law Wikipedia pages would probably qualify as a work of joint authorship, pursuant to Section 10(1) CDPA. However, as noted,[40] as each page is constantly changed, ordinarily by small increments by a myriad of contributors, it is difficult to discern which participants to the project have made the relevant contribution to identify the 'collective author' under Section 10(3) CDPA. A related issue concerns the identification of the work's contours and the relevant originality requirement. While the work is constantly in progress, so as to raise the question whether copyright law is compatible with a work which is not intended to be finished, the increments added to each pre-existing page (or line of code in the OS example) by contributors might also be seen as derivative works. However, the often minimal size of each increment throws into question the actual 'materiality of the change' that would be required in order for a derivative work to meet the originality requirement.

What we can infer from the above is that the ethics of sharing knowledge and fine grained contributions of dispersed authors – otherwise defined as 'social production'[41] – have little in common with the consolidated norms of copyright, let alone with the idea of the solipsistic genius forged in the Romantic age. Such examples of social production vividly show how the current global copyright regime is unfit for new forms of mass creativity, which obfuscate several pillars of traditional copyright tenets, beginning with the very concept of authorship. As has been argued, multi-author mass collaboration with no hierarchical organization is a game changer and the ensuing creativity is likely not authorship.[42]

Legal literature also offers examples of cases concerning 'automatic writings' – also known as psychography or texts written by persons (such as spiritualist mediums at séances) – claiming to be acting under the influence of a spirit. The psychography cases involve issues that are analogous to those arising in connection with CGWs, as in both instances there is no creative human author and nonetheless, in the event of psychography,

---

[40] See Daniela Simone, Copyright or Copyleft: Wikipedia as a Turning Point for Authorship, 25 Kings L. J. 102, 104 (2014).

[41] Niva Elkin-Koren, Tailoring Copyright in Social Production, 12 Theoretical Inquiries in Law 309, 313 (2011).

[42] Cass Sunstein, *Infotopia: How Many Minds Produce Knowledge* (OUP, 2006) 153.

copyright was found to subsist in favour of the person acting as the medium.[43] This is an original attempt at analogizing two inherently different situations. Indeed, psychography, unlike CGWs, is an esoteric practice which a court of law may not take as anything more than what it actually is, that is to say a wholly human work, even though it seems to be dictated by non-terrestrial entities (whose existence, however, may or may not be believed but certainly may not be scientifically demonstrated). Yet the analogy is still interesting, because it shows that copyright law can disengage from the concept of human author, in spite of the person to whom protection is granted denying his own creative contribution in the protected work.

The question about the role of authorship in copyright has certainly been made more pressing by the rise of CGWs. The advent and development of CGWs raise the question of whether it is time to statutorily and openly recognize author-less productions within the realm of copyright laws.

4. THE CONUNDRUM OF COMPUTER GENERATED WORKS: DESERVING PROTECTION BUT NOT BEING PROTECTED – THE UK EXCEPTION

The issue is here two-fold: whether it is desirable to afford CGWs protection from a policy perspective; and, in the affirmative, how such protection can be reconciled with current copyright law principles that seem to require a 'human' author.

Take the Next Rembrandt project as an example of where a machine came up with a creative output using the analysis of large data sets extracted from the meticulous observation of a number of original Rembrandt's paintings. The machine learnt from the data and made the artistic choice that can now be admired in the painting, meaning that the machine is effectively the 'author'. The characterization of originality and creativity as well as authorship as essentially human attributes leaves no room for granting CGWs protection because what the law demands is the exercise of 'creative freedom'[44] whereas the author(s) – intended to be human beings – would exercise none in CGWs.

Should copyright laws be less strict in terms of 'human' authorship by allowing the protection of CGWs? I believe so, primarily as such protection

---

[43] See Bridy, supra note 2, at 18.

[44] Case C-604/10, *Football Dataco Ltd. v. Yahoo! UH Ltd.* (2012).

would accommodate important policy goals. In particular, the emphasis here should be placed on the desirability of protecting a work which more often than not is unique, the result of considerable technological effort and investment, and increasing the existing stock of knowledge. As has been observed, 'given the low standard of intellectual effort required to confer originality, copyright policy does not so much encourage great mental exertion but its productive result – a new and potentially valuable work'.[45] The need to incentivize the creation of new works is particularly felt in the digital economy where heavily computer-dependent creative businesses that contribute to the accumulation of knowledge continue to ask for some forms of protection for their investments.

The case for protecting CGWs confirms that the concept of authorship is losing importance and needs to be replaced by a concept of 'deemed authorship'. In those jurisdictions that do not provide for the protection of CGWs, including in the EU following *Infopaq*, works like the Next Rembrandt or the music album 'Song of the Neurons' would remain part of the public domain.[46] This is so due to a lack of any specific provision covering CGWs in both the common and civil law legal frameworks based on principles that are at odds with the possibility of affording protection where no creative human input in the ultimate result exists.

The UK's approach demonstrates a potential way forward. Section 178 of the CDPA contains a definition of CGWs which can be borrowed to establish a universally accepted meaning of CGWs for copyright purposes: a CGW 'is generated by computer in circumstances such that there is no

---

[45] See McCutcheon, supra note 27, at 954. This position – which I endorse – places greater emphasis on the incentive to create as one of the two pillars on which copyright reposes, the other one being the protection of creativity as human attribute (which translates into the protection of the authors' moral rights in the civilian systems' parlance). Admittedly, however, this position is not readily reconcilable with the Berne Convention, which to the opposite shows a strong humanist footprint, not permitting the outright expulsion of the human authors – see Jane C. Ginsburg, supra note 38, at 134.

[46] For an opposite opinion see Guadamuz, Do Androids Dream of Electric Copyright?, supra note 24, at 179, (noting that as the creation of the Next Rembrandt was based on the selection of portrait elements that were fed by humans into the computer, then even under *Infopaq*'s standards, it would be difficult to hold that the portrait does not represent the personality of the authors and therefore the computer generated portrait would deserve copyright). I believe this is a flawed argument, to the extent that it would conflate the idea (i.e., the feeding of data in the system) with the expression, and while the former may be human, the latter is certainly not, not least because the outcome was not predictable or even wanted by the humans contributing information and – as is known – copyright only protects expressions.

human author of the work'. This definition pinpoints the feature of CGWs that is more striking and problematic from a copyright perspective – the absence of a human author. As mentioned in the introduction of this chapter, when computers are instead only mere tools in the hands of human authors, as in the case of CAWs, no issue would arise under the existing copyright theoretical framework.

As far as CGWs are concerned, UK law has adopted the solution that basically grounds protection in the concept of deemed authorship, whereby the author of a CGW is considered the person 'making the necessary arrangements' for the creation of the work. Under the CDPA, depending on the circumstances, the owner of the copyright in CGWs can alternatively be the programmer or the user of the program. In either case, there will be no personal creative input from the copyright holder; as Lord Beaverbrook (House of Lords) commented on the act during its passage, the person undertaking the arrangements for a CGW to be created 'will not himself have made any personal, creative effort'.[47] Therefore, while Section 9(3) CDPA probably marks an exception to the creativity and originality requirements,[48] it remains unclear how these requirements should be assessed in relation to CGWs.[49] At the same time, the provision of Section 12(7) of the CDPA for a term of 50 years from the end of the calendar year in which the work was made, hints at the qualification of CGWs as entrepreneurial works and the relevant right as a related or neighbouring right.

As copyright laws in most countries – excepting the UK and the few others that followed suit[50] – are not receptive to making copyright available for CGWs, the solution would require a reformulation, by way of exception, of the thresholds requirements for copyright, thus stripping away the standards of originality and creativity and instead embracing an industriousness oriented approach (that would reward skill, labour and effort or somehow focus on the creation of incremental knowledge). Therefore, as far as CGWs are concerned, the originality/creativity requirement might be replaced by a non-obviousness standard to be interpreted less stringently

---

[47] Copyright, Designs and Patent Bill, Hansard, HL, Vol. 493 (February 25, 1988), cited in Emily Dorotheou, Reap the Benefits and Avoid the Legal Uncertainty: Who Owns the Creations of Artificial Intelligence? 21 Computer and Telecomm. L. Rev. 85, 87 (2015).

[48] See Guadamuz, supra note 24, at 176.

[49] See Dorotheou, supra note 47, at 87. The question may be for example whether under UK law it would be sufficient for CGWs to be original, in the sense that they have not been copied from pre-existing works.

[50] Hong Kong, India, Ireland, New Zealand.

than in patent law and taking into consideration whether the works in question are not constrained, for example, by utilitarian[51] or expressive factors.[52] All other requirements for copyright protection should remain unchanged. For instance, CGWs should not be considered protectable if they constitute copies of pre-existing works. In this specific regard, however, it might be sensible for any reform to exclude the defence of independent creation for CGWs. Indeed, independent creation is a defence available to those accused of copyright infringement, and can be used when it is demonstrable that the respondent had no potential access to the plaintiff's pre-existing work. As a matter of fact, access implies the cognitive exposure (even inadvertent) of an individual to earlier works. However, when it comes to CGWs, this cannot be proven. Computers at most can collect and process information, but as such cannot be held to have 'knowledge' or be aware of pre-existing works. On the other hand, because computers can in fact mine huge amounts of data and information, a rebuttable presumption could be established, as an alternative, that pre-existing works have in fact been mined (or 'known' for that matter) by the computer.

CGWs could also be protected under a related right, with a shorter term of protection than the standard term (70 years after the death of the author), such terms being computed from the first publication of the work. Requirements of originality and creativity should, I believe, be abandoned for the reasons mentioned above. Similar to what is provided by the UK's CDPA, the rights could be vested in the person making the arrangements for the creation of the work without any predetermined choice in favour of the programmers or of the users of the computer generating the work.[53]

---

[51] The computational capacity of computers is increasing at such a pace that computer generated works might for example easily aim at completeness rather than selection. For instance, while an imaginary dictionary, listing all words ever used by humans since the advent of writing and elaborated by computers as the result of the processing of all the existing published writings, may be useful, nevertheless it may not deserve copyright protection (although it might still deserve the sui generis database protection).

[52] Similar to what happens with the idea-expression merger doctrine, if there is a limited number of ways to express the message expressed by the CGW, then such a work shall not be protected.

[53] In the US the CONTU Commission (National Commission on New Technological Uses of Copyrighted Works) in 1978, when dealing with computer-created works, concluded in its final report that copyright in computer created works should be vested in the program's users and not in the programmers. This conclusion, however, seems biased by the assumption that computers in no way contribute authorship, while works are produced through their use. In other words, the Commission compared computers to inert tools of creation, just as it happens with photography. As mentioned, this position is no longer tenable

An alternative solution could be to vest the exclusive rights of the CGWs in the person or entity making substantial investments in the making of CGWs, such as those financing the development of the AI systems creating the CGWs, particularly when the investment is purposively made to that end. Such an approach would be similar to that followed by the EU Database Directive in relation to the sui-generis rights, where the requirement for protection is the substantial investment in obtaining, verifying and presenting data. This would be a sensible approach, as it would ensure that the making of AI machines capable of creating literary and artistic works is not discouraged.

5. CONCLUSION

The argument has been made that offering CGWs copyright protection might lead to allocative inefficiencies from an economic perspective and stifle, rather than foster, innovation.[54] However, it cannot be denied that the developments of computer science are not predictable. After all, their pace justifies the assumption that in the not so distant future computers might rival, if not wholly overtake, human ingenuity in the production of valuable knowledge. The examples of current CGWs discussed in this chapter demonstrate that there may be no perceptible qualitative difference between a human and a computer generated work. The justification for the persisting copyright discrimination against CGWs in most jurisdictions is not attuned with a radically changed and changing contemporary scenario where the production of knowledge is no longer the preserve of a few elected intellectuals. Indeed, as they presently stand, copyright laws may disincentivize investments in technologies capable of resulting in valuable contributions to the accumulated stack of knowledge. This outcome is even less justified considering that the primary spark of any progress remains wholly human.

---

nowadays: see Evan H. Farr. Copyrightability of Computer-Created Works, 15 Rutgers Computer & Tech. L.J. 63, 66 (1989).

[54] See Perry and Margoni, supra note 32, at 628.

PART IV

Illegality and immorality

19. Copyright protection of illegal works

Eldar Haber

1. INTRODUCTION

Accounting for jurisdictional differences, copyright protection is generally granted for any work that is both original and fixed in a tangible medium of expression. There is no examination of the nature of a work as long as it is creative enough, does not infringe upon another work and as long it is fixed. Under these narrow requirements, copyright law is arguably indifferent to questions of morality and illegality. It does not presume to take a side. Child pornography, revenge porn, snuff films,[1] crush films,[2] or any other original works of authorship that involve criminal activities are copyrightable as long as they are original and fixed in a tangible medium of expression.

But that was not always the case, at least as the history of the American copyright regime teaches us. Since the formation of the first U.S. Copyright Act in 1790, many courts have interpreted copyright law to explicitly bar the registration of illegal, immoral, fraudulent or blasphemous works.[3] Until 1979, it was rather clear that copyright law will not protect illegal works. Since then, however, it has become less clear whether copyright law is content neutral or not. While the Fifth and Ninth Circuits have held that all original works of authorship fixed in a tangible medium are copyrightable, district courts in the Second Circuit have disagreed, and

---

[1] Snuff films depict the actual murder of a person for the enjoyment of the audience, often for the express purpose of increasing distribution and entertainment or financial exploitation. See Arnold H. Loewy, Obscenity, Pornography, and First Amendment Theory, 2 Wm. & Mary Bill Rts. J. 471, 476 (1993) (defining 'snuff films').

[2] Crush films usually depict stepping on small animals. See Crush Videos Law and Legal Definition, US Legal (April 2010), https://definitions.uslegal.com/c/crush-videos.

[3] See infra section 2.

other circuits have not directly addressed this matter yet.[4] Thus, much uncertainty still lies within the U.S. regarding the neutrality of copyright protection.

Aside from the current vagueness under the American copyright regime, the general perception of copyright protection on the international level is that of a content-neutral regime.[5] Even works whose mere creation or content is considered as a criminal offence could be protected by copyright and could benefit their makers. However, even if most jurisdictions worldwide require merely originality and fixation (or other narrow requirements) for copyright protection, the broader normative question at stake is whether copyright law should limit protection only to certain works such as those that benefit society.

Generally, granting legal protection in the form of copyright to some types of works could be highly questionable.[6] Examples could include works that are either conceived as immoral, like pornography, or works whose mere creation is considered a criminal offence like child pornography, snuff films and crush videos. These types of illegal works – among other types, further analyzed in the next part of this chapter – could be perceived as not merely not advancing the rationales behind granting copyright protection, such as promoting knowledge, but rather also be proven as harmful for their victims, their families and other members of society. Should the content of a work change its eligibility for protection under copyright law? What are the benefits and drawbacks of granting protection to works that do not necessarily benefit society, but are also capable of causing harm?

Incentivizing the making of illegal works by granting them legal rights is questionable and requires further examination. The purpose of this chapter is to serve as an introduction to the concept of illegal works under copyright law. While it focuses mainly on the American legal system as a key example, it is not generally limited to a single jurisdiction as it strives to raise a normative debate. It will begin by conceptualizing the types of works that could be deemed 'illegal' by offering a taxonomy to distinguish between them. The next part will analyze the American copyright regime as a key example of how a legal system treats illegal works. Then, this

---

[4] Ibid.

[5] See infra section 2 (discussing the requirements for copyright protection on the international level).

[6] For a discussion on another type of works whose protection could be questionable, for example copyright in the works of Nazi leaders, see Chapter 21 in this volume by Marc Mimler, On How to Deal with Pandora's Box – Copyright in Works of Nazi Leaders'.

chapter will proceed to a normative discussion of whether the law – and more specifically, copyright law – should place limitations on works based on their illegal nature. This part will be followed by the proposal of a number of pragmatic solutions to the problem of illegal works. The final part will conclude the discussion and suggest that under certain circumstances, the law – whether copyright law or the law in general – should not protect the makers of works that are both harmful for society and that do not promote knowledge.

2. ILLEGAL WORKS

What is the definition of illegal works? This terminology could vary between individuals and could have different meanings within various legal jurisdictions. To offer a taxonomy, one must first define the term illegality in this context. While illegality could refer to any conduct that is explicitly prohibited by the state, for the purpose of this taxonomy the term illegality would be confined to prohibitions of conduct that is linked to criminal liability. The types of works which this chapter considers 'illegal' could be further divided into three categories:[7] (1) works whose mere creation or content is a criminal offence – child pornography would be an obvious example of such works; (2) works that are directly linked to a criminal activity or made while engaging in unlawful conduct, while the content of the works or their making is not considered unlawful – for example, the filming of a bank robbery, even if carried out by one of the robbers, is not illegal per se; and (3) reputational illegal works – namely, works made by individuals who were convicted of a crime and produced works that are either directly related to a crime or not, while asserting profits from their reputation deriving from the criminal activity; that is, if a convicted felon, prior or after release, publishes a book that relates to his crimes or simply, say, an unrelated poetry book, he could potentially earn profits from copyright protection due to his misconduct. Using this taxonomy, this chapter will refer to any of the categories as illegal works, while in some instances limiting its analysis to a specific category.

---

[7] In a rather similar context, I broadly referred to illegal works as any works of authorship whose content is criminal, such as snuff; that are directly linked to harmful crimes, including works that promote criminal activity; or that are otherwise legal works made by a felon that profits from their infamous reputation. See Eldar Haber, Copyrighted Crimes: The Copyrightability of Illegal Works, 16 Yale J.L. & Tech. 454, 456 (2014).

Illegal works are thereby works whose content is illegal or that are directly related to a crime or an individual convicted of a criminal offence. Other than that, the moral nature of a work or a conduct is not part of the categorization of illegal works, unless it is embedded within an existing criminal offence. Pornography or other obscene materials, for example, if legally made and their content is legally permissible, are not labeled illegal works, but rather (potentially) immoral works.[8]

Differentiating between illegal and immoral works is crucial for the debate on their legal status.[9] Generally, considering immoral works which were legally made non-copyrightable could be dangerous for the protection of free speech. Granting legislatures, copyright registers (under the American approach), or courts the ability to decide what is immoral could lead to undesired outcomes.[10] For one thing, morality, or its lack thereof, is an ever-changing standard which varies according to time, communities, and even within communities.[11] Determination of whether a work is immoral will differ between different jurisdictions and even in different rulings within the same jurisdiction.

Even assuming that some jurisdictions could consent to a unified immoral standard, what will the legal status of works be when that standard changes and some types of material are considered morally justified in the future? Will works that were deemed non-copyrightable in the past because of immorality suddenly receive copyright protection? Consider

---

[8] For more on the protection of pornography under copyright law, see Ann Bartow, Copyright Law and Pornography, 91 Ore. L. Rev. 1 (2012).

[9] The existing academic debate in the literature regarding certain types of works, such as pornography, must be highlighted. For example, Ann Bartow argues that copyright law should not grant copyright protection when little originality is involved or there is a high risk of harm resulting from the distribution and consumption of a work. See generally, ibid. and Ann Bartow, Pornography, Coercion, and Copyright Law 2.0, 10 Vand. J. Ent. & Tech. L. 799 (2008). For more academic literature on this issue, see, e.g., Robert C. Summers, Constitutional Protection of Obscene Material Against Censorship as Correlated with Copyright Protection of Obscene Material Against Infringement, 31 S. Cal. L. Rev. 301 (1958) (discussing copyright protection of obscene material); Franklin Wallahan, Immorality, Obscenity and the Law of Copyright, 6 S.D.L. Rev. 109 (1961) (examining immorality and obscenity in copyright law); Jeremy Phillips, Copyright in Obscene Works: Some British and American Problems, 6 Anglo-Am. L. Rev. 138 (1977) (examining the legal reasoning courts give when refusing to protect obscene works).

[10] See e.g., *Bleistein v. Donaldson Lithographing Co.*, 188 U.S. 239, 251–2 (1903).

[11] See Jennifer E. Rothman, Sex Exceptionalism in Intellectual Property, 23 Stan. L. Pol'y Rev. 119, 122 (2012) (arguing that immorality is difficult to assess).

the example of a song titled 'Dora Dean' from the 19th century which was deemed by the Northern District of California as non-copyrightable as it contained the word 'hottest'.[12] The usage of the word 'hottest' was conceived by the district court as having an indelicate and vulgar meaning, and thus was denied copyright protection.[13] Clearly, even advocates for placing content-restrictions on the copyrightability of immoral works would most likely not support denying its makers copyright protection. It thus exemplifies the potentially problematic nature of categorizing works as immoral. All in all, even if certain types of immoral works should not normatively receive copyright protection, it is beyond this chapter's analysis to make such assessment. Rather, this chapter's goals are more modest. It seeks to scrutinize whether illegal works deserve legal protection – by copyright or other legal remedies – and what are the benefits and drawbacks of such legal protection.

Using the taxonomy of illegal works, this chapter will scrutinize the normative arguments for and against granting legal protection for these works. While this chapter acknowledges jurisdictional differences in both penal and copyright law regimes, which could affect the outcome of the analysis, it seeks mainly to shape a normative framework which could be universal to a great extent. But prior to such normative evaluation, it is important to learn how illegal works were treated – and might still be treated – under copyright law. As the general international framework for copyright protection does not advance this discussion much, the next part will focus on the American copyright regime as a key example of how illegal works have been treated by policymakers and courts.

3. COPYRIGHT PROTECTION FOR ILLEGAL WORKS: A TEST-CASE

The international community does not generally mandate any requirements on the copyrightability of works. Neither the Berne Convention nor the Agreement on Trade-Related Aspects of Intellectual Property Rights (TRIPs Agreement)[14] expressly mandate that works must be original or fixed in a tangible medium of expression. More specifically, the Berne

---

[12] See *Broder v. Zeno Mauvais Music Co.*, 88 F. 74 (N.D. Cal. 1898).

[13] Ibid.

[14] See Agreement on Trade-Related Aspects of Intellectual Property Rights, art. 27, Apr. 15, 1994, Marrakesh Agreement Establishing the World Trade Organization, Annex 1C, 1869 U.N.T.S. 299, 33 I.L.M. 1197 (1994).

Convention leaves signed members with the discretion to decide whether they wish to impose a fixation requirement in their jurisdictions.[15]

In many jurisdictions, like the U.S., the requirements for obtaining copyright protection for most types of works are rather simple. They require that (1) the work is in a material form (the fixation requirement);[16] and (2) the work is original (the originality requirement).[17] In addition to those two requirements, a work should generally be the result of human conduct, while some entities, like corporations, could also be considered as authors.[18] Hence, original works of authorship fixed in any tangible medium of expression are copyrightable. Other than that, the Berne Convention generally prohibits conditioning protection of copyright by any formalities like registration.[19] It should be noted that while registration is not a condition for copyright protection in the U.S., copyright registration is still necessary in order for the author to accrue certain rights and benefits.[20]

While formally it appears that copyright only requires fixation and originality to be eligible for protection, and perhaps registration to accrue certain rights and benefits, it is still unclear whether the illegal nature of a work could also place a restriction on copyrightability. The U.S. is a good example of such lack of clarity. In the past, illegal works (and immoral ones) were considered non-copyrightable under U.S. law, both by the copyright registry and by various courts. The U.S. Register of Copyrights reviews applications for copyright registration and determines whether the material deposited constitutes copyrightable subject matter.[21] Allegedly, the Copyright Office considers only whether the material falls within the subject matter of copyright and whether it represents a sufficient amount

---

[15] See Berne Convention for the Protection of Literary and Artistic Works, art. 5 (2), July 24, 1971, 1161 U.N.T.S 3 (as revised in 1971).

[16] 17 U.S.C. § 102(a) (2012). The exception in the U.S. is the protection of live musical performances that are not 'fixed' but where there is an obligation to protect this type of work under the requirements of the TRIPs Agreement.

[17] Ibid.

[18] Animals, however, cannot currently own copyright. See, e.g., *Naruto v. David John Slater et al.*, No. 3:2015cv04324 – Document 45 (N.D. Cal. 2016). For more on non-human entities in the realm of copyright protection, see Derek E. Bambauer, Copyright = Speech, 65 Emory L.J. 199, 208 (2015).

[19] Berne Convention for the Protection of Literary and Artistic Works, art. 5 (2), July 24, 1971, 1161 U.N.T.S 3 (as revised in 1971).

[20] See Erin Hogan, Survey, Approval Versus Application: How to Interpret the Registration Requirement Under the Copyright Act of 1976, 83 Denv. U. L. Rev. 843, 843 (2006).

[21] 17 U.S.C. § 410(a) (2012).

of original, creative authorship to justify registration.[22] However, for a long time, illegal works were not copyrightable if their nature was brought to the attention of the examiner.[23] But in 1959, following a decision by the Attorney General,[24] the refusal was not mandatory. This decision established a practice of limiting examination to statutory formalities,[25] which turned into an official policy in 1974.[26]

While currently the registry does not consider the nature of a work as part of its examination, courts are still somewhat inconsistent in their view of illegal copyright. Prior to 1979, many, if not most, courts interpreted copyright law to explicitly bar the registration of illegal works. Courts consistently held that illegal works were not copyrightable and denied most illegal (and immoral) works judicial remedies.[27] When the copyright holders attempted to assert their rights against infringers, most courts found that the works' connection to an illegal activity even granted alleged infringers a valid defense.[28]

Since 1979, however, the question of whether copyright subsides in illegal works became less clear. The transition began in the Fifth Circuit. In *Mitchell Bros. Film Grp. v. Cinema Adult Theater*,[29] the court held that

---

[22] The copyright examiner is mainly guided by the Code of Federal Regulations (37 C.F.R. § 202 (2012)) and a manual of examining practices entitled the Compendium of Copyright Office Practices (Library of Congress Copyright Office, II Compendium of Copyright Office Practices (1984)). For a general description of the available guidelines of the examiners, see Marybeth Peters, The Copyright Office and the Formal Requirements of Registration of Claims to Copyright, 17 U. Dayton L. Rev. 737, 739–40 (1992).

[23] See 44th Annual Report of the Register of Copyrights, Copyright Office 29 (1941), http://www.copyright.gov/reports/annual/archive/ar-1941.pdf.

[24] Walter J. Derenberg, Copyright Law, 35 N.Y.U. L. Rev. 650, 654 (1960) (describing the Librarian of Congress request from the Attorney General).

[25] Ibid. at 654.

[26] See Dan W. Schneider, Authority of the Register of Copyrights to Deny Registration of a Claim to Copyright on the Ground of Obscenity, 51 Chi.-Kent L. Rev. 691, 704 n.79 (1975).

[27] See, e.g., *Barnes v. Miner*, 122 F. 480 (C.C.S.D.N.Y. 1903); *Broder v. Zeno Mauvais Music Co.*, 88 F. 74 (C.C.N.D. Cal. 1898); *Richardson v. Miller*, 20 F. Cas. 722 (No. 11791) (C.C.D. Mass. 1877); *Bullard v. Esper*, 72 F. Supp. 548 (N.D. Tex. 1947); *Hoffman v. le Traunik*, 209 F. 375, 379 (N.D.N.Y. 1913); *Dane v. M. & H. Co.*, 136 U.S.P.Q. 426 (N.Y. Sup. Ct. 1963); *Shook v. Daly*, 49 How. Pr. 366 (N.Y. Sup. Ct. 1875); Eaton S. Drone, *A Treatise on the Law of Property in Intellectual Productions* (Little, Brown, 1879) 185.

[28] See Bonnie Wilkinson, Recent Development, 46 Fordham L. Rev. 1037, 1038–40 (1978).

[29] *Mitchell Bros. Film Grp. v. Cinema Adult Theater*, 604 F.2d 852 (5th Cir. 1979).

a motion picture titled *Behind the Green Door* was copyrightable, and its infringement enforceable, whether or not it was obscene.[30] While this decision was governed by the 1909 Copyright Act,[31] it marked a shift in the courts' perception regarding the copyrightability of illegal (and immoral) works. In 1982, under the 1976 Act (the current copyright act), the Ninth Circuit followed the Fifth Circuit's approach. In *Jartech, Inc. v. Clancy*, the court held that the fact that content is obscene – and in the context of this chapter, usually considered as illegal[32] – is not a defense to copyright infringement.[33] Since then, this content-neutral approach has seemed to prevail in a few other circuits as well.[34]

But there is still much ambiguity regarding the copyrightability of illegal works. While few circuits have not directly addressed this matter yet,[35] the Second Circuit took an opposite approach. In 1998, *Devil Films, Inc. v. Nectar Video*,[36] the Southern District of New York ruled that, 'Once a court has determined that copyrighted material is obscene, there seems no reason to require it to expend its resources on behalf of a plaintiff who it could as readily be trying for a violation of the federal criminal law'.[37] The district court repeated this view in 2012.[38]

All in all, while illegality is no longer explicitly a bar to copyright registration in the U.S., it is still unclear whether illegal works will be

---

30 *Mitchell Bros. Film Grp.*, 604 F.2d at 854.

31 Act of Mar. 4, 1909, Pub. L. 60-849, 35 Stat. 1075 (1909) (codified as amended at 17 U.S.C. §§ 1–216 (1970)) (repealed 1976).

32 See Haber, supra note 7, at 466 fn. 62.

33 *Jartech, Inc. v. Clancy*, 666 F.2d 403, 408 (9th Cir. 1982) (discussing the obscene nature of adult movies in a copyright infringement case).

34 For example, the Seventh Circuit, without deciding on the matter, noted that 'the prevailing view is that even illegality is not a bar to copyrightability'. See *FlavaWorks, Inc. v. Gunter*, 689 F.3d 754, 755 (7th Cir. 2012).

35 Cf. in the first circuit, where the court mentioned that the issue of whether pornography is copyrightable is 'unsettled in many circuits'. See *Liberty Media Holdings, LLC v. Swarm Sharing Hash File*, 821 F. Supp. 2d 444, 447 n.2 (D. Mass. 2011).

36 *Devil Films, Inc. v. Nectar Video*, 29 F. Supp. 2d 174 (S.D.N.Y. 1998).

37 Ibid. at 176. However, it should be noted that this ruling only addressed an application for preliminary relief, and the court did not eventually decide whether the obscenity of content could constitute a valid defense against allegedly infringing the content.

38 *Next Phase Distrib., Inc. v. Does 1–27*, 284 F.R.D. 165, 171 (S.D.N.Y. 2012) ('[T]he Court recognizes that, if the Motion Picture is considered obscene, it may not be eligible for copyright protection'). But see *Nova Prods., Inc. v. Kisma Video, Inc.*, No. 02 Civ. 3850(HB), 2004 WL 2754685, at *3 (S.D.N.Y. 2004) ('[E]ven if the videos were ultimately proven to be obscene, following the Fifth and Ninth Circuits holdings, this would not be a defense to copyright infringement').

deemed copyrightable when examined by courts. Without a Supreme Court decision or Congressional clarification on this matter, the status of illegal works in the U.S. will remain unclear. But beyond the current legal status of this specific jurisdictional test-case, a normative evaluation of whether copyright law should place restrictions on illegal works is necessary.

4. NORMATIVE EVALUATIONS OF ILLEGAL WORKS

Any normative evaluation of whether copyright should subside in illegal works must begin with a broader question: should copyright law examine the content of a work – any work for that matter – as part of granting protection? In other words, should copyright law be neutral to the substance upon which protection is granted, beyond mere creativity and originality?[39]

A content-neutral approach to copyright could have many benefits. First, imposing content restrictions could impede the creation of certain types of works.[40] It could lead to a chilling effect on creativity and on free speech,[41] and could stifle the creation and dissemination of expression.[42] Second, if we impose restrictions on content, we could unintentionally increase the circulation of troubling materials to the public since more people might be able to use the work without permission or paying fees.[43] Third, under a content-based restriction regime, the state – whether by registrars (for example, in the U.S.) or courts – would obtain an undesired power to decide on which works to protect and which not. These powers

---

[39] It should be further clarified that copyright law is hardly content-neutral. For a work to be copyrightable, its content must be examined, to some extent. One example is that of the examination of creativity, whereas a work must exhibit independent creation plus a modicum of creativity. See Haber, supra note 7, at 461–2.

[40] See, e.g., Schneider, supra note 26, at 719 (arguing that '[d]enial of Registration could work to discourage the development of the Arts'); *Mitchell Bros. Film Grp.*, 604 F.2d at 856 (5th Cir. 1979) (citing Schneider).

[41] *Mitchell Bros. Film Grp.*, 604 F.2d at 857; Ronald L. Green, The Obscenity Defense to Copyright Revisited, 69 Ky. L.J. 161, 182 (1980).

[42] See Jim Gibson, Copyright as Censorship, The Media Institute (Dec. 22, 2009), http://www.mediainstitute.org/IPI/2009/122209_CopyrightasCensorship.php.

[43] See Schneider, supra note 26, at 719; Jeffrey M. Ferguson, The Obscenity Defense Denied: The Rise of a Rational View of Copyright, 9 W. St. U. L. Rev. 85, 95 (1981); Rothman, supra note 11, at 156.

could be viewed as an undesirable form of government censorship.[44] Finally, if the content-based approach is not internationally accepted and implemented, and other jurisdictions chose a content-neutral approach to copyright protection, it could lead to discrepancies which would in turn negatively affect the harmonization of copyright law worldwide.

A content-neutral approach also has a few drawbacks. First, denial of copyright protection to works that do not benefit society or could even be harmful may de-incentivize their creation and potentially reduce criminal activity by removing their economic incentives.[45] Illegal works, for example, could in some instances possess an additional negative impact on society by offending unwilling onlookers, inducing criminal conduct, eroding moral standards, and harming the social fabric.[46] Second, imposing restrictions on certain types of works like illegal works could aid in the compensation of crime victims. If we reallocate the right to sue for infringement from the offender to the victim or proxy of her choice, we could grant the victim an additional means of compensation from the crime or from the additional harm caused by the work. Third, granting copyright protection to undesirable works could be viewed as government endorsement and approval of such activity.[47]

If we accept the notion that, under limited instances, copyright law could limit copyright protection based on the content of a work (and arguably this form of content-restriction is highly problematic), then we should continue the normative evaluation of whether illegal works should fall under this exception. One way of examining such a hypothesis is by securitizing the rationales behind granting copyright protection. Generally, copyright's main theoretical frameworks worldwide are the incentive/utilitarian theory, the personality theory and the labor-desert theory.[48] These frameworks led to the formation of modern copyright

---

44 Schneider, supra note 26, at 720.

45 On the economic motive of advertising and selling child pornography, see, e.g., *New York v. Ferber*, 458 U.S. 747, 747 (1982) ('[T]he advertising and selling of child pornography provides an economic motive for and is thus an integral part of the production of such materials, an activity illegal throughout the nation'). Cf. Green, supra note 41, at 183 (arguing that the 'existence of copyright would not hinder criminal prosecutions').

46 Kathleen M. Sullivan & Gerald Gunther, *First Amendment Law* (2nd edn, Foundation Press 2003) 121–2.

47 Christopher Thomas McDavid, I Know It When I See It: Obscenity, Copyright, and the Cautionary Tale of the Lanham Act, 47 U. Louisville L. Rev. 561, 562 (2009).

48 For these and other theories of copyright law see Peter S. Menell, Intellectual Property: General Theories, in Boudewijn Bouckaert & Gerrit D. Geest, eds,

law to a great extent. A brief look at the first modern copyright act, the 1709 Statute of Anne in Great Britain, suggests from its title 'An Act for the Encouragement of Learning', that its main goal is to encourage learning.[49] Subsequently, in the U.S., the copyright act was born out of the Constitution which empowers Congress to enact copyright laws in order to 'promote the progress of science and useful arts'.[50] Do illegal works encourage learning? Do they promote the U.S.'s constitutional goals? It would most likely depend on the nature of the work. Hence, looking at the literal meanings of both the Statute of Anne and the U.S. Constitution, could support that some works – most likely including illegal works – should not be granted copyright if they do not encourage learning or do not promote the progress of knowledge (science).

But arguing that all illegal works would not benefit society would be misleading. It goes much deeper than the argument that most works could advance the knowledge of specialists in the field.[51] Works by convicted felons like Mohandas Karamchand Gandhi ('Mahatma')[52] and Nelson Mandela[53] could clearly benefit society. On the other side, it is fairly evident that works like child pornography are not beneficial for society. But those are the easy cases, and the grey area could be enormous as there is a whole world of works between those two examples. In many instances, it would be very difficult to draw a clear line between what an illegal work is and what is not. Moreover, some works might be partially illegal, for instance containing only a small portion of illegality. Consider, for example, a two-hour-long film which contains ten seconds of an illegal activity such as child pornography. Would the entire film be categorized as an illegal work, only those ten seconds, or not at all? Could we even grant

---

Encyclopedia of Law and Economics (Liberty Fund Inc, 2000) 129; Neil W. Netanel, Copyright and a Democratic Civil Society, 106 Yale L.J. 283 (1996).

49 See Copyright Act 1709, 8 Ann., c. 19 (Gr. Brit.).

50 U.S. Const. art. I, § 8, cl. 8. For more discussion on the progress clause in the U.S. Constitution, see generally Orrin G. Hatch & Thomas R. Lee, 'To Promote the Progress of Science': The Copyright Clause and Congress's Power to Extend Copyrights, 16 Harv. J.L. & Tech. 1, 7 (2002).

51 See, e.g., Sean J. Kealy, A Proposal for a New Massachusetts Notoriety-For-Profit Law: The Grandson of Sam, 22 W. New Eng. L. Rev. 1, 27 (2000).

52 On March 18, 1922, a British colonial court convicted Gandhi of sedition after a protest march led to violence and sentenced him to six years imprisonment. See, On This Day: Gandhi Imprisoned for Civil Disobedience, Findingdulcinea (Mar. 18, 2011, 06:00 AM), http://www.findingdulcinea.com/news/on-this-day/March-April-08/On-this-Day--Gandhi-Imprisoned-for-Civil-Disobedience-.html.

53 Nelson Mandela was sentenced to life in prison for sabotage in 1964. See 1964: Nelson Mandela Jailed for Life, BBC, http://news.bbc.co.uk/onthisday/hi/dates/stories/june/12/newsid_3006000/3006437.stm (last visited June 15, 2018).

protection to some parts of a work, but not to others? And finally, who will decide whether this work is eligible for protection?

Clearly, there could be a wide variety of works that could be considered illegal under this chapter's taxonomy, and their contribution to knowledge or their potentially negative (harmful) effects on society could vary. The type of illegal works that are most likely to cause harm are works that fall under the first categorization of illegal works, namely work whose mere creation or content are illegal (criminal). The illegality in this type of works arises mainly from the content and not merely the act itself. Should a child pornographer, whether convicted or not, profit from her work and be even entitled to file lawsuits for copyright infringement and receive damages? The answer seems rather obvious. While the rationales of keeping copyright law content-neutral are evident, in this type of work it would seem that the drawbacks will outweigh any potential benefits. No substantive social benefit subsides when granting protection to these types of works. More broadly, criminal offences that harm individuals are undesirable for society and any knowledge gained from creative work connected to the crime does little to enrich the public domain – if at all. Such works do not benefit society but rather harm it, along with imposing further harm on the victim and their families. Furthermore, the law should not condone such works as a normative matter. These works are undesired, and granting them protection would aid in creating social injustice. Policymakers must suppress their creation and dissemination by blocking legal incentives such as profits.

The second type of illegal works are those that are directly linked to a criminal activity or made while engaging in unlawful conduct, while the work itself is not considered unlawful. The illegality in this type of works arises mainly from the act itself – but not from the content. For example, graffiti making – analyzed by Bonadio in this volume[54] – could be considered as an unlawful activity – particularly when it is connected to vandalism – and thereby the work itself could be labeled as illegal. The rationale to deny protection of copyright would be on grounds that granting protection could incentivize this harmful behavior. Another example is the filming of a murder. A snuff film, for that matter, will fall under the first categorization as the work itself will most likely be illegal. However, if someone incidentally filmed a murder without taking a part in the activity (and was unable to prevent the crime), his work could fall under this second category.

---

[54] See generally Enrico Bonadio, Street Art, Graffiti and Copyright, Chapter 4 in this volume.

Why should we differentiate between the two categories? From a social perspective there is a huge difference. Works that fall under the first category are not beneficial in social terms. They do not generally advance knowledge or creativity, and thus should not be incentivized by copyright law. But the second categorization is different. No internal considerations of copyright law are present there, as these activities are not harmful due to the scope of their creativity or their advancement of knowledge, but rather only due to their potential destructive nature. Thus, in terms of copyright protection, this type of illegal works should be considered as a legal form of free speech and should be protected by copyright. It does not imply that other laws, like penal codes, should not sanction the conduct itself – but that the work should be granted protection.

The final type of illegal works are reputational works, made by individuals who were convicted of a crime and who have produced works that are either directly related to a crime or are not, while usually assuming to increase profits from their reputation. This category is probably the most difficult to assess out of the three. Surely, when criminal offenders profit from their misconduct, it leads to injustice that should not be promoted by the law. The content, however, could be highly beneficial to society, depending on the nature of the work.

On the normative level, it is highly complex to evaluate whether the law – whether copyright law or not – should even interfere here and when. To properly evaluate these complex questions, we first need to distinguish between different criminal conducts. Criminal offences such as murder or rape should not be treated similarly to traffic violations (for example speeding or driving without insurance). It would be ridiculous to limit creations, or profitability from them, from individuals convicted of speeding. However, a convicted serial killer, writing a book about her victims, is probably a different story. While bearing in mind that this type of illegal works are based on reputation, and thus the traffic offender, unlike the serial killer, will not likely be famous due to her misconduct, a pragmatic evaluation of illegal works is necessary in order to identify how the legal system should treat different types of works.

5. PRAGMATIC SOLUTIONS TO ILLEGAL WORKS

Some forms of legal remedies already exist in many jurisdictions, and could perhaps aid in facing the complex issue of illegal works. These legal remedies could include, for example, equitable doctrines that could be invoked in copyright lawsuits as a remedy or defense. A key example of an equitable remedy is that of unjust enrichment. Individuals could potentially invoke the unjust enrichment rationale that, under the American example, state that '[a] person who is unjustly enriched at the expense of another is subject to liability in restitution'.[55] There could also be other equitable defenses which could also potentially be invoked in copyright lawsuits. These would include equitable defenses like unclean hands, whereas a plaintiff's misconduct 'affect[s] the equitable relations between the parties in respect of something brought before the court for adjudication'.[56] But these equitable doctrines are not applicable to all types of illegal works, and in some instances could not be invoked.[57] Moreover, they are incapable of untangling all matters relating to illegal conduct.

Policymakers and scholars have suggested various methods to address this problem. The first type of suggestion relates directly to copyright law, and relies on setting 'compliance conditions' to copyright protection.[58] Under this suggestion in academic literature, policymakers should invalidate copyright registration (where it exists) when the author or her agent has 'violated specific criminal laws in the immediate production of the material for which the protection is sought'.[59] Copyright protection would be conditioned on the merits that the creator or their agents have not violated criminal laws (or a statutorily defined subset). In such instances, copyright protection would be invalidated and the government could destroy physical attributes of the copyrighted work.

While applying compliance conditions could aid in resolving the problem of reputational profitability, it would still be insufficient. Generally, it only applies to legal systems which oblige the registration of a work (even if only to acquire certain rights and benefits). But more closely, it applies merely to violations of criminal law in the immediate production of the material. If criminal acts ensued after the immediate production of the material, copyright protection will not be invalidated. It does not apply to criminals who benefit from their reputation, but do not create works that are related to their criminal activity. Thus, they could still make profits indirectly from their activity, and compliance conditions will not apply. In

---

[55] See Restatement (Third) of Restitution and Unjust Enrichment § 1 (2011).

[56] *Keystone Driller Co. v. General Excavator Co.*, 290 U.S. 240, 245, 54 S.Ct. 146, 148, 78 L.Ed. 293, 297 (1933); see also *Mitchell Bros. Film Grp.*, 604 F.2d at 863. For a broader discussion and analyses of the unclean hands doctrine in copyright law, see James R. Alexander, Evil Angel Eulogy: Reflections on the Passing of the Obscenity Defense in Copyright, 20 J. Intell. Prop. L. 209 (2013).

[57] For a summary of equitable doctrines, see Haber, supra note 7, at 486–90.

[58] See Note, Can Intellectual Property Law Regulate Behavior – A Modest Proposal for Weakening Unclean Hands, 113 Harv. L. Rev. 1503, 1504–5 (2000).

[59] Ibid. at 1503.

addition, these compliance conditions could be perceived as over-inclusive as to include activities which are beneficial for society. Graffiti would fall under this suggestion, while, as previously stated, the work could be socially beneficial and should be protected under copyright law. Thus, these conditions should be limited to criminal acts that directly harm another person, either physically or mentally. In addition, the fact that the copyright in a criminal work is invalidated and even seized and destroyed by the government does not erase the early distribution of the work. As a result, the 'undesired' criminal work could be further distributed even though its copyright protection has been invalidated. Finally, victims are not compensated in any way under this proposal.

The problem of illegal works, and mainly reputational ones, was not overlooked by U.S. policymakers. In response to this problem, U.S. policymakers both at the federal and state levels enacted notoriety-for-profit laws, sometimes referred to as 'Son of Sam' laws. They are named after the serial killer David Berkowitz (nicknamed 'Son of Sam') who was offered a relatively large sum of money for the rights to his story.[60] The first Son of Sam law was enacted in the state of New York in 1977,[61] providing, *inter alia*, that an 'entity' contracting with a person 'accused or convicted of a crime' for the production of a movie, book, magazine article, tape recording, or the like, which re-enacts, or describes the person's thoughts, feelings, opinions or emotions regarding such crime, must pay any money owed to that person under the contract to a Crime Victims Board instead.[62] This law was struck down by the Supreme Court in 1991 (as violating the First Amendment due to its significant over-inclusiveness),[63] but was later revised and re-enacted. Similar Son of Sam laws throughout most of the U.S. still exist,[64] including a federal version that applies to convictions for gathering or delivering defense information to aid a foreign government, and to convictions for federal crimes resulting in physical harm to an individual.[65]

The notoriety-for-profit approach addresses the illegal works problem externally to copyright law. It does not invalidate protection, but rather limits the profitability from such works. As I previously suggested

---

60 See David L. Hudson Jr., 'Son of Sam' Laws, First Amendment Center (June 4, 2004), http://www.firstamendmentcenter.org/son-of-sam-laws.

61 N.Y. Exec. Law § 632-a (McKinney 1982).

62 *Simon & Schuster, Inc. v. State Crime Victims Bd.*, 502 U.S. 105 (1991) (describing the 'Son of Sam' law in New York).

63 Ibid.

64 N.Y. Exec. Law § 632-a (McKinney Supp. 1994).

65 18 U.S.C. § 3681 (2012).

elsewhere, policymakers should treat the problem of illegal works both internally and externally, by taking the Son of Sam approach a step forward.[66] Under this illegal copyright framework,[67] I offer three suggestions that rely on the concept that criminals who have committed certain types of criminal offences should not be able to profit from their illegal copyright or their reputation. First, the copyright of a work that is directly and substantially connected to a crime will be expropriated from the person who was convicted of, pled guilty to, or voluntarily admitted to the crime (hereinafter the copyright felon) and will be reallocated to a 'Crime Victims Board' (CVB). If such material exists but was never registered, the CVB will be able to register as its copyright owner. Second, the CVB will be able to sue the copyright felon for any existing profits from any work directly and substantially connected to the crime. Third, the CVB will be able to sue the copyright felon for any other profits he or she reaped from the notoriety he or she gained from the crime. The CVB will deposit any profits obtained from the three provisions in an escrow account, which the board will manage and use for the benefit of the victim, her living relatives, and/or any victims of similar offences throughout the country. To distribute the money, the CVB will publish legal notices to potential victims of the crime.

This proposal should be carefully examined through the lens of any legal system which seeks to implement it. In the U.S., for example, it might be perceived as unconstitutional as it potentially violates the first amendment by imposing restrictions on content which might not pass strict scrutiny. Even if it does pass constitutional or other barriers, it is notable that there are some types of works that might not be created, lacking an economic incentive to make them, and thus would undermine the rationales of copyright protection. Nevertheless, the fact that the law would reduce the profitability of some works does not necessarily mean that most would not be created as there could be many other different incentives to create other than the financial aspect.

All in all, illegal works pose difficulties for the legal system. It is crucial for any policymaker striving to place limitations on creative works due to their illegality, to first analyze the various forms of illegal works as described in this chapter. Only after differentiating between various forms of works and conduct could policymakers decide on whether or not copyright law is the proper tool to censor such works, or even simply reduce their profitability by external legislation. Overall, the law should

---

66 See Haber, supra note 7, at 494–500.

67 Ibid.

not be silent regarding the problems of at least some types of illegal works like child pornography. Even if we embrace a content-neutral approach to copyright law, it is evident that legislatures should not generally promote such criminal activity.

6. CONCLUSION

The uniqueness of illegal works raises important questions which extend far beyond the rationales behind granting copyright protection. Should the law incentivize wrongdoers to profit from their wrongdoing? Within copyright law, how can we strike a proper balance between ensuring the continued promotion of knowledge through creative works and correcting the injustice of profits from illegal works? Whether the protection of illegal works should be part of the copyright system or set by other legal mechanisms remains an open question. Policymakers, however, must acknowledge the problematic nature of neutrality in copyright law as long as other legal statutes do not provide proper victim compensation, and attempt, at the very least, to reduce the incentive for the creation of works which not only do not benefit society, but rather cause further harm.

20. Copyright and pornography

Enrico Bonadio and Nicola Lucchi*

1. INTRODUCTION

Since ancient times, a wide range of creative people have achieved great fame through sexually explicit works. Some of the most famous painters, directors, film actors and writers in history are known for their sexually provoking works, which have become controversial because of their erotic, pornographic and (what many consider) indecent or offensive content.

One may think, for example, of the *Maja desnuda* painting by Francisco Goya, which was censored by the Spanish inquisition in 1815 because of its obscene content;[1] Gustave Courbet's painting *L'Origine du monde*;[2] the novel *Madame Bovary* by Gustave Flaubert, who was prosecuted for immorality; and, more recently, Jeff Koon's sculpture depicting explicit sexual poses as well as the 1972 hardcore porn movie *Deep Throat*, now considered the most influential adult movie, capable of repositioning pornography from the underground to the mainstream.

But can one claim copyright over a sexually explicit work? In order to answer this question, this chapter reviews the developments in the case law of jurisdictions like the US, the UK and France. It also briefly expands on whether the creators of adult and sexually explicit materials – such as pornographic movies – should be able to claim rights over the result

* Both authors contributed equally to this manuscript and are listed alphabetically.

1 See Douglas Edward LaPrade, Francisco Goya, in *Censorship: A World Encyclopedia* (Derek Jones ed., Routledge, 2001) 975, 976.

2 The painting is still considered provocative and triggers censorship, see, Facebook to French Court: Nude Painting Did Not Prompt Account's Deletion (Feb. 1 2018, 08:14 PM), https://www.theguardian.com/technology/2018/feb/01/facebook-nude-painting-gustave-courbet (reporting the news of a teacher saying that his Facebook account was shut down because he had posted a picture of Gustave Courbet painting of a woman's genitals).

of their intellectual efforts and thus rely on the protection offered by copyright laws.[3]

In order to fully explore this issue, it is first necessary to have clear the evolving nature as well as the economic dimension of pornography. Pornography – understood as the representation of explicit sexual subject matter for the purpose of sexually exciting the viewer – is a very lucrative business. This industry has an estimated revenue of about $100 billion worldwide,[4] with most media conglomerates being directly or indirectly involved in the relevant market. There are hundreds of production companies and studios scattered around the world 'that produce a range of films, videos, and print porn that reflect particular genres and

---

3 These issues have already been looked at by scholars. In general, on the intersection between pornography and copyright see also Vincenzo Zeno-Zencovich, *Sex and the Contract: From Infamous Commerce to the Market for Sexual Goods and Services* (2nd edn, Martinus Nijhoff Publishers, 2015) 68; Robert C. Summers, Constitutional Protection of Obscene Material Against Censorship as Correlated with Copyright Protection of Obscene Material Against Infringement, 31 S. Cal. L. Rev. 301 (1958); Franklin J. Wallahan, Immorality, Obscenity and the Law of Copyright, 6 S.D. L. Rev. 109 (1961); Morris D. Forkosch, Obscenity, Copyright, and the Arts, 10 New Eng. L. Rev. 1 (1974); Dan W. Schneider, Authority of the Register of Copyrights to Deny Registration of a Claim to Copyright on the Ground of Obscenity, 51 Chi.-Kent L. Rev. 691 (1975); Jeremy Phillips, Copyright in Obscene Works: Some British and American Problems, 6 Anglo-Am. L. Rev. 138 (1977); Kurt L. Schmalz, Problems in Giving Obscenity Copyright Protection: Did Jartech and Mitchell Brothers Go Too Far?, 36 Vand. L. Rev. 403 (1983); Dawn A. Edick, Note. Regulation of Pornography on the Internet in the United States and the United Kingdom: A Comparative Analysis, 21 B.C. Int'l & Comp. L. Rev. 437 (1998); Ann Bartow, Pornography, Coercion, and Copyright Law 2.0, 10 Vanderbilt J. Ent. & Tech. L. 799 (2008); Ann Bartow, Copyright Law and Pornography: Reconsidering Incentives to Create and Distribute Pornography, 39 U. Balt. L.F. 75 (2008); Jennifer Rothman, Sex Exceptionalism in Intellectual Property, 23 Stan. L. & Pol'y R. 119 (2012); Ann Bartow, Copyright Law and Pornography, 91 Or. L. Rev. 1 (2012); Ned Snow, The Regressing Progress Clause: Rethinking Constitutional Indifference to Harmful Content in Copyright, 47 U.C. Davis L. Rev. 1 (2013); Ariel E. Ronneburger, A Response to Ann Bartow's Copyright Law and Pornography, 91 Or. L. Rev. Online 39 (2013); Amanda Levendowski, Note. Using Copyright to Combat Revenge Porn, 3 N.Y.U. J. Intell. Prop. & Ent. L. 422 (2014); Kaitlan M. Folderauer, Not All Is Fair (Use) in Love and War: Copyright Law and Revenge Porn, 44 U. Balt. L. Rev. 321 (2015); Ann Bartow, Copyright Law and the Commoditization of Sex, in *Diversity in Intellectual Property: Identities, Interests, and Intersections* (Irene Calboli & Srividhya Ragavan, eds, CUP, 2015) 339; Pamela Samuelson, Evolving Conceptions of Copyright Subject Matter, 78 U. Pitt. L. Rev. 17 (2016).

4 See Chris Morris, Things Are Looking Up in America's Porn Industry, NBC News (Jan. 20, 2015), https://www.nbcnews.com/business/business-news/things-are-looking-americas-porn-industry-n289431 (last accessed July 20, 2018).

subgenres'.[5] In addition, this industry 'has its own trade journals, media networks, organizations, a wide range of corporations, small businesses, news outlets, conventions, industry awards, and trade associations'.[6]

Since its very creation, pornography has also been constantly evolving to keep up with technological advances. Thus, when photography was developed in the early 19th century 'some of the first pictures taken were of people naked or having sex, which were then printed on postcards for mass-consumption'.[7] The rise of cinema at the beginning of the 20th century, offered to pornography a new medium of artistic expression and popular entertainment: 'by the end of the 1960s Hollywood was producing widely-distributed films with explicit sex scenes'.[8] Then, with the advent of the home videotape player in the 1970s and later of the DVD, pornography became highly popular. Most recently, the Internet had brought a Copernican revolution to porn's 'production, distribution and consumption',[9] with sexual media content becoming very affordable, accessible, anonymous and increasingly socially acceptable.[10]

Nowadays, online pornography is a multibillion dollar industry, where every second over 3,000 dollars are spent,[11] and is generating nearly the 30 percent of all Internet traffic.[12] It is Internet distribution that has massively contributed to the growth of pornography, and also normalized its content to the extent that copyright holders in digital works of pornography nowadays enforce their copyright against both competitors and consumers (who, for example, may easily download or exchange files incorporating adult material).[13] Porn movie producers indeed regularly send notices requesting that porn tube websites take down their material. The Internet also allows the production and distribution of pornography made by individuals in their private homes with their own recording devices. Amateur and homemade porn is frequently made by individuals

---

5 Kassia R. Wosick, Pornography, in *Handbook of the Sociology of Sexualities* (John De Lamater and Rebecca F. Plante, eds, Springer 2015) 413, 424.

6 Ibid.

7 Ibid., at 416.

8 Ibid.

9 Ibid.

10 See Pamela Paul, *Pornified: How Pornography Is Transforming Our Lives, Our Relationships, and Our Families* (Henry Holt and Company, 2007).

11 See the webpage http://www.toptenreviews.com/internet-pornography-statistics.

12 See Mark Ward, Web Porn: Just How Much Is There?, BBC News (June 30, 2013), http://www.bbc.co.uk/news/technology-23030090.

13 Ann Bartow, User-Generated Confusion: The Legal and Business Implications of Web 2.0, 10 Vand. J. Ent. & Tech. L. 799, 835 (2008).

performing without pay, but with the purpose and intent of distributing – freely or for profit – the final product to the public through the Internet.

This chapter just deals with lawful and consensual pornography, and thus excludes illegal (and hateful) material such as child pornography and other sex-related illegal content. Copyright protection of such subject matter is addressed in another chapter of this book.[14]

2. COURT CASES DEALING WITH PORNOGRAPHY AND COPYRIGHT

The copyright laws of most countries do not require copyrightable works not to be immoral or indecent, even though some of them may limit enforceability of copyright in certain circumstances.[15] International copyright treaties[16] do not specifically exclude from protection immoral or indecent works, either.[17] In many jurisdictions, pornography is simply treated like any other creative content. Also, copyright protection is considered in most countries to be content-neutral and free speech principles seem to be in contrast with content-based restrictions on copyrightability.[18]

What is required for a work to be protectable is, depending on the jurisdiction, fixation and/or originality, and (for example in the US) that

---

[14] See Chapter 19 by Eldar Haber.

[15] See for instance Section 171(3) of the UK Copyright, Designs, and Patents Act 1988, which in principle may apply to pornography. It provides that '[n]othing in this Part affects any rule of law preventing or restricting the enforcement of copyright, on grounds of public interest or otherwise'.

[16] See Berne Convention for the Protection of Literary and Artistic Works, Sept. 9, 1886, S. Treaty Doc. 99-27, 1161 U.N.T.S. 3 (amended Sept. 28, 1979) (hereinafter: Berne Convention); Agreement on Trade-Related Aspects of Intellectual Property Rights, Art. 27(2), Apr. 15, 1994, Marrakesh Agreement Establishing the World Trade Organization, Annex 1C, 1869 U.N.T.S. 299.

[17] See also Vincenzo Zeno-Zencovich, *Sex and the Contract: From Infamous Commerce to the Market for Sexual Goods and Services* (2nd edn, Martinus Nijhoff Publishers, 2015) 68.

[18] See Dominika Bychawska-Siniarska, Protecting the Right to Freedom of Expression under the European Convention on Human Rights 12 (Council of Europe 2017) (noting that the protection given by Article 10 extends to any expression, notwithstanding its content, disseminated by any individual, group or type of media and that the only content-based restriction applied by the Court has dealt with the dissemination of ideas promoting racism and the Nazi ideology, denying the Holocaust, and incitement to hatred and racial discrimination); See Eric Barendt, Freedom of Speech 51–53 (2d ed. 2005) (noting that, in the United States, courts normally invalidate content-based restrictions on freedom of speech).

the latter be registered at the copyright office in order to add certain rights and benefits. Pornography may use a wide variety of media, including books, magazines, photos, sculptures, drawings, paintings, animation, sound recordings, movies, short-videos and videogames. All these forms of expression certainly satisfy the fixation requirement, and probably many of them may be considered sufficiently original, whether we apply the originality test under US law (as affirmed in the seminal case *Feist v. Rural*),[19] the UK's 'skill, labour and judgment' standard,[20] or other originality requirements, such as the European Union (EU) 'intellectual creation' test.[21] Doubts may remain on whether certain basic sex-related scenes depicted in these media are original enough to attract protection. American scholar Ann Bartow casts doubts about whether banal erotic positions (such as heterosexual intercourse in the missionary position) could ever be protected; and hints that any sex position that is predominant in pornography as well as in the real life of most of the population may belong to the public domain due to the doctrines of 'merger' and *scènes à faire*.[22] Even where a porn movie or a sexually explicit work is protected by copyright, therefore, such a right would probably be very thin when it comes to ordinary sexual scenes: which entails that, if someone replicates these scenes without also appropriating dialogues, plots or story arcs, it would be rather difficult for a judge to find copyright infringement.[23]

That said, could sufficiently original pornographic works be denied copyright protection on morality grounds? In the United States, for example, the Copyright Office does not refuse registration simply because the work submitted for registration may be obscene. There are two leading cases supporting this view. The first is *Mitchell Bros. Film Group v. Cinema Adult Theatre*, where the owners of the copyright on a porn movie sued the defendant that had obtained and exhibited copies of the plaintiff's porn

---

19 In *Feist Publications, Inc., v. Rural Telephone Service Co.*, 499 U.S. 340 (1991) the US Supreme Court explained that the requirement of originality is not particularly stringent and is comprised of two elements: that the work be independently created by the author (as opposed to copied from other works) and that it possesses at least some minimal degree of creativity.

20 See for example *Ladbroke v. William Hill* [1964] 1 All E.R. 465, 469 (Eng.).

21 See Case C-5/08, *Infopaq Int'l A/S v. Danske Dagblades Forening*, 2009 E.C.R I-6569 (setting out the EU originality standard for copyright protection).

22 See Ann Bartow, Copyright Law and Pornography, 91 Or. L. Rev. 1, 9–10 (2012).

23 Ibid.

movie *Behind the Green Door*.[24] The second case is *Jartech v. Clancy*,[25] where the defendants copied the plaintiffs' movies by using a camera photographing images from the films, and reproduced such images in a book with the dialogue placed below the pictures (this was done to bring evidence in nuisance abatement proceedings aimed at eliminating adult movie theatres in the Californian city of Santa Ana). Both cases rejected the idea that obscenity could be a defense to copyright infringement. In particular, the Fifth Circuit in *Mitchell Bros.* found that:

> There is not even a hint in the language [of the Copyright Act] that the obscene nature of a work renders it any less a copyrightable 'writing'. There is no other statutory language from which it can be inferred that Congress intended that obscene materials could not be copyrighted . . . The history of content-based restrictions . . . is the result of an intentional policy choice and not simply an omission.[26]

A similar approach seems to have been taken by the UK Copyright, Designs and Patents Act 1988.[27] The Act indeed does not preclude the protection of scandalous or indecent works,[28] although the case law appears rather unclear as to the status of pornography.[29] In the old case *Glyn v. Weston Feature Film Company*,[30] for example, the court incidentally recognized that potentially obscene works might be protected, but eventually refused to grant an injunction for infringement of copyright in the claimant's

---

[24] 604 F.2d 852 (5th Cir. 1979), cert. denied, 445 U.S. 917 (1980).

[25] 666 F.2d 403 (9th Cir. 1982). Here the Ninth Circuit confirmed the *Mitchell Bros.* court's reasoning.

[26] See 604 F.2d 852, 854 (5th Cir. 1979). Yet, in a subsequent similar case – *Devils Films Inc. v. Nectar Video*, 29 F. Supp. 2d 174 (S.D.N.Y. 1998) – the court refused to follow *Mitchell v. Cinema Adult Theatre* and did not grant the plaintiff an injunction or seizure order. It did so by relying on the 'unclean hands' doctrine, which entails that he who seeks legal redress must not be tarnished by his own deeds. This finding is probably due to the fact that the defendant had illegally distributed the obscene movies through interstate commerce, making the whole behavior not just immoral but also unlawful (which distinguishes it from *Mitchell*). *Mitchell*'s teaching was later reinstated in *Nova Prods., Inc. v. Kisma Video, Inc.*, Nos. 02 Civ. 3850(HB), 02 Civ. 6277(HB), 03 Civ. 3379(HB), 2004 WL 2754685, at *3 (S.D.N.Y. Dec. 1, 2004) (holding that even where the films are considered obscene, that would not preclude copyright protection).

[27] Copyright, Designs and Patents Act, 1988 (Eng.).

[28] See however (again) Section 171(3) of the UK Copyright, Designs, and Patents Act 1988, mentioned supra note 15.

[29] See also Jeremy Phillips, Copyright in Obscene Works: Some British and American Problems, 6 Anglo-Am. L. Rev. 138, 140–4 (1977).

[30] *Glyn v. Weston Feature Film* [1916] 1 Ch. 261, 261.

dramatic work because the 'episode described in the plaintiff's novel [focusing on a "sensual adulterous" love story between a young man and a mysterious lady], and which she alleges has been pirated by the defendants, is in my opinion grossly immoral in its essence, in its treatment, and in its tendency'. In more recent times, a British court recognized that 'in principle public policy can prevent copyright enforcement completely',[31] admitting that the courts still have the power to deny enforcement on public policy grounds. Yet, arguably, societal norms have changed immensely since these decisions (especially, *Glyn v. Weston Feature Film Company*), and immorality through sexual acts is not seen as wholly good or evil, particularly in a more entertainment-based medium such as pornography.

In another recent case in France, the Tribunale de Grande Instance de Lyon argued that even pornographic works can be characterized by the personality of the author and of the director and therefore be individualized and recognizable. Even if the case was decided on trademark infringement grounds, the Court also had the chance to clarify the conditions under which a porn movie might be protected under copyright law. The case was as follows. A producer of porn movies ('Le Dauphin Pirate'), found out that the website Vodx.fr was offering and distributing – in video on demand – eight movies taken from their catalogue without authorization. Once warned by the plaintiff, the website removed the movies, claiming however that it had obtained permission to use them from the director's company and that in any case such movies – being of a pornographic nature – were not protected by copyright. The Court rejected the defendant's argument by noting that pornographic works are 'based largely on scenes of fornication between different individuals . . . in various places and positions . . . However, the choices that underlie a movie, the conception and the design of such work can be original and, as such, be protected under copyright'.[32]

Finally, another interesting copyright case on adult material was decided in Taiwan. Here, the Taiwan Intellectual Property Court – revers-

---

[31] See *Hyde Park Residence Ltd. v. Yelland* [1999] R.P.C. 655. Also, see Section 171(3) of the UK Copyright, Designs, and Patents Act 1988, mentioned above note 15.

[32] 'Il sera observé que la protection au titre du droit d'auteur ne saurait être déniée à une oeuvre audiovisuelle du seul fait du genre auquel elle se rattache. Il est convenu que les oeuvres à caractère pornographique, comme le souligne la société défenderesse, reposent en grande partie sur *« des scènes de fornication entre plusieurs individus, ( ) dans divers endroits et positions »*. Néanmoins, les choix ayant orienté la conception de telles oeuvres peuvent s'avérer originaux au sens du droit d'auteur.' See TGI de Lyon, ch. 10 cab. 10 J, jugement du 7 février 2017.

ing previous decisions of the Taiwan Supreme Court – extended protection to pornography following the view of the Taiwanese Intellectual Property Office that a work is copyrightable as long as it is original, even if it is deemed to be obscene.[33] In particular, producers of Japanese pornographic videos claimed that some Taiwanese online video portals and their content suppliers were making profits by making such videos available for online viewing and downloading. The Court ruled that foreign pornographic videos – being protected in their country of origin – must be considered as under copyright protection also in Taiwan and that public order or morality are not preconditions for protection.[34]

These cases teach us something, namely that copyright laws tend to treat the author of obscene works not differently from the creators of other (more ethically acceptable) content, whether it is the *Maja desnuda* by Francisco Goya or an explicit hardcore sex movie.[35] This does not necessarily entail that the dissemination of obscene works cannot be restricted by governments or parliaments. It obviously can be, as is also confirmed by the Berne Convention, whose provisions cannot in any way affect the right of countries to control or to ban the circulation, presentation, or exhibition of works that the competent authority may find it necessary to target.[36] But this has nothing to do with the copyrightability of such works. As has been noted, such limitations often 'come after the work has been created and are justified by reasons of public policy, mostly laws against obscenity'.[37] This broader point has also been reinforced by the WTO Panel in the US–China dispute over TRIPS enforcement, [38] which found that China's denial of copyright protection to works that do not meet China's 'content review' standards is not in

---

[33] See Zhang Zhongxin, Pornographic Writings Should Be Able to Enjoy Copyright, Taiwan L. 1 (May 2010).

[34] See Taiwan Intellectual Property Court Criminal Judgment 101 Xing-Zhi-Shang-Yi-Zi No. 74 (2012). For a detailed analysis of the case, see Chen, Ping-Hsun, An Unresolved Question in the First Adult Video Copyright Case of the Taiwan Intellectual Property Court, 3 NTUT J. of Intell. Prop. L. & Mgmt. 56 (2014).

[35] Vincenzo Zeno-Zencovich, *Sex and the Contract: From Infamous Commerce to the Market for Sexual Goods and Services* 68 (2nd edn, Martinus Nijhoff Publishers 2015).

[36] See Article 17 of the Berne Convention.

[37] See Jeremy Phillips, Copyright in Obscene Works: Some British and American Problems, 6 Anglo-Am. L. Rev. 138, 140–4 (1977).

[38] See WTO case DS362 China – Measures Affecting the Protection and Enforcement of Intellectual Property Rights, report of 26 January 2009.

line with the TRIPS Agreement.[39] The Panel in particular held that copyright and government censorship address different rights and interests, and that while copyright protects private rights, government censorship addresses public interests. Therefore, censorship regulations – the Panel further noted – cannot eliminate rights that are inherent in a copyrighted work.

3. SHOULD PORNOGRAPHY BE PROTECTED BY COPYRIGHT? ARGUMENTS AND COUNTER-ARGUMENTS

Opposing opinions can be voiced as to whether explicit sexual content should be copyrightable, or the relevant copyright enforceable.

First, one may note that this category of works does not deserve copyright protection, as the latter would reward immoral activities and promote sex addiction and therefore would not promote 'the progress . . . of useful arts', as requested by the Copyright clause of the US Constitution.[40] The economic exploitation of obscenity and indecency – the argument goes – should thus be opposed rather than tolerated or even promoted, and denying copyright protection for pornographic material would reduce the incentive to create these works.

Difficulties in addressing the subject matter in question should also be taken into account. Judges and lawyers for example might find themselves in an uncomfortable position when it comes to assessing whether this category of works reaches the level of originality required by copyright laws or whether, in case of alleged infringement, the defendant's work is substantially similar to the protected one. Yet, in the case of disputes focusing on the literal copying of porn movies (which happen to be the most common category of copyright litigations involving pornography),[41]

---

[39] Agreement on Trade-Related Aspects of Intellectual Property Rights Art. 27(2), Apr. 15, 1994, Marrakesh Agreement Establishing the World Trade Organization, Annex 1C, 1869 U.N.T.S. 299 (hereinafter: TRIPS).

[40] See Article 1, Section 8, of the US Constitution. The above is not what the Fifth Circuit stressed in *Mitchell v. Cinema Adult Theatre*, as it noted that considering obscene works copyrightable furthers the pro-creativity aims of copyright (604 F.2d 852, 854 (5th Cir. 1979).

[41] See Bartow, Copyright Law and Pornography, supra note 3, at 11–12 (also noting that literal copying has either been held infringing, or found to be fair use, and citing relevant cases).

judges' and lawyers' embarrassment might be less pronounced, as the need to focus on lurid details is probably avoided or anyway minimized.

What about the arguments supporting the case for the copyrightability of pornography? In addition to free-speech-related concerns (copyright is indeed considered an 'engine of free expression',[42] and denying it even in relation to obscene works may have chilling effects on this fundamental freedom), what some commentators consider problematic in relation to the 'no copyright for pornography' approach is the establishment of a standard or definition by which judges or copyright offices can be guided in assessing whether a sexually explicit content is ethically unacceptable.[43] The assessment here – the argument continues – would inevitably become subjective, and may vary depending on whether the person called to carry out that task is particularly sensitive to sex-related subject matter.

Also, moral standards towards sex change over time. Notions of obscenity 'evolve from a constant interaction of individual needs, social mores, cultural forces, religious beliefs, political ideologies and sociological realities'.[44] What was considered morally unacceptable fifty years ago might nowadays, due to changes in social attitudes, be considered less scandalous. As was noted in *Mitchell v. Cinema Adult*,[45] 'denying copyright protection to works adjudged obscene by the standards of one era would frequently result in lack of copyright protection (and thus lack of financial incentive to create) for works that later generations might consider to be not only non-obscene but even of great literary merit'. Therefore – as has been correctly observed – 'limitations to the rights of the author or the copyright holder might be expressly indicated by legislation and – as with any exceptional rule – must be interpreted stricto sensu'.[46]

Moral standards moreover change over space. What is deemed ethically admissible in one country could be considered outrageous in another, given the cultural and social differences between nations and people.

---

[42] *Harper & Row v. Nation Enterprises*, 471 U.S. 539 (1985) (finding that First Amendment rights cannot be invoked to use copyrighted works of third parties).

[43] Ariel E. Ronneburger, A Response to Ann Bartow's Copyright Law and Pornography, 91 Or. L. Rev. Online 39, 40 (2013).

[44] See Jeremy Phillips, Copyright in Obscene Works: Some British and American Problems, 6 Anglo-Am. L. Rev. 138, 140–4 (1977).

[45] 604 F.2d 852 (5th Cir. 1979), cert. denied, 445 U.S. 917 (1980).

[46] See Vincenzo Zeno-Zencovich, *Sex and the Contract: From Infamous Commerce to the Market for Sexual Goods and Services* (2nd edn, Martinus Nijhoff Publishers, 2015) 73.

Denying copyright for what is perceived obscene in a country, but protecting the same work in another, might thus lead to the fragmentation of copyright regimes. This concern was highlighted in *Jartech v. Clancy*,[47] where the court stressed that 'obscenity is a community standard which may vary to the extent that controls thereof may be dropped by a state altogether. Acceptance of an obscenity defense would fragment copyright enforcement, protecting registered materials in a certain community, while, in effect, authorizing pirating in another locale'.[48] In addition, it has been also argued that a lack of copyright protection can 'allow unfettered piracy of such an immoral work, thus increasing multifold the number of such works in circulation'.[49]

4. CONCLUSION

Copyright law often ends up being a useful financial tool in the hands of producers of pornographic works. Is that fair? In other words, is it acceptable that pornographers are given a legal tool to maximize their profits and are therefore encouraged to create and disseminate content that many consider ethically unacceptable?

These questions lead us to the very heart of a morality-related debate that has kept intellectual property scholars busy for a while. Should patents be granted in relation to potentially unethical and harmful inventions? (Many have stressed that for example bioengineered seeds are immoral and pose risks for human beings, animals and the environment.) Also, should trademark offices register signs that are offensive, scandalous or disparage ethnic minorities or vulnerable people? One of the authors of this chapter has argued elsewhere that both patent and trademark protection should be denied if the relevant subject matter is morally

---

[47] 666 F.2d 403 (1982).

[48] Ibid.

[49] Robert C. Summers, Constitutional Protection of Obscene Material Against Censorship as Correlated with Copyright Protection of Obscene Material Against Infringement, 31 S. Cal. L. Rev. 301, 309 (1958) (noting that 'it is here that the censorship laws against obscene literature must come into play to prevent the increased circulation of such a work. If, however, such censorship laws do not have a co-equal standard of obscenity in common with the copyright limitation, it is manifest that the denial of relief to the author in such a case does not in any way benefit the public by keeping from them the things which the court thinks they should not have, but has exactly the opposite effect: it multiplies the copies and increases the circulation of something that the court thinks immoral').

unacceptable.[50] Patent and trademark regimes – the argument goes – do not exist in a vacuum and should not neglect aspects and concerns related to the exploitation of dangerous or unethical subject matter. After all, that brands and inventions should not be immoral is an explicit requirement enshrined in many trademark and patent laws,[51] and the relevant offices are given, during the registration procedures, the task of assessing and deciding whether the subject matter is morally unacceptable.[52]

This is not the case with copyright, though. Indeed, no copyright statutes in major jurisdictions expressly refer to a morality requirement. This does not come as a surprise as copyright is a rather peculiar intellectual property right. It springs into action upon creation of the work, and does not require a registration procedure carried out by a public body, as happens with patents and trademarks. There is therefore no official endorsement or control by the state vis-à-vis the content and the process of creation of the work at the very moment rights are granted.

Also, introducing such a governmental control, as happens with trademarks and patents in many jurisdictions, may not be a wise policy as it is accepted that copyright regimes stimulate creation and the vibrant exchanges of opinions, especially in the fields of the arts, literature and

---

[50] Enrico Bonadio, Patents and Morality in Europe, in *Diversity in Intellectual Property: Identities, Interests, and Intersections* 149 (*Irene Calboli & Srividhya Ragavan*, eds, CUP, 2015); Enrico Bonadio, Brands, Morality and Public Policy: Some Reflections on the Ban on Registration of Controversial Trademarks, 19 Marq. Intellectual Property L. Rev. 39 (2015).

[51] See amongst many others Article 53(a) of the European Patent Convention; Article 6(1) of the EU Biotech Directive (Directive 98/44/EC of the European Parliament and of the Council of 6 July 1998 on the legal protection of biotechnological inventions); Article 4(1) Directive (EU) 2015/2436 of the European Parliament and of the Council of 16 December 2015 to approximate the laws of the Member States relating to trade marks; Regulation (EU) 2015/2424 of the European Parliament and of the Council of 16 December 2015); Article 7(1)(f) Council Regulation (EC) No 207/2009 of 26 February 2009 on the Community trade mark (codified version); Section 3(3)(a) of the UK Trade Marks Act; Section 1(3) of the UK Patents Act.

[52] This is not the case for the US Patent and Trademark Office (USPTO), at least as far as immoral and illegal inventions as well as disparaging, contemptuous and disreputable marks are concerned. The US Patent Act indeed has not incorporated a statutory ban on the patentability of inventions contrary to morality or *ordre public*. Also, while the USPTO in theory has still the authority to refuse registration of immoral and scandalous signs, on the other hand, disparaging, contemptuous and disreputable marks are now available for, and cannot be refused, registration as a consequence of the US Supreme Court First-Amendment-focused decision in *Matal v. Tam*, 582 US, No. 15-1293, Slip Op. (June 19, 2017).

audio-visual material.[53] And, as mentioned, explicitly giving copyright offices and judges the power to refuse registration of a work they deem obscene may have chilling effects on the fundamental right to freedom of expression (whether some might consider a work ethically acceptable or others might not, even at different times and in different geographical areas, does not really make a difference).[54] Furthermore, the idea that pornography might not be copyrightable challenges the very intent and meaning of international copyright laws (which do not mention any restriction on the copyrightability of subject matter on morality grounds),[55] with the concrete risk of creating a content-based restriction 'which would most certainly invite more content-based exceptions' in other fields.[56]

We obviously understand and accept that pornography appears disturbing to many, but as long as it is produced with the consent of willing and aware performers and within the boundaries of the law, it is still a form of lawful speech, whether the majority of people like it or not. It should also be reminded that pornography does not just attract criticism.

---

[53] Although accepted by many scholars and commentators, the belief that copyright stimulates the creation of works has been empirically questioned, especially when it comes to early-career artists and creators (see amongst other scholars Jessica Silbey, *The Eureka Myth – Creators, Innovators, and Everyday Intellectual Property* (Stanford University Press, 2014)).

[54] In a previous paper on trademarks and morality Enrico Bonadio argued that trademark offices' decisions to refuse to register immoral or offensive signs may often not infringe the right to freedom of expression: see Enrico Bonadio, Brands, Morality and Public Policy: Some Reflections on the Ban of Registration of Controversial Trademarks, 9 Marq. Intellectual Property L. Rev., 39 (2015), pp. 5–60. That scenario could be distinguished from a situation where copyright judges or offices would deny copyright for what they may consider an immoral or obscene work. Indeed, as also confirmed by the Strasbourg-based European Court of Human Rights in several cases, less weight should be accorded to commercial communication than to other forms of speech, with the result that states should enjoy a wider margin of appreciation when it comes to restricting such speech, including when trademark offices refuse to register morally controversial signs (which should therefore be subject to a more relaxed free speech scrutiny). As has been interestingly noted, the mere application of a trademark to products or services cannot be compared to other more complex forms of communication such as artistic speeches: see Jonathan Griffiths, Is There a Right to an Immoral Trade Mark?, Soc. Sci. Research Network 1, 3 (2009), http://ssrn.com/abstract=1492117 (last accessed on June 23 2018).

[55] See e.g. the Berne Convention and TRIPS.

[56] See Marc J. Randazza, Freedom of Expression and Morality-based Impediments to the Enforcement of Intellectual Property Rights, 16 Nevada L. J. 10 8, 130 (2016).

It is sometimes praised as it may have some positive effects, including making people less likely to commit sexual crimes and in general decreasing aggressive tendencies, relieving stress as well as being a source of safe sex education and a symbol of freedom.[57]

[57] See also Brian McNair, 'Porn Doesn't Lead to Rape Culture', The Conversation, 26 November 2012, available at http://theconversation.com/porn-doesnt-lead-to-rape-culture-10957 (last accessed on June 23, 2018).

21. On how to deal with Pandora's box – copyright in works of Nazi leaders

Marc Mimler*

1. INTRODUCTION

January 1st does not just commonly mark the beginning of the New Year according to the Gregorian calendar, it is often celebrated as Public Domain day. January 1, 2016 saw a particularly controversial work entering the public domain: Adolf Hitler's *Mein Kampf*. The lapse of copyright protection in this particular work means that anyone is able to reproduce or disseminate the work without being liable for copyright infringement. Before this date the Free State of Bavaria, who held the ownership of the copyright, vigorously challenged any form of further reproduction and dissemination by applying the tools of copyright law.

The loss of copyright law as a mechanism to restrict the use and dissemination of *Mein Kampf* – 'the unread bestseller'[1] – triggered the question as to what regulatory measures could replace copyright to restrict the further diffusion of its contents. In this regard, a formal query in the German Parliament, the Bundestag, has been posed by the Parliamentary Group of the Social Democratic Party in Germany (SPD) to the German Government as to what options remain available.[2] A starkly different approach to address this particular text was taken by the Munich-based Institute of Contemporary History (Institut für Zeitgeschichte). It

* I would like to express my gratitude for the long discussions, fruitful comments and helpful advice to Professor Guido Westkamp, Professor Jonathan Griffiths, Professor David Llewelyn, Dr. Manfred Mimler, Dr. Luke McDonagh, Dr. Gaetano Dimita, Dr. Jose Bellido, Dr. Martin Husovec, Dr. Emre Cumalıoğlu and Rainer Dresen. Finally, I would like to thank the editors, in particular Dr. Enrico Bonadio, for giving me the opportunity to conduct this piece of research on this timely topic.

1 See, Othmar Plöckinger, *Geschichte eines Buches: Adolf Hitlers 'Mein Kampf'* (2nd edn, Oldenbourg Wissenschaftsverlag 2011) 1.

2 See, Kleine Anfrage der SPD Bundestagfraktion, BT-Drucksache 17/12426, 1 (Feb. 20, 2013).

published an annotated version of *Mein Kampf* in 2016,[3] which was controversially discussed in the domestic and international media.[4] The head of the Jewish community in Munich, Charlotte Knobloch, stated that the re-publication of the book would constitute the opening of Pandora's box that 'once opened cannot be closed'.[5]

Mein Kampf is arguably the most prominent example of such 'literature', but other works of leaders of the 'Third Reich', like Joseph Goebbels' diaries and Alfred Rosenberg's *Der Mythus des zwanzigsten Jahrhunderts*, have to be mentioned in this context as well. The controversies surrounding the treatment of such works with their inhuman and racist contents are obvious. Should such works be freely available in order to educate the public? Or should such contents be erased from the face of the earth to eradicate their inflammatory rhetoric and contents – a rather futile attempt in the days of the internet. Another important aspect of this debate – wonderfully demonstrated by the discussions surrounding the Goebbels diaries – is how the income generated from such works enabled through copyright is distributed. This chapter will trace how copyright operates with regards to such works; in particular, how it has been used to restrict their access and dissemination.[6] The focus here will lie on analysis of the issues surrounding *Mein Kampf* due to the current debates surrounding its recent entry into the public domain. The chapter will then outline which other regulatory responses are currently applied to counteract the further dissemination of such works with a focus on German criminal law. Finally, the chapter will discuss whether copyright law is a suitable regulatory tool to restrict access to and the dissemination of such works and their contents.

---

[3] See, Hitler, *Mein Kampf. Eine kritische Edition*, Institut für Zeitgschichte, http://www.ifz-muenchen.de/publikationen/einzelwerke/ea/publikation/hitler-mein-kampf/ (last accessed Oct. 19, 2017).

[4] See Neil Gregor, Mein Kampf: Eine Kritische Edition review – Taking the Sting out of Hitler's Hateful Book, *The Guardian* (Feb. 4, 2016, 7:30 PM), http://www.theguardian.com/books/2016/feb/04/mein-kampf-hitler-eine-kritische-edition-review?CMP=fb_gu; Björn Hengst, 3700 Fußnoten gegen Hitlers Hass, Spiegel Online (Dec 21, 2015, 4:20 PM), http://www.spiegel.de/kultur/gesellschaft/mein-kampf-in-kritischer-edition-3700-fussnoten-gegen-adolf-hitlers-hass-a-1068562.html.

[5] See Anthony Faiola, *Mein Kampf*: A Historical Tool, or Hitler's Voice from Beyond the Grave?, *Washington Post* (Feb. 24, 2015), https://www.washingtonpost.com/world/europe/mein-kampf-a-historical-tool-or-hitlers-voice-from-beyond-the-grave/2015/02/24/f7a3110e-b950-11e4-bc30-a4e75503948a_story.html?utm_term=.121e606c9632.

[6] For a more general discussion on copyright protection of illegal works see the analysis offered by Eldar Haber in Chapter 19 of this volume.

2. A SHORT STORY OF *MEIN KAMPF*

Adolf Hitler's *Mein Kampf* is divided into two volumes. The first volume was developed and finalised during Hitler's prison sentence in Landsberg am Lech. The sentence of high treason was awarded due to Hitler's leadership in the unsuccessful Munich or Beerhall Putsch of November 1923 in which he sought to overthrow the Bavarian government. The first volume was published after Hitler's release in 1925 and contains mainly autobiographical elements and elaborates on the beginnings of the National-Socialist party in Germany. The second volume, which contains more programmatic statements, was written after Hitler's release from prison and was published in 1926. Both volumes were published by Franz-Eher Verlag, the publishing house of the National Socialist German Workers' Party (NSDAP) and were eventually combined into one tome in 1930.[7]

While the book sold reasonably well before the Nazis seized power in January 1933,[8] the sales significantly increased to around 1,080,000 copies in that year alone.[9] Various editions of the book were available – among them a wedding edition which was given to couples on their nuptials from 1936 onwards.[10] Around 10 million copies were circulating in Germany in the years from 1933 and 1945.[11] Hitler's stake in the sales of the book was considerable: until 1932, he earned royalties of 10% and from 1 January 1933 15%.[12] In order to maintain the stream of income from the tome, the President of the 'Reichsschrifttumskammer', the state chamber for literature, instructed German booksellers only to sell new copies of *Mein*

---

7 See Plöckinger, supra note 1, at 182.

8 The book outsold those of other 'Nazi' authors such as Ernst Röhm's *Geschichte eines Hochverräters* or Rosenberg's *Mythus des zwanzigsten Jahrhunderts*. Only Joseph Goebbels' booklet 'Wege ins Dritte Reich' was able to keep up with the sales of *Mein Kampf*; see, Plöckinger, supra note 1.

9 See Plöckinger, supra note 1, at 185.

10 See, Jürgen Seul, Hitlers 'Mein Kampf' im Spiegel der Justiz – Zum juristischen Umgang mit dem wohl umstrittensten Buch deutscher Sprache, 14 Medien und Recht Int. 131, 134 (2014).

11 See, Sascha Sebastian & Robert Briske, Die Verwertung von Hitlers 'Mein Kampf' – Eine urheber- und strafrechtliche Analyse, Zeitschrift für Medien- und Kommunikationsrecht 101, 101 (2013); see also, Andreas Heldrich, Das Territorialitätsprinzip im Internationalen Urheberrecht: Ein Freibrief für ein entsetzliches Buch, in *Festschrift für Claus-Wilhelm Canaris zum 70. Geburtstag*, Vol 2, 645, 645 (Andreas Heldrich ed., Beck, 2007).

12 See, Seul, supra note 10, at 134. It has been said that the income from the royalties were used to acquire Hitler's Mountain residence on the Obersalzberg near Berchtesgaden; see Heldrich, supra note 11, at 645–6.

Kampf.[13] Additionally, Hitler's income from the royalties became tax free since he generally did not pay any taxes on his income from spring 1935 on.[14]

Along with the editions sold in Germany, *Mein Kampf* was also published abroad. Interestingly, copyright law was either used as a tool to facilitate its dissemination or used as a method to restrain unauthorised translations. On the one hand, the Franz-Eher Publishing House promoted the book in English-speaking nations since Hitler sought to establish an alliance with England against the Soviet Union.[15] Therefore, an English translation was published in the UK at the publishing house of Hurst & Blackett in October 1933.[16] This version was heavily abridged as it excluded certain references to anti-Semitism[17] and to some of Hitler's foreign policy ambitions.[18] The English publisher was positively surprised by the volume of the sales; around 90,000 copies were sold until the end of 1938.[19] A slightly altered version was published in the United States in 1933.[20] Its apparent abridgements were noticed within the highest circles of the US government.[21]

A starkly different approach was applied in France, where copyright was used to restrict an unauthorised translation of *Mein Kampf.* The French publisher Fernand Sorlot published an unauthorised, full French translation at his publishing house 'Editions Nouvelles Latines'. Hitler's publishers, however, wanted to avoid a full French translation of *Mein*

---

[13] See, Roland Aegerter, Ein Machwerk und seine Wirkung im Spiegel der Zeiten – Vor 75 Jahren erschien Hitlers 'Mein Kampf', Neue Zürcher Zeitung (Jan. 5, 2002), https://www.nzz.ch/article7vitg-1.354185.

[14] See, Sven Felix Kellerhoff, 'Mein Kampf' brachte Hitler Millionen. Steuerfrei, *Die Welt* (Sep. 25, 2015), https://www.welt.de/geschichte/zweiter-weltkrieg/article146837543/Mein-Kampf-brachte-Hitler-Millionen-Steuerfrei.html.

[15] See Plöckinger, supra note 1, at 461.

[16] See, Heldrich supra note 11, at 645, 652; see also, Plöckinger, supra note 1, at 462.

[17] See Plöckinger, supra note 1.

[18] Dan Stone, The 'Mein Kampf Ramp': Emily Overend Lorimer and Hitler Translations in Britain, 26 German Hist. 504, 508 (2008).

[19] See Plöckinger, supra note 1, at 463.

[20] See Plöckinger, supra note 1, at 462; see also, Martin Husovec, A Copyright Story of An Infamous Book (Mar. 9, 2016),) http://www.husovec.eu/2016/03/a-copyright-story-of-infamous-book.html.

[21] A hand-written note within President Franklin D. Roosevelt's copy of *Mein Kampf* stated: 'this translation is so expurgated as to give a wholly false view of what Hitler really is or says – The German original would make a different story'; see Plöckinger, supra note 1, at 499.

Kampf[22] and successfully sued for copyright infringement in the French courts. Hitler himself sanctioned this since he did not want to have a complete version available in France, which he harshly attacked in his book.[23]

The publication history of *Mein Kampf* in Germany shared a similar fate to its author at the end of the Second World War: its publication forbidden by the Allied Commission;[24] its printing plates held by the Franz Eher Verlag melted and destroyed.[25] Some of the published editions were used for heating purposes along with other devotional material of the Nazis, as Heldrich suggests.[26] Abroad, the book was still being published and disseminated,[27] while the Free State of Bavaria has eagerly fought to restrict publication.[28] Based on the already wide distribution of the book and the inception of the internet – the world's largest copy machine – this resembled a Sisyphean task. Now, after the lapse of copyright protection, it appears that the 'genie' cannot be put back into the bottle again, that Pandora's box has irrevocably been opened. This then begs the question whether the lapse of the copyright term will be of much relevance with all these tangible and intangible versions already in circulation.

3. COPYRIGHT IN NAZI LITERATURE

The existence of copyright is commonly based on natural law approaches or on its utilitarian function to incentivise the creation of new works.[29]

---

[22] See Plöckinger, supra note 1.

[23] See, *Le Monde*, Pour une édition critique de 'Mein Kampf' (Oct. 6, 2011, 4:30 PM), http://www.lemonde.fr/livres/article/2011/10/06/pour-une-edition-critique-de-mein-kampf_1582887_3260.html.

[24] See Caesar C. Aronsfeld, Mein Kampf, 1945–1982, 45 Jewish Soc. Stud. 311, 311 (1983).

[25] The melted printing plates were allegedly reused to make the printing plates of the *Süddeutsche Zeitung*, a very reputable broadsheet published in Munich since October 1945; see, Marita Krauss, Amerikanische Besatzungskultur und 'konstruktive' Transformation im Jahr 1945. Das Beispiel Bayern, in *Die lange Stunde Null: Gelenkter sozialer Wandel in Westdeutschland nach 1945*, 59, 80 (Hans Braun, Uta Gerhardt & Everhard Holtmann, eds, Nomos, 2007); see also, Herbert Hess, *50 Jahre Süddeutsche Zeitung* 4 (Süddeutscher Verlag, 1995).

[26] See Heldrich, supra 11, at 645, 652–3.

[27] See generally, Aronsfeld, supra note 24.

[28] See, Owen Bokott, Charity Returns £250,000 Royalties for Hitler's Credo, *The Guardian* (Jun. 18, 2001, 2:14 PM), https://www.theguardian.com/uk/2001/jun/18/books.humanities.

[29] See Gehard Schricker & Ulrich Loewenheim (eds), *Urheberrecht: Kommentar, Einleitung* 9 (7th edn, Beck, 2017).

Both of these justifications are often challenged for various reasons.[30] But, as this chapter will outline and discuss, copyright may serve yet another purpose: it was used to inhibit any further dissemination of the Nazi propaganda contained within *Mein Kampf* – a legislative aim enshrined within the preamble of the constitution of the Free State of Bavaria.[31] Thus, copyright with its exclusionary function can be applied to restrict access to and dissemination of works. With regards to the Goebbels diaries, a more classical justification for copyright and its term, however, may prevail – to provide income for the estate of the author[32] – though the question arises whether income should be generated from such works.

The following analysis will assess the protectability of works of Nazi leaders by German copyright law. It will discuss copyright subsistence, duration, ownership as well as the infringement of copyright in such works as well as possible exceptions and limitations which may enable an otherwise unauthorised use of such works. The discussion of the copyright aspects will be led from the perspective of German copyright law as codified within the Urhebergesetz (UrhG).

4. COPYRIGHT SUBSISTENCE

Any work within the literary, artistic or scientific domain can qualify for copyright protection under German law.[33] § 2(1) UrhG specifies the categories of works that may be protected by copyright law. The catalogue is, however, non-exhaustive,[34] meaning that the German Copyright Act provides an open list system. Additionally, only works that contain the author's own intellectual creation can qualify for copyright protection according to § 2(2) UrhG. While the level of originality may be higher than

---

30 Ibid., 10.

31 'In the face of the scene of devastation into which the survivors of the 2nd World War were led by a godless state and social order which lacked any conscience and respect for human dignity, with the firm intention of permanently securing for the future generations the blessings of peace, humanity and justice and mindful of its history of more than a thousand years, the Bavarian people herewith bestows upon itself the following Democratic Constitution'; Bayerischer Landtag (Dec. 15, 1998), https://www.bayern.landtag.de/fileadmin/Internet_Dokumente/Sonstiges_P/BV_Verfassung_Englisch_formatiert_14-12-16_neu.pdf.

32 See Eugen Ulmer, Urheber- und Verlagsrecht 340–1 (3rd edn, Deutscher Taschenbuch Verlag, 1980).

33 § 1 UrhG.

34 See Artur-Axel Wandtke & Winfried Bullinger, *Praxiskommentar zum Urheberrecht* § 2 [2] (4th edn, Beck, 2014).

in other jurisdictions, a work like *Mein Kampf* will generally overcome this threshold.[35]

Copyright subsistence for diary entries such as those by Joseph Goebbels, however, may be questionable under German copyright law. Such personal records often display trivial daily occurrences which may not constitute the author's own intellectual creation pursuant to § 2(2) UrhG,[36] even if their author has some form of prominence.[37] Bogedain, however, holds that the Goebbels diaries would surpass this threshold: while Goebbels refers to occurrences such as the landing of the Allies in Normandy in 1944, Goebbels often displays his sentiments in a stylistic manner.[38] Hence, such diary entries may qualify as original works and be protected under German copyright law. Furthermore, the disputes surrounding the copyright of Goebbels diaries underline that courts acknowledged that copyright subsisted in them.[39]

In comparison to other intellectual property rights, considerations of ordre public or morality do not affect copyright subsistence.[40] Patent and trade mark law,[41] on the other side, provide for provisions that enable the competent authorities to reject an application for a patent or trade mark based on morality or ordre public considerations.[42] Such thresholds would seem unreasonable due to copyright being an IP right that comes into existence without examination. This means that even works containing

---

35 This was confirmed by the Higher Regional Court of Munich that held that excerpts of *Mein Kampf* would qualify as literary works pursuant to § 2 (1) Nr. 1, (2) UrhG; see, OLG München ZUM-RD 2012, 479, 480.

36 See Gehard Schricker & Ulrich Loewenheim (eds), *Urheberrecht: Kommentar*, § 2 [136] (7th edn, Beck, 2017).

37 Furthermore, Goebbels increasingly dictated his remarks for his diary which leads Winkler to speculate whether they actually originated from him or were rather produced by this staff; see, Willi Winkler, *Der Schattenmann: Von Goebbels zu Carlos: Das mysteriöse Leben des François Genoud* (Rowohlt Digitalbuch, 2011).

38 See, Clemens Bogedain, 'Mein Kampf', der 'Mythus des 20. Jahrhunderts' und die 'Goebbels-Tagebücher': Werke früherer NS-Größen im Spannungsfeld von Strafrecht, Urheberecht und künftiger Gemeinfreiheit, 59 Die Zeitschrift für Urheber- und Medienrecht 205, 208 (2015).

39 That is, LG München I, Urt. v. 19.09.2014, Az: 21 O 28238/13.

40 BGH, Urt. V. 23.02.1995 – I ZR 68/93; see also, Bogedain, supra note 38, at 435.

41 See, Lisa P. Ramsey, A Free Speech Right to Trademark Protection, 106 Trademark Rep. 797 (2016).

42 § 2(1) of the German Patent Act; § 8 (2) Nr. 5 of the German Trade Mark Act.

national-socialist contents can qualify for protection[43] as long as the criteria of copyright subsistence are met.

5. COPYRIGHT DURATION

The German Author's Right Law applies a 70-year post mortem auctoris rule, according to § 64 UrhG. In addition, § 69 UrhG clarifies that the calculation of the 70-year period commences 'at the end of that calendar year in which the event giving rise to them [i.e. the death of the author] occurs'. Previously, the German Copyright Act for Literary Works (LUG)[44] provided for a 30-year post mortem auctoris rule within § 20.[45] This rule applied until 1934 when the term for copyright protection was extended to 50 years post mortem auctoris.[46] In 1964, § 64 UrhG introduced the present rule, which also applied to works that were created before 1965 according to § 129(1) UrhG.

With regards to *Mein Kampf*, copyright protection lapsed on January 1st 2016, as already mentioned. Similarly, Joseph Goebbels' diaries fell into the public domain on the same day, since Goebbels committed suicide with his wife Magda on 1st of May 1945, a day after Hitler's death. Alfred Rosenberg, the author of *Der Mythus des zwanzigsten Jahrhunderts* was sentenced to death during the Nuremberg Trials of 1946. The sentence was carried out in the morning of 16 October of that year. This means that copyright in *Der Mythus des zwanzigsten Jahrhunderts* lapsed on 31st December 2016 at midnight. Due to the fact that the Nazi dictatorship ended in 1945, most leading protagonists have deceased already. Thus, most of these works written by Nazi leaders are in the public domain.

The fact, that copyright was used to restrict access and further dissemination of *Mein Kampf* leads to the question whether the term could be extended to maintain control over such work. Other jurisdictions

---

[43] Artur-Axel Wandtke & Winfried Bullinger, *Praxiskommentar zum Urheberrecht* § 2 [31] (4th edn Beck, 2014); Clemens Bogedain, 'Mein Kampf', der 'Mythus des 20. Jahrhunderts' und die 'Goebbels-Tagebücher': Werke früherer NS-Größen im Spannungsfeld von Strafrecht, Urheberecht und künftiger Gemeinfreiheit, 59 ZUM 205, 208 (2015).

[44] See Gesetz betreffend das Urheberrechts an Werken der Literatur und Tonkunst vom 19. Juni 1901 (LUG).

[45] See Sebastian & Briske, supra note 11 at 104.

[46] Gehard Schricker & Ulrich Loewenheim (eds), *Urheberrecht: Kommentar, Einleitung* [135] (7th edn, Beck, 2017).

provide significantly longer copyright terms. Mexico, for instance, provides for a 100-year post mortem auctoris rule within its Copyright Act.[47] European Union law, however, prohibits an extension.[48] Article 1 of the Term Directive[49] stipulates a 70-year post mortem auctoris rule. The provision is mandatory, as Recital 12 of the Term Directive suggests, thus making a unilateral extension of copyright law protection by Germany not possible.[50] In any event, any form of extension usually begs the question how long copyright should extend for such works which would become an arbitrary exercise and politically difficult to deliver.

6. COPYRIGHT OWNERSHIP

Copyright ownership played a crucial role in relation to the works discussed in this chapter. On the one hand, the works of Nazi leaders were used to generate income, as the example of the diaries of Joseph Goebbels show,[51] on the other, copyright was used to restrict the dissemination of such works, as seen with *Mein Kampf*, where the Free State of Bavaria acted more like a custodian of the work, containing its contents. Copyright protection was therefore applied in totally different ways depending on who had control over the work in question.

Under German Law, copyright is originally vested with the creator of the work[52] and is generally not transferable[53] but is inheritable.[54] Ownership of the copyright of deceased authors is held by their estates according to

---

47 Ley Federal del Derecho de Autor, Art, 29(1), Diario Oficial de la Federacion [DO], 12 de diciembre de 1996 (Mex.).

48 Hannes Ludyga, Der Vertrieb von Adolf Hitler's Hetzschrift 'Mein Kampf' – Urheber- und strafrechtliche Aspekte, 59 Die Zeitschrift für Urheber- und Medienrecht 435, 436 (2015).

49 Directive 2006/116/EC of the European Parliament and of the Council of 12 December 2006 on the term of protection of copyright and certain related rights, 2006 O.J. (L 372) 12.

50 As held by the German Government in response to the query of the Parliamentary Group of the Social Democratic Party in the German Parliament (see note 2) – Antwort der Bundesregierung auf die Kleine Anfrage der Fraktion der SPD – BT Drucksache 17/12426 (Mar 11.2013).

51 Peter Maxwill, Steinreich dank Goebbels' Tagebüchern, *Spiegel Online* (Jan. 10, 2014), http://www.spiegel.de/einestages/fran-ois-genoud-und-joseph-goebbels-tagebuecher-a-953258.html.

52 § 7 UrhG.

53 § 29(1) UrhG.

54 § 28(1) UrhG.

§ 30 UrhG.[55] Primarily, the successors in title of the author are determined through the last will of the deceased. In the absence of such a will, statutory succession will apply under German law. Since the Third Reich ended in 1945, the works of many Nazi authors would have been held by the estates of the deceased. The copyright in Hitler's *Mein Kampf*, however, was held by the Bavarian Free State. This begs the question how this was possible.

After Hitler's suicide in the bunker of the Reichskanzlei on the 30th of April 1945, the question of ownership of the copyright of *Mein Kampf* became an issue. In the absence of a testament, the copyright would have been transferred to Hitler's next of kin.[56] Hitler's private testament, however, stipulates that any of his assets would become part of the Party's, i.e. the NSDAP's, possessions.[57] He added that in the case that the Party no longer existed, copyright would fall to the state.[58] At Hitler's point of death, the NSDAP still existed. This means that if this testament had been held to be valid, all of Hitler's possessions and the copyright in *Mein Kampf* would have fallen into the Party's possession. The Allied Command, however, issued Control Council Law No. 2. Its Article 2 prescribed that that all possessions of Nazi organisations, such as the NSDAP, were to be confiscated, which would include the copyright over *Mein Kampf*.[59]

Ownership of the Free State of Bavaria of the copyright for *Mein Kampf*, however, took another route. This requires some explanation of the particular historical context of the aftermath of the Second World War. After the defeat of Germany, the Allied forces sought to eradicate Nazism from German society. For this purpose, many instruments were applied. While many of these tasks were completed by the Allied forces themselves, within their respective zones of occupation, some tasks were given to the self-governance of the German people. One of these measures was the passing of the Act Nr. 104,[60] which was signed by the three German states of the US occupied zone (Bavaria, Württemberg-Baden and Greater Hesse) in the city hall of Munich on 5 March 1946. All persons who actively supported the Nazi dictatorship and its sympathisers or profiteers

---

55 'The author's successor in title holds the rights to which the author is entitled according to this Act, unless otherwise provided for'.

56 Hitler's sister made claims with regards to the copyright of *Mein Kampf* – Spruchkammer München I, Spruch v. 15.10.1948 – I-3568/48.

57 *Der Spiegel*, Verlagsrechte – Hilters Erben (Dec 3. 1952), http://www.spiegel.de/spiegel/print/d-21978350.html.

58 Heldrich, supra note 11, at 645, 652–6.

59 Ibid., at 645, 652–6.

60 Gesetz Nr. 104 zur Befreiung von Nationalsozialismus und Militarismus vom 5. März 1946 (bayerisches Gesetz – und Verordnungsblatt – GVBl. – 1946 S. 145).

were required to be registered and classified within five groups pursuant to Article 4 of the Act: Major Offenders; Offenders; Lesser Offenders; Followers; and Exonerated Persons. The Act also provided for measures of atonement which could be imposed on such individuals. For instance, the sanction for major offenders foresaw the confiscation of all their assets according to Article 15 Nr. 2 of Law 104.

The process of qualifying an individual into the categories of Article 4 of Act Nr. 10 was determined by a judicial sentence. Hence, special panels were established to adjudicate on this matter and on the corresponding sanctions. Since Hitler had his last domicile in Munich's Prinzregentenplatz 16, the judicial panel in Munich, Bavaria, was competent pursuant to Article 29(a) of Act Nr. 104. In its decision from the 15 October 1948, the panel unsurprisingly declared that Hitler was a major offender, holding that this would not require any further explanation.[61] Since Hitler was already deceased, Article 37 of Law 104 was applicable. This provision states that proceedings against a deceased person to confiscate their assets could be initiated by the competent Minister. Such confiscations were only available against major offenders and would overrule any rules on testamentary disposition or statutory succession. The confiscation would include all assets, rights and other assets which are situated within the State of Bavaria according to Article 3 of the Regulation on the confiscation, administration and exploitation of assets according to Law 104. This included the copyright over *Mein Kampf*.[62] Hence, the Free State of Bavaria became the owner of Hitler's assets and the copyright for *Mein Kampf*. This enabled it to exercise the economic rights deriving from the copyright over the book. The Bavarian Ministry of Finance was declared as the competent authority to exercise the rights in practice.

The situation with regards to the diaries of Joseph Goebbels is a different one. Goebbels originally wanted his diaries to be released 20 years after his death and the copyright to be exercised by the Franz Eher publishing house.[63] After his death, the diaries remained unpublished, although some excerpts which were discovered in the courtyard of Goebbels' ministry of propaganda in 1945 were published in the US and Switzerland in 1948.[64]

---

[61] Spruchkammer München I, Spruch v. 15.10.1948 – I-3568/48; see also, Martin Husovec, A Copyright Story of An Infamous Book (Mar. 9, 2016),) http://www.husovec.eu/2016/03/a-copyright-story-of-infamous-book.html.

[62] See Sebastian & Briske, supra note 11 at 102.

[63] Willi Winkler, *Der Schattenmann: Von Goebbels zu Carlos: Das mysteriöse Leben des François* Genoud (Rowohlt Digitalbuch, 2011).

[64] Ibid.; *Der Spiegel*, Goebbels-Nachlass – Der Stenograf muß es wissen (Jan 14. 1951), http://www.spiegel.de/spiegel/print/d-29191941.html.

With regards to the situation of ownership of the copyright, it needs to be pointed out that in contrast to Hitler's possessions, the possessions of Goebbels were not confiscated.[65] Hence, the diaries of Joseph Goebbels apparently fell to his heirs.[66] Goebbels' estate subsequently authorised the Swiss businessman François Genoud with the administration of exclusive rights over the diaries. Genoud, who had been an enthusiastic admirer of Hitler since he met him as a teenager in the 1920s,[67] agreed with Goebbels' estate that half of the revenues created from exploiting the copyright would fall to them while he could keep the other half after the administrative costs were deducted.[68] Additionally, Genoud's position as administrator of the exclusive rights in Goebbels' published and unpublished literary works was confirmed by contractual arrangement with the executor of Goebbels' estate in 1955.[69]

While Genoud's position with regards to his entitlement over the diaries has been disputed – it has been suggested that he could not be owner of the copyright because of the expropriation of Nazi criminals by the laws of the Allied command[70] – he succeeded in exercising his claimed ownership of copyright in relation to Goebbels' works. For instance, he secured payment from the Hoffmann und Campe publishing house, which published fragments of the diaries in relation to its last entries before the downfall of the Third Reich. Additionally, he was also able to provide a post-script in this edition which was criticised for being apologetic.[71] Genoud also secured a settlement with the Federal Archive and the Institute for Contemporary History who received all records of the diary which the Hoffman and Campe publishing house held and did not seek to publish.[72] After committing suicide in 1996, Genoud left the rights as

---

65 Winkler, supra, note 63.

66 Martin Broszat, Zur Edition der Goebbels-Tagebücher, 37 Vierteljahrshefte für Zeitgeschichte 156, 157 (1989).

67 Ibid.

68 Winkler, supra note 63.

69 The Local Court of Zehlendorf in Berlin appointed the attorney and notary Kurt Leyke as executor of the estate of Goebbels in 1945; Winkler, supra note 63.

70 It was said that Goebbels left the rights to the Franz Eher Publishing House which would mean that the rights would have fallen to the Free State of Bavaria – *Der Standard*, Prozess über Tantiemen für Goebbels-Nachlass (Apr. 22, 2015) http://derstandard.at/2000014642242/Gericht-entscheidet-ueber-Tantiemen-fuer-Goebbels-Nachlass.

71 Martin Broszat, Zur Edition der Goebbels-Tagebücher, 37 Vierteljahrshefte für Zeitgeschichte 156, 157 (1989).

72 Ibid., 156.

literary executor over Goebbels' works to Cordula Schacht,[73] the daughter of the former President of the Reichsbank and Minister of Economics during Hitler's rule. The diaries continued to provide a considerable but also arguably tainted stream of income. For instance, the German actress Iris Berben toured Germany, Austria and Switzerland in 2002 reciting passages of Goebbels' diary along with those from Anne Frank's diary. While Goebbels' estate received a payment negotiated by Cordula Schacht, no payment was made to the heirs of Anne Frank.[74]

Recently, Cordula Schacht sued the publishing house Random House before the Courts of Munich over the use of extracts of Goebbels's diaries within a biography by the historian Peter Longerich of the former minister of propaganda, published in 2010 by a subsidiary.[75] Schacht, a German attorney, demanded payment for the use of the excerpts based on a settlement agreement between her and the publishers. Random House, on the other hand, was adamant it would not pay any amount, inter alia, disputing Schacht's right to exercise the exclusive rights over the diaries.[76] Another point raised by the defendants was the fact that the settlement agreement, which the publishers only committed to in order not to jeopardise the publication of the biography on Goebbels, was invalid. This was because it related to a legal transaction which would be void pursuant to § 138 of the German Civil Code since it would enable the commercialisation of the thoughts and intentions of one of the greatest criminals of the Nazi regime.

The District Court of Munich, however, acknowledged Schacht's position in relation to the diaries and held that the agreement between the parties would not be void as it would enable a critical evaluation of the texts within the diaries.[77] Furthermore, the request for remuneration by Schacht would not render the agreement void since she would not commercialise the ideas of Goebbels but rather their critical examination.[78] The subsequent appeal to the Higher District Court of Munich was rejected based on formal reasons.[79] The Court, inter alia, found that the District Court

---

[73] *Der Spiegel*, Cordula Schacht (Jul. 14, 1997), http://www.spiegel.de/spiegel/print/d-8743624.html.

[74] Winkler, supra, note 63.

[75] Willi Winkler, Geld der Vergangenheit, *Südeutsche Zeitung* (Jul. 31, 2014), http://www.sueddeutsche.de/kultur/verhandlung-um-goebbels-tantiemen-geldder-vergangenheit-1.2069707.

[76] LG München I, Urt. v. 19.09.2014, Az: 21 O 28238/13.

[77] Ibid.

[78] Supra, note 76.

[79] OLG München, Urteil vom 09.07.2015, Az. 6 U 3707/14.

was correct in not having declared the appeal of Random House as admissible pursuant to § 511(4) Nr. 1 of the German Civil Procedure Rules. This provision mandates that a court of first instance shall admit an appeal which is of 'fundamental significance' or serves 'the further development of the law or the interests in ensuring uniform adjudication'. The Court held that the case was not of fundamental significance as it would only relate to a dispute between the parties. The historical interest in Goebbels and his diaries would not render the dispute as having fundamental significance due to the lack of connection with the case at hand – the obligation stemming from the settlement agreement. Furthermore, since copyright protection would lapse at the end of 2015, it could not be expected that many courts would have to adjudicate similar cases.

7. COPYRIGHT INFRINGEMENT

German copyright law provides the right holder with the exclusive rights of reproduction (§ 16 UrHG) and distribution (§ 17 UrhG) of the protected work. While this generally means that copyright holders may commercialise the work, they are also at liberty not to commercialise the work.[80] Furthermore, they can restrict these uses by third parties where no authorisation is provided and no exception or limitation provision applies. Any unauthorised use therefore may constitute copyright infringement and may entail a claim for damages according to § 97(2) UrHG. Additionally, the copyright owner may also seek injunctive relief pursuant to § 97(1) UrhG.

The Free State of Bavaria used its position as right holder to forbid any reproduction or dissemination of *Mein Kampf*. This, for instance, led to the removal of a pdf version of *Mein Kampf* which was hosted on a website of a higher education institution in Berlin, at the request of the Ministry of Finance of the Free State of Bavaria.[81] Another example of the firm and litigious stance taken by the Bavarian Free State can be seen in a case before the Higher District Court of Munich in 2012.[82] The British publisher Peter McGee sought to publish parts of the book with some commentary in his journal *Zeitzeugen* ('contemporary witnesses'). McGee's publishing house advertised a supplement to the journal with the title 'Das unlesbare Buch' ('The unreadable book'). This publication placed excerpts of *Mein*

80 OLG München ZUM-RD 2012, 479, 484.
81 VG Berlin: Beschluss vom 19. Dezember 2013, Az. 3 L 1220.13.
82 OLG München, ZUM-RD 2012, 479.

Kampf next to some commentary by a professor of journalism. The Free State of Bavaria successfully applied for a preliminary injunction before the District Court of Munich, whose decision was upheld by the Higher District Court of Munich.

8. LIMITATIONS AND EXCEPTIONS

The firm stance of the Free State of Bavaria meant that seeking for authorisation for any otherwise restricted use would have been futile. Hence, anyone seeking to reproduce or distribute excerpts of *Mein Kampf* would have to rely on exceptions to and limitations of copyright which would permit an unauthorised use of the work. In relation to *Mein Kampf*, the quotation rule enshrined in § 51 UrhG is particularly relevant:

> It shall be permissible to reproduce, distribute and communicate to the public a published work for the purpose of quotation so far as such exploitation is justified to that extent by the particular purpose. This shall be permissible in particular where
>
> 1. subsequent to publication individual works are included in an independent scientific work for the purpose of explaining the contents,
> 2. subsequent to publication passages from a work are quoted in an independent work of language,
> 3. individual passages from a released musical work are quoted in an independent musical work.

This provision was central to the discussion of the already introduced decision of the Higher District Court of Munich in 2012. The purpose of the use of the third party is of particular relevance for determining whether the limitation is applicable. The citation used must be necessary for the analysis of the cited work and must be distinguishable from the work it is inserted into.[83] The Higher Court of Munich outlined previous case law on this issue: the citation would need to be accompanied with remarks from the citing person, while a merely internal connection between the original and the cited work would not suffice.[84] Furthermore, the citation should not become the main part of the citing work and, by this, replace the work

[83] Clemens Bogedain, 'Mein Kampf', der 'Mythus des 20. Jahrhunderts' und die 'Goebbels-Tagebücher': Werke früherer NS-Größen im Spannungsfeld von Strafrecht, Urheberecht und künftiger Gemeinfreiheit, 59 ZUM 205, 209 (2015).

[84] OLG München ZUM-RD 2012, 479, 480.

in which the citation is incorporated.[85] This means that the citation must serve as the basis for elaborations by the person citing the work; it cannot simply be the aim to make the cited work more accessible.[86] Importantly, the Federal Court of Justice has held that an entire citation would fall outside the scope of the limitation where only parts of it were not covered by the purpose of the citation.[87]

The Higher District Court of Munich examined some of the passages of the defendant's publication and held that the exception would not apply here. While the commentary part which included passages from *Mein Kampf* in inverted commas would be covered by the exception,[88] the Court found that this was not given for the passages of 'Mein Kampf' next to the commentary. These were rather being inserted to provide the reader with access to passages and not to support the commentary.

9. OTHER REGULATORY RESPONSES

Copyright law is not the only regulatory response available. The obvious tool is criminal law, which has already been used in this context. Criminal law has become even more relevant now due to the lapse of the copyright in *Mein Kampf* and the possible unrestricted dissemination of the book. Hence, criminal law has been mooted as a possibility by the Social Democratic Party in the German Parliament, which also inquired whether new legislation would be required.[89] Particularly, three provisions from the German Criminal Act are being applied against right-wing extremist propaganda, as follows.[90]

Firstly, § 86(1) of the German Criminal Code sanctions, inter alia, the dissemination and production of propaganda material of political parties or organisations held to be unconstitutional, or such organisations which aim to usurp the constitutional order within Germany. The German Federal Constitutional Court has held that works, like *Mein Kampf* or Rosenberg's *Mythus* that propagate a 'Führerstate' based on racist

---

85 Ibid.
86 Ibid.
87 BGH ZUM 2012, 981 [29].
88 See Sebastian & Briske, supra note 11 at 103.
89 The Parliamentary Group specifically mentioned Sections 318 and 319 of the Canadian Criminal Code that sanction hate propaganda; Kleine Anfrage der SPD Bundestagsfraktion, BT-Drucksache 17/12426, 1. (Feb. 20, 2013).
90 See Sebastian & Briske, supra note 11 at 104.

parameters and which discard liberty, equality and democracy would easily fit within this category of such illegal works.[91]

The provision, however, only applies to works devised after the German Basic Law entered into force. In a decision of 1979, the German Federal High Court (Bundesgerichtshof; BGH) held that the possession, purchase, and sale of antiquarian version of *Mein Kampf* would not violate § 86 of the German Criminal Code.[92] The Court held that *Mein Kampf* could not violate the constitutional order of the federal Republic of Germany since it predates the German Basic Law. This decision was criticised as being questionable from a political point of view.[93]

Criminal liability could, however, still occur where someone would adapt or edit such a pre-constitutional work to create references to current affairs.[94] Such references could, for instance be created in the text of the blurb.[95] Annotated academic editions would, however, escape criminal liability since they usually do not have the intention of furthering the aims of the quoted text or advertising for unconstitutional organisations but rather provide a critical discussion of the work.[96] In such a scenario, the provision would not be applicable due to the lack of criminal intention. Furthermore, such editions could be considered to fall under the proviso of § 86(3) since they 'serve civil education, to avert unconstitutional movements, to promote art or science, research or teaching, the reporting about current or historical events or similar purposes' and would therefore escape criminal liability.[97]

Secondly, § 86(a) of the Criminal Code sanctions the use of symbols of unconstitutional organisations. Such symbols could be Hitler's image and the use of the Imperial or Party Eagle along with the oak-leaved swastika.[98] A book, however, cannot be such a symbol so this provision only gains relevance where the books in question show such symbols.[99] On many occasions, even the use of such symbols on a copy of *Mein Kampf*

---

[91] Clemens Bogedain, 'Mein Kampf', der 'Mythus des 20. Jahrhunderts' und die 'Goebbels-Tagebücher': Werke früherer NS-Größen im Spannungsfeld von Strafrecht, Urheberecht und künftiger Gemeinfreiheit, 59 ZUM 205, 206 (2015).

[92] Bundesgerichtshof, Urt. v. 25.07.1979, Az.: 3 StR 182/79 (S).

[93] Caesar C. Aronsfeld, Mein Kampf, 1945–1982, 45 Jewish Social Studies 311, 312.

[94] Bogedain, supra note 91; Timo Handel, Mein Kampf – Gilt ein Verbreitungsverbot auch nach 2015?, Juristische Rundschau 433, 434 (2016).

[95] Timo Handel, ibid., 434.

[96] Ibid. 435.

[97] § 86 (3) German Criminal Act.

[98] Bogedain, supra note 91, at 207.

[99] Handel, supra note 94, at 437.

has not as such been penalised. Here, the proviso of subsection 3 of § 86, which also applies to § 86(a), could help again. The BGH has held that the small-scale sale of original editions of *Mein Kampf* which contained the sanctioned symbols could fall within that proviso. Hence, both §§ 86 and 86(a) of the Criminal Act would not be applicable to restrict republications of *Mein Kampf*.[100]

Finally, § 130 of the Criminal Code could be applicable in the discussed context. This provision sanctions the incitement of hatred against, inter alia, 'a national, racial, religious group or a group defined by their ethnic origins'. The public display or dissemination of written material which incites such hatred is specifically mentioned as a sanctionable action. Passages of *Mein Kampf* would easily qualify as such written material.[101] In comparison to the situation under Article 86, Article 130 also covers such works which predate the German constitution, which means that a new edition of *Mein Kampf* and its digital dissemination could be penalised under this provision.[102] However, scientific, and educational editions would not be sanctioned pursuant to § 130(6) with Sec. 86(3) StGB.[103]

10. HOW SUITABLE IS COPYRIGHT AS A TOOL TO REGULATE HATE SPEECH?

The manner in which the Free State of Bavaria exercised its ownership over *Mein Kampf* demonstrates that copyright can not only be used as a tool to disseminate works but also as a tool to restrict the dissemination of works and the information they contain.[104] This raises the question whether copyright law is an adequate tool for this purpose. While this question is of an academic nature with regards to works of the Nazi dictatorship due to the lapse of copyright, works containing hate or discriminatory speech are still being produced and still may be inflammatory. The analysis within this chapter, however, has demonstrated that copyright has only a limited

---

100 Karl-Nikolaus Peifer, Urheberrechtlich Probleme um Hitlers 'Mein Kampf', (Issue 11 Anm. 2) jurisPR-WettbR 1, 3 (2012).

101 See Sebastian & Briske, supra note 90, at 106; Hannes Ludyga, Der Vertrieb von Adolf Hitler's Hetzschrift 'Mein Kampf' – Urheber- und strafrechtliche Aspekte, 59 ZUM 435, 437 (2015).

102 See Sebastian & Briske, supra note 90, at 109; Bogedain, supra note 91, at 210.

103 Karl-Nikolaus Peifer, Urheberrechtlich Probleme um Hitlers 'Mein Kampf', (Issue 11 Anm. 2) jurisPR-WettbR 1, 3 (2012).

104 Ibid., 2.

use in restricting the creation and dissemination of works of a racist or discriminatory nature.

The works of Nazi leaders and similar works created in present times will attract copyright protection in many jurisdictions. This leads to the question whether copyright should not be available for such works. As seen, Hitler accumulated substantial wealth through the royalties of the sales of *Mein Kampf*. Additionally, the disputes over the Goebbels diaries have generated criticism as to whether the estate of Goebbels should be monetising this particular work. Abolishing copyright protection for these type of works could substantially diminish such streams of income. Indeed, some jurisdictions have renounced copyright protection for illegal or immoral works.[105] But this begs the question how such works are qualified as immoral and by whom? Since copyright arises automatically, an ex-ante exclusion or ground of refusal similar to patent or trade mark law appears not to be possible due to conflict with the Berne Convention. Even an ex-post assessment by courts in infringement proceedings would not appear to be a straightforward matter. Similar to the situation where the judge cannot easily evaluate the artistic value of a work to assess its protectability, the question arises whether works containing hate speech could be refused protection by copyright.

Aside from the doctrinal and practical problems with such approach, the denial of copyright just might lead to the opposite of its proposed purpose of restricting the creation and dissemination of such works. The absence of copyright protection does not hinder the dissemination of the information the work contains. Additionally, the economic argument behind copyright protection based on an incentive to create might not be of importance for the creator of such a work.[106] Often, the aim of hate speech is its expansive dissemination to reach a wide audience. Control of the work, i.e. the essence of copyright protection, may often not be what the 'author' seeks.

Finally, the term for copyright protection is limited, hence, the works created by the Nazi leader being in the public domain. A retroactive extension of copyright law over these works is no longer viable nor possible. Furthermore, for similar contemporary works, extending copyright protection would prove to be counterproductive. This is because the copyright would initially be vested in the authors of the work and after their deaths

---

105 See the discussion on US law in Chapter 19 of this volume, 'Copyright Protection of Illegal Works' by Eldar Haber.

106 It should be borne in mind that Hitler generated substantial income from his book – see supra 434–5.

their estates, which could exploit the work for the extended period unless the copyright were confiscated, as seen with *Mein Kampf*. This, however, might lead to problems relating to the expropriation of the property and having to justify this.

On a wider regulatory scale, possible conflicts with fundamental or human rights need to be mentioned here. On the one hand, restricting the dissemination of discriminatory or hate speech through copyright law can lead to a conflict with the fundamental right of freedom of expression or the freedom of science.[107] Freedom of expression is protected in various international and regional human or fundamental rights frameworks[108] and is also envisaged within the German Basic Law.[109] Importantly, the fundamental right does not only protect the freedom to actively express information but would encompass the right of accessing information.[110] While freedom of expression is a highly treasured right, it is not provided without limits. The right can be restricted by law, and particularly where this conflicts with constitutionally guaranteed interests of other parties. In this regard, the Higher District Court of Munich has held that the copyright act is such a piece of legislation.[111] Additionally, the exception provisions, in this particular case the quotation rule (see discussion above), would aim at balancing the interests of the right holders and those of third parties.[112]

It appears that copyright is not the best mechanism to regulate discriminatory speech.[113] So should copyright law be content-neutral? In Chapter 19 of this volume Haber provides a good overview of the benefits of such an approach. In any event, it is the criminal law which can be applied with regards to such works and their dissemination, and this is the appropriate

---

107 Article 13, EU Charter on Fundamental Rights; Article 5(3) German Basic Law.

108 Article 10, ECHR; Article 11, EU Charter on Fundamental Rights.

109 Article 5(1) Basic Law. The German Federal Constitutional Court has held that the dissemination of National Socialist ideology is not as such discarded from the scope of the fundamental right of freedom of expression but may be curtailed by law – BVerfG, NJW 2010, 47, 49.

110 With regard to the ECHR: ECtHR, *Youth Initiative for Human Right v Serbia*, no. 48135/06 25 June 2013, [20], [24].

111 OLG München ZUM-RD 2012, 479, 484.

112 OLG München ZUM-RD 2012, 479, 484.

113 European Copyright Society, Limitations and Exceptions as Key Elements of the Legal Framework for Copyright in the European Union – Opinion on the Judgment of the CJEU in Case C-201/13 *Deckmyn*, [33] (Nov. 1, 2014). http://www.create.ac.uk/wp-content/uploads/2014/11/Limitations-and-Exceptions-as-Key-Elements-of-the-Legal-Framework-for-Copyright-in-the-EU-Opinion-of-the-European-Copyright-Society-on-Case-C-201-13-Deckmyn-final.pdf.

tool to tackle the task. An interesting approach that infuses the copyright discussion with that of criminal law is provided by Rehbinder. He argues for limiting the exercise of exclusive rights where the public performance and dissemination of works containing illegal contents would be considered as criminal activities.[114] Only the negative rights targeting piracy should be provided, while the positive rights of exploitation and seeking revenues would not be available to the right holder.[115] Rehbinder states that otherwise the legal order would be contradictory where criminal law would forbid an action on the one hand, which would be permissible by copyright law, on the other.[116]

An example that highlights such an approach is – to some extent – the case of Goebbels' diaries, which has attracted considerable news coverage.[117] The defendant in the case asserted that the claimant should not benefit from the royalties created by such a book. Alternatively, the claimed sum should at least go towards an association that alleviates the atrocities of the Nazi dictatorship.[118] While no criminal action was argued on behalf of the defendant, it was suggested that the exploitation of Goebbels's works would be immoral. A criminal action would also be difficult to argue with regards to the diaries, as the discussion on the criminal law within this chapter has shown. But perhaps, not strictly legal measures could have been applied in this scenario in form of 'shaming'.[119] But this discussion is now moot due to the lapse of copyright. For contemporary works of hate speech whose dissemination would entail criminal

---

114 Manfred Rehbinder & Alexander Peukert, *Urheberrecht* § 10 [209] (17th edn, Beck, 2015).

115 Ibid.

116 Ibid.

117 See for instance: *Spiegel Online*, Streit um 6507,87 für Goebbels-Zitate (Sept 9, 2014), http://www.spiegel.de/kultur/gesellschaft/joseph-goebbels-random-house-muss-auskunft-ueber-gewinn-geben-a-992702.html; Dayla Alberge, Joseph Goebbels' Family Win Legal Battle to be Paid Royalties for Diary Extracts, *The Guardian* (Jul 9. 2015), https://www.theguardian.com/world/2015/jul/09/joseph-goebbels-family-royalties-diaries-nazi-propaganda-minister; Ofer Aderet, Goebbels' Heiress Wins Legal Battle to Receive Royalties From Biography of Nazi Minister (Jul 11. 2015), http://www.haaretz.com/israel-news/.premium-1.665364.

118 Again, this may prove to be counterproductive, as seen in the case relating to the royalties in the UK received by the German Welfare Council from the publisher Random House. In 2001, the charity decided to return the royalties to the publisher after its source was disclosed. Owen Bokott, Charity Returns £250,000 Royalties for Hitler's Credo, *The Guardian* (Jun 18. 2001), https://www.theguardian.com/uk/2001/jun/18/books.humanities.

119 Elizabeth L. Rosenblatt, Fear and Loathing: Shame, Shaming, and Intellectual Property, 63 DePaul Law Review, 1, 4 (2013).

sanctions, Rehbinder's approach could be a feasible way to counteract the enforcement of such works by their right holders, while bearing in mind the difficulties in its practical application.

11. CONCLUSION

The treatment of works such as *Mein Kampf* is a timely discussion. Heldrich states that anyone who read the book and then voted for Hitler in 1933 cannot claim to be totally innocent of the occurrences that led to the Holocaust.[120] Alas, works containing hate speech are still being produced, concerts with Nazi rock music were still able to attract thousands of Neo-Nazis in 2017.[121] However, as this chapter has argued, copyright is not the ideal regulatory tool to tackle hate speech – in the past, in the present and in the future – simply, because it has not been designed to do so. In Germany and other jurisdictions, criminal law, will remain applicable to the extent discussed above. Along these lines, the Bavarian Minister of Justice Bausback said that the 'present tools of criminal law would be applied – we owe this to the victims of the Holocaust and their relatives'.[122] Seul, on the other hand, welcomes the publication of the annotated version of *Mein Kampf*.[123] Not only could the annotated version provide a historically-critical alternative to all the already available versions of the book but it would highlight its anti-human ideology and perverted rationality to the unknowing reader.[124] And hopefully thus educate the reader so that history will never repeat itself – principiis obsta!

---

[120] Heldrich, supra note 11, at 645.

[121] *Deutsche Welle*, Hundreds of Neo-Nazis Band Together for Concert in Germany (Jul. 30, 2017), http://www.dw.com/en/hundreds-of-neo-nazis-band-together-for-concert-in-germany/a-39896740.

[122] Bayerisches Staatsministerium der Justiz (Jun. 25, 2014), https://www.justiz.bayern.de/presse-und-medien/pressemitteilungen/archiv/2014/91.php.

[123] Seul, supra note 10.

[124] Ibid.

Economic epilogue

22. Non-conventional copyright: an economic perspective

Tim W. Dornis*

1. INTRODUCTION

Since its inception, copyright law has worked in pursuit of a system of monolithic rights protection. This made sense a long time ago, when the number of works to be protected was limited to a single category of creative works – printed books. Today, however, creative production continues to be protected under uniform and solid entitlements, even though the scope of protectable subject matter has been extended to many different categories of works. This is not surprising considering that the rationale for protection has always been based on rather simplistic theoretical fundaments, notably ideas of protecting the author's natural right, her 'sweat of the brow', or other variants of if-value-then-right paradigms.[1] But economists recently seem to have dissolved these monolithic fundaments. Although they still prescribe the protection of individual entitlements, their rationale has become more consequence-oriented. Individual rights are no longer an end in and of themselves; they have become a means to the end of welfare maximization. Indeed, modern thinking – particularly in US law – has increasingly begun to center on more economic explanations.

Despite economists' turn away from solid and self-contained rights and toward an elaborate – and, as one might expect, more flexible – cost-benefit analysis, they, too, fall into the trap of over-simplification. Of course, econo-legal theory now admits that marketplaces are heterogene-

* This chapter is based in part on the author's prior articles 'Wigmorian Copyright: Law, Economics, and Socio-Cultural Evolution' (IPG 2018, pp.159–180) and 'Urheberrecht, Ökonomik und Evolution: Kreativität im Gravitationsfeld multipler Einflussfaktoren' (forthcoming in ZGE/IPJ 2018).

1 For an overview on the theoretical landscape, see William Fisher, Theories of Intellectual Property, 168, 169–173, in *New Essays in the Legal and Political Theory of Property* (Stephen R. Munzer ed., CUP, 2001).

ous and volatile, no longer claiming that there is a single, all-encompassing economic model or ideal structure of copyright protection. Nonetheless, most economic theorists still approach the field with specific products or marketplaces in mind. Inevitably, therefore, their analyses provide for individually, locally, and industrially limited insights only. Moreover, while many theorists address the question whether an actual extension of rights is justified, they are surprisingly silent on the issue of copyright homogeneity. In other words, they seldom explore the fact that copyrights are widely homogeneous and provide their owners with rather monolithic entitlements.

At the center of this problem – and, in essence, touching on the very fundaments of both legal doctrine and economic theory – is the matter of non-conventional copyright. Protecting or not protecting works of non-conventional stature is ultimately an issue of the speciation of protection standards – and such a diversification is woefully needed.

2. ECONOMIC THEORY

Over time, legal doctrine has continuously broadened copyrights.[2] The color of the argument, of course, has varied significantly: Natural-rights claims have been brought forward next to reward-based and incentive-based theories, as well as neoliberal economic and even constitutional-democratic arguments.[3] Indeed, no matter what the advance – printing press, photography, player piano, gramophone, photocopier machine, radio, television, videocassette recorder, MP3 file, or other means of digitization – whenever new technology has arrived, copyright has been extended in order to uphold the allegedly endangered protection-creativity mechanism.[4] It is therefore no surprise that today we have an almost all-encompassing and solid doctrine of copyright protection. It is all-

---

[2] For the so-called proprietarian creed in intellectual property law, see Peter Drahos, *A Philosophy of Intellectual Property* 200–210 (Routledge, 1996); Gillian K. Hadfield, The Economics of Copyright: An Historical Perspective, 38 Copyright L. Symp. 1, 11 (1988).

[3] For a concise overview of the different justifications and their malleability for the sake of rights extension, see L. Bently & B. Sherman, *Intellectual Property Law* 35–40 (4th edn, OUP, 2014).

[4] Paul A. David, The End of Copyright History?, 1 Rev. Econ. Research Copyright Issues 5 (2004); Mark A. Lemley, Is the Sky Falling on the Content Industries?, 9 J. on Telecomm. & High Tech. L. 125 (2011); for a concise overview of the evolution in the UK, see, e.g., William Cornish, David Llewelyn & Tanya Aplin, *Intellectual Property*, paras 10-01–10-41 (8th edn, OUP, 2013).

encompassing since it covers a virtually unlimited variety of products of the human mind. And it is solid since it prescribes a monolithic concept of right creation and protection. Unlike what we might expect and what is often claimed, economic theory has usually subscribed to this creed rather than providing an effective corrective.

Economists' basic reasoning centers on the general characteristics of so-called public goods. When we hear the term 'public goods' we usually think of government-provided infrastructure and public security or defense – but they may also include intangible objects such as information and knowledge. The fundamental problem with public goods is that rational actors must be expected to free ride on their production. Since the use of public goods is both non-excludable and non-rival, once a good is created, the creator can no longer stop others from making use of it – with or without payment.[5] In addition, the utilization of public goods by one person does not limit their availability to others.[6] Ultimately, such a panorama leads to a state of under-production by private actors, which is why public goods are usually provided by the state. With respect to creative works, these same characteristics of non-exclusivity and non-rivalry threaten an oft-enunciated market failure. Copiers can replicate products of the mind at virtually zero cost. In a competitive market, prices for such objects will be driven down to the cost of supplying the marginal unit of the product. Accordingly, the original creator's revenue will not even cover her production costs, resulting in a lack of incentive to produce.[7] Intellectual property law is thought to provide an escape from this dilemma by offering exclusive rights (i.e., monopolies). The downsides

---

[5] See, e.g., Kenneth J. Arrow, Economic Welfare and the Allocation of Resources for Invention, 609, 616–622, in *The Rate and Direction of Inventive Activity: Economic and Social Factors* (National Bureau Committee for Economic Research et al. eds., Princeton University Press, 1962); Harold Demsetz, Toward a Theory of Property Rights, 57 Am. Econ. Review 347, 348, 359 (1967); François Lévêque & Yann Ménière, *The Economics of Patents and Copyright*, 4–5 (Berkeley Electronic Press, 2004).

[6] See generally Paul A. Samuelson, The Pure Theory of Public Expenditure, 36 Rev. of Econ. & Statistics 387, 387 (1954); for intellectual property rights, see, e.g., François Lévêque & Yann Ménière, *The Economics of Patents and Copyright* 4–5 (Berkeley Electronic Press, 2004) (for patents and copyrights). For the different character of trademarks, see Tim W. Dornis, *Trademark and Unfair Competition Conflicts: Historical-Comparative, Doctrinal, and Economic Perspectives*, 321–325 (Open-Access-Source on Cambridge Core, 2017).

[7] For copyright specifically, see, e.g., L. Bently & B. Sherman, *Intellectual Property Law* 37–38 (4th edn, OUP, 2014).

of monopolies are accepted for the sake of incentivizing innovation and creativity.[8]

Depending on the perspective, intellectual property monopolies are seen either as a catalyst for or as a hindrance to marketplace competition. Harold Demsetz is credited as having laid the foundation for a neoliberal approach when he explained that '[a] primary function of property rights is that of guiding incentives to achieve a greater internalization of externalities'.[9] Accordingly, the creation of clearly determined and solid rights is explained as the *sine qua non* of a functioning marketplace. If property rights are defined clearly, so the neoliberal dogma goes, the market's invisible hand will do the rest.[10] This idea has also prospered in copyright theory.[11] As Paul Goldstein has famously explained, 'The logic of property rights dictates their extension into every corner in which people derive enjoyment and value from literary and artistic works. To stop short of these ends would deprive producers of the signals of consumer preference that trigger and direct their investment'.[12]

But, on the contrary, a much less optimistic perspective has always existed. Arnold Plant was the first to formulate such a theory. He questioned the idea that creativity would perish in the absence of full-fledged value appropriation. He was concerned that excess rights protection would spur the wrong incentives and that scarce resources might be

---

[8] See, e.g., Jim Leitzel, *Concepts in Law and Economics*, 46–51 (OUP, 2015) ('Static efficiency – optimal use of existing inventions – is sacrificed for the purpose of enhanced dynamic efficiency, the production of more inventions over time'); Robert Cooter & Thomas Ulen, *Law and Economics*, 104–105 (6th edn, Pearson, 2014); Richard A. Posner, *Economic Analysis of Law*, § 3.1, § 11.1 (9th edn, Aspen, 2014).

[9] Harold Demsetz, Toward a Theory of Property Rights, 57 Am. Econ. Rev. 347, 348, 359 (1967).

[10] For the normative Coase theorem expressed in this sentence, see Robert Cooter & Thomas Ulen, *Law and Economics*, 78 (6th edn, Pearson, 2014).

[11] See, e.g., Wendy J. Gordon, An Inquiry into the Merits of Copyright: The Challenges of Consistency, Consent, and Encouragement Theory, 41 Stan. L. Rev. 1343, 1389 (1989); Robert P. Merges, The End of Friction? Property Rights and Contract in the 'Newtonian' World of On-Line Commerce, 12 Berkeley Tech. L.J. 115, 131 (1997); Frank H. Easterbrook, Cyberspace versus Property Law?, 4 Tex. Rev. L. & Pol. 103, 111–112 (1999); Wendy J. Gordon & Robert G. Bone, Copyright, 189, 192–194, in 2 *Encyclopedia of Law and Economics* (B. Bouckaert & G. de Geest, eds, Edward Elgar, 2000). For a similar perspective but with a more natural rights flavor, see David Ladd, The Harm of the Concept of Harm in Copyright, 30 J. Copyright Soc'y U.S.A. 421, 431 (1982/83).

[12] Paul Goldstein, *Copyright's Highway: From Gutenberg to the Celestial Jukebox*, 146 (rev. edn, Stanford University Press, 2003).

misallocated toward creativity and away from more efficient uses.[13] In addition, then-professor (now Justice) Stephen Breyer and S.J. Liebowitz suggested distinguishing between copyright subject matter that would be created even without legal protection (and that, accordingly, should not be protected) and subject matter where creators are responding to a rights-based incentive only (and that, accordingly, should be protected).[14] More recently, a more extended concept of cost–benefit balancing has been brought forward. According to this perspective, not all 'spillovers' of benefits to others (i.e., positive externalities not appropriated by the right owner) will drive a wedge between private and public welfare. In particular, there is no risk of disincentivizing in many scenarios where third parties appropriate the benefit of others' creativity without compensation.[15] More simply put, one does not forego planting flowers in one's front garden simply because one cannot deny passersby the 'benefits' from looking at those flowers.[16] In essence, what is suggested, therefore, is a return to a more Coasean sobriety: The costs and benefits of rights creation must be balanced, and 'the gain from preventing the harm [must be] greater than the loss which would be suffered elsewhere as a result of stopping the action which produces the harm'.[17]

---

[13] Arnold Plant, The Economic Aspects of Copyright in Books, 1 Economica 167, 183–184 (May 1934); see also Robert M. Hurt & Robert M. Schuchman, The Economic Rationale of Copyright, 56 Am. Econ. Rev. 421, 422–423 (1966). For the characterization of such theories as 'copyright pessimism,' see Paul Goldstein, *Copyright's Highway: From Gutenberg to the Celestial Jukebox*, 10–11 (rev. edn, Stanford University Press, 2003).

[14] Stephen Breyer, The Uneasy Case for Copyright: A Study of Copyright in Books, Photocopies, and Computer Programs, 84 Harv. L. Rev. 281, 285–286 (1970); S.J. Liebowitz, Copyright Law, Photocopying, and Price Discrimination, in 8 *Research in Law and Economics: The Economics of Patents and Copyrights*, 181, 183–188 (J. Palmer & R. Zerbe, eds, Jai Press, 1986).

[15] See, e.g., Lisa N. Takeyama, The Welfare Implications of Unauthorized Reproduction of Intellectual Property in the Presence of Demand Network Externalities, 42 J. Indus. Econ. 155 (1994); Jeffrey L. Harrison, A Positive Externalities Approach to Copyright Law: Theory and Application, 13 J. Intell. Prop. L. 1, 14–15 et passim (2005); Brett M. Frischmann & Mark A. Lemley, Spillovers, 107 Colum. L. Rev. 257, 262 et passim (2007).

[16] Mark A. Lemley, Property, Intellectual Property, and Free Riding, 83 Tex. L. Rev. 1031, 1048 (2005).

[17] Ronald H. Coase, The Problem of Social Cost, 3 J. L. & Econ. 1, 27 (1960). See also H.R. Rep. No. 60-2222, 60th Cong., 2d Sess., To Amend and Consolidate the Acts Respecting Copyright (Currier), at 7 (February 22, 1909) ('In enacting a copyright law Congress must consider . . . two questions: First, how much will the legislation stimulate the producer and so benefit the public; and, second, how much will the monopoly granted be detrimental to the public?').

As we can see, copyright is no longer focused on an absolute conception of private property.[18] But our understanding is still far from complete. Not only are analyses often limited to specific products or marketplaces, but the real-world mechanics of protection–creativity correlations are still widely unexplored.

3. MONOLITHIC COPYRIGHT

Finding the incentive that is 'just enough' to spur the optimal degree of creativity is a complex task. Neoliberal theory avoids the pain of resolving the question, instead letting market mechanics do the work. Incentive-based approaches, by contrast, have the difficult job of comparing and balancing static and dynamic efficiency effects. Since overall welfare depends on these time-shifted efficiency correlations, one cannot limit economic analysis to a single production period. If one were to do so, the level of protection (z) would have to be 0 in order to provide the maximum amount of aggregate consumer and producer welfare in production period 0. In this case, however, there would be insufficient incentives to spur creative activity and to provide for sufficient production in future production periods (1, 2, . . . ∞). We must not attempt to provide a comprehensive and detailed catchall equation; this is simply impossible. Nonetheless, as in the formal economic model suggested in 1989 by William M. Landes and Richard A. Posner, a look at the interrelation between copyright protection, creativity, and welfare helps highlight several relevant characteristics.[19]

Most basically, overall welfare in a certain creative product's marketplace can be explained as:

$$W = f\,(N, w_{p/c}, c_{creation}) \qquad (1)$$

In this equation, overall welfare (W) is a function of the number of existing creative works (N), consumers' and producers' combined welfare (w_p and w_c), and the costs of creation ($c_{creation}$). N concerns the number of all creative works in a specific marketplace. Consumers' and producers' combined welfare ($w_{p/c}$) concerns their combined surplus – without a deduction of the costs of creation – with respect to N. Accordingly, the

[18] William M. Landes & Richard A. Posner, *The Economic Structure of Intellectual Property Law*, 9–10 (Harvard University Press, 2003).

[19] The following account relies on the classic model explained in William M. Landes & Richard A. Posner, An Economic Analysis of Copyright Law, 18 J. Legal Studies 325 (1989).

costs of creation cover only the cost of actually creating the work, not the cost of reproduction.[20] The number of creative works (N) is a function of the protection level (z). Further, both producer and consumer welfare directly depend not only on the level of protection but also on the number of works and are thus also a function of z. This is also the case with regard to the costs of creation.

We can therefore rewrite overall welfare (W) as a function of z as:

$$W = N(z) * w_{p/c}(z) - c_{creation}(z) * N(z) \qquad (2)$$

The fact that all elements of the equation are functions of the protection level (z) complicates the analysis. But complications aside, the equation allows for some insightful explanations for the variation of the protection level. Starting with the two extreme points on the scale of protection level ($z = 0$ and $z = 1$),[21] we can explain the following scenarios.

First, as the formal economic model posits, if there is no copyright protection ($z = 0$), there will be no incentive to create. The result is a scarcity of creative works (N) in future production periods (1, 2, . . .∞). The lack of incentives will keep N at 0 (note that this is the consequence of the classic rational-author paradigm). Accordingly, overall welfare will remain close to zero.

Second, with all-comprehensive and eternal protection of creative activities ($z = 1$), producer surplus in production period 0 is maximized. Overall welfare suffers an immediate deadweight loss as a result of the creative monopoly. Yet monopoly gains offer incentives for creativity. Over time, the maximum extension of rights will drive N upward. However, the consequence of this apparently desirable development of increasing the output of creativity is shrinkage of the public domain. Let us recall that protection is *all-encompassing* and *eternal.* The public domain is the residual pool of preexisting works of creativity that anybody can avail herself of at zero cost. Considering now that creators' revenue is a function of the costs of creativity, a rise in the protection level (z) and in the number of protected works (N) makes the production of new creative works increasingly expensive.[22] As a result, welfare will ultimately be pushed back again toward a minimum of creative productivity.

---

20 William M. Landes & Richard A. Posner, An Economic Analysis of Copyright Law, 18 J. Legal Studies 325, 341–343 (1989).

21 If $z = 0$, there is no right in any creation. If $z = 1$, protection is all-encompassing and eternal.

22 Wendy J. Gordon & Robert G. Bone, Copyright, 189, 195, in 2 *Encyclopedia of Law and Economics* (B. Bouckaert & G. de Geest, eds, Edward Elgar, 2000);

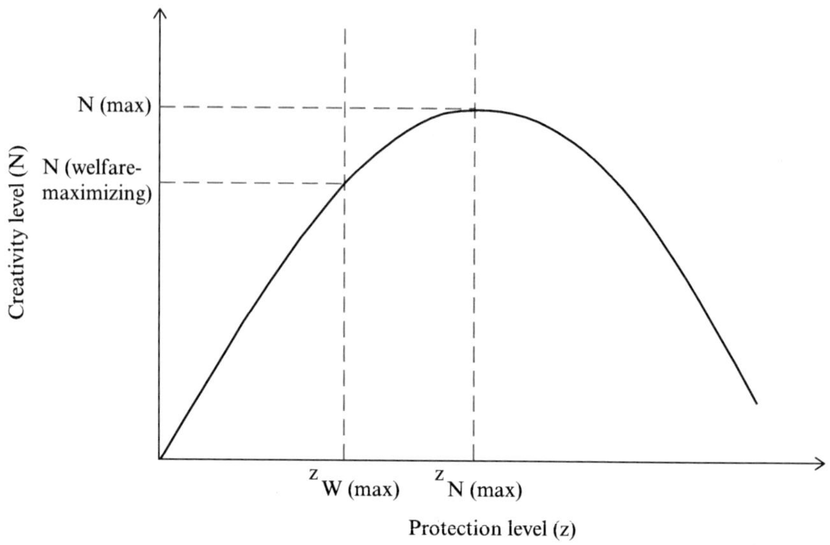

Figure 22.1 Formal protection–creativity model

Of course, none of these extremes reflects the optimum protection level. It is the in-between (i.e., $0 < z < 1$) that is determinative for a realistic economic concept. The formal law and economics assumption of what happens if we move upward from a starting point of no copyright protection ($z = 0$) is straightforward. Yet it needs clarification with respect to the effects that protection levels have on the number of creative works (N) and overall welfare (W). Figure 22.1 illustrates the movement of N in relation to z in the formal model of law and economics.

Initially, starting at $z = 0$, the number of creative works (N) rises with any increment (Δz). The stock of preexisting and protected works thereby continually grows. So does the cost of creation for subsequent producers of creative works. As long as the protection level does not overcome a critical threshold ($z_{N(max)}$), the number of new works will continue to grow. In other words, new producers will enter the market until their marginal costs exceed their revenue. Once the threshold is overcome, however, N starts to decline.[23]

Peter S. Menell & Suzanne Scotchmer, Intellectual Property Law (Chapter 19), 1473, 1500, in 2 *Handbook of Law and Economics* (A. Mitchell Polinsky & Steven Shavell, eds, North Holland, 2007).

[23] Beyond the point of maximum creativity, N_z' is negative. William M. Landes & Richard A. Posner, An Economic Analysis of Copyright Law, 18 J. Legal Studies

Similarly, producer and consumer surplus rises as the protection level (z) moves up from 0. As the formal economic model explains, with a higher level of protection, incentives to create are larger, and the number of creative works (N) rises. With any increase in z, the marketplace for creative works grows. Yet the increase in N in turn raises the costs of creation, which leads to a counter-effect on overall welfare. Since new creation requires a certain amount of utilization of preexisting works, marginal overall welfare will become negative ($W_z' < 0$) beyond a certain level of protection ($z_{W(max)}$).

More simply put, we can say that while some protection generates the necessary incentive for creative activity, too much will raise the costs of creation and, accordingly, stifle creativity in future production periods. The key is how to set the protection level according to the dimensions of the legal framework in order to generate optimal incentives for creativity. This perspective – by assuming away the possibility that an optimal copyright could exist without protection or at significantly lower levels of protection[24] – risks neglecting the realities of actual protection–creativity correlations.

4. COMPLEXITY OF (NON-CONVENTIONAL) COPYRIGHT REALITIES

In certain scenarios, legal protection may be less important for the existence of creative activities. Common examples are cases where copying cannot provide a perfect substitute for the original; where copying takes time, giving the original producer a head start without competition; where price discrimination by the right owner is possible; and where technology (e.g., DRM and TPM devices) can exclude copying.[25] Some of these examples have been taken as reasons for demanding limitations to – and occasionally even a full jettisoning of – copyright.[26] But due to the fact that

325, 335 (1989); William M. Landes & Richard A. Posner, *The Economic Structure of Intellectual Property Law*, 73 (Harvard University Press, 2003).

[24] See William M. Landes & Richard A. Posner, *The Economic Structure of Intellectual Property Law*, 73 n. 6 (Harvard University Press, 2003).

[25] For these and further examples, see, e.g., Michael O'Hare, Copyright and the Protection of Economic Rights, 6 J. Cultural Econ. 33 (1982); S.J. Liebowitz, Copying and Indirect Appropriability: Photocopying of Journals, 93 J. Pol. Econ. 945 (1985); Hal R. Varian, Copying and Copyright, 19 J. Econ. Perspectives 121, 134–136 (2005).

[26] See, e.g., Arnold Plant, The Economic Aspects of Copyright in Books, 1 Economica 167 (May 1934); Stephen Breyer, The Uneasy Case for Copyright: A

most of these scenarios deal with specific products or marketplaces with an accordingly limited validity, the formal economic model has remained largely unaltered. In other words, although it is clear that creativity is not immediately dependent on legal rights protection, the basic assumption remains that – at least in principle – more protection spurs more creativity.[27]

This assumption is not unproblematic. Indeed, a closer look at the correlation between copyright protection (z) and creativity (N) shows that the legal regime of formal rights not only may be irrelevant for a marketplace to evolve but also may even be detrimental when seen in the light of welfare maximization as the ultimate goal. The different aspects of non-conventional copyright that are explored in this volume provide an illustrative picture of these real-world complexities with respect to protection levels and actual creativity.

In the real world, as many of the non-conventional copyright examples explain, correlations between protection and creativity levels are not inverse-U shaped. A few examples are illustrative and can be explained on the basis of Figure 22.2:

1. Beginning with the curve in alternative 1, we can imagine a number of different industries where information and knowledge are protected by a purely norms-based system. Although there may be interrelations between legal and social norms and their efficacy,[28] in such a system, prima facie there is no need for legal rights protection. It is the system of social norms that directly provides for sufficiently effective appropriation and incentive structures. Accordingly, copyright law will be largely ineffective with respect to the creativity level and, very likely, also with respect to overall welfare.

Study of Copyright in Books, Photocopies, and Computer Programs, 84 Harv. L. Rev. 281 (1970); Tom G. Palmer, Intellectual Property: A Non-Posnerian Law and Economics Approach, 12 Hamline L. Rev. 261, 287–300 (1989); Steven Shavell, Should Copyright of Academic Works Be Abolished?, 2 J. Legal Analysis 301 (2010).

[27] See, e.g., Glynn S. Lunney, Reexamining Copyright's Incentives–Access Paradigm, 49 Vand. L. Rev. 483, 485, 553, 557 (1996); Richard A. Posner, Misappropriation: A Dirge, 40 Hous. L. Rev. 621, 638 (2003); Mark A. Lemley, Property, Intellectual Property, and Free Riding, 83 Tex. L. Rev. 1031, 1058–1065 (2005). For an illustrative summary, see, e.g., Giovanni B. Ramello, Pelle Sub Agnina Latitat Mens Saepe Lupina: Copyright in the Marketplace, Liuc Papers n. 141, Serie Economia e Impresa, 35 (02/2004), at 3–5.

[28] This question is still rarely explored. See, e.g., Georg von Wangenheim, Evolutionary Law and Economics, 161, 167 et seq., in *Oxford Handbook on Law and Economics* (Francesco Parisi ed., OUP, 2017).

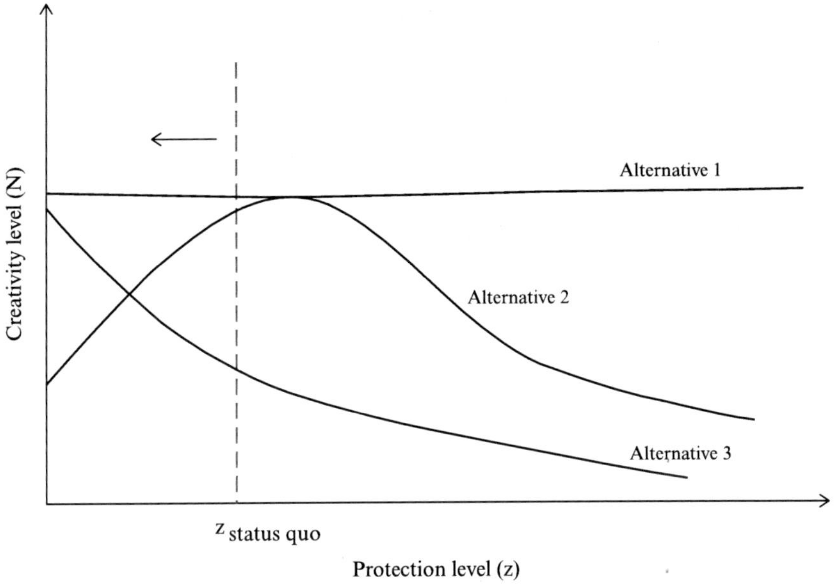

Figure 22.2 Market realities of protection–creativity mechanics

As Luke McDonagh explains, for instance, traditional musicians in Ireland highly value free sharing.[29] This seems to be a situation akin to the environment for creativity in the fashion industry and also resembles the social-norms-based systems of regulation in high-cuisine circles.[30] Furthermore, a similar social-norms system also seems to exist among comedians, as Trevor Gates explains by reference to earlier research by Dotan Oliar and Christopher Sprigman.[31] Sanctions for violations in such circles are usually informal (e.g., shaming, status loss, or blackballing) and may be so effective that they make legal

---

[29] See Chapter 7 in this volume, p. 151 et seq.

[30] For high-cuisine chefs' social norms, see Emmanuelle Fauchart & Eric von Hippel, Norms-Based Intellectual Property Systems: The Case of French Chefs, 19 Org. Science 187 (2008). For the fashion industry, see, e.g., Kal Raustiala & Christopher Sprigman, The Piracy Paradox: Innovation and Intellectual Property in Fashion Design, 92 Va. L. Rev. 1687, 1776 (2006).

[31] See Chapter 10 in this volume, p. 219 et seq.. For earlier research on the comedy industry and its handling of 'copying issues', see Dotan Oliar & Christopher Sprigman, There's No Free Laughs (Anymore): The Emergence of Intellectual Property Norms and the Transformation of Stand-Up Comedy, 94 Va. L. Rev. 1787 (2008).

protection of copyright virtually unnecessary.[32] Accordingly, in principle, there exists no significant correlation between legal protection and actual creativity levels. If the actors do not care about the law (or care very little about the law), their normative framework and their incentives for creativity are virtually isolated from the legal system. Hence the flat curve in alternative 1.

2. In other industries, as the curve for alternative 2 illustrates, although creativity levels may grow with the introduction of an effective regime of copyright protection, the growth may still be different from what can be found in the inverse-U-shaped ideal.

Protection–creativity correlations of this kind notably exist for industries where members are motivated mostly by non-copyright-related benefits. One oft-mentioned example is academic publishing. Scholarly authors gain peer recognition and career opportunities by producing books and articles; earning copyright royalties is not their primary motivation.[33] In this regard, the field of non-conventional copyright offers fascinating new insights and examples. As to the question of whether disk jockey (DJ) sets should be protected by copyright, for instance, Tom Iverson makes a forceful claim in favor of DJ gratification.[34] Yet, regardless of whether we want to find sufficient creativity as well as other preconditions for copyright protection involved in DJ productivity, one cannot disagree that a DJ's service of 'rocking a party' is, in most cases (at least with respect to less famous members of the DJ community), a rather ephemeral affair. In other words, the common DJ gets paid for the event, and nothing is left after the party is over and the music has been played. We should, therefore, expect a protection–creativity curve similar to alternative 2. This means that there exists a basic level of non-copyright-related incentives. Legal copyright protection only complements the overall picture by an additional bounty. In other words, compensation for the DJs' investment in creativity – and, accordingly, minimum incentives that are sufficient to spur creativity – will, in most cases, be provided for by payment of the contractual remuneration. Copyright protection may enhance the incentive, but it is not essential for sufficient

---

32 For French chefs, see Emmanuelle Fauchart & Eric von Hippel, Norms-Based Intellectual Property Systems: The Case of French Chefs, 19 Org. Science 187, 192 (2008).

33 Arnold Plant, The Economic Aspects of Copyright in Books, 1 Economica 167, 168–169 (May 1934); Steven Shavell, Should Copyright of Academic Works Be Abolished?, 2 J. Legal Analysis 301, 305, 315 et passim (2010).

34 See Chapter 9 in this volume, p. 201 et seq.

appropriability of 'DJ investment'. The marketplace already exists (and seems to work well) without extensive copyright protection.

3. Finally, the introduction of a legal copyright regime may reduce a preexisting level of creativity, as illustrated by alternative 3. In these scenarios, legal rights protection may reduce rather than enhance actual creativity levels.

   Scenarios of this kind may come up particularly when effective norms-based systems are thrown off by the introduction of legal rights or the enhancement of enforcement options. As just explained, the fashion industry, similar to circles of *haute cuisine* chefs and Irish musicians, can be described as a system that provides for a substantial degree of creativity that is not served by the legal system of intellectual property protection.[35] Yet the introduction or enhancement of legal protection and rights enforcement may still create a rather undebated risk: even though the actors may not generally rely on copyright law, their motivation may change once their awareness of its potential for strategic competition increases. More concretely, legal action may become an element of competition by, for instance, seeking nuisance rents from better-established market participants. At any event, the introduction of legal protection, as Cathay Smith points out, will increase uncertainty and ultimately reduce motivation on the side of the creators.[36] If this is true, the protection–creativity curve has a constant downward slope, as illustrated in alternative 3.

5. NON-CONVENTIONAL COPYRIGHT AND THE UNIFORMITY TRAP

In light of its economic downsides, the uniform treatment of divergent scenarios of protection–creativity correlations must be acknowledged as detrimental. As a closer economic analysis reveals, the so-called uniformity of rights creates costs of both under- and over-protection. Our aim must thus be to diversify the law: First, we need to determine how to adapt the law to reality. Second, we need to identify the potential problems inherent in such an adaptation – i.e., the traps we may fall into when attempting to break up monolithic copyright structures. Third, we need to explain how to modify the mechanics of lawmaking in order to provide for a more adaptive system. On this basis, we can then return to the issue

---

35 See supra n. 30 (and accompanying text).

36 See Chapter 6 in this volume, p. 128 et seq, in particular at pp. 148–149.

of non-conventional copyright and illustrate economics 'in action' with regard to the numerous examples given in this book.

5.1 What Is the Economic Downside of Uniformity?

A closer look at the costs involved in the uniform protection model helps us understand that in many scenarios, such a doctrine is either under-inclusive or over-inclusive, for the uniform status quo of the law will rarely fit the varying market conditions for all products and industries.[37]

The divergence between actual and optimal protection levels has been explained using the idea of 'uniformity costs'.[38] Uniformity costs exist if the actual protection level is above or below the optimal protection level ($z_{W (max)}$). If actual copyright protection – i.e., the status quo of legal protection – is higher than necessary to provide for optimal welfare-maximizing incentives, the marketplace is distorted through over-monopolization and over-investment. In addition, there are more administrative and enforcement-related costs than would exist under an optimal protection level.[39] Conversely, when the legal status quo is below the optimal protection level ($z_{W (max)}$), the public-good problem surfaces. Here, the effects are under-investment and dynamic inefficiency. Figure 22.3 illustrates how uniformity costs play out.

Different products and marketplaces are listed along the horizontal axis in accordance with the actual protection level that each product and marketplace requires in order to yield a welfare-maximizing equilibrium. The status quo (the dashed horizontal line) illustrates the uniform and monolithic rights protection under current copyright law. Since some products and marketplaces require more protection while others require less, the hypothetical optimum can be explained as a fluctuating line. The divergence between actual and optimal protection levels illustrates the marketplace distortion. The striped areas above and below the optimal protection levels do not directly reflect the uniformity costs, but they illustrate how uniformity leads to a virtually ubiquitous failure of the copyright incentive mechanism.

Looking at these costs might tempt one to conclude – somewhat radically

---

[37] See, e.g., Glynn S. Lunney, Patent Law, the Federal Circuit, and the Supreme Court: A Quiet Revolution, 11 Sup. Ct. Econ. Rev. 1, 39–56 (2003) (for patents); Michael W. Carroll, One for All: The Problem of Uniformity Cost in Intellectual Property Law, 55 Am. U. L. Rev. 845, 852–861 (2006) (for patents and copyrights).

[38] See extensively Michael W. Carroll, One for All: The Problem of Uniformity Cost in Intellectual Property Law, 55 Am. U. L. Rev. 845 (2006).

[39] Ibid., 847.

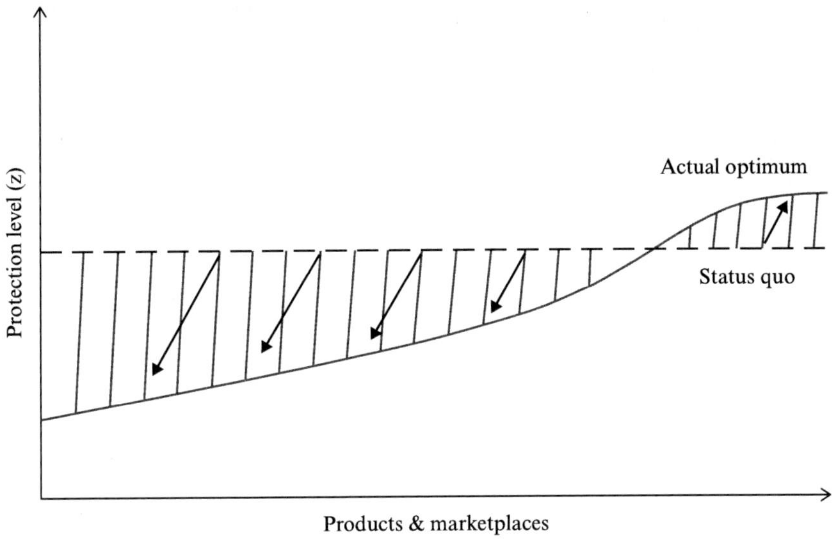

Figure 22.3 *Over-/under-protection across different products and marketplaces*

– that if different products and industries require differing protection levels, a custom-made copyright statute should be required for each industry and marketplace, if not each product.[40] After all, only tailor-made copyright policies can avoid the Scylla and Charybdis of over- and under-protection. At the same time, however, too elaborate a fine-tuning may accumulate costs as well – notably with respect to rights administration and enforcement.[41] Any kind of complexity in adjudication creates costs of differentiation. If the scope of protectable subject matter, the range of defenses and exceptions, or even the duration of copyright varies with the product and industry, private parties – both would-be and actual creators, as well as users and consumers – have to spend more time and effort on their decision-making. It is then difficult to determine what kind of creation constitutes a good 'investment' in the first place. In addition, it is hard to identify the demarcation line between owner rights and user freedom. Courts must also invest more time and effort in adjudication, for litigation is more fact-sensitive, less reliably

---

[40] Ibid., 848.

[41] Wendy J. Gordon & Robert G. Bone, Copyright, 189, 194–196, in 2 *Encyclopedia of Law and Economics* (B. Bouckaert & G. de Geest, eds, Edward Elgar, 2000); Mark A. Lemley, Property, Intellectual Property, and Free Riding, 83 Tex. L. Rev. 1031, 1058–1059 (2005).

founded on a 'standing doctrine', and more dependent on non-legal experts (in economics and other areas). As a result, a more diversified copyright doctrine is hard for the judiciary to handle.[42]

Considering these two sides of the equation, it is evident that there is no easy solution. Instead, we need a balancing of the costs and benefits of increased heterogeneity.

5.2 What Must Be Done?

Several decades ago, Friedrich von Hayek analyzed issues of concurrent technological, social, and legal change. His starting point has become known as the 'impossibility theorem'. Most simply put, this theorem contends that human design and planning are always and inevitably incomplete for want of information and cognitive capacities. This means that nobody is ever able to construct a rational – in other words, 'perfectly efficient' – economic order.[43] And this also holds true for other aspects of human culture, notably the legal order. Indeed, no lawmaker is ever able to deliberately design or construct an optimal system of legal norms.[44] As a consequence, the law is not a 'made' institution. Rather, it evolves spontaneously out of the actions of many – notably out of the myriad decisions by courts and judges.[45] Necessarily, any legal order is in a permanent state of evolution. Such limitations apply to all areas of the legal order, including intellectual property in general and copyright more specifically.[46]

---

42 See, e.g., Heimo Schack, *Urheber- und Urhebervertragsrecht*, paras 518 and 516 (6th edn, Beck, 2013) (on a diversified copyright duration). On the uniformity of rights in general, see Albrecht Schmidt-Bischoffshausen, 'One size fits all': Wie viel Einheitlichkeit verträgt das Urheberrecht?, 15, 21 et seq., *in Überprotektion durch Geistiges Eigentum?* (Horst-Peter Götting & Anja Lunze, eds, Nomos, 2009).

43 See especially Friedrich A. von Hayek, The Use of Knowledge in Society, 35 Am. Econ. Rev. 519 (1945).

44 See Friedrich A. von Hayek, *Law, Legislation and Liberty, Vol. 1: Rules and Order*, 11–15, 55 et seq. (Routledge and Kegan Paul, 1973), particularly 102.

45 See, e.g., Friedrich A. von Hayek, *Law, Legislation and Liberty, Vol. 1: Rules and Order*, 5, 36–37, et passim (Routledge and Kegan Paul, 1973); see also Oliver E. Williamson, Economic Institutions: Spontaneous and Intentional Governance, J. Law, Econ. & Organization 7, (1991) 159; James S. Coleman, Constructed Organization: First Principles, J. Law, Econ. & Organization 7, (1991), 7.

46 See, e.g., Richard A. Posner, Intellectual Property: The Law and Economics Approach, J. Econ. Perspectives 19 (2005) 57, 59; Thraínn Eggertson, The Unbearable Lightness of A: Useful Knowledge and Economic Growth, 54, 64, in

Although the situation of an inevitable imperfection and failure is unfortunate, the state of affairs is not too dismal since we can rely on a most basic institutional-theoretical concept for the reconceptualization of regulatory design. In essence, as von Hayek further explained, what is required is an institutional framework that allows for the law to 'evolve'.[47] Such a framework must create spaces of freedom for individual activity and decision-making on the basis of concrete cases rather than provide for a dense and rigid system of bright-line statutory norms. While the framework may incorporate some deliberate institutional and constitutional choices, it must leave the evolutionary process *as such* untouched. Only then can the 'invisible hand' of legal evolution do its creative and innovative work.[48]

In practice, and particularly for copyright law, this means that context-sensitivity and flexibility must be provided for by a process-oriented modification of lawmaking. Only if evolution in the sense of creative chaos and competition is possible, will the legal order be able to adapt quickly and adequately to the realities of complex and constantly changing circumstances. The statutory lawmakers may still provide for the fundamental framework of norms, but it is the practical application of the law that ultimately generates the regulatory policy 'in action'. In essence, the catalyzer of such an evolution is the unhindered competition of legal ideas and concepts. As in biological evolution, the mechanism of competition is founded on a cyclical process of information and knowledge creation, replication, transmission, selection, and retention.[49] This process best evolves in the courtrooms and on the basis of concrete cases. And one more aspect is important: different from statutory lawmaking, the promulgation of legal doctrine by means of a constantly iterating, court-crafted norm-generation process provides for a more transparent and multi-faceted landscape of regulatory alternatives. In such a system, the

---

Law, Economics and Evolutionary Theory (Peer Zumbansen & Gralf-Peter Calliess, eds, Edward Elgar, 2011).

[47] Viktor Vanberg, Cultural Evolution, Collective Learning, and Constitutional Design, 171, 190, in Economic Thought and Political Theory (David Reisman, ed., Springer, 1994).

[48] See Friedrich A. von Hayek, Die Theorie komplexer Phänomene, 21 et seq., in *Vorträge und Aufsätze*, vol. 36 (Walter Eucken Institut, ed., 1972); E. Ullmann-Margalit, Invisible-Hand Explanations, Synthese 39 (1978), 263.

[49] Friedrich A. von Hayek, Law, *Legislation and Liberty, Vol. 1: Rules and Order*, 102 (Routledge and Kegan Paul, 1973); Friedrich A. von Hayek, *The Collected Works of Friedrich August von Hayek, Vol. 1: The Fatal Conceit – The Errors of Socialism* 26 (W.W. Bartley, III, ed., Routledge 1988).

selection of concepts can build on a much more varied and better tested stock of concepts to choose from.[50]

Of course, the implications that result from peculiarities of creativity and innovation regulation are wide-reaching and provide a fertile playground for further studies.[51] At this point, however, we must return to the specific issue of non-conventional copyright: seen in light of the Hayekian legal-evolutionary theory, it is evident that a number of changes are necessary to overcome the defects of a statutory copyright law that consists of monolithic rights and inflexible rules rather than context-sensitive standards.

5.3 What Does This Mean in Practice?

Clearly, a fundamental reconceptualization of copyright law is impossible. International agreements on intellectual property law prevent us from jettisoning existing categories and structures; the same holds true for national statutory laws.[52] In other words, we have a definite number of different intellectual property rights ('categories') and a vast homogenization within these categories. Thus, modifications are possible only at the fringes of each category. Yet even a re-differentiation at the fringes is often discouraged, especially because it would allegedly lead to a loss of legal certainty and, ultimately, a reduction in creative incentives.[53] But we must not be too pessimistic. While a number of changes are necessary, they do not require us to turn copyright upside down. Roughly summarized, modifications to the status quo must seek to make the law more controllable, flexible, and corrigible.[54] A few aspects are particularly important for such an adaptation, as follows.

First, current copyright doctrine is driven by statutory lawmaking.

---

[50] In essence, this process of juridical collective learning reflects the Popperian paradigm of trial-and-error elimination and conjecture and refutation in action. See Karl R. Popper, *Objective Knowledge: An Evolutionary Approach*, 255, 341 et seq. (OUP, 1972).

[51] With respect to copyright law in general see Tim W. Dornis, Wigmorian Copyright: Law, Economics, and Socio-Cultural Evolution, IPQ 2018, 159–180.

[52] See Michael W. Carroll, One for All: The Problem of Uniformity Cost in Intellectual Property Law, 55 Am. U. L. Rev. 845, 878 (2006).

[53] Wolfgang Kerber, Zur Komplexität der Anwendung des ökonomischen Anreizparadigmas bei geistigen Eigentumsrechten, Zeitschrift für Geistiges Eige ntum (ZGE/IPJ) 5 (2013), 245, 271.

[54] With respect to decision-making concerning technology in general, see David Collingridge, *The Social Control of Technology*, 32–42 (Frances Pinter, 1980).

The problem may be more drastic in civil law jurisdictions, notably in Europe,[55] but even common law systems like that of the United States have witnessed an increase in statutory lawmaking.[56] In this regard, the most critical problem is the fact that statutory law is rather static and inflexible by definition, and thus not an optimal instrument for addressing the peculiarities of a heterogeneous reality of copyright.

In addition, as we have seen, lawmakers' information capacities and knowledge are limited.[57] While this problem is not endemic to statutory lawmaking, it is particularly problematic when a larger-scale conception or reconceptualization of legal doctrine is at issue. This defect is due to the fact that it is impossible to collect comprehensive and all-encompassing information and to analyze the peculiarities of creative industries with sufficient precision. Fritz Machlup expressed a similar concern for patent law reform when he explained that if one does not know whether a system 'as a whole' is good or bad, the 'safest "policy conclusion"' is to 'muddle through'.[58] This also holds true for copyright. We have a better understanding and position – in other words, a lower risk of being wrong – when taking small steps away from existing doctrine as opposed to trying to overthrow and rebuild the whole cathedral. Of course, this does not mean balking at more wide-ranging legal reform if such change is indicated. Rather, simply being aware of the risks that come with large-scale modifications helps us avoid drastic pitfalls. In other words, a reactive lawmaker is preferred over a proactive one in many cases.

Even if we wanted to trust statutory lawmakers' information-collection and decision-making capacities, we would still have to be leery of their democratic and political integrity. Copyright often is a question of interest-group will. In other words, the diversification of copyright doctrine at the level of statutory law invites opportunities to capture the process of norm promulgation.

Furthermore, Machlup's recommendation that we make incremental amendments highlights that the issue is not related to substantive law policy alone. In light of the complexities and dynamics of copyright

---

[55] See, e.g., Tim W. Dornis, Urheberrecht und Evolution: Kreativität im Gravitationsfeld multipler Einflussfaktoren, forthcoming 2018 in ZGE/IPJ (Zeitschrift für Geistiges Eigentum/Intellectual Property Journal).

[56] See, e.g., Michael W. Carroll, One for All: The Problem of Uniformity Cost in Intellectual Property Law, 55 Am. U. L. Rev. 848 (2006).

[57] See supra n. 44 (and accompanying text).

[58] Fritz Machlup, An Economic Review of the Patent System, Study No. 15 of the Subcomm. on Patents, Trademarks, and Copyrights of the Committee on the Judiciary United States Sen., 85th Cong., 2d Sess., at 80 (1958).

marketplaces, changes to legal doctrine must take place as part of a process-oriented transformation. And if change comes in small steps rather than large leaps, courts' function must be upgraded substantially. This does not mean that statutory law loses its relevance. It does, however, require acknowledging that there is no prerogative of statutory lawmaking. On the contrary, if flexibility is the order of the day, courts are predestined to fine-tune legal doctrine.

Such a downward shift of policymaking responsibilities from statutory lawmaking institutions to the courts introduces more decision-making discretion and flexibility. Accordingly, a turn from clear-cut rigid rules to more open-textured standards of adjudicating is necessary.[59] If a 'rule' is understood as the complete or almost complete *ex ante* determination of the outcome of a particular case, a 'standard' provides for the counter-concept of an *ex post* specification.[60] Rules are cheap to apply in practice, for they are bright-line and straightforward. Standards, by contrast, are cheap to promulgate at the level of lawmaking, but they create more costs when actually applied. In reality, the *ex ante/ex post* distinction is not always an easy one to make. Rather, rules and standards lie on a continuum that – moving from the 'rules end' to the 'standards end' – increases in flexibility and discretionary elements, providing decision-makers with more and more leeway to determine what the law actually is.[61]

In sum, what is required is less statutory lawmaking and more court-made doctrine. More leeway for the courts is another important aspect. In this regard, the law must change from a mostly rule-driven to a more standard-oriented state of aggregation. Ultimately, with more judge-made law comes the need for faster review, as well as the need for flexibility with respect to case-law systems' built-in tendency to create an ever-denser network of rules (and, accordingly, to weed out standards).[62] This does present

[59] See generally Cass R. Sunstein, Problems with Rules, 83 Cal. L. Rev. 953, 972 (1995). For copyright, see, e.g., Michael W. Carroll, One for All: The Problem of Uniformity Cost in Intellectual Property Law, 55 Am. U. L. Rev. 845, 890–900 (2006); Matthias Leistner, Der Beitrag ökonomischer Forschung zum Urheberrecht, 1 Zeitschrift für Geistiges Eigentum (ZGE/IPJ) 403, 423–434 (2009); Axel Metzger, Urheberrechtsschranken in der Wissensgesellschaft: 'Fair use' oder enge Einzeltatbestände?, 101, 118–122, in *Europäische Perspektiven des Geistigen Eigentums* (Matthias Leistner, ed., Mohr Siebeck, 2010).

[60] See, e.g., Louis Kaplow, Rules Versus Standards: An Economic Analysis, 42 Duke L. J. 557, 568 et seq. (1992); Cass R. Sunstein, Problems with Rules, 83 Cal. L. Rev. 953, 972 (1995).

[61] Cass R. Sunstein, Problems with Rules, 83 Cal. L. Rev. 953, 984 (1995).

[62] See, e.g., Arthur L. Goodhart, Precedent in English and Continental Law, 50 L. Q. Rev. 40, 50 (1934) ('Where a rule has once been decided, even though

a critical defect of court-based lawmaking and thus provides an interesting object of analysis.[63] Our special interest in non-conventional copyright, however, requires us to focus on several concrete aspects. With respect to the need for process-oriented modifications of copyright law, the area of non-conventional copyright provides a number of illustrative insights.

5.4 How Does This Apply to Non-Conventional Copyright?

The necessary diversification and tailoring of copyrights may concern all doctrinal levels of rights protection – beginning with the question of subject matter, extending to exceptions and defenses, and also including issues of duration. A closer look at the topics discussed in this volume helps us understand how a more context-sensitive and flexible doctrine of copyright law 'in action' can work:

1. To start with the issue of the 'subject matter' of copyright protection, many of the chapters in this volume address the demarcation line between non-protectable and protectable substances. It is in this regard that a careful handling of non-conventional copyright issues proves to be a question of fundamental economic policymaking. As a look at all these areas illustrates, current copyright offers numerous ways to 'thin out' – and thereby economically fine-tune – rights protection.

   One example is Yolanda King's lucid analysis of body art, which may not necessarily be considered protectable subject matter per se. As King notes, although the artistic content of such works is seldom disputed, there are a number of hurdles to protection, particularly regarding issues of functionality.[64] Similar problems, namely with respect to the issue of originality, must be overcome for artists seeking protection for typefaces and with regard to the protection of CAD files needed in the process of 3D-printing.[65] The idea/expression

---

wrongly, it is difficult or impossible to depart from it . . . when a space has once been closed by a precedent, then no further development is . . . possible'). More recently, see, e.g., Oona Hathaway, Path Dependence in the Law: The Course and Pattern of Legal Change in a Common Law System, 86 Iowa L. Rev. 601 (2001).

63 For a more extended process-oriented perspective and analysis of necessary modification of copyright law, see Tim W. Dornis, Wigmorian Copyright: Law, Economics, and Socio-Cultural Evolution, IPQ 2018, 159–180.

64 See Chapter 5 in this volume, p. 111 et seq..

65 See Chapter 13, p. 299 et seq and also Chapter 16, p. 355 et seq..

distinction, another complex issue of fine-tuning, is splendidly analyzed in Shane Burke's chapter on conceptual art.[66] Furthermore, as F. Jay Dougherty illustrates, it requires no magic to see that the 'useful article conundrum' may hit the spot with respect to magicians' props, devices, costumes, and masks.[67] It is particularly in the performing arts – the area of music improvisation is an illustrative example provided by Giuseppe Mazziotti[68] – where numerous open-textured standards give the courts discretion to determine copyrightability. With regard to sport moves, as Péter Mezei explains, similar dynamics of copyright protection evolve in the gravitational field of originality and functionality.[69] Evidently, it is the specific type of sport and its peculiarities – American football, soccer, wrestling, ice-skating, etc. – that determines whether 'creative' movements and choreographies are awarded copyright protection.

In essence, all these doctrinal instruments used in the larger context of 'copyrightability' are 'adjusting screws' that the courts can use to fine-tune the scope of protection. In economic terms, these instruments are the crossroads where express policymaking is required. Although this is often considered the statutory lawmakers' domain, it is the courts that are in charge at this point. Consequently, when confronted with issues regarding the subject matter of protection, originality, functionality, the non-protection of ideas, or any other of the numerous micro-doctrines, decision-makers have to openly consider the economic peculiarities and the evolutionary nature of copyright.

It is in this regard that another aspect becomes evident: Hardly any of these questions of copyrightability can be answered without solid information on marketplace circumstances. In other words, what we tend to discuss in normative terms requires an empirical foundation. Therefore, we must begin to seek more information on the economic backgrounds, particularly concerning the relevant actors' (artists') state of mind and pecuniary-incentive orientation, as well as users' (consumers') valuation of the subject matter.

2. Another doctrinal-level question of copyright protection is whether a work actually 'infringes' on an existing copyright.

---

66 See Chapter 2, p. 44 et seq. And related issues surface in Jani McCutcheon's discussion on bio art. See Chapter 3, p. 62 et seq.

67 See Chapter 11, p. 237 et seq.

68 See Chapter 8, p. 174 et seq.

69 See Chapter 12, p. 271 et seq.

The issue of infringement is regularly the most fact-specific of them all and, accordingly, gives courts wide discretion to adjust the outcome in accordance with the case at hand. One example is copyright protection for jokes and comedy. As Gates explains, it is the question of whether substantial similarity exists in order to warrant a finding of actual infringement. He fittingly notes that courts tend to find a rather 'thin' copyright in jokes.[70] Hence, we once again find an instrument of economic fine-tuning. And the case of dramatic works (including magic tricks) explored by Dougherty lucidly explains how delicate the analysis of 'substantial similarity' in infringement analysis can be.[71] Under US law (and the standard is similar in other regimes), in order to prove that the defendant copied the protected work, the plaintiff must show not only that the infringer had access to her copyrighted work but also that 'the works at issue are substantially similar in their protected elements'.[72] In one of Dougherty's example cases, the district court – in accordance with the Ninth Circuit's doctrine – asked 'whether the ordinary, reasonable audience would find the works substantially similar in the total concept and feel of the works'.[73] This provides for a very fitting – and economically reasonable – open-ended standard to adjust the law to real-world circumstances.

3. For cases where all obstacles of protectability have been overcome and the defendant's work is found to be infringing, the most oft-discussed issue in theory and practice is fair use (or fair dealing).

   Here, Enrico Bonadio provides an inspiring insight into modern street artists' realities. One of the many aspects of graffiti copyright protection is the question whether the incorporation of street art into other kinds of creative works that depict the public space – e.g., a video shoot or a movie scene – constitutes fair use.[74] This point invites further elaboration on the introduction of open-ended standards at the expense of rigid rules. Fair use in US doctrine is not only an obvious open-textured standard (unlike UK doctrine and that of other jurisdictions with limited lists of narrow exceptions),[75] but also the subject of intense discussion, notably with respect to its economic

---

[70] See Chapter 10, p. 219 et seq.
[71] Dougherty, Chapter 11, p. 237 et seq.
[72] *Cavalier v. Random House, Inc.*, 297 F.3d 815, 822 (9th Cir.2002).
[73] *Teller v. Dogge*, 8 F. Supp. 3d 1228, 1235 (D. Nev. 2014).
[74] See Chapter 4, p. 83 et seq.
[75] For the UK and the 'rigidity' of its exceptions, see, e.g., L. Bently & B. Sherman, *Intellectual Property Law*, 221 et seq. (4th edn, OUP, 2014).

underpinnings and ratio.[76] In a sense, therefore, it provides a *pars pro toto* for courts' balancing of the costs and benefits of copyright protection – including open-endedness and context sensitivity with respect to the heterogeneity of case scenarios and the quick and constant variability of circumstances.

One more example of real-world heterogeneity: with respect to Irish musicians, McDonagh paints a picture of a rather diverse sharing culture. There may be generosity toward fellow musicians and composers, but not toward television or film producers.[77] In doctrinal terms, this difference suggests that inter-musician copying might be seen as covered by a context-specific doctrine of fair use. External groups' commercial use of traditional music, by contrast, may still require licensing. As this empirical insight suggests, the more we know about the real-world circumstances, the better we can create a more context-sensitive and tailored system of copyright protection.

4. It is in this regard, finally, that we must emphasize that particular care is required when interference between copyright law and other areas of intellectual property law, private law, and regulation of free speech is the issue.

   To begin with a crossroads between intellectual property categories, one must be careful with extending copyright to areas of DNA engineering. As Nicola Lucchi explains, copyright does appear flexible and useful to handle contemporary technologies of DNA engineering – possibly more useful than patent law.[78] Yet, the risk of over-regulation is enormous. Not only is it likely that copyright protection might 'amend' patent protection, rather than 'substitute' it, but over-protection under a dual regime rather than copyright flexibility would result. In addition, it is far from clear whether the system of copyright protection as it currently exists would be a suitable substitute at all. To take just one example: can the period of rights protection (70 years p.m.a.) be a good thing considering the fast pace of development in the field? Evidently, as Nicola Lucchi correctly points out, a reframing of the contours of copyright is required. This must not neglect the economic underpinnings.

   Moreover, it is also essential to understand that copyright law is not an instrument for regulating the publication, much less the crea-

---

[76] See, e.g., William W. Fisher III, Reconstructing the Fair Use Doctrine, 101 Harv. L. Rev. 1659 (1988).

[77] Similar traits – as McDonagh further explains – can be found in other forms of traditional music.

[78] See Chapter 17, p. 367 et seq.

tion, of illegal or pornographic works. In this regard, Eldar Haber as well as Enrico Bonadio and Nicola Lucchi provide inspiring insights into the debate on whether copyright doctrine should be employed as an instrument to regulate creative 'content'.[79] Under an economic perspective, copyright norms must not be used to moralize or even censor. On the contrary, economics describes copyright as concerned with the incentives for creativity – the balancing does not include the *content* of such creativity. There is no 'guiding' of the creative process into the 'right' direction. In the same vein, there is also no copyright policy with respect to regulating the rights in the works of Nazi leaders. This point is convincingly made by Marc Mimler. He concludes that copyright law has not been designed to tackle hate speech.[80] Here, criminal laws or confiscation laws are what should be applied.

And there are also limitations with respect to the question of extending protection levels at the crossroads to other areas of private law where individual entitlements are at stake. Here, too, however, an economic caveat is in place: as Xiyin Tang suggests, 'copyright' in land art should be seen not only in the light of copyright policies but also with respect to other private law doctrines – particularly trespass and nuisance.[81] No doubt, this is an attempt to present a 'creative solution'.[82] Yet it is questionable whether such a solution shows us the right way. On the contrary, using nuisance law 'in lieu of' copyright is a perfect example of academic meandering that suggests broader protection for the worse. The reason is because 'in lieu of' will very likely mean 'in addition to'. In economic terms, such an extension of copyright's domain by means of private law doctrine ultimately risks transformation into a more comprehensive and over-extended doctrine.

Finally, a similar critique is warranted with respect to the areas of protection for press publications as well as perfumes and fragrances. Stavroula Karapapa convincingly explains risks of over-protection with respect to the hotly debated right to press publications.[83] As to perfume producers, Charles Cronin points out that the 'paucity' of intellectual property protection—notably the denial of copyright protection—may ultimately affect the industry's sustainability.[84] Yet here, as with press

---

79 See Chapter 19, p. 401 et seq. and Chapter 20, p. 418 et seq.
80 See Chapter 21, p. 432 et seq.
81 Chapter 1, p. 23 et seq.
82 See Tang, Chapter 1, p. 23, p. 24.
83 See Chapter 14, p. 316 et seq.
84 See Chapter 15, p. 340 et seq.

publications, land art, and other sectors 'at the fringes', a clear look at the overall picture of protection structures advises us to be careful. Since the production and marketing of perfumes and fragrances is neither fully unprotected (as suggested by the protection of perfume trademarks and brands in many legal regimes) nor in danger of withering (as suggested by empirical data),[85] any attempt to extend copyright protection may end up distorting the delicate cost-benefit equilibrium that has come to be established under current marketplace conditions. Again, the key is to avoid both over-extension and over-unification of copyright.

85 See supra p. 352.

Index